AF736064

Subgroup Complexes

Mathematical Surveys and Monographs

Volume 179

Subgroup Complexes

Stephen D. Smith

American Mathematical Society
Providence, Rhode Island

EDITORIAL COMMITTEE

Ralph L. Cohen, Chair
Jordan S. Ellenberg
Michael A. Singer
Benjamin Sudakov
Michael I. Weinstein

2010 *Mathematics Subject Classification.* Primary 20D05, 20D06, 20D08, 20D30, 20J05, 20C33, 20C34, 05E18, 55Pxx, 55Uxx.

For additional information and updates on this book, visit
www.ams.org/bookpages/surv-179

Library of Congress Cataloging-in-Publication Data

Smith, Stephen D., 1948–
Subgroup complexes / Stephen D. Smith.
p. cm. — (Mathematical surveys and monographs ; v. 179)
Includes bibliographical references and index.
ISBN 978-0-8218-0501-5 (alk. paper)
1. Finite groups. 2. Group theory. I. Title.

QA177.S65 2012
512′.23–dc23

2011036625

Copying and reprinting. Individual readers of this publication, and nonprofit libraries acting for them, are permitted to make fair use of the material, such as to copy a chapter for use in teaching or research. Permission is granted to quote brief passages from this publication in reviews, provided the customary acknowledgment of the source is given.

Republication, systematic copying, or multiple reproduction of any material in this publication is permitted only under license from the American Mathematical Society. Requests for such permission should be addressed to the Acquisitions Department, American Mathematical Society, 201 Charles Street, Providence, Rhode Island 02904-2294 USA. Requests can also be made by e-mail to `reprint-permission@ams.org`.

© 2011 by the American Mathematical Society. All rights reserved.
The American Mathematical Society retains all rights
except those granted to the United States Government.
Printed in the United States of America.

♾ The paper used in this book is acid-free and falls within the guidelines
established to ensure permanence and durability.
Visit the AMS home page at `http://www.ams.org/`

10 9 8 7 6 5 4 3 2 1 16 15 14 13 12 11

To my mother, Anna Elizabeth Yust Smith Kirn

Contents

Preface and Acknowledgments

As will be indicated in a moment in the Introduction, this book is primarily intended as an exposition—which hopes to bring a wider audience into contact with an area of research that I have enjoyed working in, over many years.

But of course during those years, I gained much of my own experience by benefiting from the knowledge of very many colleagues. So in this preface, I would first like to take the opportunity to thank them—apologizing in advance to anyone I may have left out. (Of course the reader will see the work of these experts emerging, as the later exposition in the book proceeds.)

Some personal acknowledgments. My introduction to the methods of finite geometry dates mainly to my collaboration with Mark Ronan, beginning around 1979. I also learned a great deal about geometries from Bill Kantor, Jon Hall, Don Higman, Ernie Shult, Francis Buekenhout, and Bruce Cooperstein.

During the 1980s, many experts in finite group theory, motivated partly by the work of Tits on buildings, became interested in geometries underlying simple groups. I particularly benefited from long-term contact with Michael Aschbacher, Franz Timmesfeld, and Geoff Robinson.

Discussions with Peter Webb and Jacques Thévenaz were instrumental in leading me into the more specifically topological methods underlying subgroup complexes; and in effect led to my later collaboration with Dave Benson. Many other topologists helped educate me in their area; particular Alejandro Adem, Jim Milgram, Bill Dwyer, Bob Oliver, and Jesper Grodal. Especially in recent years it has been a pleasure to discuss developments made by John Maginnis and Silvia Onofrei.

Also during the 1970s and 1980s, many combinatorialists (notably Stanley) were also developing similar techniques for the combinatorics of posets (partially ordered sets). Some of my initial contacts with that area were around 1981 with Jim Walker and Bob Proctor. Soon thereafter I began a particularly valuable ongoing correspondence with Anders Björner. Over the years I have also profited from discussions with other experts—notably Volkmar Welker, Michelle Wachs, and John Shareshian.

And of course we also learn from our students: It was a pleasure to work with Peter Johnson, Andrew Mathas, Matt Bardoe, Kristin Umland, and Phil Grizzard—who wrote their theses with me at the University of Illinois at Chicago (UIC), in aspects of this general research area. I also had some involvement in the thesis work of Tony Fisher under George Glauberman, and of Paul Hewitt under Jon Hall.

In a similar vein, it was a pleasure to work in this area with several postdoctoral scholars at UIC: namely Alex Ryba, Satoshi Yoshiara, and Masato Sawabe; and indeed with Yoav Segev, even before completion of his Ph.D.

The more specific history of this book. I first collected much of the present material while on sabbatical at Notre Dame, in preparation for a Fall 1990 graduate course there: Math 671, *Subgroup Complexes*.

During Fall 1994, I revised and expanded those old notes, to use as the text for the UIC graduate course Math 532 (Topics in Algebra): *Subgroup Complexes*. I would like to thank the students in that course for their questions and corrections, and for their general interest: Matt Bardoe, Joe Fields, Venketraman Ganesan, Julianne Rainbolt, and Kristin Umland.

A preliminary draft of the book was provisionally accepted for *Surveys of the AMS* in 1995. At that time, I received many detailed and very helpful suggestions from various colleagues, particularly Satoshi Yoshiara and Jacques Thévenaz, which strongly influenced the overall structure of the final version.

However, the book went to the back burner for some years, when I was involved in more urgent collaborations on books with Michael Aschbacher, Dave Benson, Richard Lyons, and Ron Solomon; and I have only managed to complete this book recently. (I particularly thank Sergei Gelfand and his staff at the American Mathematical Society, for their patience with me during this lengthy delay.)

During July 2005, the material of the book was again used as a text—for the summer graduate seminar Math 593 at UIC. Again I thank the students in the course for their willingness to assist me in the final revision process: Hossein Andikfar, Chris Atkinson, Chris Cashen, Phil Grizzard, Jason Karcher, Dean Leonardi, Jing Tao, and Klaus Weide. Their suggestions in particular led me to try to make a clearer distinction between the more elementary exposition, and the more advanced examples. This essentially resulted in the "optional tracks" for reading the book, described below in the Introduction.

I received helpful suggestions on the final (2011) draft of the book from a number of colleagues, including Matt Bardoe, Anders Björner, Jesper Grodal, Jon Hall, Bill Kantor, Ian Leary, Silvia Onofrei, Geoff Robinson, Masato Sawabe, Jacques Thévenaz, Rebecca Waldecker, Satoshi Yoshiara, and Peter Webb. I also thank the anonymous referees contacted by the AMS.

Institutional acknowledgments. Parts of this book were developed during several sabbatical periods at Caltech, as well as at Notre Dame and U. Illinois–Urbana. I am also grateful to All Souls College-Oxford, for a Visiting Fellowship during Hilary Term 2009, when some of the final work was carried out.

My overall work has been partially supported over the years by summer grants, first from NSF and more recently from NSA.

Dedication. Of course the support and encouragement of my wife Judy Baxter have been unflagging.

Finally I'd like to formally dedicate this book to my mother, Anna Elizabeth Yust Smith Kirn: who at various times earlier in my career asked when I was going to write a book (as opposed to the usual journal articles).

So, although several other books have actually appeared since I started this one, I'm finally in a position to say: Well, Mom—here it is.

Introduction

Aims of the book

Much of the material in this book originated as a set of lecture notes for a graduate course titled *Subgroup Complexes*, given in Fall 1990 at Notre Dame, and in Fall 1994 at the University of Illinois at Chicago (UIC); a later draft was used for the summer graduate seminar at UIC during Summer 2005.

This final version is also intended—at least in part—for possible use as a graduate text for such a course; especially in the chapters in the first half or so, where usually reasonably full details are given. The level of the course would typically be second-year: the exposition should ideally be accessible to students who have taken first-year graduate algebra—but who may have seen at most some basics of algebraic topology and combinatorics. The choice of the material aims to give a rough overview of a modern research area, from its historical origins through some directions of current research problems. The treatment includes proofs of many basic and intermediate results, as well as examples and exercises.

On the other hand, especially in the later chapters, more typically proofs are just sketched, or omitted; and fuller references are given, which ideally the more motivated reader can follow up. Thus in these areas, the material is perhaps more appropriate for an advanced reading course.

Finally, the book is also intended to be of at least some use as a reasonably general reference for this research area: So as indicated above I have also included statements of various more advanced results; typically without proof, since the proofs would involve details beyond the fairly elementary setting of the book.

In selecting the topics, I have made an effort to collect and compare material from various sources (some of it apparently in the "folklore" category); and to cover various areas reasonably well known to me. Of course I cannot hope to be truly encyclopedic about the many research directions related to subgroup complexes—so I apologize in advance for omissions of any favorite material of potential readers. (In particular because of my various other commitments after this book was started, in various areas I have fallen behind developments that have taken place since about 1994; so I thank numerous colleagues who have supplied more up-to-date references in many of these areas.)

Optional tracks (B,S,G) in reading the book

An additional feature of the treatment is that I have included many examples of subgroup complexes related to finite simple groups, which is my own particular area of interest. It seemed sensible to present this material as "optional" reading—since the details of the structure of all but the smallest simple groups may not be so easily accessible to students, or even to more advanced readers not specializing in group theory.

Accordingly I begin, in Chapter 1 and the earlier part of Chapter 2, with elementary examples—related to more familiar groups, such as symmetric groups and matrix groups. Then in the latter part of Chapter 2, I introduce more advanced examples from simple groups, explicitly separating them off into two optional tracks—which do *not* need to be read in order to understand the main material in the rest of the book. These tracks are:

• Option B—examples related to the *building* geometries for groups of Lie type, which may be of interest to readers in the many areas of mathematics where some version of the Lie theory plays a role; and:

• Option S—"sporadic" geometries, mostly related to *sporadic* simple groups, which will develop certain analogies with the material in Option B—but will likely be of more restricted interest, typically to specialists in areas closely related to finite group theory. After the initial material on Options B and S has been laid down in Chapter 2, in later chapters any examples related to that material will be flagged via an explicit reference to the corresponding option.

A further optional track arises mainly in some of the more technical material in Chapters 3 and 4, and especially in Chapter 7; I have called it:

• Option G—on *G-homotopy* equivalences. This is not needed for the more elementary part of our material in this book, which has to do with the standard notion of homotopy equivalence in topology (including material on contractibility). However, the material in Option G mainly concerns some further arguments, which extend homotopy equivalence results to the stronger condition of equivalence of *G-equivariant homotopy* of a space (defined using the action on that space of a group G). Much of this optional material begins with the observations of Thévenaz and Webb in [**TW91**]. These extended equivalences are in turn used in several of our more advanced applications, notably for *G-equivariant cohomology* in Section 7.3. But these extensions will not be essential for the main flow of the exposition in this book.

In a somewhat similar vein, some of the numbered remarks in the text are given the more explicit descriptive label "Digression" or "History"; in particular, this will indicate that they are also optional—that is, they are not required for understanding the subsequent main material.

A preview via some history of subgroup complexes

The remainder of this Introduction is devoted to a quick summary of some of the history of the research area of "subgroup complexes". More detailed historical information and references will be given at various later points in the main text; but the summary below will at least indicate some of the overall motivation, and introduce a number of the main themes to be developed in the book.

The exposition in this brief history will be informal, intended to convey the general flavor of the book: the more formal definition and properties of various terms and concepts mentioned here will appear later in the main text. In particular, see the Index for corresponding page references; typically the page number for the definition of an indexed term is given in **boldface**.

Much of the work in the modern research area of "subgroup complexes"[1] can be traced back to the influence of one very striking result: namely Corollary 2 in K. S. Brown's 1975 paper [**Bro75**], sometimes called the *Homological Sylow Theorem*, which we now quickly review.

For comparison we first recall the numerical part of the usual Sylow Theorem: If G is a finite group, and p is a prime dividing $|G|$ (i.e. the order of G), then the cardinality of the set $\mathrm{Syl}_p(G)$ of Sylow p-subgroups satisfies the congruence:

$$|\mathrm{Syl}_p(G)| \equiv 1 \pmod{p}.$$

We turn to the restriction of Brown's result [**Bro75**, Cor. 2] to finite groups G: below $\mathcal{S}_p(G)$ denotes the poset (partially ordered set) of all nontrivial p-subgroups of G under inclusion, in which the inclusion-chains define a simplicial complex—for whose Euler characteristic $\chi(\mathcal{S}_p(G))$ we obtain an analogous congruence:

$$\chi(\mathcal{S}_p(G)) \equiv 1 \pmod{|G|_p}, \tag{0.0.1}$$

now modulo the full p-part $|G|_p$ of the group order. The similarity of this numerical congruence with Sylow's Theorem led to the nickname Homological Sylow Theorem.

To briefly summarize the subsequent history, Brown's result then led to many further applications of topological techniques to the investigation of subgroup complexes such as $\mathcal{S}_p(G)$. These developments have been pursued by researchers from a number of different parts of mathematics; in particular such complexes can be viewed as lying in the intersection of the three areas in the following diagram (imitating a diagram of Webb below), and providing a natural link among them:

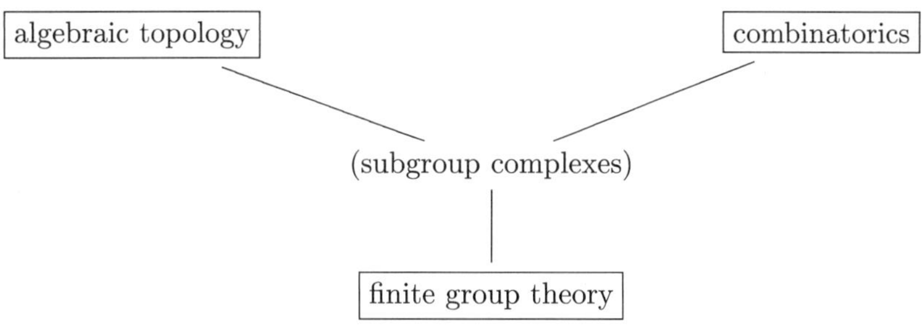

The interplay of these areas is a major theme of this book; and indeed many of the results to be discussed later are not easily restricted to just one of these areas. We now indicate a few aspects of the subsequent history of subgroup complexes in each of the areas.

It is natural to begin from the viewpoint of algebraic topology: Probably the most influential early development from Brown's result was Quillen's 1978 paper [**Qui78**], on the homotopy properties of $\mathcal{S}_p(G)$ and related posets. The results in this paper extended many of Brown's ideas, and laid down a great deal of foundational material for the topological study of posets of subgroups; and in particular the paper initiated a number of major directions of future research, including:

---

[1] But we emphasize also that the more specific research area of geometries for simple groups is largely inspired by the fundamental work of Tits on buildings (which will be another main theme throughout this book, starting especially in Section 2.2).

- contractibility
- homotopy equivalences
- homological construction of G-modules
- relations with Tits buildings (and other finite geometries for groups)
- spherical and Cohen-Macaulay subgroup complexes
- computation of group cohomology $H^*(G)_p$ of G at the chosen prime p.[2]

These and other research directions will be discussed in corresponding chapters of this book.

Next from the viewpoint of combinatorics: The study of more general posets—that is, posets not necessarily arising from subgroups of a group G—was already established in combinatorics by the time of the Brown–Quillen work. But during the 1970s and 1980s, there was an especially vigorous parallel development of topological methods in poset combinatorics, pioneered by Rota, and pursued by Stanley, Garsia, Baclawski, Björner, Walker, and others; this included the study of possible properties of a poset such as Cohen-Macaulayness and shellability. In particular, the 1982 paper [**Sta82**] of Stanley focused on the combinatorics of a poset with an action by a group G (which of course includes the case of suitable posets of subgroups of G). There have been numerous further developments in poset combinatorics; see for example Chapter 3 in Stanley's *Enumerative Combinatorics I* [**Sta86**]. (We will sketch some of these developments in later chapters.)

Finally from the viewpoint of finite group theory: It seems that researchers trained in finite groups did not immediately take up the study of the Brown–Quillen topological methods from the late 1970s; perhaps work in the early- to mid-1980s of Kratzer–Thévenaz, Bouc, and Webb was among the first. Possibly the most influential introduction for group theorists to those methods was provided by the expository paper [**Web87b**] of Webb, developed for the 1986 AMS Summer Symposium at Arcata. Webb's paper reviewed the background, and also indicated his own contributions in several of the abovementioned directions; but additionally, the paper emphasized the role of subgroup complexes in terms of inter-relating the study of certain sub-areas of finite group theory, summarized in the diagram on p 350 of [**Web87b**]:

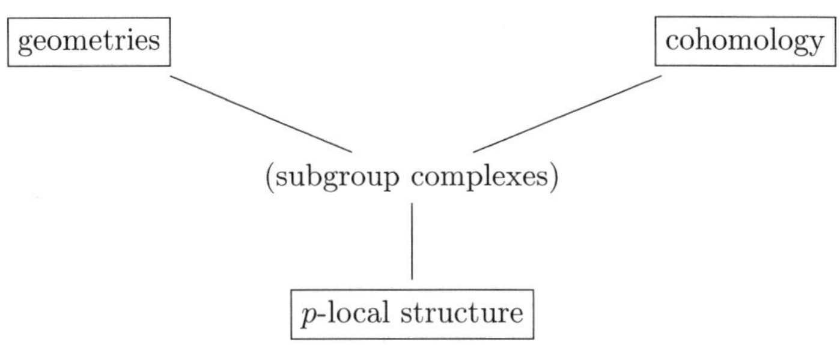

In particular, this interaction afforded a natural confluence of work on subgroup complexes with work on "groups and geometries"—the study of finite geometries

[2]Actually group cohomology was not really pursued explicitly in Quillen's paper; but it was prominent in the motivation for Brown's original work—see e.g. [**Bro75**, Thm 1]. And group cohomology has continued to be an important application of the subgroup-poset viewpoint in the literature since that time.

defined primarily for *simple* groups (largely inspired by Tits buildings), which had been strongly developing especially from the early 1980s. For example: for many of the sporadic simple groups, partial analogues of Tits buildings called *p-local geometries* [**RS80**] had been discovered—and these geometries have also been profitably investigated from the Brown–Quillen viewpoint. (The geometries are determined by *p-local subgroups*, namely the normalizers of nontrivial p-subgroups.) Indeed applying that general viewpoint to the particular case of geometries for simple groups has been a major interest of mine; and such applications provide a main theme in numerous examples throughout the text, for readers who wish to follow Options B and S.

To bring this brief history up to the present: During the 1990s, the application of subgroup–poset methods to group cohomology led to generalizations of the earlier Brown–Quillen–Webb results, using the viewpoint of the homotopy colimit over a category; this study has come to be known as "homology approximations" or "homology decompositions"—with notable contributions from researchers such as Jackowski, McClure, Oliver, Benson, Wilkerson, Dwyer, and Grodal. Such decompositions have in turn been one of the tools used in the program of Adem and Milgram and their collaborators, on computation of cohomology of simple groups; and they are the focus of the project of Benson and Smith in [**BS08**] which comprehensively treats certain aspects of sporadic geometries, a number of which appear as examples in Option S of this book. Chapter 7 on group cohomology will outline some of the above development of decompositions—though it is beyond the scope of this book to treat that work in detail; further exposition on decompositions is available in [**Dwy01**], [**Gro02**], and [**BS08**].

Finally at a more speculative level about the future: The study of suitable p-subgroup posets and decompositions seems to be one component of some emerging work in algebraic topology, and of related work in group representation theory—on such topics as "fusion systems" and "p-local finite groups". We mention this topic briefly at later Remark 7.1.3; important developments appear in work of Oliver, Levi, Broto, Grodal, Linckelmann, Kessar, and others. These topological structures, generalizing certain features of groups, are of course very interesting in their own right; and also one can hope that they will cast new light on outstanding open problems about groups and their representations such as the Alperin and Dade Conjectures. Furthermore Aschbacher is now investigating how results from the classification of finite simple groups can be pursued in this context. Again we will mention some of this work in the text, though without being able to go into great detail.

This concludes our brief and highly selective history; the main text itself will now try to provide some development of the interactions among the research areas indicated above.

Part 1

Background Material and Examples

The first two chapters are devoted first to reviewing background material on poset topology; and then to giving examples of these topological constructions—by introducing a number of geometries for simple groups, considered from the viewpoint of subgroup posets and complexes.

CHAPTER 1

Background: Posets, simplicial complexes, and topology

This initial chapter reviews some background material on the fundamentals of "poset topology", which will be our basic context throughout this book.

Many of the concepts and results mentioned in the chapter are elementary—such as that of a poset (partially ordered set); and so the corresponding definitions and proofs will often be given only informally, with sources given for further reference. But the review will in particular allow us to establish some further notation, and to provide some initial examples.

More experienced readers will of course be able either to skim through this background chapter, or to skip over it entirely.

Chapter overview: the viewpoint of "the three levels". One advantage of using the context of poset topology is that we can start with just the elementary setup of a poset—but then we will be able to obtain conclusions at the two successively deeper levels of simplicial complexes and topological spaces; and furthermore we can do so fairly easily, by using just the elementary combinatorics of the original poset.

Correspondingly the first three Sections 1.1— 1.3 introduce these levels of analysis. We mainly focus on our usual special case of a poset of subgroups of some fixed group G, so that we will be considering :

- subgroup posets;
- the associated simplicial complexes, namely subgroup complexes; and
- some basics of algebraic topology, applied to subgroup posets and complexes.

And in particular our discussion will emphasize how we make the *transition* from one level to the next deeper level.

The two subsequent Sections 1.4 and 1.5 then continue the theme of transition—by comparing, at these three levels, the corresponding notions of:

- mappings; and
- G-actions.

These interrelations will be fundamental throughout our later analysis.

The final Section 1.6 indicates one further such transition between levels—this time in the "reverse" direction, from deeper to more elementary—by introducing:

- the barycentric subdivision, described in the viewpoint of face posets,

along with some further notions for complexes such as joins and links.

1.1. Subgroup posets

The fundamental objects of our study will be posets consisting of subgroups of a group; but sometimes we will also consider other kinds of posets. So we begin by recalling the wider context:

General posets. We will assume the standard notion of a *partially ordered set*, or *poset*. Recall that this is a set equipped with a reflexive, transitive, antisymmetric order relation; see e.g. [**Hun80**, p 13]—or for further development, also Section 3.1 of [**Sta86**].

NOTATION 1.1.1 (Some notation for posets). We typically use $\mathcal{P}$ to denote some poset, with the inequality symbol $\leq$ (or $\leq_{\mathcal{P}}$) giving its order relation; more formally we may write $(\mathcal{P}, \leq)$ to indicate both the set and the relation. Often we write x, y, ... for the elements of the set $\mathcal{P}$. Of course $<$ denotes the *strict* order relation: that is, $x < y$ means that $x \leq y$, but $x \neq y$.

Except where explicitly stated otherwise, all posets considered in this book are *finite;* that is, they have a finite number of elements. We emphasize that *this restriction allows us to use simplified forms* of the statements of various standard results in the literature on general posets. ♢

Initial examples of posets. A sufficiently small finite poset is often represented visually by means of a diagram,[1] as in [**Sta86**, p 98]: namely a graph whose ascending edges indicate the "unrefinable" strict relations, as in:

EXAMPLE 1.1.2 (The diagram of a small poset). Here is an initial example of a poset, chosen more or less at random:

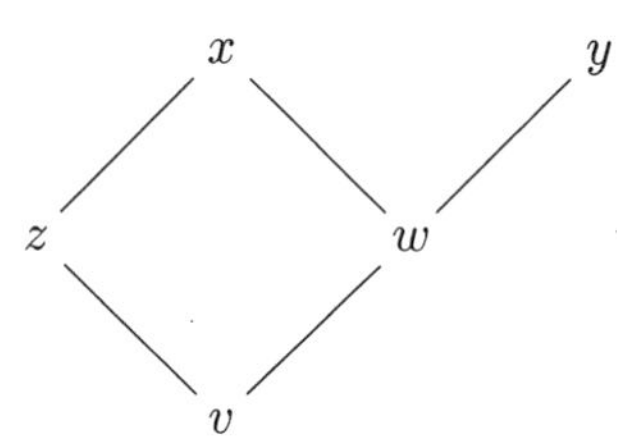

From the diagram we see (for example) that $v < w$ and $w < y$ are unrefinable relations, with their composition giving a refinable relation $v < y$. ♢

EXAMPLE 1.1.3 (Poset diagrams in Galois theory). Most readers have already encountered poset diagrams—typically in a course on Galois theory (e.g. [**DF99**, Sec. 14.2]): namely the diagrams for subextensions of a Galois field extension, and the corresponding diagrams for subgroups of a Galois group. ♢

Indeed in analogy with the above example given by subgroups of a group, many of the posets that we will consider in this book will have the members of the poset given by the substructures of some fixed structure—and the order relation is given just by inclusion $\subseteq$. Of course the most general such inclusion-poset is defined by the subsets of a set:

[1]Often called a *Hasse* diagram in the literature

EXAMPLE 1.1.4 (The power set as a poset). Given some set S (which we normally assume to be finite), we get a natural poset structure on the standard *power set* of S: Namely we take $\mathcal{P} := 2^S$,[2] so that $\mathcal{P}$ has elements given by the subsets of S; and the role of the order relation "$\leq_{\mathcal{P}}$" is played by the set-inclusion $\subseteq_S$ among those subsets.

In particular, if we choose $S := \{1, 2, 3\}$, then the power-set poset 2^S has the diagram:

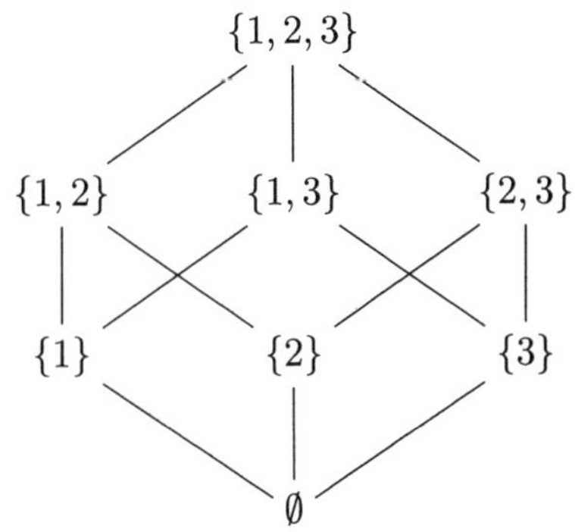

Posets closely related to power sets will turn up at various points later, e.g. 1.1.6 and 1.2.7. ♢

DIGRESSION 1.1.5 (lattice theory). We mention that the power set 2^S of Example 1.1.4 has the special property of being a *lattice* (see e.g. [**Sta86**, Sec 3.3])—since each pair of members has a (unique) least upper bound and greatest lower bound: in this example, given respectively by the union and the intersection of the pair as subsets of S. However, most of the posets in this book will not have this lattice property; cf. later Example 1.1.6.

Nonetheless we emphasize that lattices are important in a number of areas, such as the logical foundations of quantum mechanics (see e.g. Mackey [**Mac63**]); and there has been a considerable corresponding literature, especially since Birkhoff's book [**Bir40**], on the combinatorics of lattices. For further reference on that material, see e.g. Sections 3.3–3.5 in Stanley [**Sta86**]; and also see the historical discussion and references in the Notes (p 150) following Chapter 3 there. ♢

Subposets. We will also assume the standard notion of *subposet*; Namely given a subset $\mathcal{Q}$ of elements from a poset $\mathcal{P}$, we simply take the order relations in $\mathcal{Q}$ to be all[3] of the relations which are defined (in the original poset $\mathcal{P}$) between members of the subset $\mathcal{Q}$.

That is, passing from $\mathcal{P}$ to a subposet $\mathcal{Q}$ amounts to removing all the elements in $\mathcal{P}$ which are not in $\mathcal{Q}$—and then also removing the order relations on those elements being removed.

---

[2]Notational convention: We will use := to denote a definition—or more generally, a possibly temporary choice of notation. By contrast, we will use $\overset{\text{bydef}}{=}$ to mean "by definition", in the sense of a reference back to a definition or choice made earlier.

[3]as opposed to just some: i.e. we normally mean subposet in the sense of *induced* subposet (e.g. [**Sta86**, p 98]).

For example in the poset $\mathcal{P}$ of 1.1.2, the subposet $\mathcal{Q} := \mathcal{P} \setminus \{x, z\}$ is given by the linear subdiagram corresponding to the three remaining elements, namely the chain $v < w < y$. (But since this chain implies $v < y$ in $\mathcal{P}$, we would still have $v < y$ in $\mathcal{P} \setminus \{w\}$.)

Also our use of simplicial complexes, beginning in the next section, is based on the notion of a simplex—and the simplex arises as a subposet in the following way:

EXAMPLE 1.1.6 (The simplex poset). We will see (cf. later Definition 1.2.1) that a simplex σ, defined by vertices which are indexed by a finite set S, has its faces indexed by the *nonempty* subsets of S. Thus the poset of faces of σ (cf. later Definition 1.6.5) is given by the subposet $2^S \setminus \{\emptyset\}$ of the power-set poset 2^S of Example 1.1.4 above.

In particular for a 2-simplex σ based on $S := \{1, 2, 3\}$, the face-poset of σ is obtained from the diagram in Example 1.1.4 by removing the "bottom" element $\emptyset$:

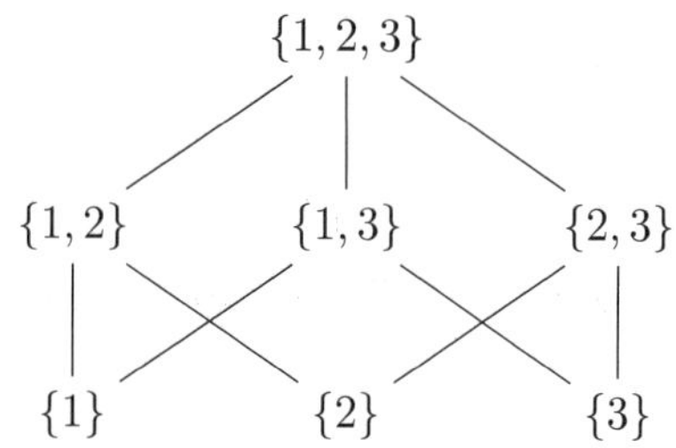

Note that pairs of minimal elements here do not have a greatest lower bound—so the simplex poset is not a lattice (cf. 1.1.5).

Simplex posets will often arise for us as indexing sets, because of flag-transitive group actions—cf. later Remark 2.1.4.4. ◊

Another class of subposets which will arise naturally at many points in our later analysis (e.g. starting at 1.6.13) can be thought of as arising from a "single generator", roughly the analogue of a principal ideal in a ring:

DEFINITION 1.1.7. Given $x \in \mathcal{P}$, we define the subposet

$$\mathcal{P}_{\leq x} := \{y \in \mathcal{P} : y \leq x\}.$$

We also define $\mathcal{P}_{<x}$, $\mathcal{P}_{\geq x}$, and $\mathcal{P}_{>x}$ in the obvious analogous ways. ◊

Thus for x as in Example 1.1.2, the subposet $\mathcal{P}_{\leq x}$ is the diamond-subdiagram on the elements other than y. And the simplex poset of Example 1.1.6 can be described in the context of the power set $\mathcal{P} := 2^S$ from Example 1.1.4 as $\mathcal{P}_{>\emptyset}$.

Posets of subgroups. We now turn to our main situation of interest, namely posets defined by suitable subgroups of a finite group.

We will assume the most basic notions of finite group theory—material normally covered in an undergraduate algebra course, such as Chapter 2 of Herstein's book [**Her75**]. In addition we will sometimes quote material more usually covered in a first-year graduate course in algebra, from texts such as Dummit and Foote [**DF99**] and Jacobson [**Jac80**]. For more detailed reference on finite group theory, see classical sources such as the books of Hall [**Hal59**], Huppert [**Hup67**], and Gorenstein [**Gor80**]; or the more recent treatment of Aschbacher [**Asc00**].

NOTATION 1.1.8 (Some notation for groups). We often use G to denote some finite group of interest. We use lower-case letters such as $g, h, \ldots$ for elements of G; with 1 used for the identity element of G. We use upper-case letters $A, B, \ldots$ for subgroups of G. We use $\leq$ to denote inclusion of subgroups; of course this has more content than just inclusion of subsets (which we will denote by $\subseteq$, whenever that distinction might be needed). $\diamondsuit$

DEFINITION 1.1.9 (The subgroup lattice $L(G)$). The *subgroup lattice* of a group G is given by:[4]

$$L(G) := \text{the set of all subgroups of } G,$$

where the role of the poset order relation "$\leq_{L(G)}$" is played by the subgroup-inclusion relation $\leq_G$. This poset is a lattice in the sense of Digression 1.1.5—since each pair A,B of subgroups has a greatest lower bound, given by $A \cap B$; and has a least upper bound, given by $\langle A, B \rangle$.

We refer to a subposet $\mathcal{P}$ of $L(G)$ as a *poset of subgroups*, or just a *subgroup poset*. Of course arbitrary subgroup posets $\mathcal{P}$ will typically not be lattices. $\diamondsuit$

For instance, Example 1.1.3 includes the full subgroup lattice $L(G)$ of a Galois group G.

In fact more explicitly, most readers will probably recall having seen a Galois group G represented via permutations of the roots of the relevant polynomial. So we will present some initial examples of subgroup posets which are taken from the symmetric group.

NOTATION 1.1.10 (Some notation for the symmetric group). For a natural number n, the symmetric group S_n is the group of permutations of the "letters" $\{1, 2, \ldots, n\}$. We write elements of S_n in the standard *cycle notation*, as in Herstein [**Her75**, 2.10.1]; indeed typically for brevity we will omit commas between letters making up a cycle—writing $(123)(45)$ rather than $(1,2,3)(4,5)$ for the corresponding permutation in S_5. In particular we will follow the usual algebraists' convention[5] as in [**Her75**] of multiplying group elements *left to right*: so that $(12)(13) = (123)$. We also recall here the useful standard formula for *conjugation*[6] of a cycle by some permutation g: writing $(a_i)g$ for the image of the letter a_i under the group element g, we have

$$g^{-1}\,(a_1\ a_2\ \ldots\ a_k)\ g \;=\; \big(\ (a_1)g\ (a_2)g\ \ldots\ (a_k)g\ \big) \tag{1.1.11}$$

(Cf. [**DF99**, Ch 4, Prop 10]—but adjust for our left-to-right convention.) $\diamondsuit$

---

[4]The notation L_G also appears in the literature.

[5]Some other standard sources such as Dummit–Foote [**DF99**, Sec 1.3] instead multiply permutations from right to left—that is, following the order of composition for functions in the conventions of analysis.

[6]Notational convention: We normally write out a conjugation $g^{-1}Ag$ in full; that is, we avoid using the fairly common group theorists' convention of A^g for this conjugation—because we want to reserve the use of a superscript g instead for the topologists' convention of fixed points in A under action of g (as in Definition 1.5.8).

EXAMPLE 1.1.12 (The subgroup lattice of S_3). We consider first the small permutation group S_3. One checks easily (e.g. [**DF99**, p 548]) that the subgroup lattice $L(S_3)$ has the form

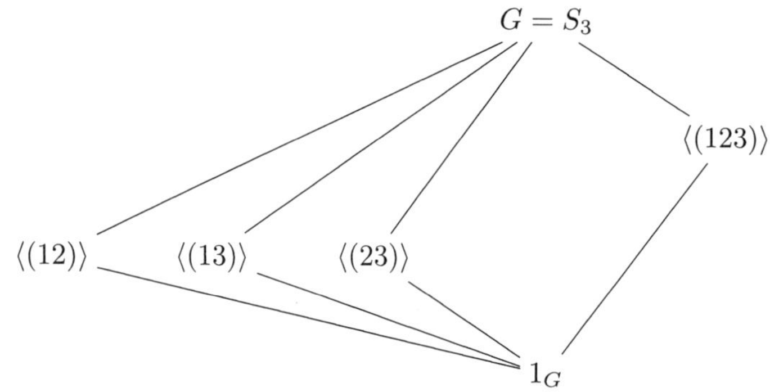

where for emphasis we place the subgroups of the descending orders 6,3,2,1 at distinct levels. ♢

The group S_3 of the example above can be viewed as the dihedral[7] group D_6 of order 6. We get a somewhat richer pattern of containments if we instead consider the next larger dihedral group:

EXAMPLE 1.1.13 (The subgroup lattice of D_8). We may realize $G := D_8$ of order 8 inside the permutation group S_4, using generators (13) and (1234). This time we find (cf. [**Hun80**, p 275]) the following subgroups, of orders $8, 4, 2, 1$:

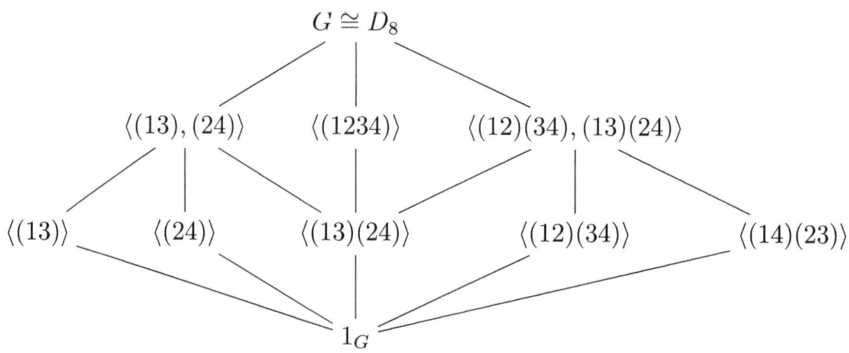

Notice that the center $Z(G)$ (the subgroup whose elements commute with all elements of G)—which is generated by $(13)(24)$ of order 2)—is contained in all three of the subgroups of order 4. (An element of order 2 is often called an *involution*.)

We can of course realize D_8 "concretely" as the symmetries of a square, i.e. a regular 4-gon; where the above elements permute the vertices—with the two generators giving a reflection and a rotation. (A more abstract notation for such generators, in the case of a regular 6-gon, is exhibited in later Example 1.1.19; and that notation could also be used in Exercise 1.1.14 below.) ♢

[7] We follow the group theorists' convention of writing D_{2n} for the dihedral group of order $2n$; some more classical sources write just D_n for that group, especially as the group of a regular n-gon.

EXERCISE 1.1.14 (The subgroup lattice of D_{12}). Draw the subgroup diagram for the full subgroup lattice $L(G)$ of the dihedral group $G := D_{12}$ of order 12.

This more complicated diagram combines various features from the previous two examples: For example, there are three subgroups of index 2 (and hence order 6)—one cyclic, with the others isomorphic to S_3. There are seven subgroups of order 2, one of which again generates the center $Z(G)$. A subposet of this lattice will be studied in later Example 1.1.19. ♢

Another larger example, again related to those above, is provided by the quasidihedral (also called semidihedral) group of order 16; its subgroup lattice can be found at [**DF99**, p. 560].

DIGRESSION 1.1.15 (Theory of the subgroup lattice). We continue the theme of Digression 1.1.5; we have seeb that $L(G)$ is indeed a lattice in that earlier sense.

Though most of the subgroup posets considered in this book are not lattices, we emphasize that there is a long tradition of research in combinatorics and group theory on the subgroup lattice $L(G)$; and we will make occasional references to that literature in later chapters. For further reference on that material, see for example Section 1 of the Hawkes-Isaacs-Özaydin paper [**HIÖ89**]: this provides a brief history of the area of subgroup lattices, and discusses some of the connections of that area with the Brown–Quillen stream of research on more general subgroup posets which is the main theme of this book. Other references include Chapter 19 in Hall [**Hal59**], and Suzuki [**Suz56**]. ♢

The p-subgroups poset $\mathcal{S}_p(G)$. As suggested in the Introduction (at our discussion of the Homological Sylow Theorem (0.0.1)), the subgroups in our posets will very often be p-subgroups, for a particular prime p dividing the order of a group G of interest. We first establish some corresponding conventions:

NOTATION 1.1.16 (Terminology for G relative to a prime p). We let p denote a prime dividing $|G|$ (which in turn denotes the order of G). Then $|G|_p$ denotes the p-part of $|G|$. Of course we assume Sylow's Theorem (see e.g. Section 4.5 in [**DF99**]): that is, the p-subgroups of G (namely those of p-power order) are contained in *Sylow p-subgroups* of G—which have order $|G|_p$, and which are all conjugate in G. We let $\mathrm{Syl}_p(G)$ denote the set of Sylow p-subgroups, and we have the standard congruence

$$|\mathrm{Syl}_p(G)| \equiv 1 \pmod{p}$$

(which we mentioned as motivation for the Homological Sylow Theorem (0.0.1)). ♢

Sylow's Theorem suggests an obvious corresponding poset of p-subgroups:

DEFINITION 1.1.17 (The nontrivial[8] p-subgroups poset $\mathcal{S}_p(G)$). Set:

$$\mathcal{S}_p(G) := \text{the poset of non-trivial } p\text{-subgroups of } G.$$

This poset was crucial in the seminal paper [**Bro75**] of Brown; and is also fundamental in the subsequent literature—where it is sometimes informally referred to as the *Brown poset*. ♢

[8]The identity p-subgroup 1 (whose order is just p^0) is conventionally excluded when we define this poset; for example, so that the poset will not always be contractible—cf. later Lemma 3.2.3.1.

In due course we will study many properties of $\mathcal{S}_p(G)$.

Of course by Sylow's theorem (cf. 1.1.16), the set $\mathrm{Syl}_p(G)$ of Sylow p-subgroups of G provides the members of $\mathcal{S}_p(G)$ which are maximal in the order relation $\leq$.

Next we recall from elementary group theory (e.g. part (3) of Theorem 1 in section 6.1 of [**DF99**]) that a finite nontrivial p-group P contains at least one maximal (indeed normal, i.e. preserved by conjugation) subgroup M, which is of index p. Using this inductively, we see that if $|P| = p^n$, then P contains a chain:

$$1 =: P_0 < P_1 < \cdots < P_n := P, \text{ with } P_i \trianglelefteq P_{i+1} \text{ and } |P_{i+1}/P_i| = p \tag{1.1.18}$$

for each $i \leq n-1$. It follows from (1.1.18) that subgroups of each possible order occur; and in particular that the minimal members of $\mathcal{S}_p(G)$ have order exactly p.

For $G := S_3$ in Example 1.1.12, with $p := 2$, the subposet $\mathcal{S}_2(G)$ of $L(G)$ consists just of the three subgroups of order 2—with no (strict) partial-order relations, so that the diagram consists of three isolated points. For $G := D_8$ in Example 1.1.13, with $p = 2$: since G itself is a 2-group, the subposet $\mathcal{S}_2(G)$ consists of the diagram for $L(G)$ with only the identity 1 removed.

Next we consider an example which is only slightly larger, and which will be of further interest later (e.g. for contractibility at Example 3.3.6).

EXAMPLE 1.1.19 ($\mathcal{S}_2(D_{12})$). Take $G = D_{12}$, the dihedral group of order 12, and $p = 2$; that is, we consider the subposet $\mathcal{S}_2(G)$ of the subgroup lattice $L(G)$ which was the subject of Exercise 1.1.14. We could for instance realize G in cycle notation as the subgroup of S_5 generated by the permutations $a := (34)$ and $b := (12)(345)$. However it will be briefer notationally to work with the abstract presentation of G given by the generators and relations:

$$G = \langle a, b : a^2 = b^6 = 1, a^{-1}ba = b^{-1} \rangle;$$

where these represent a reflection and a rotation, when G is regarded concretely as the group of a regular hexagon.

We find that the maximal members of the poset are three Sylow 2-subgroups of order 4; and the minimal members are generated by 7 involutions. Thus the poset diagram for $\mathcal{S}_2(D_{12})$ is given by:

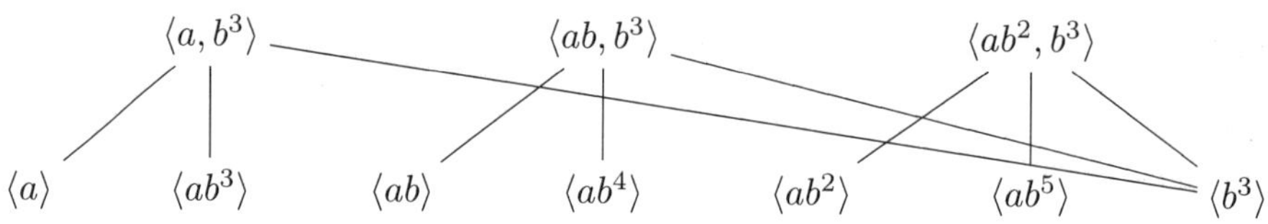

Observe, as in Example 1.1.13, that the center $Z(G)$ of order 2 (generated by b^3) is contained in all three (Sylow) subgroups of order 4. The other involutions (in pairs) lie in just one Sylow subgroup each. ◊

EXERCISE 1.1.20 ($\mathcal{S}_2(S_4)$). Draw the poset $\mathcal{S}_2(G)$ for the permutation group S_4. (This diagram is in fact given on p. 351 of [**Web87b**].) Observe that there are 3 Sylow 2-subgroups isomorphic to D_8; hence there are 3 copies of the picture in Example 1.1.13, except that we avoid repetitions whenever a 2-subgroup appears in more than one Sylow group—for example the normal subgroup $\langle (12)(34), (13)(24) \rangle$ of order 4 is common to all three Sylow groups. ◊

Next we will see how an obvious property of chains in the order relation allows us to view posets as simplicial complexes.

1.2. Subgroup complexes

In this section, we will introduce the *order complex* of a poset: this is a standard way of associating a simplicial complex to a poset—which then allows us to apply, to the poset, the viewpoint and techniques of classical algebraic topology.

We begin by recalling the context of more general simplicial complexes:

Simplicial complexes. Below we will review the standard notion of a simplicial complex; many readers should be able to skim or skip this material.

For more detailed reference on simplicial complexes, see for example Chapter 1 (especially Section 3) in the standard algebraic topology text of Munkres [**Mun84**]. We mention that Munkres' book originates from the period of vigorous development of poset topology circa 1975–1985—especially at M.I.T., around Quillen in algebraic topology, and around Rota and Stanley in poset combinatorics; the viewpoint in Munkres's text is particularly natural for the material in the present book. For a more specifically combinatorial viewpoint on complexes in poset topology, see for example sections 9 and 10 of Björner [**Bjö95**], or Walker [**Wal81b**]. A more group-theoretic viewpoint can be found for example in Chapter 6 of Benson [**Ben91**]. Also, various individual papers (including many cited in later chapters) develop some of this material as needed.

In broad overview, a simplicial complex is a geometric object that can often provide a useful "approximation" of a more general topological space—especially for the purpose of computing homology and other topological invariants.

DEFINITION 1.2.1 (Simplices and complexes). Viewed geometrically, for $n \geq 0$ a *simplex* of dimension n (or just an *n-simplex*) is just the convex hull of a set of $n+1$ points in general position:

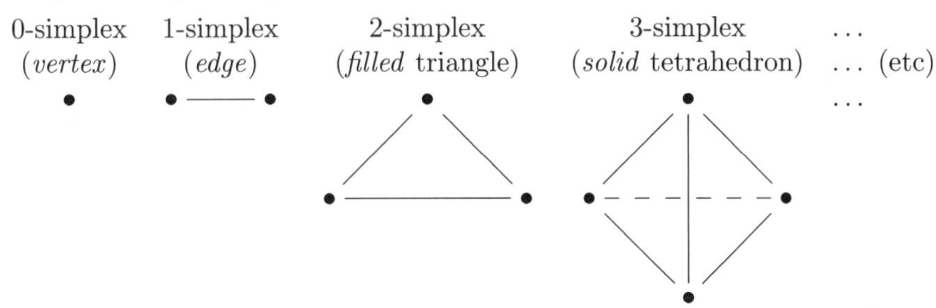

In this section, we will emphasize the property that in a simplex σ, *any (nonempty) subset of vertices also determines a simplex*—such a sub-simplex is called a *face* of σ.

A *simplicial complex* (usually we will just say *complex*) is a union of simplices of various dimensions, which typically share some of their faces. The *dimension* of a complex is the maximum of that of its simplices. ◊

EXAMPLE 1.2.2 (Surface of a tetrahedron as simplicial complex). As an initial example of a simplicial complex (which is not just a single simplex), consider the surface of a tetrahedron: for example, from the 3-simplex σ (on four vertices) drawn in Definition 1.2.1 just above, we can just *remove the interior*. Then the resulting complex K is determined by four 2-simplices (filled triangles), corresponding to the $\binom{4}{3} = 4$ triples of vertices—and so K has dimension 2. Of course the $\binom{4}{2} = 6$ pairs of vertices determine edges, each of which is shared by some pair of maximal faces. ◇

EXERCISE 1.2.3 (Surface of an octahedron as simplicial complex). Exhibit the surface of an octahedron as a 2-dimensional simplicial complex K, on a set of 6 vertices; with 12 edges, and 8 faces of dimension 2 .

Note by contrast that the surface of a cube has 6 "faces", which are squares rather than 2-simplices. However, bisecting each square along a diagonal does result in a 2-dimensional simplicial complex (a *triangulation*—cf. 1.3.3—of the surface of the cube); now with 12 rather than 6 maximal simplices of dimension 2. ◇

Our geometric viewpoint on complexes so far has been naive, in the sense of implicitly assuming some kind of embedding in a Euclidean space—with the complex being a collection of certain closed subsets. But we can in fact given an "intrinsic" viewpoint on complexes, by focusing on purely set-theoretic aspects. For observe in the surface of the tetrahedron in Example 1.2.2 that

(i) each simplex σ is determined by its (nonempty) set $V(\sigma)$ of vertices;

(ii) each nonempty subset of $V(\sigma)$ determines a sub-simplex of σ.

Observation (i) says, in the poset language introduced earlier, that the poset given by the simplices of K (under the sub-simplex relation) is isomorphic to a suitable subposet of the power set 2^V on the set V of all vertices of K. Indeed (i) suggests that the closed set σ in Euclidean space could in effect be identified with the set $V(\sigma)$ given by the vertices of σ.

And then observation (ii) gives a set-theoretic expression of the basic property of a simplex:

(1.2.4) each nonempty subset of a simplex is also a simplex.

(Cf. Example 1.1.6 for the case of a 2-simplex.) So we will use this condition, to in effect replace the classical geometric viewpoint on a complex (as a collection of suitable closed sets in Euclidean space) by a more abstract set-theoretic characterization. Thus as in standard treatments (e.g. on pages 15ff of [**Mun84**]), the condition (1.2.4) can be formalized as:

DEFINITION 1.2.5 (Abstract simplicial complex). An *abstract* simplicial complex is a collection K of finite nonempty sets σ, called the *(abstract) simplices* of K—satisfying the subset-closure property:

if τ is a nonempty subset of the set σ, then the set τ is also a simplex of K.

The sets in K of size 1 are the *vertices* of K. If we write V for the set of vertices of K, notice that a simplex σ is then determined by its set $V(\sigma)$ of vertices. ◇

This set-theoretic axiomatization of simplicial complexes sometimes turns out to be more convenient, in formal settings such as notation and many proofs, than the more geometric view of complexes in Euclidean space.

And the two indicated viewpoints are equivalent: For first, from our discussion above, it should be clear that a classical geometric complex determines an abstract complex—when we replace a geometric simplex σ by its vertex set $V(\sigma)$. And conversely, we will see shortly at Remark 1.3.3 that an abstract complex K can be embedded in a suitable subspace $|K|$ of a Euclidean topological space, in such a way that the simplices σ of K are realized as corresponding closed sets $|\sigma|$ of $|K|$.

So in order to pursue our preferred abstract formulation of complexes, we introduce:

NOTATION 1.2.6 (Some notation for abstract complexes). Often we write K for some simplicial complex, and $\sigma, \tau, \ldots$ for the simplices of K, with $\sigma \subseteq \tau$ for the *face relation* (i.e., sub-simplex containment). If σ consists of $(n+1)$ vertices, we say that σ is an *n-dimensional simplex* or *n-simplex*, and denote this by $\dim(\sigma) = n$; and we write K_n for the subcollection of n-simplices of K. In particular, a 0-simplex is also called a *vertex*, so that K_0 denotes the set of vertices of K; and a 1-simplex is also called an *edge*. We write $\dim(K)$ for the maximum dimension of a simplex in K.

A *subcomplex* of K is a subcollection L which still satisfies the subset-closure property in Definition 1.2.5. For example, the *boundary* $\partial\sigma$ of a simplex σ is the subcomplex given by the union of the proper faces of σ.

Furthermore much as for subposets (as in our discussion prior to Example 1.1.6), we often have the case of an *induced* subcomplex—determined by a subcollection L_0 of the vertices K_0 of K, along with *all* simplices which are defined (in the original complex K) on vertices of L_0. (However, note that the boundary $\partial\sigma$ above is *not* an induced subcomplex.)

Note that the simplices of an abstract complex K are just certain distinguished subsets of the set of vertices K_0 of K. We may refer to such a simplex σ, regarded just as one of the specified subsets[9] of the vertices K_0, as an *abstract* simplex.

Following Tits (e.g. in [**Tit81**]), we often use the term *incidence* for the relation on vertices, given when a pair of vertices defines an edge of the complex—or more generally, when a set of vertices defines a simplex. Thus a simplicial complex K may be given by specifying the vertex set K_0, and the incidence relations—subsets of *mutually* (i.e. subsetwise, as opposed to just pairwise) incident vertices. $\diamondsuit$

Now just as for posets in Notation 1.1.1: Except where explicitly stated otherwise, all simplicial complexes K considered in this book will be *finite*; that is, will have a finite number of simplices. Again this provides some simplifications in the foundations; cf. also the discussion at the end of Section 2 of [**Mun84**].

EXAMPLE 1.2.7 (Surface of tetrahedron again, now as abstract simplicial complex). For the surface K of the tetrahedron in Example 1.2.2, let us now choose notation $K_0 := \{1, 2, 3, 4\}$ for the vertices. In our set-theoretic realization of K, we declare its maximal simplices to be the four triples from K_0—so that the full set of simplices of K is given by their nonempty subsets. Hence in this view, K is the collection of all nonempty *proper* subsets of K_0. In particular this choice satisfies (1.2.4) from our definition—and so is an abstract simplicial complex as in Definition 1.2.5.

An example of a 2-simplex in K is given by $\{1, 2, 3\}$; of an edge, by $\{1, 2\}$; and of a vertex, by $\{1\}$. (In dealing with such small set-theoretic examples, for economy

[9] But see our later remark 1.2.17 for a mention of some more general topological contexts such as CW-complexes, which are based on structures which are *not* identified with a set of vertices.

we may sometimes omit the uses of braces and even commas—e.g. writing 123 for the above 2-simplex.)

Also we have face relations given by inclusions such as $\{1,2\} \subseteq \{1,2,3\}$, for simplices of different dimensions—where $\{1,2\} \in K_1$ and $\{1,2,3\} \in K_2$. But we do not have inclusions for the full collections of simplices of different dimensions—e.g. we do *not* have $K_1 \subseteq K_2$, since these collections have members of different sizes.

(Of course this surface complex K also arises as the boundary subcomplex $\partial\sigma$ of the simplex σ given by the solid tetrahedron—which is defined by *all* nonempty subsets of the four vertices.) $\diamondsuit$

EXERCISE 1.2.8 (Surface of octahedron as abstract complex). In the surface of an octahedron in Exercise 1.2.3, list the proper nonempty subsets of the set V of 6 vertices which occur as simplices of the complex K. In contrast with the tetrahedron in Example 1.2.7, this time not all proper nonempty subsets of the 6 vertices arise as simplices. $\diamondsuit$

We will develop further properties of simplicial complexes primarily in later sections and chapters, when they are needed. Since our primary focus will be on posets, the main purpose of this subsection has been to introduce the definition of an abstract simplicial complex via the subset-condition above; so that the next subsection can show how posets naturally give examples of complexes.

The order complex of a poset. To see how the ordering in a poset leads to the simplices of an abstract complex, we recall the standard notion:

DEFINITION 1.2.9 (Chains in a poset). A *chain* in a poset $\mathcal{P}$ (or more specifically, a $<_{\mathcal{P}}$*-chain*) is a nonempty sequence of elements $X := (x_0 < x_1 < \cdots < x_r)$ of $\mathcal{P}$, with the successive pairs (when $r > 0$) satisfying the strict order relation $<$; the *length* of X is just the number of such relations—here r, and we may refer to X as an r*-chain*. A *subchain* of X is any nonempty subsequence $x_{i_0} < x_{i_1} < \cdots < x_{i_s}$ of the elements in X.

Notice in particular that

a subchain (of a chain X in $\mathcal{P}$) is also a chain in $\mathcal{P}$.

since the partial order relation $<$ of the poset is transitive. $\diamondsuit$

The above inheritance by subsets of the chain property shows that taking the set of chains in $\mathcal{P}$ as "K" in Definition 1.2.5 gives an abstract simplicial complex—usually called the *order complex* of $\mathcal{P}$ as in Definition 1.2.11 below.

In fact this elementary but fundamental observation was not always obvious:

HISTORY 1.2.10 (Some history of complexes and topology for posets). Rota (cf. [**Rot64**]) was perhaps most influential, especially during the 1960s, in developing the early foundations of topology for posets, via methods such as "crosscuts". Björner indicates after 9.3 in [**Bjö95**], that the order complex construction goes back at least to Aleksandrov [**Ale37**]; but seemingly it was Folkman [**Fol66**] who drew wider attention to the construction of a simplicial complex from a poset, which makes the connection of posets with topology transparent. In particular, Folkman showed the equivalence of his viewpoint (which has since become standard) with the somewhat more complicated approach of Rota. For further history and references, see Sections 9 and 10 of [**Bjö95**], and the discussion on p 150 of Stanley [**Sta86**]. ◊

We now establish further notation for the use of this construction:

DEFINITION 1.2.11 (The order complex of a poset). For a poset $\mathcal{P}$, the *order complex* is:

$K(\mathcal{P}) :-$ the simplicial complex of inclusion-chains from $\mathcal{P}$.

In particular, the elements of $\mathcal{P}$ give the set $K(\mathcal{P})_0$ of vertices in the order complex $K(\mathcal{P})$. ◊

Notice also that if $\mathcal{Q}$ is a subposet of $\mathcal{P}$, then $K(\mathcal{Q})$ is an induced subcomplex of $K(\mathcal{P})$, defined by the vertices given by the members of $\mathcal{Q}$.

In this book, usually $\mathcal{P}$ will be a subgroup poset determined by some group G; and then we will say that $K(\mathcal{P})$ is the corresponding *subgroup complex*.

We make the notational comment that the "mapping" $K(-)$ should not be regarded as an ordinary function, from the poset $\mathcal{P}$ to the complex $K(\mathcal{P})$; e.g. the chains are subsets rather than just elements of $\mathcal{P}$. More precisely, $K(-)$ *includes* an ordinary set mapping, from (the set of elements of) $\mathcal{P}$ to the vertex-set $K(\mathcal{P})_0$ of $K(\mathcal{P})$.

REMARK 1.2.12 (The abbreviated notation $|\mathcal{P}|$). Some authors, notably Quillen in [**Qui78**], use the notation $|\mathcal{P}|$ for the order complex of $\mathcal{P}$. In effect, this makes no distinction between the order complex $K(\mathcal{P})$ and the "underlying topological space" $|K(\mathcal{P})|$ of that complex—a construction which we will introduce at later Definition 1.3.1. That is, Quillen's usage is essentially extending the customary usage of $|K|$ in the literature for the topological space of a complex K. But for clarity, we will typically *maintain* that distinction: That is, we will use the notation $K(\mathcal{P})$ when referring to the order complex of a poset; and reserve the notation $|K|$ for the topological space of a simplicial complex K. However, it will occasionally be convenient to adopt a variant of the abbreviated notation: see our later Notation 1.3.4, concerning writing $|\mathcal{P}|$ as an abbreviation for $|K(\mathcal{P})|$. ◊

Notice that if the chains in a poset $\mathcal{P}$ are of length at most 1, then the order complex $K(\mathcal{P})$ is of dimension 1; so the complex $K(\mathcal{P})$ is just a graph, and it coincides with the graph given by the poset diagram. This is the case for example in $\mathcal{P} := \mathcal{S}_2(D_{12})$ in Example 1.1.19. But just as in our earlier remark on the case of posets, most order complexes even of moderate size are too complicated to be easily drawn, or even visualized. We can, however, return to an earlier small example giving a complex of dimension 2 which can be fairly easily pictured:

EXAMPLE 1.2.13 (The order complex $K(L(S_3))$). Recall the diagram for the poset $\mathcal{P} := L(S_3)$ of Example 1.1.12. To extend that diagram to a picture of the simplicial complex $K(L(S_3))$: We first denote by $E := (1 < G)$ the edge of the complex which arises from the various chains of length 2 in the poset; and then we denote by F_1, F_2, F_3, F_4 the four filled triangles which are determined by each of those 2-chains with the new edge E:

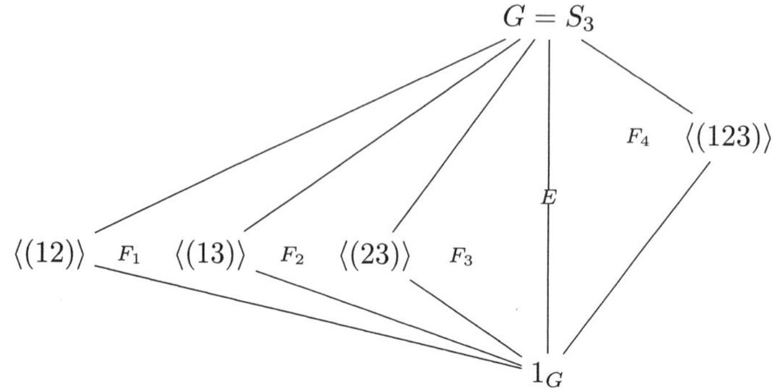

Thus $K(\mathcal{P})$ consists of four 2-simplices, all on the common edge E. ◊

EXERCISE 1.2.14 ($K(\mathcal{S}_2(D_8))$). Draw the 2-dimensional complex $K(\mathcal{S}_2(D_8))$. Recall that the poset $\mathcal{S}_2(D_8)$ is obtained from the subgroup lattice $L(D_8)$ in Example 1.1.13 by removing the identity subgroup 1. There are 7 maximal (that is, 2-dimensional) faces, so the picture will be somewhat more complicated than that in the example above. ◊

REMARK 1.2.15 (Dimension of $K(\mathcal{S}_p(G))$). Observe, using the definition of chain length in 1.2.9 and of dimension in 1.2.6, that $\dim(K(\mathcal{P}))$ is the length of the longest chain in $\mathcal{P}$. For a more general poset of subgroups in a general group G, say the full subgroup lattice $L(G)$, the length of such chains can be far from obvious. And there is a corresponding literature on the problem of subgroup-chain length; see for example [**CST89**].

However in the case of $\mathcal{S}_p(G)$, we get:

$$\dim(\ K(\mathcal{S}_p(G))\ ) = \ \log_p(\ |G|_p\ ) - 1,$$

using the standard fact (1.1.18). ◊

For example, in $\mathcal{S}_2(D_{12})$ of Example 1.1.19, the dihedral group D_{12} of order 12 has $|D_{12}|_2 = 2^2$. Hence by Remark 1.2.15, the order complex $K(\mathcal{S}_2(D_{12}))$ is of dimension 1—and so as a simplicial complex is just a graph. However for $\mathcal{S}_2(D_8)$, which is obtained by removing the trivial subgroup from $L(D_8)$ in Example 1.1.13, notice since $|D_8|_2 = |D_8| = 2^3$ that the order complex $K(\mathcal{S}_2(D_8))$ is of dimension 2.

We close the section on simplicial complexes with several final comments:

DIGRESSION 1.2.16 (Viewing a single poset or complex as a category). We mention that some authors prefer to regard posets and complexes within the context of categories and functors. (For general background on categorical language, see e.g. [**Mun84**, Sec 28], or [**Jac80**, I, Ch 1].)

For a single poset $\mathcal{P}$ can be regarded as a very easy example of a category—whose objects are given by its elements, and whose morphisms are given by the order relation $\leq$ for pairs. Similarly a single complex K can be regarded as a category whose objects are its simplices, and whose morphisms are given by the face relation $\subseteq$ for pairs.

Furthermore our later discussion in Section 1.5 of action by a group G can also be subsumed into this categorical viewpoint: namely the actions $\phi(g)$ for $g \in G$ will relate the pairs of objects in $\mathcal{P}$ of the form $(x , (x)(\phi(g)))$; so these action-relations can also be added to the collection of morphisms.

However, in this book we will not normally use the language of categories—instead we will remain in the elementary context of posets or complexes. Occasionally, we may add remarks relevant to the viewpoint of categories: For example, we had mentioned after Definition 1.2.11 that we should not regard the order complex construction $K(-)$ as an ordinary function (e.g. from the poset $\mathcal{P}$ to the complex $K(\mathcal{P})$.) However (here we preview our later more formal observation in Remark 1.4.8), we can regard $K(-)$ as in particular inducing a funct*or*—from the category of all posets to the category of all simplicial complexes: for a morphism of posets (such as an inclusion $\mathcal{P} \subseteq \mathcal{Q}$) is mapped to a morphism of complexes (here, an inclusion $K(\mathcal{P}) \subseteq K(\mathcal{Q})$). $\diamondsuit$

REMARK 1.2.17 (CW-complexes and simplicial sets). We mention also that the more classical topological literature often states standard results in the context of Whitehead's *CW-complexes* (see e.g. [**Mun84**, p. 214]), in which simplices are replaced by *cells*—which are *not* necessarily determined just by a set of vertices. These structures are sometimes convenient, for example in describing quotients modulo a G-action; cf. our discussion of later Example 3.2.9.

We mention furthermore that many of the constructions of modern homotopy theory are instead performed in the category of *simplicial sets*—in which the object corresponding to a simplex has an appropriate "semi-simplicial" structure in every dimension, allowing in particular for sets with *repeated* copies of vertices. We refer the reader for example to Section 3 of [**Dwy01**] for a discussion of the technical difficulties of simplicial complexes, and how they can be resolved using simplicial sets. For an introduction to simplicial sets more relevant to our context in this book, see Section 3.1 of [**BS08**] and the references it gives; Section 3.6 there discusses Quillen's result on the equivalence of the approaches via CW-complexes and via simplicial sets.

However, our convention in this book will be to provide treatment whenever possible at the elementary level of simplicial complexes; but when it is necessary to mention more general contexts such as CW-complexes or simplicial sets, we will just provide references (and assume results as needed). In many results stated in those more general contexts, it will suffice for our purposes in this book to simply quote those results in the special case of simplicial complexes. $\diamondsuit$

1.3. Topology for subgroup posets and complexes

We will see as we proceed that at least some topological notions for a poset $\mathcal{P}$ (for example, the Euler characteristic in 5.1.1) can be adequately described just using the context of the previous section—that is, at the level of the simplicial complex given by the order complex $K(\mathcal{P})$ of the previous section. However, a number of other important topological concepts are no easier to define for simplicial complexes than for arbitrary topological spaces; and so often we may as well work in the more general topological context.

So in this section, we briefly review the standard construction of a topological space for a simplicial complex, and some of the other fundamental concepts of more general algebraic topology that we will require. For further references on algebraic topology, as mentioned earlier, Munkres [**Mun84**] is particularly natural for our subgroup-complexes viewpoint. Some other standard references include Hilton-Wylie [**HW60**], Spanier [**Spa81**], and Hatcher [**Hat02**].

Once this material is in place, we will be able to make transitions fairly freely among the related contexts of posets, simplicial complexes, and topological spaces.

The underlying topological space of a complex. The set-theoretic definition 1.2.5 of an abstract simplicial complex K does not in any obvious way include topology—that is, open and closed sets.[10] But by contrast, a natural topology *is* visible in the usual geometric viewpoint of a complex embedded in a Euclidean space (such as in the surface of a tetrahedron in Example 1.2.2 in $\mathbb{R}^3$). Indeed there is a standard construction (which we will describe in a moment; cf. page 8 of [**Mun84**]) that associates to an abstract complex K a corresponding topological space, for which we use the standard notation $|K|$:

NOTATION 1.3.1 (The underlying topological space of a complex). For a simplicial complex K, we let $|K|$ denote the *(underlying) topological space* of K. (See also the terminology of *geometric realization* in 1.3.3 below). ◇

For our purposes in this book, the details of the construction of $|K|$ are not really important: for we will normally use only the *existence* of $|K|$, in order to invoke the usual definitions of further topological concepts for $|K|$ (with corresponding implications for the original abstract complex K).

However, for use in some later expository discussions, we do now mention a few further details: In the construction of $|K|$, the simplices of the abstract complex K (which are defined just as subsets of the vertex set K_0 in 1.2.6) are realized as standard "Euclidean simplices" on those vertices—with the topology induced from the Euclidean topology.

Indeed since we always consider a finite complex K, we can indicate a straightforward embedding of $|K|$ in the finite-dimensional Euclidean space $E := \mathbb{R}^{K_0}$ indexed by the vertices K_0 of K: First, we map each vertex v in K_0 to the endpoint $e(v)$ of the unit coordinate vector in the v-th direction. Then we map each abstract simplex σ in K_0 (regarded just as a distinguished *subset* of vertices in K_0) to the usual Euclidean *convex hull* $|\sigma|$ in E of the corresponding discrete points $e(v)$ for $v \in \sigma$, namely:

$$|\sigma| = \{ \textstyle\sum_{v\in\sigma} t_v e(v) : 0 \leq t_v \leq 1, \sum_{v\in\sigma} t_v = 1 \}.$$

We may refer to $|\sigma|$ as a *Euclidean simplex* or *geometric simplex*, in contrast with the abstract simplex σ regarded set-theoretically.

Then the topology of $|K|$ is that inherited from the Euclidean space $E = \mathbb{R}^{K_0}$; cf. page 9 of [**Mun84**]. For example, since the convex hull $|\sigma|$ is closed in E, we

[10] Indeed since our abstract complexes are *finite* sets, considered just set-theoretically they could only have the *discrete* topology—and that is not of much interest. So the process discussed here—of tying our abstract complex back to the original Euclidean-simplex viewpoint, by "filling in" continuous structures among the discrete points defining simplices in our abstract complex—definitely represents proceeding to a deeper level of analysis.

see that:

(1.3.2) For a simplex σ of K, the image $|\sigma|$ is closed in $|K|$.

REMARK 1.3.3 (The underlying space and geometric realizations). We have introduced $|K|$ as a topological space—and by that space we will normally mean the collection of *all* open sets in $|K|$. (Or equivalently, all closed sets in $|K|$.)

But we should mention that there is a concept which is intermediate between the abstract set-theoretic complex K, and the full topological space $|K|$: We could for example consider the "classical" geometric simplicial complex which is defined using *only* those closed subsets of $|K|$ which are of form $|\sigma|$ for some simplex σ of K (as opposed to taking all closed sets in $|K|$ arising from these). Sometimes this intermediate construction is referred to in the literature as a *geometric realization* of K. It has the set-theoretic content of the abstract complex K; but the geometric content of simplices arising from the space $|K|$.

However (especially because our K is always finite), we have no real need to distinguish this intermediate construction from the full $|K|$. As a result, we will also use the language of geometric realization to refer to the topological space $|K|$; we can always recover from $|K|$ the particular closed sets corresponding to simplices.

(1) We may refer to our realization of the topological space $|K|$ via the above embedding in the *particular* Euclidean space $\mathbb{R}^{K_0}$ as our *standard* geometric realization of K. But more generally, a geometric realization of K can be given using an embedding of the vertices K_0 in another Euclidean space (that is, of dimension not necessarily equal to the size of K_0)—such that the simplices of K are obtained as the appropriate convex hulls. (For example, some variant constructions are indicated within Definition 2.7.2 of [**BS08**].) Such superficially different realizations of a fixed K are in fact linearly isomorphic (cf. page 15, especially Theorem 3.1, and also page 16, in [**Mun84**]); so we usually just refer informally to "the" geometric realization of K, rather than distinguishing among various possible realizations. In particular the topological space $|K|$ is, up to homeomorphism (i.e. continuous isomorphism, cf. Remark 1.4.1), independent of the particular geometric realization.

(2) Some sources in the literature prefer to avoid the topological terminology, and refer only to the geometric realization—cf. Section 9 of Björner [**Bjö95**]; one also sometimes sees the hybrid term *topological realization*. And some authors use the double-bar notation $\|K\|$ rather than $|K|$. Ideally these minor variations should not cause the reader any serious confusion when consulting various sources.

(3) We can now be more specific with regard to our earlier mention of simplicial complexes used to "approximate" topological spaces: If X is a topological space, and K a simplicial complex such that $|K|$ is homeomorphic to X, we say K is a *triangulation* of X. (See e.g. p. 118 of [**Mun84**]). In particular, K is a triangulation of $|K|$.

(4) Much as for the earlier order complex construction $K(-)$, the realization construction $|-|$ is not an ordinary function from K to $|K|$: for example, the "object" σ of K is mapped not to a point but to a sub*set* of points $|\sigma|$ in $|K|$. (Instead, again previewing Remark 1.4.8, $|-| : K \mapsto |K|$ defines a functor from the category of all simplicial complexes to the category of all topological spaces.) $\diamondsuit$

Topology for a poset, via its complex. As just remarked, typically our simplicial complexes will in fact be order complexes—namely they will have the form $K(\mathcal{P})$ for some poset $\mathcal{P}$; so that $|K(P)|$ would be our full notation for the topological space associated with $\mathcal{P}$. However, partly following Quillen (as mentioned in our earlier Remark 1.2.12), it may sometimes be convenient to use the following briefer form:

NOTATION 1.3.4 (The shorthand notation $|\mathcal{P}|$ for the space of a poset $\mathcal{P}$). For $\mathcal{P}$ a poset, we may write $|\mathcal{P}|$ as an abbreviation for $|K(\mathcal{P})|$. However, we will *not* use $|\mathcal{P}|$ to also refer to the order complex; instead we will always write $K(\mathcal{P})$ for that complex. ◊

Of course if L is a subcomplex of K, then $|L|$ is a topological subspace of $|K|$. Hence if $\mathcal{Q}$ is a subposet of $\mathcal{P}$, $|\mathcal{Q}|$ is similarly a subspace of $|\mathcal{P}|$.

In the remainder of the chapter, we will be able to examine various standard concepts—comparing them at the related levels of poset, simplicial complexes, and topological spaces.

1.4. Mappings for posets, complexes, and spaces

We assume the usual elementary notions related to mappings of sets, such as $1:1$, onto, and isomorphism (denoted by $\cong$). We will often use mappings as a way of measuring the similarity of various posets, or complexes, or spaces.

In fact most readers will already be familiar with the notion of continuous maps—which are defined at our deepest level, namely topological spaces. The main purpose of this section will be to introduce the natural mappings at our more elementary level of complexes, and then of posets; and to further observe that such mappings do indeed lead to continuous mappings for the corresponding topological spaces.

Continuous maps for topological spaces. We first quickly review:

REMARK 1.4.1 (Continuous maps). Recall the standard notion of a *continuous* map $f : X \to Y$ of topological spaces: namely for A open in Y, $f^{-1}(A)$ is open in X.

Recall that an isomorphism of spaces (that is, when both f and f^{-1} are continuous) is called a *homeomorphism*. ◊

In fact in addition to isomorphism of spaces, we will often use weaker notions of equivalence; notably homotopy equivalence—which we will develop further in Chapters 3 and 4 (see especially Remark 3.1.11).

Simplicial maps for complexes. Now in actual practice in this book, we will rarely need to consider open and closed sets as such: For usually our topological space will come from a complex, as described in the previous section; so that the topology will be determined by the face relation. Thus it should seem plausible that mappings which preserve the simplicial face relation should also preserve the induced topology; we indicate some of the relevant details:

REMARK 1.4.2 (Simplicial maps). (Cf. 2.7 in [**Mun84**]). Assume K and L are simplicial complexes. A set mapping $f : K_0 \to L_0$ of vertices is called a *simplicial map* (in which case we write $f : K \to L$) if for each simplex σ of K, the image $f(\sigma)$

is correspondingly a simplex of L. In particular then f preserves the face relation; that is, we have:

$$\sigma \subseteq_K \tau \text{ implies } f(\sigma) \subseteq_L f(\tau).$$

for simplices $\sigma, \tau \in K$. ◊

For example, the inclusion of a sub-simplex in a larger simplex is a 1:1 simplicial map; more generally, so is an inclusion of a subcomplex in a complex.

REMARK 1.4.3 (Collapsing a 2-simplex to an edge). Here is an example of a simplicial map which is not 1:1 (and in particular reduces dimensions): The non-injective set map on $\{1, 2, 3\}$ given by $f : (1, 2, 3) \mapsto (2, 2, 3)$ induces a simplicial map on the 2-dimensional simplex defined by the vertices $\{1\}$, $\{2\}$, $\{3\}$—which "collapses" that simplex down onto its edge $\{2, 3\}$. (We will revisit this example later at Remark 3.1.3; and indeed starting at Definition 3.4.3, we will routinely apply the term *collapse*, in a formal topological sense, to generalizations of this example.) ◊

REMARK 1.4.4 (The induced map $|f|$ of spaces). If we are given a simplicial map $f : K \to L$, then we can make a natural extension of the language of Notation 1.3.1, writing $|f| : |K| \to |L|$ for the map induced by f on the underlying topological spaces.

But we emphasize, much as in our discussion before Definition 1.3.1, that $|f|$ adds some non-trivial geometric content to the information provided by f, by "filling in" the convex hulls in the discussion after Definition 1.3.1: That is, $f(\sigma)$ just records the images of the individual vertices which define σ, as an abstract simplex in the language of Notation 1.2.6. But $|\sigma|$ is instead the full convex hull, in a Euclidean space, of the discrete points corresponding to the vertices of σ; and then all these points are mapped by $|f|$ to those of the full convex hull $|f(\sigma)|$—of the discrete points corresponding to the image $f(\sigma)$. This is the fuller meaning of the notation $|f|(|\sigma|)$ for the image $|f(\sigma)|$ of the induced map on an abstract simplex. ◊

We see from Remark 1.4.2 that $|f|$ maps closed sets to closed sets; that is, $|f|$ is a *closed map* (cf. [**Mun84**, p. 111]). Indeed standard arguments, e.g. 2.7 in [**Mun84**], show that the map is continuous:

LEMMA 1.4.5. *If $f : K \to L$ is a simplicial map of complexes, then f induces a continuous map $|f| : |K| \to |L|$ of topological spaces.*

Poset maps. Our complexes will most often arise in the form $K := K(\mathcal{P})$ of the order complex for some poset $\mathcal{P}$. Hence we wish to make an analysis analogous to that above, for maps which preserve the order relation of the poset. In order to allow for the possibility of non-injective maps, in the obvious definition we use $\leq$ rather than the strict relation $<$:

DEFINITION 1.4.6 (poset map). A set mapping $f : \mathcal{P} \to \mathcal{Q}$ of posets is a *poset map* if

$$x \leq_{\mathcal{P}} y \text{ implies } f(x) \leq_{\mathcal{Q}} f(y) .$$

We may also use the term *poset morphism*. ◊

For example, for $\mathcal{P}$ a subposet of $\mathcal{Q}$, the natural inclusion $i : \mathcal{P} \to \mathcal{Q}$ is a 1:1 poset map. For a non-injective example, we can adapt our earlier Remark 1.4.3: the set mapping $(1, 2, 3) \mapsto (2, 2, 3)$ collapses the poset $1 < 2 < 3$ down to the subposet $2 < 3$.

Since the order relation $\leq$ on $\mathcal{P}$ determines the $<$-chains which are the simplices of $K(\mathcal{P})$, we see that the induced map $K(f)$ takes simplices to simplices. This immediately gives us the natural analogue below of Lemma 1.4.5. (In particular we use the natural extension of the shorthand notation $| - |$ in 1.3.4, namely $|f|$ for $|K(f)|$.)

LEMMA 1.4.7. *Assume* $f : \mathcal{P} \to \mathcal{Q}$ *is a poset map. Then*
(1) $K(f) : K(\mathcal{P}) \to K(\mathcal{Q})$ *is a simplicial map.*
(2) $|f| : |\mathcal{P}| \to |\mathcal{Q}|$ *is a continuous map.*

PROOF. Note that (2) follows from (1), using Lemma 1.4.5. □

REMARK 1.4.8 (The category of all posets (or complexes)). We can now more formally justify an earlier informal observation we had made, using the language of categories. (As suggested by the use of the term of "morphism" in Definition 1.4.6.)

Namely instead of viewing a single poset as a category as in Remark 1.2.16, one can instead take the collection of *all* posets as the objects of a category; and then the poset maps give the relevant morphisms. Similarly the collections of all complexes gives a category, in which simplicial maps are the morphisms.

And in this language, Lemma 1.4.7 essentially shows that our notation of K for the induced map (more precisely, $K : f \mapsto K(f)$) defines K as a *functor* between the two categories just described; and further that $| - |$ defines a functor—from either of those categories, into the analogous category of topological spaces. ◇

1.5. Group actions on posets, complexes, and spaces

In this book, our posets and complexes will usually come from some collection of subgroups of a group G; and then the permutations of those subgroups coming from the conjugation action by the elements $g \in G$ will provide an important special case of the natural mappings described in the previous section.

The appropriate definitions for general group actions are similar, for all three of our usual levels of analysis:

DEFINITION 1.5.1. A *G-action* on a poset $\mathcal{P}$ is given by a group homomorphism ϕ of G into $Aut(\mathcal{P})$—the group of poset maps which are isomorphisms of $\mathcal{P}$ with itself. In that case we say $\mathcal{P}$ is a *G-poset*.

Similarly for a simplicial complex K, a G-action is a group homomorphism of the form $\phi : G \to Aut(K)$, into the group of simplicial self-automorphisms of K; and we say K is a *G-complex*.

Finally a G-action on a topological space X is a group homomorphism of the form $\phi : G \to Homeo(X, X)$, into the group of self-homeomorphisms of X (for which we may also use the notation $Aut(X)$); and we say X is a *G-space*. ◇

NOTATION 1.5.2 (Our conventions for group actions). For mappings, we usually follow the standard calculus-style functional notation of writing mappings on the left—for example, $f(x)$ in Definition 1.4.6, and also $\phi(g)$ in Definition 1.5.1.

However, when we want to then consider the above image $\phi(g)$ as a member of $Aut(\mathcal{P})$—namely regarding $\phi(g)$ in turn being a mapping, now from $\mathcal{P}$ to $\mathcal{P}$—then we typically instead follow the more algebraic convention of writing this latter action *on the right*: This is to conform with the standard convention for group multiplication, from left to right, as in Notation 1.1.10; in particular this avoids any necessity to use inverses of group elements, when we are expressing the general group-homomorphism property for ϕ. That is, if for $x \in \mathcal{P}$ we write the image as $(x)(\phi(g))$, then the homomorphism condition takes the form:

$$(x)(\phi(g_1g_2)) = ((x)\phi(g_1))\ \phi(g_2) \quad \text{for } x \in \mathcal{P} \text{ and } g_1, g_2 \in G. \tag{1.5.3}$$

Furthermore when we are given a G-action ϕ, we will often simplify notation by letting the action of $\phi(g)$ be denoted just by the element g itself. And then for example the above homomorphism condition appears instead in the briefer form:

$$(x)(g_1g_2) = ((x)g_1)g_2.$$

This convention is useful when ϕ is clear, so that it can just be left implicit. ◇

We can now return to the conjugation-action suggested at the start of the section:

EXAMPLE 1.5.4 (G-action on subgroup lattices). For our poset $\mathcal{P}$, we first take the full subgroup lattice $L(G)$ of our group G. And we let ϕ be the conjugation map c: namely for $g \in G$, the map $c(g)$ on $H \leq G$ is given by $(H)(c(g)) := g^{-1}Hg$—using the action on the right indicated in Notation 1.5.2.

It is elementary that conjugation preserves subgroup inclusions, so that $c(g)$ is a poset map on $L(G)$; and also that $c(g)$ and $c(g)^{-1} = c(g^{-1})$ are 1:1 and onto, so that $c(g) \in Aut(\mathcal{P})$. And the group homomorphism property, for c as in (1.5.3) above, is just the elementary observation that $(gh)^{-1}H(gh) = h^{-1}(g^{-1}Hg)h$. Thus c is indeed a G-action, and $L(G)$ is a G-poset.

Indeed note that the same observations apply to any subposet of $L(G)$ which consists of a union of G-conjugacy classes of subgroups—such as $\mathcal{S}_p(G)$ of Example 1.1.17. For this reason, most of the posets we consider will even be G-posets, essentially as a subcase of the present Example. ◇

Furthermore we next observe, using the results of the previous section, that these G-poset examples in turn give rise to examples of G-complexes and G-spaces.

LEMMA 1.5.5. *Let G be a (finite) group.*

(1) For ψ a G-action on a complex K, the map $|\psi|$ is a G-action on $|K|$.

Let ϕ denote a G-action on a poset $\mathcal{P}$. Then

(2) The induced map $K(\phi)$ is a G-action on the order complex $K(\mathcal{P})$.

(3) The induced map $|\phi|$ is a G-action on the space $|\mathcal{P}|$.

PROOF. Since the images $\psi(g)$ are simplicial maps on K, (1) follows from Lemma 1.4.5. Using poset maps in a similar way, (2) follows from Lemma 1.4.7.1; and then (3) follows from (2) by applying $|-|$ and then quoting (1). □

Of course in addition to using the above Lemma, we can also directly obtain examples of G-complexes and G-spaces. We mention:

EXAMPLE 1.5.6 (Action of S_4 on the surface of the tetrahedron). Consider the complex K given by the surface of the regular tetrahedron—as realized abstractly on vertices $K_0 = \{1, 2, 3, 4\}$ in Example 1.2.7. The natural action of an element g, from the symmetric group S_4, induces a permutation of the subsets of K_0 which are the simplices of K—that is, this permutation defines a member $\phi(g)$ of $Aut(K)$. Clearly this association $g \mapsto \phi(g)$ satisfies the homomorphism condition in (1.5.3); and so ϕ defines a G-action on K. (Indeed $Aut(K) \cong S_4$ in this particular case.) $\diamondsuit$

REMARK 1.5.7 (Type-preserving actions on complexes). Note in Example 1.5.6 above that the subgroup S_3 stabilizes the 2-simplex defined by the vertices $\{1, 2, 3\}$—and in fact permutes the vertices of that simplex nontrivially.

For many of the G-complexes K in our later chapters, it will be more common to have the situation where the stabilizer G_σ of a simplex σ *must in fact also fix each vertex of* σ; and then it also acts trivially on all faces of σ. (This property turns out to be crucial in some later arguments; see e.g. the discussion in Example 4.4.5.)

For such a G-action, we will use the terminology of *type-preserving*:[11] since as we will see in the discussion around later Remark 2.3.3, it often arises when a simplex is composed of vertices of distinct "types" which are preserved by the G-action.

Indeed a case that arises very frequently for us is when $\mathcal{P}$ is some G-poset—and then the action of G on the order complex $K(\mathcal{P})$ is type-preserving: since an element $g \in G$ stabilizing a (finite!) chain of proper inclusions in $\mathcal{P}$ must in fact stabilize each term of the chain. That is,

G-action on a subgroup complex is order-preserving.

(One could think of subgroup order as the "type" being preserved.) $\diamondsuit$

We end the section with several concepts which arise naturally within the context of G-actions.

DEFINITION 1.5.8 (Fixed-point subposets). If $\mathcal{P}$ is a G-poset, and A is a subgroup (or even subset) of G, then the *fixed-point subposet* is:

$$\mathcal{P}^A := \{x \in \mathcal{P} : (x)a = x \quad \text{for all } a \in A\}.$$

where we use the shorthand a rather than $\phi(a)$ as indicated at the end of Notation 1.5.2. In particular if $\mathcal{P}$ is a G-subposet of the subgroup lattice $L(G)$ with the usual G-conjugation action, then $\mathcal{P}^A$ consists of the subgroups in $\mathcal{P}$ which are *normalized* by A.[12]

There are obvious analogous notions of the fixed-point subcomplex K^A of a G-complex K, and the fixed-point subspace X^A of a G-space X.

Furthermore the fixed-point subposet $\mathcal{P}^A$ determines in the order complex $K(\mathcal{P})$ the fixed-point subcomplex $K(\mathcal{P}^A) = K(\mathcal{P})^A$; and in the space $|\mathcal{P}|$, determines the fixed-point subspace $|\mathcal{P}^A| = |\mathcal{P}|^A$. $\diamondsuit$

---

[11]There seems to be no universal terminology for this concept: e.g. one also sees *admissible* and *regular*. Our use of type-preserving is mainly for convenience, as often our complex will have some clear notion of type. (For a general G-complex, the property could always be obtained by passing to a barycentric subdivision.)

[12]For certain actions, we may also denote fixed points using the group theory *centralizer* notation: for example in the conjugation action of g on elements of a group G, we may write G^g as $C_G(g)$; and similarly for action on elements of a vector space V, we may write V^g as $C_V(g)$.

For example if $G := S_3$ and $\mathcal{P} := L(S_3)$ as in Example 1.1.12, with $A := (12)$, then $\mathcal{P}^A$ is the subposet defined by 1, G, $\langle(12)\rangle$, and $\langle(123)\rangle$. (The subgroups $\langle(13)\rangle$ and $\langle(23)\rangle$ are interchanged by (12), and so do not lie in the fixed-point subposet.)

In our discussion in the previous section, we introduced the mappings which are natural in the sense of preserving the defining relations of posets and complexes. But we also observed in Example 1.5.4 that we can usually expect our posets and complexes to support G-actions; and when we regard those actions also as part of the fundamental structures, we typically wish to restrict attention to mappings which respect that additional structure:

DEFINITION 1.5.9 (G-equivariant map). A mapping $f : \mathcal{P} \to \mathcal{Q}$ of G-posets is said to be *G-equivariant* if it commutes with the G-actions:

$$f(xg) = \big(f(x)\big)g \quad \text{for all } x \in \mathcal{P} \text{ and } g \in G.$$

There are obvious analogous definitions of G-equivariance for maps between G-posets, and between G-spaces. ◇

For example, the natural inclusion of a G-subposet in a G-poset will be a G-equivariant poset map—as in the case of $\mathcal{S}_p(G) \subseteq L(G)$ in Example 1.5.4.

Finally we mention that there is a considerable combinatorial literature relevant to group actions on posets; see for example Section 1 of Stanley [**Sta82**]; another reference for general background is Section 1 of Curtis-Lehrer [**CL81**].

1.6. Some further constructions related to complexes

We close the chapter with a quick review of some standard basic constructions for simplicial complexes, which we will soon need as we continue our development: namely the barycentric subdivision, the join, and the link. As usual, the reader familiar with these concepts can probably just skip over this material.

The barycentric subdivision for complexes and posets. In overview, the subdivision constructs a certain "refinement" of a simplicial complex—which at the underlying topological level remains homeomorphic with the realization for the original complex. In particular, this begins our study of various types of equivalence: namely relations at the level of complexes or posets, which involve some suitable notion of similarity weaker than isomorphism. (At the topological level, these relations will typically induce at least a homotopy equivalence; this theme is continued later especially in Chapters 3 and 4—see e.g. Remark 3.1.11.)

We will also be able to examine the subdivision from our poset viewpoint.

But we will begin just with a fairly informal geometric description of the *barycentric subdivision* $\mathrm{Sd}(K)$ of a simplicial complex K. (For details, see e.g. Section 15 of [**Mun84**].) In the intuitive geometric view of K (and hence in its geometric realization), in order to form $\mathrm{Sd}(K)$, we simply place an additional vertex at the barycenter (i.e. center of mass) of each simplex of K; and then we form "new" simplices on sets of these new vertices, whenever the corresponding original simplices were mutually incident in the face relation. For instance:

EXAMPLE 1.6.1 (The barycentric subdivision of a 2-simplex). We will subdivide a 2-simplex K with vertex set $K_0 = \{1, 2, 3\}$: First we indicate these vertices below, with a solid node •. Since simplices in the abstract complex K are just subsets of vertices in K_0, we use each subset larger than a singleton to index a hollow node ∘—indicating the corresponding "new" vertex of $\mathrm{Sd}(K)$. And we use dotted lines to indicate the new edges—which connect pairs of vertices that satisfy a face relation, of form $\{i\} \subseteq \{1,2,3\}$ or $\{i,j\} \subseteq \{1,2,3\}$. Finally these edges, along with those arising from having subdivided the old edges, complete the boundary of each of the 6 new 2-simplices, each corresponding to a set of mutual face relations of the form $\{i\} \subseteq \{i,j\} \subseteq \{1,2,3\}$. We sketch this (with the new nodes only roughly centered):

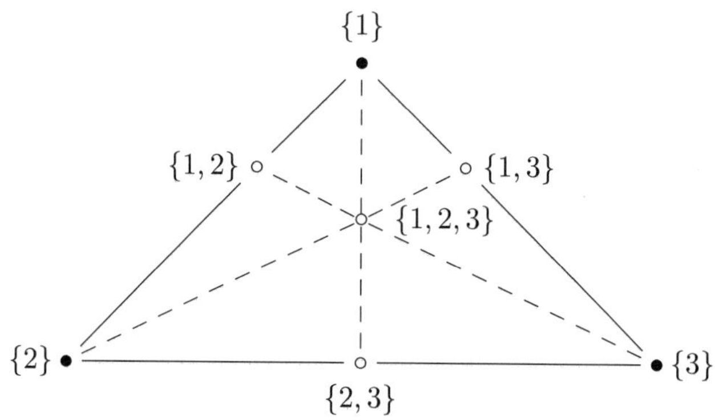

◇

EXERCISE 1.6.2 (barycentric subdivision of a 3-simplex). Similarly analyze the barycentric subdivision of a 3-simplex K. Here we get $4! = 24$ possible choices for an inclusion-chain of faces of successive dimensions $0, 1, 2, 3$; and so $\mathrm{Sd}(K)$ has 24 maximal (i.e. 3-dimensional) simplices. (The structure is a bit complicated to try to sketch.) ◇

The reader has probably noticed that in the above construction, we are in effect considering the set of simplices of a complex as a poset (cf. Example 1.1.6)—where the simplicial face relation defines the ordering; and indeed we will return to this viewpoint shortly, at Definition 1.6.5. But for the moment, we will remain in the "classical" viewpoint of complexes. And indeed in our set-theoretic context 1.2.5 of abstract simplicial complexes, the standard definition (cf. 15.3 in [**Mun84**]) is given by:

DEFINITION 1.6.3 (The barycentric subdivision of a complex). The *barycentric subdivision* $\mathrm{Sd}(K)$ of an abstract simplicial complex K has vertices given by the original simplices of K; and simplices of higher dimension are given by chains of original simplices under the proper face relation $\subsetneq$. ◇

We emphasize that the map $\mathrm{Sd} : K \to \mathrm{Sd}(K)$ is not completely "natural", if we adopt the strictest viewpoint of simplicial complexes: In particular, it is not a simplicial map in the sense of Remark 1.4.2: since for example the vertices of an edge of K (for instance, $\{1\}$ and $\{2\}$ in Example 1.6.1) are mapped to vertices which are not on an edge of $\mathrm{Sd}(K)$. Hence K and $\mathrm{Sd}(K)$ are not isomorphic at the level of complexes.

However, we do still obtain a homeomorphism of the underlying topological spaces; that is, of $|K|$ with $|\mathrm{Sd}(K)|$: Indeed in Example 1.6.1, the obvious geometric realization of $|\mathrm{Sd}(K)|$ is even the *same* filled-in triangle defined by $|K|$—only the triangulation, i.e. the "marking" via simplices, differs. For a general K, it is a consequence of Definition 1.6.1 (cf. p. 83 of [**Mun84**]) that a particular geometric realization of the new complex $\mathrm{Sd}(K)$ can be exhibited on the same subset of $\mathbb{R}^{K_0}$ which is provided by the standard geometric realization of the original complex K. And then the realization $|K|$ and the realization $|\mathrm{Sd}(K)|$ of the subdivision again coincide as sets of points in Euclidean space. So in particular, $|K|$ and $|\mathrm{Sd}(K)|$ inherit the same Euclidean topology. And finally: Since by remark 1.3.3.1, the topological space $|\mathrm{Sd}(K)|$ is up to homeomorphism independent of *any* particular geometric realization, we have the general fact:

LEMMA 1.6.4. *For a simplicial complex K, $|\mathrm{Sd}(K)|$ is homeomorphic to $|K|$.*

Furthermore if K is a G-complex in the sense of Definition 1.5.1, then it also follows easily that the above homeomorphism is G-equivariant in the sense of Definition 1.5.9

From this homeomorphism, we see that if we wish to investigate the topological properties of $|K|$, we may instead investigate those properties in the barycentric subdivision $|\mathrm{Sd}(K)|$. And we will see later (e.g. Lemma 2.2.20) that sometimes it is indeed advantageous to instead work with the latter. In particular we will see, notably among the examples developed in the following chapter (e.g. (2.1.22)), that for some ostensibly-different complexes, one may arise naturally as the subdivision of another. Thus the homeomorphism, at the level of the underlying topology for K and $\mathrm{Sd}(K)$, provides one useful notion of equivalence for complexes, which is weaker than isomorphism as complexes.

We now return to an idea mentioned earlier, of regarding the set of simplices of a complex, with the inclusions among them, from the viewpoint of a poset. This will allow us to re-phrase the barycentric subdivision in language which is very natural for the poset viewpoint in this book.

And also, since more usually we proceed in the direction from posets to the deeper level of complexes (via the order complex construction of Definition 1.2.11), this will provide an instance of how we can also move in the opposite direction, from complexes to the more elementary level of posets. We start with:

DEFINITION 1.6.5 (Face poset). For a simplicial complex K, the *face poset* of K is defined by

$\mathcal{P}(K) :=$ the poset of simplices of K, ordered by the face relation $\subseteq$.

(Some authors denote this construction by $S(K)$, the poset of *simplices* of K). $\diamondsuit$

Of course the only real difference between K and $\mathcal{P}(K)$ is that $\mathcal{P}(K)$ takes "official" notice of the inclusions among the various subsets which define K as an abstract complex. So often, it is not really necessary to distinguish between K and $\mathcal{P}(K)$; and then we may refer informally to the complex K as a poset—meaning of course the face poset $\mathcal{P}(K)$ of K.

EXAMPLE 1.6.6. Let K be the edge on vertices $\{1\}$ and $\{2\}$. Then $\mathcal{P}(K)$ has its "bottom" level given by these two 0-simplices—both are contained in the unique 1-simplex $\{1,2\}$ at the "top" level:

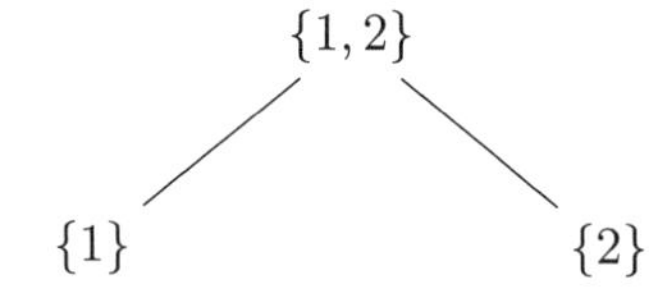

Note that the order complex $|K|$ of $\mathcal{P}(K)$ has two edges—giving the barycentric subdivision of the single edge K. $\diamondsuit$

As this example suggests, we can follow Definition 1.6.5 of the face poset by Definition 1.2.11 of the order complex, to obtain an equivalent viewpoint on Definition 1.6.3 of the subdivision:

LEMMA 1.6.7 (Cf. 1.4 in Quillen [**Qui78**]). *For a simplicial complex L,*

$$\mathrm{Sd}(L) = K(\mathcal{P}(L)).$$

This observation helps to render our two standard viewpoints, of posets and of complexes, roughly interchangeable. Furthermore the observation also suggests a natural extension of the notion of subdivision to posets—by reversing the order of composition of the constructions involved:

DEFINITION 1.6.8 (The barycentric subdivision of a poset). For a poset $\mathcal{Q}$, the *barycentric subdivision* is

$$\mathrm{Sd}(\mathcal{Q}) := \mathcal{P}(K(\mathcal{Q})).$$

That is, $\mathrm{Sd}(\mathcal{Q})$ is the poset of all inclusion-chains in $\mathcal{Q}$, ordered by the subchain relation. $\diamondsuit$

For example, if $\mathcal{Q}$ is the poset $1 < 2$, then $K(\mathcal{Q})$ is the 1-simplex on those two vertices; hence the face poset of this edge gives the subdivision poset $\mathrm{Sd}(\mathcal{Q})$, namely the poset in Example 1.6.6. (Notice that the asymmetry of 1 and 2 in the ordering of $\mathcal{Q}$ has disappeared in the subdivision poset—whose ordering corresponds only to the simplex face relation, i.e. *inclusion of chains* in the original ordering on $\mathcal{Q}$.)

The join of complexes. For simplicial complexes (in contrast to the deeper level of topological spaces), it is standard that the natural notion of direct product is not the usual Cartesian product, but instead the *join*. We will see (for example at later Lemma 8.1.2) that the join is the right product to use, when we apply standard subgroup-poset and subgroup-complex constructions to the direct (Cartesian) product of groups. So we now briefly review the notion of join; for fuller detail, see e.g. Section 62 of [**Mun84**].

Roughly, to join two simplices, we "make all possible connections": Thus the join of a vertex v and an edge E will put v on an edge with every vertex w of E; so that the join $v * E$ "fills out" a 2-simplex—whereas the Cartesian product of v and E would be just a single edge again. Similarly the join $E * E'$ of two edges gets the structure of a 3-simplex on the four vertices involved (rather than the "square" structure of their Cartesian product).[13] Geometrically, the simplices in such a join are given by the convex hulls of all the nonempty subsets taken from the *union* of

[13]The relation of the join with the Cartesian product is a little indirect—cf. [**Qui78**, 1.9].

the vertices of the factors; and this is very naturally expressed within our context of abstract simplicial complexes:

DEFINITION 1.6.9 (Join of simplicial complexes). Assume that σ and τ are abstract simplices, defined by (disjoint) vertex sets σ_0 and τ_0. Then the *join* $\sigma * \tau$ is the abstract simplex defined on the vertex set $(\sigma * \tau)_0 := \sigma \cup \tau$.

Thus the "typical" faces of $\sigma * \tau$ have the form $\sigma' * \tau'$, where $\sigma' \subseteq \sigma$ and $\tau' \subseteq \tau$. (But in order to allow for the fact that σ and τ themselves are faces of $\sigma * \tau$, we implicitly include the cases where one of σ' or τ' could be empty; this essentially abuses the standard assumption for simplicial complexes that faces are nonempty.)

More generally, assume that K and L are two abstract simplicial complexes (on disjoint vertex sets K_0 and L_0). Then the join $K * L$ is the complex (on vertex set $(K * L)_0 := K_0 \cup L_0$) whose simplices are the joins $\sigma * \tau$ for $\sigma \in K$ and $\tau \in L$—where again we allow for the "abusive" case that one of σ or τ might be empty. We saw above that $\sigma * \tau$ is an abstract simplex, and so satisfies the subset-closure property in Definition 1.2.5; so $K * L$ is indeed an abstract simplicial complex. $\diamondsuit$

In particular we can regard an n-simplex as the iterated join of its $n+1$ vertices. Or more generally, the n-simplex can be decomposed as the join of the various simplices generated by any partition of its vertices. (The order is immaterial; furthermore the join is associative—cf. 62.4 in [**Mun84**].)

The star and link of a vertex (or simplex) in a complex. In our viewpoint in the next chapter on simplicial complexes as geometries for simple groups, we will often wish to consider suitable "local" subcomplexes. At the level of topological spaces, local properties are often examined in terms of the neighborhood of a point (i.e. an open set containing that point). For complexes, an analogue is provided by the *star* of a vertex—and indeed a kind of "cross section" of the star will be provided by the *link* of that vertex. We give a quick review; for fuller background on these topics, see e.g. [**Mun84**, pp 11, 371].

In overview, the (closed) star of a simplex σ is the subcomplex generated by all simplices which contain σ. And then the link of σ is obtained by roughly "removing" σ from each of those simplices generating the star; that is, the link consists of all simplices which are disjoint from σ, but which are incident with σ in the sense of Notation 1.2.6. (And it will turn out below at Lemma 1.6.12 that the star is obtained as the join of σ with its link.)

In more detail, we will first follow the traditional development, starting with the definition of the closed star, and then (independently) of the link. Again this is conveniently done in the set-theoretic context of abstract simplicial complexes:

DEFINITION 1.6.10 (Star and link of a simplex). Assume that σ is a simplex of an abstract simplicial complex K.

The *closed star* of σ is the subset of K defined by

$$\overline{\mathrm{St}}(\sigma) := \{\, \tau \in K : \sigma \cup \tau \in K\};$$

where we have written $\in$ in the expression "$\sigma \cup \tau \in K$" as a convenient shorthand for the requirement that the vertex-set $\sigma_0 \cup \tau_0$ should define a simplex of K. (Again we implicitly allow for the abusive case where one of σ or τ is empty.) Since by construction $\overline{\mathrm{St}}(\sigma)$ consists of simplices, we see that it is in fact a sub*complex* of K. In particular, $|\overline{\mathrm{St}}(\sigma)|$ gives a closed set in the geometric realization $|K|$, as suggested by the terminology of *closed* star. We may use a subscript, as in $\overline{\mathrm{St}}_K(\sigma)$, when it is desirable to specify the complex K.

Typically we will work at the level of simplicial complexes, and hence implicitly also restrict attention to the closed sets $|\overline{\mathrm{St}}(\sigma)|$. However in a few later arguments (especially in Chapter 3) which are more explicitly topological in nature, it will be convenient to work instead with the corresponding *open* star—namely a certain open set of $|K|$, whose closure gives this closed set $|\overline{\mathrm{St}}(\sigma)|$. For this, we obtain constituent open sets from the interiors of the geometric simplices containing $|\sigma|$. Thus the *(open) star* of σ in $|K|$ is the open subset defined by

$$\mathrm{St}_{|K|}(\sigma) := \bigcup_{\sigma \subseteq \tau \in K} (\ |\tau| \setminus |\partial\tau|\ ).$$

Since $\mathrm{St}(\sigma)$ is open, it is not a subcomplex of the geometric realization complex $|K|$. This definition in $|K|$ also suggests a roughly corresponding structure in the abstract complex K, namely:

$$\mathrm{St}_K(\sigma) := \bigcup_{\sigma \subseteq \tau \in K} \tau\ ;$$

here we are deviating from the standard notational convention in [**Mun84**]—in particular we are using essentially the same notation as for the open star above, though of course in the abstract complex K, we have no real notion of open sets.

Finally the *link* of σ is defined by

$$\mathrm{Lk}(\sigma) := \{\rho \in K : \rho \cap \sigma = \emptyset,\ \rho \cup \sigma \in K\},$$

where similarly "$\rho \cap \sigma = \emptyset$" is shorthand for the more precise statement $\rho_0 \cap \sigma_0 = \emptyset$ about vertices. Again $\mathrm{Lk}(\sigma)$ consists of simplices, and so defines a subcomplex of K; so $|\mathrm{Lk}(\sigma)|$ is closed in $|K|$. ◇

EXAMPLE 1.6.11. Recall the poset $\mathcal{P} := \mathcal{S}_2(D_{12})$ in Example 1.1.19: as the poset has just two levels, its order complex $K := K(\mathcal{S}_2(D_{12}))$ is 1-dimensional—and so we can picture K just as the poset diagram for $\mathcal{P}$.

If we take σ to be the vertex $\langle a, b^3\rangle$, we find that the closed star $\overline{\mathrm{St}}(\sigma)$ consists just of the three edges on σ—which are on the vertices $\langle a\rangle$, $\langle a, b^3\rangle$, and $\langle b^3\rangle$. (This star is just the order complex of the subposet $\mathcal{P}_{\leq\sigma}$ from Definition 1.1.7.)

Further $\mathrm{Lk}(\sigma)$ consists just of those three vertices $\langle a\rangle$, $\langle a, b^3\rangle$, and $\langle b^3\rangle$. (This link is just the order complex of the subposet $\mathcal{P}_{<\sigma}$.)

For the open star $\mathrm{St}(\sigma)$ in $|K|$, we in effect remove from the geometric realization $|\overline{\mathrm{St}}(\sigma)|$ just the three lower vertices $\langle a\rangle$, $\langle a, b^3\rangle$, and $\langle b^3\rangle$. In particular *we obtain* $\mathrm{St}(\sigma)$ *as* $|\overline{\mathrm{St}}(\sigma)| \setminus |\mathrm{Lk}(\sigma)|$. ◇

We can now return to the connection with the join mentioned earlier: In view of our remarks after Definition 1.6.9 of the join, we can decompose any simplex in the closed star $\overline{\mathrm{St}}(\sigma)$ in the form of a join $\sigma * \tau$, where τ is a simplex in the link $\mathrm{Lk}(\sigma)$. Thus we obtain the relation:

LEMMA 1.6.12 (e.g. 62.6 in [**Mun84**]). *For σ a simplex of a complex K,*

$$\overline{\mathrm{St}}(\sigma) = \sigma * \mathrm{Lk}(\sigma).$$

For instance in the order complex $K := K(L(S_3))$ of Example 1.2.13, for σ given by the vertex $\langle(123)\rangle$, the link $\mathrm{Lk}(\sigma)$ is the edge $E = (1 < G)$; and the closed star $\overline{\mathrm{St}}(\sigma)$ is the 2-dimensional simplex $\langle(123)\rangle * E = F_4$.

Since our complexes will often arise as the order complexes of posets, we consider the poset special case of the above lemma. Here obviously chains through a member x involve only the above- and below-subposets, as suggested in Example 1.6.11, so:

LEMMA 1.6.13. *For $x \in \mathcal{P}$ a poset, and $K := K(\mathcal{P})$ its order complex,*

$$\mathrm{Lk}_K(x) = K(\mathcal{P}_{<x}) * K(\mathcal{P}_{>x}).$$

For instance, in the poset $\mathcal{P} := L(D_8)$ in Example 1.1.13, with $K := K(\mathcal{P})$, take $x := \langle(12)(34)\rangle$. Then $\mathrm{Lk}_K(x)$ is given by the join: of the vertex $\{1\}$ below x, with the edge strictly above x given by $\{\ \langle(12)(34),(13)(24)\rangle\ ,\ G\ \}$. (And so that join is a 2-simplex.)

CHAPTER 2

Examples: Subgroup complexes as geometries for simple groups

This chapter introduces a number of examples of subgroup complexes, which are taken from the literature on geometries for finite simple groups. Since this is a main area of interest for the author, such examples will appear fairly frequently in later chapters.

Of course, readers will have varying levels of experience with simple groups—and so may wish to accordingly limit the examples that they follow through the book. In order to facilitate such a choice, the examples in the chapter are organized into sections which proceed from more elementary to greater technical depth. So in effect, these sections provide some optional tracks to follow:

For those readers acquainted only with elementary group theory, we have already provided in Chapter 1 some easier examples, typically phrased in the language just of the symmetric group—notably in Examples 1.1.12 through 1.1.19.

Section 2.1 turns to the classical matrix groups: The natural geometries for these groups are projective (and polar) spaces—which are realized essentially via elementary linear algebra.[1] Most readers likely to take up this book will probably be reasonably comfortable with this context of classical matrix groups over finite fields.

This leads naturally to a deeper optional track: For the above elementary projective examples are in turn special cases of—and indeed exhibit very natural motivation for—the more general notion of the Tits building: The building is the natural geometry for a Lie type group, and we introduce various aspects in Section 2.2. The context of buildings provides Option B in the book—it is essentially the model case for many of the results we will discuss, primarily in the sense that it affords a starting point for various generalizations. Furthermore buildings arise in various guises in many areas of mathematics involving Lie type groups; so this material may well interest a reasonably large subset of readers.

The final Option S considers geometries for sporadic simple groups, introduced in Section 2.3; this topic is a more specialized interest of the author—and the readership for this more technical option will very likely be rather smaller. Nonetheless, the inspiration for the results within this option does comes from generalizing results for the model case of buildings in Option B; and this might motivate at least some less experienced readers to sample Option S.

[1] Furthermore some of the smaller projective examples can in fact be translated back into the combinatorial language of the symmetric group. In particular (cf. later Remark 2.1.23 and the subsequent examples), one can use isomorphisms of certain small alternating groups with matrix groups, such as: $A_5 \cong L_2(4) \cong \Omega_4^-(2)$; $S_6 \cong Sp_4(2)$; and $A_8 \cong L_4(2)$.

Complexes provided by geometries. As mentioned in the Introduction, a research area of considerable interest among finite group theorists (especially starting in the 1980s) became known as "groups and geometries"—and this area developed largely independently of the area of "subgroup complexes" which is the main theme of this book, already introduced in Chapter 1. However, Quillen's paper [**Qui78**] already indicated (e.g. in Section 3 there) certain interrelations between the two areas; and Webb's paper [**Web87b**] continued such connections. In the subsequent years, there has been considerable further interaction.

Consequently the purpose of this chapter is to provide an introduction to the area of groups and geometries—including some indication of the viewpoint on those geometries afforded by subgroup complexes introduced in Chapter 1. In particular, a number of examples of such geometries will be introduced for later use.

The chapter is organized as follows: The first section gives a quick overview of the simple groups, and of some efforts to understand them via geometric structures. The second section reviews the ideal case—namely the geometry given by the Tits building for a group of Lie type; this case will be called Option B, whenever it is followed up in later chapters. The third section introduces some basics of "diagram geometries", which carry at least some ideas from buildings over to other simple groups such as sporadic groups; this is Option S, which also recurs in later chapters.

Introduction: Finite simple groups and their "natural" geometries

We first recall the general context of *simple* groups: In the usual Jordan-Hölder theory of composition series (cf. Section 3.4 of [**DF99**]), a finite group is built up from simple groups (i.e. those having no nontrivial proper normal subgroups). The abelian (i.e. commutative) simple groups are easily seen, from the usual theory of invariant factors, to be of prime order (cf. Exercise 1 on page 107, and Theorem 3 on page 160, of [**DF99**]).

For *nonabelian* finite simple groups (and this is what we will usually mean when we write "simple groups"), efforts of many finite group theorists, in hundreds of papers, culminated in the Classification of Finite Simple Groups (CFSG)—which was announced around 1980 (cf. [**GLS94**, p. 6]), although certain parts (notably [**AS04b**]) did not appear until recently. In this book, we will not need to quote[2] that immense result; instead, we will consider just the explicit structure of certain particular simple groups. Thus any reader uncomfortable with the classification can always take "the simple groups" to mean "the *known* simple groups".

And we mention that for reference on the individual simple groups, the reader may wish to consult [**GLS98**], or Wilson's recent book [**Wil09**].

The types of simple groups. The nonabelian simple groups are usually described (cf. page 6 in [**GLS94**]) with the following terminology:

REMARK 2.0.1 (The three classes of simple groups). The finite simple groups are subdivided into the following three classes (sometimes called families):

- alternating groups;
- groups of Lie type;
- the 26 "sporadic" groups. ◊

[2]Though in later chapters, for expository purposes we will sometimes *mention* certain results in the literature which do quote the CFSG.

As further background, we now briefly describe each of these classes:

The *alternating* groups A_n are of course the subgroups of even permutations in the full *symmetric* groups S_n. And so these groups are probably familiar to most mathematicians, for example as Galois groups of suitable field extensions—with A_n shown by a standard proof to be simple for $n \geq 5$; see e.g. Chapter 5 of [**GLS98**] and Chapter 2 of [**Wil09**] for fuller details on the structure of these groups.

The groups of *Lie type* are certain groups of matrices, which are analogues of the real and complex Lie groups of analysis: However, the matrix entries of a finite Lie type group G are taken instead a finite field $\mathbb{F}_q$, where q is a power of a prime p. (And then we say G is of Lie type *in characteristic* p.) A quick introduction to these groups is provided in Section 6.8 of Benson [**Ben91**], from a viewpoint similar to that of this book. For fuller reference on the structure of the finite groups of Lie type, see for example Carter [**Car89**], or [**GLS98**, Ch 2–4], or [**Asc00**, Ch 14], and Chapters 3 and 4 in [**Wil09**]. Also many smaller Lie type groups are described in the Atlas [**CCN$^+$85**]. (We will be providing some further background on Lie type structure later in this chapter, starting at Remark 2.2.2.)

There are various infinite families of finite Lie type groups; and so they are usually regarded as comprising "most" of the nonabelian finite simple groups—and hence they provide the "generic" examples of simple groups. For present purposes, we will simply mention that they are often subdivided into the following three subclasses: first:

∘ the *classical* matrix groups,

namely linear, unitary, symplectic, and orthogonal groups over finite fields; next:

∘ the *exceptional* groups,

corresponding to the Lie algebras of the exceptional types G_2, F_4, E_6, E_7, and E_8; and finally:

∘ the *twisted* groups,

which are determined as the fixed points under suitable outer automorphisms of the classical and exceptional groups. (It turns out that the classical unitary groups, as well as certain classical orthogonal groups, are also very naturally described as twisted groups.) In this book, we will often choose examples from the classical matrix groups, which should presumably be reasonably familiar to many readers.

HISTORY 2.0.2 (Some sources of the Lie-type viewpoint). It is a remarkable feature that the seemingly disparate groups in the above three subclasses can in fact be treated uniformly, by means of the Lie theory. The study of the structure of the finite Lie type groups (and especially the finite twisted types, which need not have analogues in the characteristic-0 Lie theory) dates largely to work of Chevalley and Steinberg in the 1950s and 1960s; this partly followed as a by-product of the classification around 1955 of the semisimple algebraic groups. (These algebraic groups are defined over infinite algebraically closed fields of arbitrary characteristic, and so the groups are also infinite.) For further historical remarks on finite groups of Lie type, see e.g. [**Car89**, Sec 1.1]. ◊

The final class of simple groups, namely the *sporadic* groups, is not really an abstractly-defined class at all—instead it just arbitrarily collects together the remaining 26 simple groups, which do *not* fit into any of the infinite families exhibited by the alternating and Lie type groups. The first 5 sporadic groups were discovered by Mathieu in the 19th century; while the remaining 21 sporadics were discovered during the heyday of simple-group classification in the 1960s and 1970s. Some group theorists (including the author) believe that further study of these "exotic" simple groups, via suitable analogies with properties of the generic examples given by Lie type groups, will lead to better understanding of all the simple groups. Indeed exhibiting some such analogies between Lie type groups and sporadic groups, at the level of subgroup complexes, will be a frequent theme in Option S of this book.

For details on the the various individual sporadic groups, see e.g. [**GLS98**, Ch 5], and [**CCN**$^+$**85**], and Chapter 5 in [**Wil09**]. Since these groups are less likely to be familiar to non-specialist readers, in this book we will try to provide more specific further references, whenever sporadic groups are used as examples.

Geometries for the simple groups. It has long been a standard approach to study a given group G via its action on some suitable combinatorial or geometric structure. For example in analysis, a real or complex Lie group G is often viewed in terms of its action on a natural *symmetric space*—where for a suitable subgroup H, the coset space $K := G/H$ admits the structure of a manifold. Similarly Klein's Erlangen Program of 1872 proposed to study more general groups G, in terms of more general non-Euclidean geometries K—notably projective geometry, i.e. the more purely combinatorial structure of *projective spaces*. Indeed if we restrict to finite G, the relevant projective geometries can usually be viewed in the context of a finite simplicial complex K, admitting a natural G-action.

In this subsection, we will give an initial overview of some modern efforts (say since the 1960s) to understand the finite *simple* groups from such a geometric perspective. Then in the subsequent subsection, we will give a brief overview of how such geometric research on simple groups led to connections with the area of subgroup complexes for more general groups. (Of which we had given some history earlier, in the Introduction.)

It will be natural to consider, in turn, the three classes of simple groups G described in the previous subsection.

First if G is the alternating group A_n, embedded in S_n, then we may regard S_n as the full symmetry group of the $(n-1)$-dimensional simplex K, whose vertices are the "letters" $\{1, 2, \ldots, n\}$ on which S_n acts. (Cf. our earlier Example 1.5.6 with A_4 in S_4—but note of course that this particular small alternating group happens not to be simple.) So this simplicial complex K provides at least an initial example of a geometry for the group S_n. It is perhaps not an overwhelmingly convincing example, since the geometric structure of the simplex K is so uniform and obvious: indeed the action of S_n on the various faces of K really exhibits just the combinatorics of the permutation action on the corresponding subsets of the vertex set $\{1, 2, \ldots, n\}$.

But the viewpoint produces more interesting geometric structure when we turn to the next class, provided by Lie type groups. Here a suitable analogue of the group A_n acting on vertices $\{1, 2, \ldots, n\}$ is the linear group $G = GL_n(q)$ (which again is not *quite* simple) acting on a vector space V of dimension n over the finite field $\mathbb{F}_q$. For this classical group, classical mathematics provided a natural geometry—afforded by the *projective space* $\mathbb{P}(V)$: this is the poset of nonzero proper $\mathbb{F}_q$-subspaces of V under inclusion; and we can pass to the order complex $K(\mathbb{P}(V))$. Then the geometry from that complex is given by inclusion-chains of subspaces—this is a complex of much richer structure than the single simplex of the previous paragraph. In the case of the other classical matrix groups (unitary, symplectic, orthogonal), each of which preserves a form on a space V, the classical analogue of the full projective space $\mathbb{P}(V)$ is the *polar* space, the poset of subspaces which are totally isotropic (or singular) in that form. (We will explore such spaces in Section 2.1.)

However, for the other subclasses of Lie type groups, namely exceptional and twisted groups, classical mathematics had not provided any such natural geometry. And indeed it was the search for a natural geometry for exceptional groups, generalizing the projective and polar spaces, which around 1965 led Tits (using the context of work on algebraic and finite groups of Lie type by Bruhat, Chevalley, and others) to the notion in [**Tit74**] of the *building*. (We will examine buildings in Option B of this book, starting in Section 2.2.) Very roughly, the building is a simplicial complex constructed from the poset of *parabolic* subgroups of G—which in the above example of the linear group, correspond to the subgroups stabilizing chains of subspaces; we mention that the parabolics are *p-local subgroups*, namely they normalize certain nontrivial p-subgroups of G, where p is the characteristic of the finite field $\mathbb{F}_q$ of definition of the matrix group G. Thus, just as the Lie theory provides a uniform approach to the structure of the various subclasses of Lie type groups, the building provides a uniform construction of a natural geometry for each such group. (For further history, see e.g.: the introduction in Tits's book [**Tit74**]; the introduction in Ronan's book [**Ron89b**]; or Section V.4 in Brown's book [**Bro98**]. The latter book has recently been expanded in a joint work [**AB08**] with Abramenko.)

Tits's further general structure theory for buildings became a fundamental tool in the study of Lie type groups. In the wake of this accomplishment, it was natural to ask whether there might also be a similar natural geometric approach to the third class of simple groups, namely the 26 sporadic groups; in particular, the hope might be to find some suitable weak approximation to the theory of buildings. And ideas soon began to emerge, offering extensions of certain parts of the building methodology, roughly as follows:

To a group G of Lie type is associated an underlying Lie algebra, with its corresponding *Dynkin diagram*. And one feature of Tits's original construction of buildings (a feature further emphasized in his subsequent "local approach" paper [**Tit81**]) is that the Dynkin diagram could equally well be regarded as describing, from an inductive viewpoint, some axioms for how the building geometry for G is built up from the subgeometries given by the links of vertices (recall Definition 1.6.10). Around 1975, Buekenhout in [**Bue79**] extended this inductive viewpoint: he was able to exhibit for many sporadic groups G a *diagram geometry*: namely a certain simplicial complex, arising from a poset of subgroups of G which are suitably analogous to parabolic subgroups—with axioms described by a diagram somewhat more general than a Dynkin diagram. Soon thereafter Ronan and Smith in [**RS80**] focused more specifically on sporadic diagram geometries exhibiting something like the p-local property of parabolics mentioned above—they defined *p-local geometries* based on posets of suitable p-local subgroups of G.

In the years since that time, various versions of such "sporadic geometries" have been intensively studied by many finite group theorists. Option S in this book will highlight a number of examples of such development, starting in Section 2.3.

Connections with more general subgroup-complex theory. For a general finite group G, as opposed to just the special case of simple G discussed above, the theory of buildings seems to have provided one of the motivations for various developments in the study of subgroup complexes in topology and combinatorics. For example, in Quillen's fundamental paper [**Qui78**] on subgroup complexes, his result 3.1 (in conjunction with 2.1 there) observed that for G of Lie type defined over a field of characteristic p, the building of G is homotopy equivalent to (the order complex of) the poset $\mathcal{S}_p(G)$ of p-subgroups. Similarly many further connections of subgroup complexes with group geometries such as buildings emerged in Quillen's paper, and also in work of various authors in subsequent years; and we summarize this experience roughly as:[3]

REMARK 2.0.3 (Building Principle). Buildings provide explicit group geometries which, in the context of Lie type groups G, provide particularly natural demonstrations of general topological properties of subgroup complexes like $\mathcal{S}_p(G)$ defined for arbitrary groups G. ◊

Such properties include for example projective Lefschetz modules, homology decompositions, and Cohen-Macaulayness, which are examined in later chapters of this book. So when we develop such properties for $\mathcal{S}_p(G)$ for general G, in accordance with this Principle we will usually (under Option B) consider buildings for G of Lie type among our first examples of those properties.

Continuing in the spirit of Quillen's observations, Webb's paper [**Web87b**] included (among other topics) an emphasis on the application of Quillen's techniques to various more general subgroup complexes K—notably via conditions on K sufficient to obtain many of the conclusions that Quillen had demonstrated essentially just for $\mathcal{S}_p(G)$ or for buildings.

Analysis like Webb's of more general complexes has in subsequent years proved especially successful in the case of G a sporadic group and K a p-local geometry, for example in such work as [**RSY90**], [**SY97**], and [**BS08**, Ch 6]. Indeed in summary one could add to the previous rough principle the following less well-defined observation:

REMARK 2.0.4 (Vague Sporadic Principle). The p-local geometries for sporadic groups should satisfy analogues of various properties of buildings (and of $\mathcal{S}_p(G)$). ◊

The Sporadic Principle is "vague" in several senses—especially in that, unlike buildings, the p-local geometries are not intrinsically defined, but instead are obtained essentially by experimental observation in the individual sporadic cases. However, the Principle has certainly led to many results on sporadic groups analogous to those for buildings; hence when we examine later topics for $\mathcal{S}_p(G)$ and for buildings, following the Sporadic Principle we will usually also (under Option S) indicate examples for p-local geometries.

---

[3]The terms Building Principle here, and Vague Sporadic Principle below, were coined just for use in this book; the reader will not see them elsewhere in the literature.

2.1. Motivating cases: Projective geometries for matrix groups

Our examples in this initial section will come essentially from elementary linear algebra. So we will begin by considering the classical matrix groups—for which the natural geometry is just the standard *projective* space (or the corresponding *polar* subspace, when the group preserves a quadratic form).

These geometries are perhaps the easiest examples that we could begin with. In particular, they will afford an elementary introduction to the inductive "local approach" viewpoint mentioned in the introductory section above: we will see that these projective spaces are in turn involved, as links of vertices, in the construction of similar but larger projective geometries—and indeed in more complicated geometries of other types. This inductive viewpoint will be a central theme when we proceed to the remaining sections of the chapter, which introduce Option B on buildings, and Option S on sporadic geometries.

Projective and polar spaces are defined using a vector space V, which affords the natural module for a classical matrix group G. For these groups and modules we will typically use the following conventions:

NOTATION 2.1.1 (Notation for classical groups). For a prime p, with $q := p^m$ a power, let $F := \mathbb{F}_q$ denote the finite field[4] of order q.

Let V denote a vector space of finite dimension n over F; the vectors of V are naturally permuted under multiplication by the matrices in the general linear group $GL(V) \cong GL_n(q)$.

In explicit computations of this group action, we continue the convention mentioned in Notation 1.5.2—which here gives the opposite of the convention of undergraduate linear algebra: Namely we write elements of V as *row* vectors; and matrices in G will act naturally by *right* multiplication on those vectors.

We will view our classical group G as a subgroup of $GL(V)$; when G is not the full linear group $GL(V)$, G will be the subgroup (symplectic, unitary, or orthogonal) preserving a suitable bilinear or quadratic form on V. (Cf. Chapter 1 of [**Car89**], or p 72 of [**GLS98**], or Chapter 3 of [**Wil09**]. ♢

REMARK 2.1.2 (Matrix groups as "nearly-simple" examples). We mention that a classical matrix group G as described in Notation 2.1.1 is usually not quite simple; cf. the remarks at the end of each subsection of Chapter 1 in [**Car89**]: For example if G is $GL_n(q)$, the related group which is usually simple (i.e. except for a few small values of n and q) is instead the projective special linear group $PSL_n(q)$ (the matrices $SL_n(q)$ of determinant 1, with scalar multiples identified)—usually abbreviated by just $L_n(q)$.

But for expository purposes in our examples, it will be easier to compute directly with matrices and vectors; in particular, matrix multiplication gives the group product in a classical matrix group G arising as a subgroup of some $GL_n(q)$. And usually results within $GL_n(q)$ can be easily translated into corresponding statements for the simple section which gives "most of" G. ♢

---

[4]This field is also often denoted by $GF(q)$ in the literature.

The projective space for the full linear group. As our initial example of a classical group, we consider the full linear group $G = GL_n(q)$. This matrix group is perhaps the easiest type of simple group (in the "nearly-" sense of Remark 2.1.2 above) to begin with. *Indeed we recommend that the reader should keep the group $GL_n(q)$ in mind—not just in this subsection, but as a standard example in all later developments related to groups of Lie type.*

Definitions and generalities. The natural finite geometry for $G = GL_n(q)$ is provided by the full projective space of the natural module V for G as in Notation 2.1.1; this is the combinatorial structure defined by the linear F-subspaces of V:

DEFINITION 2.1.3 (Projective space $\mathbb{P}(V)$). Let V etc. be as in Notation 2.1.1. Then the *projective space* of V is:

$\mathbb{P}(V) :=$ the set of nonzero proper F-subspaces of V, ordered by inclusion $\leq$.
We we wish specify the dimension n of V and the field $\mathbb{F}_q$, we write $\mathbb{P}^{n-1}(q)$—this uses the geometric notion of *projective dimension*.[5] As motivation for this convention, note that we are regarding each subspace as a single object. In the case of a 1-subspace, this means we are in effect identifying all the scalar multiples of any nonzero vector; hence geometrically we obtain just a point, which has projective dimension 0. Indeed if we regard all vectors as being projected to an image on a suitable "hemisphere" of radius 1,[6] then a linear subspace of dimension k is mapped to a projective subspace of dimension $k-1$. That is, linear 1-subspaces are projective *points*; linear 2-subspaces are projective *lines*; linear 3-subspaces are projective *planes*; and so on—through linear hyperplanes, of codimension 1 in V, giving *projective hyperplanes* in $\mathbb{P}(V)$.

A sequence of linear subspaces related by inclusion is called a *flag* in V (and when read projectively, a flag in $\mathbb{P}(V)$); in particular, a longest-possible such flag $V_1 < V_2 < \cdots < V_{n-1}$, namely where each $\dim_F(V_i) = i$, is called a *full* or *maximal* flag. Thus the flags are in fact just the chains in the poset $\mathbb{P}(V)$—that is, the simplices of the order complex $K(\mathbb{P}(V))$ (recall Definitions 1.2.9 and 1.2.11). In particular, the full flags are the simplices of maximal dimension; these maximal chains have $n-2$ successive $<$-relations, so that the simplicial complex $K(\mathbb{P}(V))$ has topological dimension one smaller than the dimension of the projective space, namely $n-2$. $\diamondsuit$

We single out one particular property, which will also be important for later simple-group examples:

REMARK 2.1.4 (Flag-transitivity). Observe that $\mathbb{P}(V)$ is a G-poset, in the sense of Definition 1.5.1, under the natural action of multiplication by the matrices in the linear group $G = GL(V) \cong GL_n(q)$.

(1) It is a standard fact from linear algebra (cf. [**DF99**, Example, p 393]) that any two vector-space bases of V are conjugate in G. Since an ordering on a basis determines a flag, it follows in particular that G acts transitively on the set of (full) flags in $\mathbb{P}(V)$; and we say that (the action of) G is *flag-transitive*. Hence G is transitive on the maximal faces of the order complex $K(\mathbb{P}(V))$.

---

[5] This space is also often denoted by $PG(n-1,q)$ in the geometric literature.

[6] This visualization of a hemisphere is inspired by (and perhaps only really appropriate for) the case of a vector space over the *reals*.

(2) Indeed in the buildings for the remaining Lie type groups, and even in the geometries for the other simple groups G that we study, there will be a notion of "flags"—typically as the maximal simplices of a suitable G-complex defining the geometry; and then the action of G on the complex will usually be flag-transitive (in the sense of being transitive on those maximal faces).

(3) We emphasize that a flag-transitive group is a very special feature for a general simplical complex; but we can often expect it in our examples, since typically it arises roughly from the *simplicity* of our example groups: For example, recall the n-transitive permutation action of S_n (e.g. 5.4.2 and 5.4.3 in [**Hal59**]); this implies transitivity on k-subsets for any k—and hence on the faces of the simplex on $\{1, 2, \ldots, n\}$, with the faces affording the relevant flags for that complex.[7]

(4) We often use flag-transitivity conditions to see that our geometries have only a few "types" of simplices—making it easier to describe (and count) them. (Cf. Example 2.1.9.) Notice in particular that if G is flag-transitive on a complex K, then the orbit complex K/G is just a single simplex. This gives an especially convenient form for the indexing set, when various later formulas (e.g. 7.2.5) are indexed by K/G. $\diamondsuit$

We have exhibited $\mathbb{P}(V)$ as a poset, in order to emphasize the poset viewpoint of the previous Chapter 1. We also now mention that when $m = 1$, so that $q = p$ is prime, there is a further connection of $\mathbb{P}(V)$ with the poset structures in the previous chapter: For in that case, the F-subspaces of V are just the additive subgroups of V regarded as an abelian group; and then $\mathbb{P}(V)$ is just the "interior" subposet of the full subgroup lattice $L(V)$, namely that obtained by removing the top and bottom elements V and $\{0\}$.

Because projective spaces are so fundamental for our later development, we will now devote the next several pages to examining in more detail some small examples of such spaces.

Numerical parameters in some smaller examples. In introducing examples, our initial focus will be on elementary combinatorial analysis: it will be convenient to compute some of the numerical "parameters" associated to those small geometries. Furthermore we will record those parameters, for use later on when we analyze larger geometries in which these examples arise as links.

For instance, a projective point over $F = \mathbb{F}_q$ in effect identifies the $q-1$ nonzero vectors of a linear 1-space over F. So since the whole space V has $q^n - 1$ nonzero vectors, we see that:

(2.1.5) The projective space $\mathbb{P}^{n-1}(q)$ has $\dfrac{q^n - 1}{q-1} = (1 + q + \cdots + q^{n-1})$ points.

In fact our examples will focus most often on the case of the smallest prime power $q = p = 2$—so that V is a vector space over the field $\mathbb{F}_2$ of just two elements $\{0, 1\}$. These small examples over $\mathbb{F}_2$ are particularly easy to make sample computations in; but we will also see later that examples over $\mathbb{F}_2$ also arise disproportionately often in the study of geometries for sporadic simple groups.

[7] Indeed this transitivity, for $(n-2)$-sets, is customarily used in "conversely" proving the simplicity of the subgroup A_n.

NOTATION 2.1.6 (Conventions for spaces over $\mathbb{F}_2$). We may sometimes explicitly write 0_F and 1_F for the elements of the field $F = \mathbb{F}_2$. In particular this will distinguish the field entries from our use of integers, often including 0 and 1, in certain other notational conventions such as the following:

Since $\dim_F(V) = n$, the space V is isomorphic to the power set 2^S of a set S of size n—where addition in the power set is given by symmetric difference. For notational brevity, we adopt the following conventions: We denote a basis for V by $S = \{1, 2, \ldots, n\}$—rather than by more usual notation for vectors such as $v_1, v_2, \ldots, v_n$. We regard a vector of V as a subset of S, which we write without braces, and usually without commas delimiting the elements—for example 124 for the subset $\{1, 2, 4\}$. (And this subset, as a vector, is in turn the sum of the vectors denoted by 1, 2 and 4). Of course there are $2^n - 1$ such nonzero vectors in V—and this will be our most frequent special case of (2.1.5), since each nonzero vector v is the unique nonzero vector of the linear 1-subspace (i.e. projective point) given by $\langle v \rangle$. ♢

We can now start to develop our smaller examples:

EXAMPLE 2.1.7 (The projective line $\mathbb{P}^1(2)$ over $\mathbb{F}_2$). We take $q = p = 2$, along with $n = 2$. In the convention of Notation 2.1.6, the $2^2 - 1 = 3$ nonzero vectors of V are denoted by $1, 2, 12$; each is the unique nonzero vector of a 1-subspace of V. Indeed typically we will just write 1 rather than $\langle 1 \rangle$, when we are regarding that space as a projective point.

So in the viewpoint of (2.1.5), there are 3 points in our projective line $\mathbb{P}(V)$. We will often use abbreviated notation of the form:

$$\#P = 3,$$

to record this value of the total number of points. (This is the only parameter value which we record for this small space.)

Thus $\mathbb{P}(V)$ is a poset with those three projective points; but since these are linear 1-subspaces—with the same dimension—there are no proper inclusions among them; and hence no proper order-relations $<$ in the poset. Consequently the order complex $K(\mathbb{P}(V))$ is of dimension 0; it consists just of three isolated vertices.

We mention that the corresponding automorphism group $GL_2(2)$ of the poset or complex is one of the few cases of a "small" Lie type group which is not simple—indeed the group is isomorphic to S_3, and in particular is solvable. ♢

EXERCISE 2.1.8 (The projective line $\mathbb{P}^1(4)$ over $\mathbb{F}_4$). Mimic the previous example—with $p = 2$ again, but now taking $q = 4$; use (2.1.5) to obtain $\#P = 5$ for the count of points in this larger projective line (and hence 5 isolated vertices in the order complex).

Note that the denominator of $(q - 1)$ in the calculation of (2.1.5) had in effect been concealed in earlier Example 2.1.7, where we were working with $\mathbb{F}_2$—since there $(q - 1)$ just had the value 1. This emphasizes that there can be some subtleties in generalizing from the $\mathbb{F}_2$-case to larger q. ♢

The next example moves up in dimension, from a line to a plane; our longer discussion of it reflects the richer geometric structure that begins to be available:

EXAMPLE 2.1.9 (The projective plane $\mathbb{P}^2(2)$ over $\mathbb{F}_2$—parameters). We continue to work over $\mathbb{F}_2$, namely with $q = p = 2$; but now we take $n = 3$.

This time we get $\#P = 2^3 - 1 = 7$ for the number of points in the plane $\mathbb{P}(V)$, corresponding as in 2.1.6 to the nonempty subsets of the basis $\{1, 2, 3\}$ of V.

It turns out that we get the same value $\#L = 7$ for the parameter recording the number of projective lines in $\mathbb{P}(V)$. We will exhibit one computation of this value in some detail, since it demonstrates some elementary combinatorial methods which we will be using frequently in the future: We wish to count the number of linear 2-subspaces of V. By the previous paragraph, there are 7 choices for an initial generator v of a 2-space. And a second generator v' of a 2-space must be linearly independent of v; hence there are $(7 - 1) = 6$ choices for v'. However, this total of $7 \cdot 6 = 42$ choices for the ordered pair (v, v') actually overcounts the number of 2-spaces—since certain ordered pairs (v, v') will generate the same 2-subspace. So we must adjust for this overcount, with a division by the number of such pairs. That is, fixing some such 2-subspace W, we must count how many of our above pairs (v, v') chosen from V actually have both v and v' in our fixed W. So first, using the previous Example 2.1.7, we see that there are 3 choices for v in the 2-space W. Then (arguing as above, but in W rather than V) there are $3 - 1 = 2$ choices for v' linearly independent of v in W; hence $3 \cdot 2 = 6$ choices for (v, v') taken from W. We conclude then that there are $42/6 = 7$ choices for the 2-space W; that is, there are 7 projective lines in V. In the future we may refer to this type of combinatorial enumeration in informal abbreviated language such as

$$\frac{\#(v, v') \text{ in } V}{\#(v, v') \text{ in fixed } W} = \frac{7 \cdot 6}{3 \cdot 2} = 7. \tag{2.1.10}$$

Indeed the method of (2.1.10) can be extended (exercise!), e.g. to count the number of k-spaces in an n-space: Use k-sequences of vectors, rather than pairs; and the requirement of linear independence will then give terms (in the numerator) with the form $(2^n - 1) - (2^i - 1)$.

Now that we have computed the total number of points and lines in the projective plane $\mathbb{P}(V)$, we will next determine further parameters for the plane, which correspond to the natural incidence relation between points and lines, given by containment. First, we saw during the previous paragraph, via Example 2.1.7, that there are 3 points in the projective line given by (any) linear 2-subspace W. Typically we express this value of the parameter recording the number of points per line in the abbreviated notation $\#\ P/L = 3$. Next, note that if we fix some vector like 1, then the quotient linear 2-space $V/\langle 1 \rangle$ similarly affords a projective line $\mathbb{P}^1(2)$, and so has 3 points; and the preimage of each of these is a linear 2-subspace of V containing 1, so that we also obtain 3 projective lines "above" the point 1. But this argument does not depend on the particular point 1, so we conclude that there are 3 lines above *any* point in the plane; we abbreviate this parameter value as $\#\ L/P = 3$. (This independence also follows from the property of flag-transitivity in Remark 2.1.4, and illustrates how we will often use that property to simplify computation of geometric parameters.)

Using the above parameter values, we can also now compute the number of full flags in the plane: we denote this count by the abbreviation $\#PL$, where we actually mean pairs (P, L) with "$P \in L$"—that is, a point P contained in a line L in the sense of linear-subspace inclusions such as $\langle 1 \rangle < \langle 1, 2 \rangle$. We get that $\#PL = (\#L)(\#\ P/L) = 7 \cdot 3 = 21$; of course we obtain the same result via the alternative computation $(\#P)(\# L/P) = 7 \cdot 3 = 21$. (And again, flag-transitivity in 2.1.4 provides another way of seeing that our counts are independent of any particular choice of point or line.) $\diamondsuit$

EXERCISE 2.1.11 (The projective plane $\mathbb{P}^2(4)$ over $\mathbb{F}_4$). Mimic the previous example, again with $p = 2$, but now taking $q = 4$. Obtain $\#P = 21 = \#L$, and $\# P/L = 5 = \# L/P$, so that $\#PL = 21 \cdot 5 = 105$. $\diamondsuit$

Geometric incidence structures in larger examples. We now turn from strictly numerical considerations, to the more geometric content that is determined by the set-theoretic inclusions among subspaces in a projective space $\mathbb{P}(V)$.

So we return to an earlier Example, from the new viewpoint:

EXAMPLE 2.1.12 (More on $\mathbb{P}^2(2)$—picturing incidence structures). We continue with the projective plane of Example 2.1.9. In a moment, we will draw the usual poset diagram, for the poset $\mathbb{P}(V)$. But first, we will give a terser representation, in which the projective lines are not just abstract objects (i.e. single elements in a poset), but instead "look like" lines—in something like the usual geometric sense of continuous straight lines in the Euclidean plane $\mathbb{R}^2$. (Actually the plane $\mathbb{P}^2(2)$ is essentially the *only* projective space, larger than a line, which is small enough to be visualized in this way by a picture.) We use essentially the subdivided triangle[8] of Example 1.6.1, but now with a new interpretation: The 7 vertices are our projective points; and the 7 projective lines (consisting of 3 points each) are given by the 3 edges of the triangle, and its 3 diagonals—along with the dotted circle, which provides the final projective line (since it turns out that we cannot draw all 7 of the projective lines as *straight* lines in $\mathbb{R}^2$):

(2.1.13)

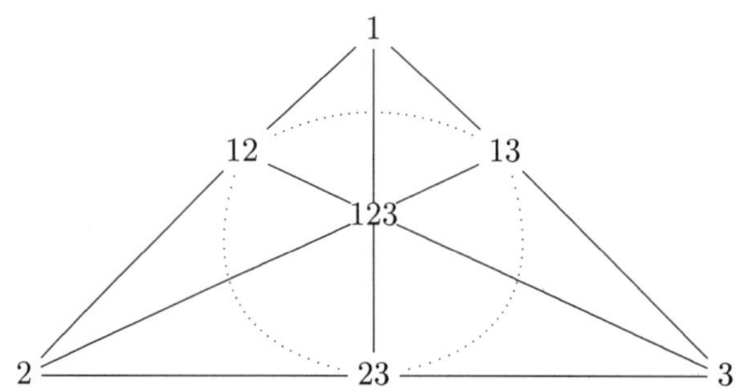

Of course the vector-space structure of V determines the picture; but to proceed in the other direction, notice for example that the three points of each line in the picture do have vector sum zero—this is required for the sum of the three nonzero vectors of a 2-subspace over $\mathbb{F}_2$.

Next, either directly from V itself, or from the picture (2.1.13) derived from V, we can obtain the usual diagram for the poset $\mathbb{P}(V)$. We position the 1-spaces below the 2-spaces, and indicate the corresponding inclusions; the result is somewhat more spread-out than (2.1.13), and in particular the picture does not especially suggest any "straight-line" feature of the projective lines:

[8]In the geometric literature, the plane over $\mathbb{F}_2$ (especially in this visualization) is often called the *Fano plane*.

(2.1.14)

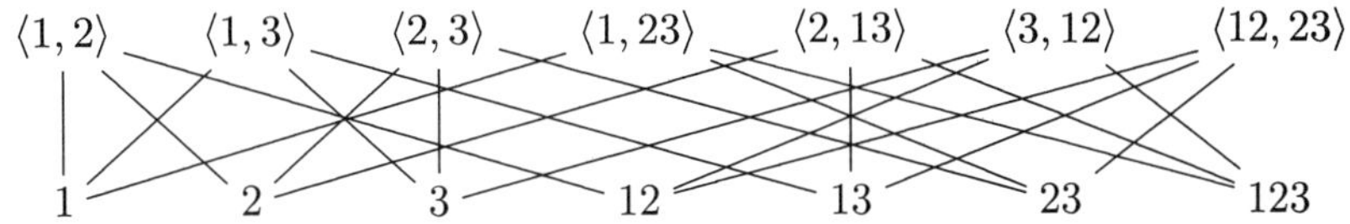

Now the maximal length of a chain in this poset is 1; so the order complex $K(\mathbb{P}(V))$ is of dimension 1. Thus (cf. our observation after Remark 1.2.12) the complex is just a graph—indeed that graph is the same as the poset diagram.

In our inductive viewpoint on geometries, we can now begin to see how projective spaces are built up, via the links of vertices, from projective spaces of smaller dimensions: First the link $\mathrm{Lk}(\langle 1,2\rangle)$ of the line-vertex $\langle 1,2\rangle$ in the above complex consists of the points 1, 2, 12; so this link has (more or less tautologically) the structure of the projective line $\mathbb{P}^1(2)$ of Example 2.1.7. And similarly (as we had essentially observed earlier), the three lines in the link $\mathrm{Lk}(1)$ of the point-vertex 1 correspond to (the points of) the projective line determined by the quotient 2-space $V/\langle 1\rangle$. Thus we regard $\mathbb{P}^2(2)$ as being inductively assembled from vertex-links having the "previous-case" structure of $\mathbb{P}^1(2)$. $\diamondsuit$

EXAMPLE 2.1.15 (The projective 3-space $\mathbb{P}^3(2)$ over $\mathbb{F}_2$). We continue with the choice of $q = p = 2$, but now take $n = 4$.

We will give a briefer derivation of the values of the corresponding parameters. For points, from 2.1.6 we get $\#P = 2^4 - 1 = 15$, corresponding to the nonempty subsets of the basis $\{1,2,3,4\}$ of V. Using the method of (2.1.10), we calculate that $\#L = \frac{15\cdot 14}{3\cdot 2} = 35$; and extending that method as indicated there, we count planes (denoting them by π, and using $n = 4$ and $k = 3$) to get $\#\pi = \frac{15\cdot 14\cdot 12}{7\cdot 6\cdot 4} = 15$. From Examples 2.1.7 and 2.1.9, we get $\#\ P/L = 3$ and $\#\ P/\pi = 7 = \#\ L/\pi$. Applying those Examples in the quotient spaces $V/\langle 1,2\rangle$ and $V/\langle 1\rangle$ gives $\#\ \pi/L = 3$ and $\#\ L/P = 7 = \#\ \pi/P$. So the number of full flags (naively "$P \in L \subset \pi$") is given by $\#PL\pi = 315$—e.g. using the product $\#P\cdot(\#\ L/P)\cdot(\#\ \pi/L) = 15\cdot 7\cdot 3$. (Actually the final factor should more precisely be written $\#\ \pi/(P \in L)$, i.e. the number of planes over a flag $P \in L$; but any plane containing L also contains P, so this value agrees with $\#\ \pi/L$.)

It will also be convenient here to give an alternative computation of $\#L$, which demonstrates another standard combinatorial technique, namely "counting two ways": For (as we observed in Example 2.1.9) the number of partial flags of form $P \in L$ is given by either of the expressions in $(\#P)(\#\ L/P) = (\#L)(\#\ P/L)$. And we obtained the values other than $\#L$ from more elementary calculations, so we can compute $\#L = \frac{(\#P)(\#\ L/P)}{\#\ P/L} = \frac{15\cdot 7}{3} = 35$. Similarly we could make the alternative computation $\#\pi = \frac{(\#P)(\#\ \pi/P)}{\#\ P/\pi} = \frac{15\cdot 7}{7} = 15$.

We can at least imagine drawing the poset diagram, in the spirit of (2.1.14): Now there will be three levels, corresponding to the three possible subspace dimensions. In particular, maximal chains in the poset $\mathbb{P}(V)$ (e.g. $1 < \langle 1,2\rangle < \langle 1,2,3\rangle$) have length 2—so that the order complex $K(\mathbb{P}(V))$ is of dimension 2. Furthermore such a chain implicitly includes a non-irreducible containment $1 < \langle 1,2,3\rangle$ in the poset—which provides the third edge in the 2-simplex corresponding to this maximal chain. Such simplices demonstrate how the simplicial complex exhibits further geometric structure, which is derived from the poset diagram.

And there is also now further content, in our inductive viewpoint via links of vertices (introduced at the end of Example 2.1.12): First the 7 points and 7 lines of $\mathrm{Lk}(\pi)$, for a plane-vertex such as $\pi = \langle 1,2,3\rangle$, have (again more or less tautologically) the structure of the projective plane $\mathbb{P}^2(2)$ of earlier Example 2.1.12. Similarly a plane structure also arises for the link $Lk(1)$ of the point-vertex 1, by working in the quotient 3-space $V/\langle 1\rangle$. Finally for a line-vertex such as $L = \langle 1,2\rangle$, we had observed that each of the 3 points in L is contained in each of the 3 planes above L: so we see that the link $\mathrm{Lk}(L)$ has the structure of the *join* of the two projective lines $\mathbb{P}^1(2)$ determined by the 2-spaces $\langle 1,2\rangle$ and $V/\langle 1,2\rangle$. (Combinatorially, this join is just a complete bipartite graph $K_{3,3}$. And we mention that this join structure also follows from poset properties using Lemma 1.6.13.) ◇

Polar spaces for groups preserving forms. Next we recall that the remaining classical groups, i.e. those other than the full linear group $GL(V)$, are defined as subgroups of $GL(V)$ which preserve a suitable nondegenerate bilinear (or sesquilinear) form defined on $V \times V$. See e.g. Chapter 1 of Carter [**Car89**], or Chapter 3 of Wilson [**Wil09**], for basic background, including various properties stated (usually without specific reference) in the next several paragraphs.

Definitions and generalities. In summary: the group is *orthogonal* if the form is symmetric; *symplectic* if the form is skew-symmetric; and *unitary* if the form is conjugate-symmetric (and actually conjugate-linear rather than bilinear)—for a "conjugation" field automorphism of order 2. Furthermore over a field of characteristic 2, an orthogonal group must also preserve an associated *quadratic* form, defined for just *one* variable from V. And for V of even dimension over a finite field $\mathbb{F}_q$, an orthogonal form can be of "plus" or "minus" type.

Now if we were working with a form in characteristic 0, the associated geometry would focus on lengths and angles, and in particular on subspaces which are nondegenerate under restriction of the form. But since our space V in this section is defined over some finite field $\mathbb{F}_q$, we get *isotropic* vectors (i.e. of "length zero", when (v,v) is divisible by p); and it turns out for the purposes of projective geometry to be more natural to study the subspaces which are instead totally degenerate in the form, in the sense described below. And the subcollection given by those subspaces will determine a *polar* space—as a natural subposet (indeed a "projective subspace") of the full projective space $\mathbb{P}(V)$.

DEFINITION 2.1.16 (Polar spaces). Let V be as in Notation 2.1.1; and assume further that V is equipped with a nondegenerate $\mathbb{F}_q$-valued bilinear form, which we denote by (v,v') for $v,v' \in V$. (When the form is orthogonal, and the field order q is a power of 2, we also assume an $\mathbb{F}_q$-valued quadratic form denoted by $f(v)$.) We let G denote the corresponding *isometry group*, namely the subgroup of $GL(V)$ preserving the form.

The vectors on which the form vanishes (i.e. $(v,v) = 0$) are called *isotropic* (or *singular*, when we further have $f(v) = 0$ in the case of an orthogonal quadratic form). Similarly the linear subspaces W of V on which the bilinear form vanishes (that is, the restriction of $(-,-)$ to $W \times W$ is zero) are called *(totally) isotropic subspaces* (or *(totally) singular* in the case of a quadratic form).

The subposet $\mathcal{P}$ of the projective space $\mathbb{P}(V)$, given by the totally isotropic (resp. totally singular) linear subspaces of V, is called the *polar space* of V. Correspondingly the *flags* in the polar space are just the inclusion-chains in this subposet; that is, the flags are the simplices of the order complex $K(\mathcal{P})$ of the polar space $\mathcal{P}$.

Furthermore, since we are primarily interested in the case where the isometry group G is essentially simple (in the sense of Remark 2.1.2), we usually restrict attention to the case of a form of *maximal Witt index*. See e.g. [**Car89**, pp. 5,6] or [**Wil09**, 3.4.6] for details; this condition says essentially that the largest subspaces which are isotropic in the form should have the maximum possible dimension, namely $\lfloor \dim_F(V)/2 \rfloor$. In particular, the dimension of the order complex $K(\mathcal{P})$ of the polar space $\mathcal{P}$ is about half the dimension of the complex $K\big(\mathbb{P}(V)\big)$ for the full projective space of V.

We mention that—in analogy with Remark 2.1.4.1 on the flag-transitivity of $GL_n(q)$ on the full projective space $\mathbb{P}(V)$—here the group G preserving the form is (usually) flag-transitive on the polar subspace $\mathcal{P}$; this is essentially a consequence of Witt's Lemma (see for example [**GLS98**, 2.7.1], or [**Jac80**, I, p. 369], or [**Wil09**, 3.4.8]). $\diamondsuit$

We begin our discussion with some general observations which will be in the background for many of our examples:

REMARK 2.1.17. For a symplectic form, the antisymmetric condition (namely that $(v, v') = -(v', v)$ for all v) in turn implies that $(v, v) = 0$ when the field $\mathbb{F}_q$ has odd characteristic; and correspondingly $(v, v) = 0$ is *assumed* in characteristic 2. So we have:

(1) For V a symplectic space, all 1-subspaces are totally isotropic.

Indeed continuing to examine the characteristic 2 case, we see since $v = -v$ that there is no difference between symmetric and antisymmetric forms; that is:

(2) In characteristic 2, an orthogonal space V is also symplectic.

An orthogonal space is in fact additionally equipped with an associated quadratic form f. In characteristic other than 2, the associated quadratic form $f(v)$ is just the diagonal (v, v) of the bilinear form (v, v'), and so in particular satisfies the condition $f(v + v') = f(v) + f(v') + 2(v, v')$. But by contrast in characteristic 2, the diagonal values (v, v) are zero by (1); so instead we define f by requiring it to satisfy the analogue of the previous equation, with the factor of 2 removed:

(3) In (2) above, the forms are related instead by $f(v+v') = f(v)+f(v')+(v, v')$.

Thus in any characteristic, if we are given the quadratic form f, we can recover the bilinear form $(-,-)$. $\diamondsuit$

Some small symplectic examples. We first examine some symplectic polar spaces; again we restrict attention to characteristic 2.

Note by Remark 2.1.17.1 that a symplectic form on a space of dimension 1 is necessarily degenerate. Indeed more or less for this reason, a nondegenerate symplectic space must have *even* dimension (cf. [**Car89**, p.3]). Correspondingly, we will replace our earlier use of "n" for $\dim V$ in Notation 2.1.1 by $2m$; so as mentioned in 2.1.16, maximal isotropic subspaces will have dimension m.

Thus the first case to consider is V of dimension 2, corresponding to $m = 1$. Here Remark 2.1.17.1 shows that all the 1-subspaces are isotropic—that is, none are excluded (or even distinguished from each other) by the symplectic structure. Hence in this case, our polar space $\mathcal{P}$ as a subposet turns out to actually be the full projective-line poset $\mathbb{P}(V)$. As before we focus on the smallest field $\mathbb{F}_2$:

EXAMPLE 2.1.18 (The polar space for $Sp_2(2)$). If under $m = 2$ we now further restrict to $\mathbb{F}_2$ (namely $q = p = 2$), then we obtain—just at the more elementary level of set theory—the same content as in Example 2.1.7: namely the polar space $\mathcal{P}$ consists of the three isolated points of the projective line $\mathbb{P}^1(2)$ (that is, $\#P = 3$); and the group $Sp_2(2)$ preserving the form is just the full linear group $GL_2(2)$ again.

However in terms of the symplectic structure, we do make the new observation (using the nondegeneracy of the form) that the three nonzero vectors are pairwise non-orthogonal: that is, they are not perpendicular in the bilinear form—instead they satisfy $(v, v') = 1$. $\diamondsuit$

We can generalize this example, to consider a nondegenerate 2-subspace in a larger space with a form in any characteristic: Such a nondegenerate 2-space is called a *hyperbolic* 2-space; and a pair of generators v, v'—namely isotropic vectors with inner product 1—is a *hyperbolic pair*.

It is standard (cf. [**Car89**, p 3]) that a nondegenerate symplectic form of maximal Witt index on V of dimension $2m$ is uniquely determined up to conjugacy; and in particular, V can be written as the orthgonal direct sum of a set of m hyperbolic 2-subspaces. This is called a *symplectic decomposition* of V. We use this in describing the next-larger case:

EXAMPLE 2.1.19 (The polar space of $Sp_4(2)$). Still with $q = p = 2$, we take V of dimension 4, corresponding to $m = 2$.

Given a basis $1, 2, 3, 4$ of V, in the literature it would be usual to take a symplectic decomposition in the above sense, in which orthogonal hyperbolic pairs are given by "adjacent" pairs $1, 2$ and $3, 4$. However, in this book we will use a different (though conjugate) choice,[9] in which the members of our hyperbolic pairs are "spread out" successively, as far as possible—namely we take the hyperbolic pairs in our *symplectic basis* instead to be $1, 4$ and $2, 3$. Recall that this choice of pairs means that our bilinear form satisfies $(1, 4) = 1_F = (2, 3)$, but we have $(i, j) = 0_F$ for all other pairs i, j from the basis.

As in the previous subsection, we record the values of the relevant geometric parameters: As all vectors are isotropic by Remark 2.1.17.1, we get $\#P = 15$ from the number of projective points in a 4-space V (which we had seen in Example 2.1.15). Since $m = 2$, the maximal isotropic subspaces here are 2-dimensional, and hence determine projective lines; so $\#\ P/L = 3$ (as in Example 2.1.9). From 2.1.15, there are 35 linear 2-subspaces of V; however, for such a 2-space over some vector v to be isotropic, it must lie in the 3-subspace $v^{\perp}$. It follows that $\#\ L/P = 3$:

[9] One reason for this choice is so that the subgroup stabilizing the maximal flag $1 < \langle 1, 2 \rangle$ will have (see later Example 2.2.16) a lower-triangular matrix form; this is in analogy with the lower-triangular form for the stabilizer of a maximal flag in projective space—which we will see in the abstract in Example 2.2.12, and more concretely in Example 2.2.5.

this uses the count of points for the polar space of Example 2.1.18, applied to the (symplectic) 2-space quotient $v^{\perp}/\langle v \rangle$.

Then using these values in the alternative method demonstrated in Example 2.1.15, we compute that $\#L = \frac{15 \cdot 3}{3} = 15$. As our final calculation, we obtain that $\#PL = \#P \cdot \# \ L/P = 15 \cdot 3 = 45$. As in the examples in the previous subsection, flag-transitivity (which we indicated in 2.1.16) provides one way of seeing that these values do not depend on any particular choice of isotropic point or line.

The polar space $\mathcal{P}$ is a subposet of $\mathbb{P}(V)$ with chains of length at most 1, for example $1 < \langle 1, 2 \rangle$. We could (somewhat tediously) draw the usual poset diagram, in the manner of (2.1.14)—with two levels, corresponding to the projective points and lines (this time of course there are 15 of each). And just as in (2.1.14), this feature implies that the order complex $K(\mathcal{P})$ is 1-dimensional, namely a graph coinciding with the poset diagram. (Here the graph is a 1-dimensional subcomplex of the 2-dimensional order complex $K\big(\mathbb{P}(V)\big)$ of the full projective space on V.)

Finally in parallel with the inductive viewpoint on links introduced in Example 2.1.12: we have essentially already indicated that the link $\mathrm{Lk}(L)$ of the line-vertex $L = \langle 1, 2 \rangle$ is the corresponding projective line $\mathbb{P}^1(2)$; and the link $\mathrm{Lk}(P)$ of the point-vertex $P = 1$ is also a projective line—in the form of the polar space of 2-dimensional $v^{\perp}/\langle v \rangle$, as in Example 2.1.18. ◇

EXERCISE 2.1.20 (The polar space of $Sp_4(4)$). Mimic the previous example—again with $p = 2$, but now with $q = 4$, as in earlier Exercise 2.1.8). For the isotropic points and lines, obtain $\#P = \frac{4^4 - 1}{4 - 1} = 85 = \#L$ and $\# \ P/L = 5 = \# \ L/P$, so that $\#PL = 425$. ◇

Some small orthogonal examples. Because we have been working in characteristic 2, we see by Remark 2.1.17.2 that we can convert the above symplectic spaces into orthogonal spaces—by adding a quadratic form.

EXAMPLE 2.1.21 (The polar space for $O_4^+(2)$). We proceed from the symplectic space for $Sp_4(2)$ in Example 2.1.19 to the orthogonal space for $O_4^+(2)$: In that earlier symplectic basis with hyperbolic pairs $1, 4$ and $2, 3$, we obtain (see [**Car89**, Sec 1.6] for further background) an *orthogonal basis* by choosing each basis vector v to be *singular* in the quadratic form f (that is, we have $f(v) = 0$). In particular this makes each of the corresponding hyperbolic 2-spaces into an orthogonal space of $+$-type; and then the full 4-space is also orthogonal of $+$-type.

Notice that when we apply the relation in Remark 2.1.17.3 to the point 14, namely calculating that $f(14) = f(1) + f(4) + (1, 4) = 0 + 0 + 1 = 1$, we conclude that this third vector 14 of $\langle 1, 4 \rangle$ is in fact nonsingular. Thus the polar space of an orthogonal 2-space of $+$-type has just 2 singular points—rather than all 3 points of the projective line, which we had in the symplectic 2-space of Example 2.1.18. (We will soon see in Example 2.1.25 that for a 2-space of minus-type, all 3 points are nonsingular.)

We can also use the relation in Remark 2.1.17.3 to determine the remaining singular vectors in V: namely 12, 13, 24, 34, 1234; so in particular $\#P = 9$. For singular 2-spaces, we apply the relation to pairs among these singular points, to check for singularity of the sum of the pair: we conclude that the singular lines are $\langle 1,2\rangle$, $\langle 1,3\rangle$, $\langle 2,4\rangle$, $\langle 3,4\rangle$, and $\langle 12,34\rangle$, and $\langle 13,24\rangle$; and so we have $\#L = 6$. Since each singular line is in fact a full projective line (just as in the symplectic case in Example 2.1.18), we have $\#\ P/L = 3$. However, this time we can check directly from the above list that $\#\ L/P = 2$—rather than 3 as in the symplectic case of Example 2.1.19: this is because when we mimic that earlier argument, the quotient space $1^{\perp}/\langle 1\rangle$ is now an orthogonal 2-space of $+$-type (indeed isomorphic to $\langle 2,3\rangle$)—and we saw above that such a space has just two rather than three singular vectors. Finally we get $\#PL = 9\cdot 2 = 6\cdot 3 = 18$.

This is another case of a geometry small enough for us to draw the projective lines as straight lines, in the spirit of (2.1.13); we have a "grid" of 3 horizontal and 3 vertical lines (with 3 points each):

(2.1.22)
$$\begin{array}{ccccc} 2 & \text{———} & 24 & \text{———} & 4 \\ | & & | & & | \\ 12 & \text{———} & 1234 & \text{———} & 34 \\ | & & | & & | \\ 1 & \text{———} & 13 & \text{———} & 3 \end{array}$$

And of course we can also draw the usual poset diagram for the polar space $\mathcal{P}$—in the spirit of (2.1.14), with two levels: the 9 points below the 6 lines. Again because of the two levels, the order complex $K(\mathcal{P})$ is of dimension 1, namely a graph agreeing with the poset diagram.

We mention that this graph can be obtained in an alternative manner: Namely first form a complete bipartite graph $K_{3,3}$—on vertex sets given by the 3 horizontal, and the 3 vertical, lines indicated above. Next note that each of the 9 edges connecting a horizontal/vertical pair is indexed by the unique point of intersection of the corresponding pair of lines. So the barycentric subdivision of $K_{3,3}$ gives the graph for $K(\mathcal{P})$—where the 9 "new" vertices from subdivided edges give the lower level (points) of the graph, and the 6 "old" edges give the upper level (lines). (This observation will be relevant later, toward the end of Remark 2.1.27.)

From our usual inductive viewpoint (cf. Example 2.1.12): The link $\mathrm{Lk}(L)$ of the line-vertex given by $L = \langle 1,2\rangle$ is (as before) the associated projective line; but the link $\mathrm{Lk}(P)$ of the point-vertex $P = 1$, consisting of two isolated vertices, is instead the polar space for an orthogonal 2-space of $+$-type, namely $1^{\perp}/\langle 1\rangle$. ♢

We next indicate an elementary construction which is particularly useful for exhibiting some further small examples of polar spaces, and especially for direct computations in them:

REMARK 2.1.23 (A set-theoretic definition of a form). Let W denote the vector space over $\mathbb{F}_2$ given by a power set—i.e. subsets of a set $\{1,2,\ldots,k\}$. Then we can define a symmetric bilinear form on W, using the size of intersections: namely by setting $(a,b) := |a\cap b| \pmod 2$. For example, we see that 1234 and 345 have even-size intersection 34, and so are orthogonal; while 123 and 34 are not orthogonal.

Next let E be the codimension-1 subspace of W given by the subsets of even size: Then we have $(e,e) = 0$ for $e \in E$; i.e. all vectors of E are isotropic. Thus the restriction of the form to E is even symplectic.

Furthermore on those even sets in E, we can define a quadratic form by setting $f(e) := \frac{1}{2}|e| \pmod 2$; for example, 4-subsets are singular, but 2-subsets are nonsingular. ♢

REMARK 2.1.24 (A variant where we regard $Sp_4(2)$ as S_6). We use the well-known isomorphism $Sp_4(2) \cong S_6$ (see e.g. the Atlas [**CCN⁺85**, p 4]); it will allow us to give an alternative approach to the polar space of Example 2.1.19, using the viewpoint of the previous Remark 2.1.23:

Take $G = S_6$, with $k = 6$ in 2.1.23. The symplectic form on the 5-dimensional space E of even subsets is degenerate—since it has radical $E^\perp$ spanned by 123456. We define our 4-space V as the quotient $E/\langle 123456 \rangle$: this amounts to identifying each subset with its complement; but we will still abusively refer to a vector of E as a subset such as 1234, where really we should refer to the partition 1234|56. The symplectic form on E is now naturally inherited by V.

And now using this abstract definition of the form, we can check directly that the pairs 1234, 1235 and 3456, 2456 afford perpendicular hyperbolic 2-spaces, and so give a symplectic basis.

The computation of parameters is also very elementary in this viewpoint: For example a vector is given by a 4-set, so we get $\#P = \binom{6}{4} = 15$. And an isotropic line is spanned by a pair of such 4-sets intersecting in a 2-subset—thus determining a $2,2,2$-partition; so that

$$\#L = \binom{6}{2,2,2} = \frac{\binom{6}{2}\binom{4}{2}\binom{2}{2}}{3!} = 15.$$

A similar analysis gives $\#\ L/P = \binom{4}{2,2} = 3$ etc. ♢

EXAMPLE 2.1.25 (The polar space for $O_4^-(2)$). This is also a convenient place to complete a theme which we had introduced for $O_4^+(2)$ in Example 2.1.21: Namely an orthogonal hyperbolic pair in characteristic 2 by contrast has minus-type, when its two vectors instead are nonsingular; and then we see, as usual via the relation 2.1.17.3, that the third vector of the 2-space is also nonsingular. Thus there are no singular vectors—and so the corresponding polar space is actually empty. Correspondingly we usually only discuss orthogonal polar spaces of minus-type for (even) dimensions at least 4: and minus-type occurs when one (or more generally an odd number) of the hyperbolic 2-spaces in a decomposition is of minus-type.

So when we apply this to V of dimension 4 over $\mathbb{F}_2$, we can take our first hyperbolic pair $1,4$ to be of $+$-type (with both vectors singular), and the second pair $2,3$ to be of minus-type (with both vectors nonsingular). Applying the relation 2.1.17.3, we find three further singular vectors, namely 124, 134, and 1234—so that $\#P = 5$. The remaining 10 nonzero vectors are of course nonsingular.

Now V does have some 2-subspaces, like $\langle 1, 2 \rangle$, which are totally *isotropic*; but such a subspace has some nonsingular vectors, and so is not totally *singular*. Hence the polar space subposet $\mathcal{P}$ consists just of these 5 points, with no edges among them; and the order complex $K(\mathcal{P})$ consists of 5 isolated vertices. (So just at the more elementary level of set theory, we get the same information as in the projective line for $GL_2(4)$ in Exercise 2.1.8; and this is perhaps suggested by the standard isomorphism $\Omega_4^-(2) \cong SL_2(4)$—though viewing this subgroup G in these two different ways determines non-isomorphic "natural" G-modules V.)

These orthogonal structures are perhaps more immediate in the alternative viewpoint of 2.1.23, when we use the standard isomorphism $O_4^-(2) \cong S_5$: Replacing V above by the 4-dimensional space E of even subsets of a 5-set with elements denoted $1, 2, 3, 4, 5$, we see that the singular vectors are given by the five 4-subsets, such as 1234; while the $\binom{5}{2} = 10$ subsets of size 2 are nonsingular. $\diamondsuit$

Some larger-dimensional examples. We remark that the small polar spaces considered so far will be involved very frequently later, in the construction of larger geometries. Indeed we will see this already in the next larger symplectic case:

EXAMPLE 2.1.26 (The polar space for $Sp_6(2)$). We continue with $q = p = 2$, and now take V of dimension 6, corresponding to $m = 3$. This time we choose our symplectic basis with hyperbolic pairs given by $1, 6$ and $2, 5$ and $3, 4$.

We give a briefer indication of parameter values: First $\#P = 2^6 - 1 = 63$ as in 2.1.6—since all nonzero vectors are isotropic by 2.1.17.1. Next the quotient $1^{\perp}/\langle 1 \rangle$ is a symplectic 4-space isomorphic to $\langle 2, 3, 4, 5 \rangle$, so from 2.1.19 we get $\#\ L/P = 15 = \#\ \pi/P$. Similarly $\langle 1, 2 \rangle^{\perp}/\langle 1, 2 \rangle$ is a symplectic 2-space isomorphic to $\langle 3, 4 \rangle$, so we get $\#\ \pi/L = 3$ by 2.1.18. Furthermore totally isotropic 2-spaces and 3-spaces determine full projective lines and planes, so we get $\#\ P/L = 3$ and $\#\ P/\pi = 7 = \#\ L/\pi$ using 2.1.7 and 2.1.9. Now we can apply the alternative method in 2.1.15 to obtain $\#L = \frac{63 \cdot 15}{3} = 315$ and $\#\pi = \frac{63 \cdot 15}{7} = 135$. Finally $\#PL\pi = 63 \cdot 15 \cdot 3 = 945$.

For our usual inductive viewpoint on links of vertices, we have essentially observed already that the link of a plane-vertex is a projective plane, and the link of a point-vertex is an $Sp_4(2)$ polar space; while the link of a line-vertex such as $\langle 1, 2 \rangle$ is the join (compare with the line-vertex situation in 2.1.15) of the projective line from $\langle 1, 2 \rangle$ with the $Sp_2(2)$ polar space for $\langle 1, 2 \rangle^{\perp}/\langle 1, 2 \rangle$—where the latter could also be considered just a projective line. $\diamondsuit$

We close this subsection by describing one situation where we use a variant of the usual polar space:

REMARK 2.1.27 (The "oriflamme" geometry for $\Omega_{2m}^+(2^a)$). We consider an orthogonal space V of dimension $2m$ and $+$-type, over a field $\mathbb{F}_q$ of characteristic 2.

It turns out that the subgroup $\Omega_{2m}^+(q)$ of the full subgroup $O_{2m}^+(q)$ of V preserving the form has two orbits (rather than just one) on the maximal singular subspaces (of dimension m) in V. Since we wish to have the flag-transitive property (recall our mention of "usually" in 2.1.16), we define a new type of polar space, by making a suitable adjustment to the general definition in 2.1.16.

For fuller details, see e.g. p 112 of [**Tit74**]; here is a rough summary: Let C_1 and C_2 denote the two classes of singular m-spaces. We take "most" of the order complex $K(\mathcal{P})$ for the full polar-space poset $\mathcal{P}$ of 2.1.16; however, we omit the vertices given by singular $(m-1)$-spaces, and "compensate" for this by now letting members $X \in C_1$ and $Y \in C_2$ be on a new edge (and suitably on higher-dimensional simplices), when $X \cap Y$ is of dimension $m-1$. In particular, this variant-complex K is the natural building for the group $\Omega_{2m}^+(q)$. (Of course buildings are a main focus for this book, especially starting in the next section.)

We also mention that the complex $K(\mathcal{P})$ of the polar space $\mathcal{P}$ can essentially be recovered from K, by a process of *partial* barycentric subdivision—where subdividing only the edges $X * Y$ indicated above leads to the recovery of the vertices for singular $(m-1)$-spaces such as $X \cap Y$. (And just as in 1.6.4, this partial barycentric subdivision gives a homeomorphism of topological spaces.)

The case of $\Omega_4^+(2)$ gives a small (and somewhat degenerate) illustration of the oriflamme geometry—which we had implicitly introduced in our "alternative" approach toward the end of Example 2.1.21: In this case the variant complex K is the complete bipartite graph $K_{3,3}$ indicated there—with vertex sets given by given by the 3 horizontal, and the 3 vertical, lines. And we indicated there that each of the 9 edges connecting a horizontal-vertical pair is indexed by the unique point of intersection of the corresponding pair of lines. Arbitrarily calling the horizontal lines by the name of points (as vertices), and the vertical ones by the name of line-vertices, we have $\#P = 3 = \#L$; and also $\#\ P/L = 3 = \#\ L/P$ with $\#PL = 9$ for this small oriflamme geometry. (Contrast this with the 6 points and 9 lines of the original $O_4^+(2)$ polar space.)

A more representative example of the oriflamme geometry K is provided by the next-larger case $\Omega_6^+(2)$: where the two different classes of singular 3-spaces give vertices, with pairs on an edge when they intersect in a 2-space (which is of course not visible as a vertex of K); but now K also has further vertices—namely for singular 1-spaces. ♢

2.2. (Option B): The model case: Buildings for Lie type groups

In this section, we give an introduction to buildings, regarded very specifically as geometries for finite Lie-type groups; this discussion is tailored to our subgroup-complex theme, and so differs from the standard axiomatic treatments of buildings in the literature.

In particular, here we will *not* much emphasize the abstract definition of Tits in [**Tit74**] via the notion of apartments. For fuller background in that more standard approach, the reader can refer to Tits's original Springer Lecture Notes volume [**Tit74**]; but there are also a number of more recent expository treatments, such as Chapter 15 of Carter [**Car89**], and the books of Abramenko-Brown [**AB08**], Ronan [**Ron89b**], and Weiss [**Wei03**]. Some briefer discussions are given in Section 42 of Aschbacher [**Asc00**], and in Section 6.8 of Benson [**Ben91**]—the latter is also partly developed in the viewpoint of subgroup complexes used in this book.

Our discussion here will eventually focus on the inductive approach in Tits's local-approach paper [**Tit81**], building up from more basic structures of links of vertices—a viewpoint we had already begun to suggest, starting at Example 2.1.12. But first we begin with some general background, on a uniform approach to buildings, using parabolic subgroups (much as in Tits's earliest viewpoint).

The desirability of a uniform approach. The projective and polar spaces of the previous Section 2.1 provide initial, and comparatively elementary, examples of buildings; indeed, the reader less well acquainted with Lie type groups can just keep in mind the easy case of a classical matrix group, throughout our coverage of Option B on buildings.

However, if we now tried to approach the exceptional Lie type groups by just continuing the viewpoint of projective space, the process would be become increasingly complicated—and in particular would require very specialized case-by-case analysis. This difficulty was in fact one important motivation for Tits, in developing the general theory of buildings—to provide a single *uniform* approach to the geometries for Lie type groups: namely a theory which includes the projective and polar spaces for the classical groups—but also covers the geometries for the exceptional Lie types.

To give just a glimpse of the difficulty mentioned above, we indicate one exceptional Lie type—in which, with some further effort, we can exhibit the geometry as a suitable subspace of a projective space:

EXAMPLE 2.2.1 (A brief description of the building of $G_2(2)$). A group of Lie type G_2 acts on an orthogonal space of dimension 7 (see e.g. [**Hum72**, Sec 19.3] or [**Wil09**, Sec 4.3]), and further preserves an anti-commutative multiplicative structure called the *Cayley algebra*. In characteristic 2, the multiplication becomes commutative, and the space has a radical of dimension 1; thus we can in fact pass to the 6-dimensional quotient V—which inherits a symplectic structure, as well as the multiplication.

Further restricting to $\mathbb{F}_2$ (i.e. $q = p = 2$ again), we obtain the building as a certain subspace of the polar space for $Sp_6(2)$ in Example 2.1.26: For given the multiplication $* : V \times V \to V$ we consider just the linear subspaces of V on which the *multiplication* now restricts to zero. In fact all 63 of the 1-subspaces have zero multiplication. (To check this more explicitly, at least on a basis for V, consider the elements denoted by c_3 through c_8 on p 105 in Humpryeys [**Hum72**]: indeed we find zero in the corresponding diagonal entries of the multiplication table given there.) Furthermore the only larger subspaces with zero multiplication are 63 of the 2-subspaces—which in fact are among the 315 isotropic 2-spaces in the symplectic 6-space of 2.1.26. (For example, one can check in the table in Humphreys that $c_3 * c_7 = 0$; so that $\langle c_3 \rangle < \langle c_3, c_7 \rangle$ gives a typical full flag of the geometry.) Thus our geometry has $\#P = 63 = \#L$, and as usual for lines over $\mathbb{F}_2$ we get $\# P/L = 3$; hence also $\# L/P = 3$ and $\#PL = 63 \cdot 3 = 189$. ◊

Now the algebra structure in this example is special to this one Lie type, namely G_2. Some of the other exceptional Lie types do preserve a suitable multilinear form on a space V, which can then lead to a partly-similar description of the building geometry, again as a subspace of the projective space on V. But because of the differing case-by-case analysis, it is not particularly easy or illuminating to try to find such specialized projective substructures, for *each* individual Lie type.

A similar difficulty complicates one other variation on extending the projective-space approach: Namely the Chevalley construction of a Lie type group G (which we sketch starting at later Remark 2.2.2) exhibits G as automorphisms of the vector space V given by the adjoint module. And the simplices of the building correspond (e.g. using [**Smi82**] as in [**RS85**, 2.1]) with certain linear subspaces of V that they stabilize, thus in turn determining a corresponding subposet of the projective space $\mathbb{P}(V)$. However, this approach again requires detailed case-by-case knowledge of the representation theory of the adjoint module V.

Ideally this discussion underlines the achievement of Tits in developing a uniform approach to the building geometry for Lie type groups G which does not depend on analysis of the particular Lie type of G.

Exhibiting the building via parabolic subgroups. To summarize our discussion below of Tits's uniform approach, we first make the following distinction:

The posets from projective and polar spaces in the previous Section 2.1 can be regarded as an "external" viewpoint on the indicated classical groups G—in that they are obtained from a vector space V, rather than from G itself. However, since we now want to discuss the case of a general Lie type group G, here we will begin instead with an "internal" view of the building—obtained from certain subgroup structures inside G:[10] namely we focus on the subposet $\mathcal{P}$ of the full subgroup lattice $L(G)$ given by the *parabolic subgroups*—which we will describe shortly.

In fact this poset $\mathcal{P}$ of parabolics in turn determines the order complex $K(\mathcal{P})$; and the order complex is the particular form (among several equivalent possibilities) which we will use as our (initial) definition of the building for G. And thereafter we will examine some explicit examples of this abstract structure.

We also mention, more or less conversely, that if we were instead initially given just the purely combinatorial structure of the above complex $K(\mathcal{P})$, along with the action of G, then (see Lemma 2.2.11 below) we could independently recover the parabolic subgroups of G—as the *stabilizers* of simplices in $K(\mathcal{P})$. Indeed when we restrict back to our earlier case of classical matrix groups G, we will see (at Lemma 2.2.20 below) that the internal building poset $\mathcal{P}$ is just the barycentric subdivision of the external projective (or polar) space poset from V.

General background on parabolic structure. We now begin to expand on the above summary. We will devote the next several pages to giving some elements of the theory describing a general Lie type group G (rather than just a classical matrix group as in the previous section).

And in doing so, we will periodically motivate the various abstract concepts needed, by illustrating them explicitly—in the particular case which leads to the classical group $G = SL_3(2)$ (which is the group for the polar space $\mathbb{P}^2(2)$ in earlier Example 2.1.9).

Before describing the structure of parabolic subgroups, we first need to provide some more general background, via a rough sketch of the standard Chevalley construction, giving the "untwisted" Lie type groups:

REMARK 2.2.2 (Some basics of Chevalley groups). To eventually construct a group G of Lie type, we in fact start with a simple Lie algebra $\mathcal{G}$ over $\mathbb{C}$. By the standard Cartan-Killing classification, this algebra has its *Lie type* given by one of:

$$A_n, B_n, C_n, D_n, G_2, F_4, E_6, E_7, E_8,$$

where the first four families (in which n varies) are the *classical* types,[11] and the last five are the *exceptional* types. The subscript in each type indicates the *Lie rank*. (In our special case leading to $G = L_3(2)$, we take $\mathcal{G}$ to be $sl_3(\mathbb{C})$, namely

[10] We should emphasize here that Tits axiomatized buildings in the abstract: that is, without reference to the main "target" example, namely the groups of Lie type. However, in this chapter we are restricting our attention to that main example; and only later, in the discussion leading up to (9.3.5) and Theorem 9.3.6, will we sketch the work of Tits which essentially characterizes abstract finite buildings as those coming explicitly from parabolic structure in Lie type groups.

[11] Cf. [**Car89**, Sec 11.3]: these correspond to groups of types SL_{n+1}, Ω_{2n+1}, Sp_{2n}, and Ω_{2n}.

the 3×3 matrices over $\mathbb{C}$ with trace 0; this $\mathcal{G}$ has classical Lie type A_2, with Lie rank 2.)

The algebra $\mathcal{G}$ is equipped with a *root system* Φ, and a corresponding *Weyl group* W, with its associated *Dynkin diagram* D (where the number of nodes in D is the Lie rank). The standard list of those diagrams, corrresponding to the above Lie types,[12] is given for example on page 43 of [**Car89**]. (Continuing our running example leading to $G = SL_3(2)$, the corresponding Lie type is A_2: here, the system Φ has *positive roots* (certain vectors in real 2-space—cf. Carter [**Car89**, p 46]) often denoted by α, β, $\alpha + \beta$, along with the negatives of these roots; these can be represented as the 6 vertices of a regular hexagon, with natural action by $W \cong D_6$, which has Dynkin diagram •—• (see e.g. pp 46, 12, 40 in [**Car89**]).)

Next we select (see e.g. [**Car89**, Sec 4.4]) a suitable integral lattice $\mathcal{G}_{\mathbb{Z}}$ within $\mathcal{G}$. We wish to work over a finite field $F = \mathbb{F}_q$, where q is a power of a prime p; and to this end we define the F-vector space $V := \mathcal{G}_{\mathbb{Z}} \otimes_{\mathbb{F}_p} \mathbb{F}_q$. (To obtain $G = L_3(2)$, we take $q = p = 2$; and then V is abstractly just some 8-space over $\mathbb{F}_2$—though in this particular case, we may explicitly identify V with $sl_3(2)$, namely the space of 3×3 matrices over $\mathbb{F}_2$ of trace 0.)

In overview, our Lie type group G is now obtained as a subgroup of $GL(V)$—via a suitable finite analogue, for operators on V, of the classical exponentiation process defined on the complex Lie algebra $\mathcal{G}$. A group G arising in this *Chevalley construction* is called an *untwisted* or *Chevalley* group. (In particular, the operators on $V = sl_3(2)$ above exponentiate to a subgroup G of $GL(V)$, which turns out to be isomorphic to $SL_3(2)$.) ♢

REMARK 2.2.3 (Adjustments for twisted Lie type groups). Before continuing our discussion of the Chevalley construction, we now digress briefly to mention that certain modifications (Steinberg's "variations") lead to the *twisted* groups G of Lie type: Very roughly, a twisted group G arises within a Chevalley group H constructed as above, as the subgroup $G := H^\sigma$ centralized by a suitable outer automorphism σ of H (see e.g. Chapter 13 of [**Car89**]). And although a twisted G is no longer directly connected with the Lie algebra that underlies H, there is still a process of assigning a "twisted Lie type", denoted by one of:

$$^2A_n, {}^2B_2, {}^2D_n, {}^3D_4, {}^2G_2, {}^2F_4, {}^2E_6,$$

where the superscript denotes the order of the relevant outer automorphism. Furthermore, in the Dynkin diagram E for H, we can identify images of nodes under σ—in order to obtain the "twisted" diagram D for G. This D then determines a suitable root system Φ and Weyl group W for the twisted group G.[13] ♢

The Chevalley and twisted constructions above produce the Lie type groups G from an abstract viewpoint. However, the classical Lie types in fact lead to the classical matrix groups—which we had examined in the previous section from the more usual explicit viewpoint. (Indeed we have just seen, in our running example,

[12] We may use the term *Coxeter diagram* or *Coxeter-Dynkin diagram* to refer to a more general diagram with nodes and bonds, which might not necessarily correspond to a Lie algebra.

[13] The number of nodes in twisted D gives the rank of the *BN-pair* for G in later Remark 2.2.22; and the term "BN-rank" thus covers this case, as well as the Lie rank in the untwisted case.

that type A_2 reproduces a linear group of type SL_3.) These standard relationships with the classical groups are listed e.g. in [**Car89**, Secs 11.3, 14.5].

The above description of G as a subgroup of $GL(V)$ will now provide the overall context, for our somewhat more specific (but still fairly informal) description of the usual "internal" definition of the parabolic subgroups of G—which, as indicated below, are built up using the root subgroups in the Chevalley construction. For fuller detail, see e.g. Chapter 8 of [**Car89**], or Section 2.6 of [**GLS98**], or [**Wil09**, pp 47, 63, 125].

REMARK 2.2.4 (Borel subgroups and parabolic subgroups). Our discussion from here on will apply to an untwisted group G in the Chevalley construction in Remark 2.2.2; some minor modifications are required for the case of a twisted group.

The Lie algebra $\mathcal{G}$ in 2.2.2 has a vector-space decomposition $\mathcal{U} \oplus \mathcal{H} \oplus \mathcal{U}^-$: where $\mathcal{U}$ and $\mathcal{U}^-$ are the sums of *root* 1*-spaces*—corresponding to the *positive* roots Φ^+ in Φ, and to the *negative* roots Φ^- (respectively); and $\mathcal{H}$ is a *Cartan subalgebra*. We get an analogous decomposition of the corresponding space V in characteristic p. (For our running example with $\mathcal{G} = sl_3(\mathbb{C}), \mathcal{U}$ is given by the subdiagonal (nilpotent) matrices: compare with the matrix picture for the corresponding group U in later Example 2.2.5—the root spaces for α, β, and $\alpha + \beta$ (respectively) are given by matrices which are nonzero only in positions $(2, 1)$, $(3, 2)$, and $(3, 1)$ (resp.); the negative root spaces are given by their transposes; and finally $\mathcal{H}$ is given by diagonal matrices of trace 0. Analogous remarks holds for spaces in $V = sl_3(2)$.)

Now in the Chevalley exponentiation mentioned in 2.2.2, for each positive or negative root γ, the root space for γ leads to a *root group* U_γ of order q.[14] Indeed the subalgebra $\mathcal{U}$ for all positive γ leads to a subgroup U; and G is generated by U and the subgroup U^- corresponding to $\mathcal{U}^-$, where U and U^- are in fact Sylow p-subgroups of G. This U is called a *(full) unipotent subgroup* of G; it is the semidirect product of the root subgroups corresponding to the positive roots, and it has order q^N where $N := |\Phi^+|$. The subgroup H corresponding to $\mathcal{H}$ is a *Cartan subgroup* or *(split) torus* (in particular a p'-subgroup normalizing U). The product $B = UH$ defines, via its G-conjugates, the set of *Borel subgroups* of G. (See e.g. Section 8.1 in [**Car89**] for fuller details.) Finally, the overgroups in G of the Borel subgroups are the *parabolic subgroups* of G. (We will examine the parabolic subgroups in our case of $G = SL_3(2)$ just below, in Example 2.2.5. Notice in particular there that the unipotent group U is lower unitriangular; where the root subgroup U_α is lower unitriangular, with zero subdiagonal entries away from position $(2, 1)$ etc.; while $H = 1$ there, since the multiplicative group of the field $\mathbb{F}_2$ is just the identity.)

In fact this lower-unitriangular behavior in Example 2.2.5 holds in general: For when we choose a basis for the space V consistent with the above algebra decomposition, and consider the resulting matrix form for the action of elements of G, then (in our convention of right-action in 1.5.2) the matrices for the Borel subgroup $B = UH$ chosen above will be *lower triangular* (with H acting as diagonal matrices); in particular, they lie in the stabilizer of a maximal flag of V. (For further detail see e.g. 5.2.2 and 6.2.1 of [**Car89**].) ◊

---

[14]These are analogues of 1*-parameter subgroups* in the characteristic 0 Lie theory.

We will next examine the parabolic subgroups in our case of $G = SL_3(2)$; this is partly a preview of our larger collection of examples later—but also it will allow us to demonstrate explicitly one further product-property of parabolics, which we will indicate in general in Remark 2.2.7 directly afterwards.

EXAMPLE 2.2.5 (Parabolic subgroups of $SL_3(2)$). We take $G = SL_3(2)$. This is the group of a space of dimension 3 over $\mathbb{F}_2$; which here we will call W, since above we have been using V for $sl_3(2)$ of dimension 8. Then W affords the projective plane $\mathbb{P}^2(2)$ of Example 2.1.12. And it turns out that we can exhibit the Borel subgroup $B = UH$ of Remark 2.2.4 above as the stabilizer of a maximal flag in the plane. Such a flag is given by an incident point-line pair PL: we can take $P = \langle 1 \rangle$ and $L = \langle 1, 2 \rangle$ (in terms of the usual basis vectors from W). Then the Borel subgroup B (namely the stabilizer G_{PL} in later Notation 2.2.10) is given by the corresponding lower triangular subgroup:

$$B = G_{PL} = \begin{pmatrix} 1 & 0 & 0 \\ * & 1 & 0 \\ * & * & 1 \end{pmatrix}.$$

Here each $*$ denotes an *independent* choice from $\mathbb{F}_2$, so that $|B| = 2^3 = 8$. In the language of Remark 2.2.4, we have diagonal subgroup $H = 1$, as we are working over $\mathbb{F}_2$; thus $B = UH = U$—and U is the semidirect product of the root subgroups U_α, U_β, and $U_{\alpha+\beta}$, given by lower unitriangular matrices which are zero away from subdiagonal positions $(2,1)$, $(3,2)$, and $(3,1)$.

Now we consider what larger parabolic subgroups lie above this particular Borel subgroup B. Since B is the stabilizer G_{PL} of P and L, clearly it lies in the separate stabilizers G_P and G_L; and it is not hard to check that there are in fact no other proper subgroups of G above B. Thus the proper parabolic subgroups above this B can be described (cf. our more general observation about partitioned matrices in later Example 2.2.12) via:

$$G_P = \left(\begin{array}{c|cc} 1 & 0 & 0 \\ \hline * & * & * \\ * & * & * \end{array}\right) \text{ and } G_L = \left(\begin{array}{cc|c} * & * & 0 \\ * & * & 0 \\ \hline * & * & 1 \end{array}\right),$$

where the lower-right and upper-left 2×2 blocks must be invertible (that is, the matrices must lie in $GL_2(2) = SL_2(2)$). The partitioning also makes it easy to exhibit a further semidirect-product property (which we state in general below at (2.2.8)). Namely we have $G_P = U_P L_P$, where:

$$U_P = \left(\begin{array}{c|cc} 1 & 0 & 0 \\ \hline * & 1 & 0 \\ * & 0 & 1 \end{array}\right) \text{ and } L_P = \left(\begin{array}{c|cc} 1 & 0 & 0 \\ \hline 0 & * & * \\ 0 & * & * \end{array}\right),$$

so that $|U_P| = 2^2$ and $L_P \cong GL_2(2)$. (Exercise: Exhibit the analogous decomposition for G_L, where the partitioning occurs after the second row and column.)

In summary: The proper parabolic subgroups are given by the G-conjugates of $B = G_{PL}$, G_P, and G_L. In other words, the parabolic subgroups are the conjugates of the stabilizers of the full flag PL, and of its nonempty subflags.

In terms of the poset diagram—and complex—for $\mathbb{P}(V)$ at (2.1.14), we see that the conjugates of G_P and G_L correspond (as stabilizers) to the vertices; while the conjugates of G_{PL} correspond to the edges. This suggests how the parabolic subgroups themselves can be used to represent the building given by the polar space. $\diamondsuit$

EXERCISE 2.2.6 (Parabolic subgroups of $SL_3(4)$). Mimic the previous Example 2.2.5—again with $p = 2$, but now with $q = 4$. This time in $B = UH$, obtain that $|U| = 4^3$ and $|H| = 3^2$; and in $G_P = U_P L_P$, obtain that $|U_P| = 4^2$ and $L_P \cong SL_2(4)$. ♢

We now indicate the general form of the product-property in the above Examples:

REMARK 2.2.7 (The Levi decomposition for a parabolic subgroup). We can in fact give a more precise view of the structure of the parabolic subgroups; see e.g. Section 8.3 in [**Car89**] for fuller details: Namely the set Φ^+ of positive roots contains a subset called a *simple system* (or *fundamental system*) Π, which in particular contains Φ^+ in its $\mathbb{Z}^+$-span. (For $G = SL_3(2)$ in Example 2.2.5, we can take $\Pi = \{\alpha, \beta\}$, since the only other positive root $\alpha + \beta$ is given by their sum.)

Furthermore for each nonempty subset $J \subseteq \Pi$, we have a parabolic subgroup G_J—which in the standard *Levi decomposition* (cf. 8.5.2 in [**Car89**]) factors as a semidirect product:

$$G_J = U_J L_J, \text{ where } G_J = N_G(U_J). \tag{2.2.8}$$

Here U_J is the *unipotent radical* of G_J, in fact giving $O_p(G_J)$;[15] and is formed from root subgroups for roots not involving $\hat{J} := \Pi \setminus J$.[16] (In Example 2.2.5, we see that the point stabilizer G_P arises as the parabolic G_α, for $J = \{\alpha\}$: in particular, note that $U_P = U_\alpha$ is the product of the root subgroups for the positive roots α and $\alpha + \beta$ which are not in the span of $\Pi \setminus \{\alpha\} = \beta$.)

And L_J is a *Levi complement*: we will see in later Remark 2.2.30 that this is a subgroup formed "inductively" from (H and) the groups for positive and negative roots which are linear combinations of roots from $\hat{J}$. (In Example 2.2.5, we see that $L_P = L_\alpha$ is generated by the root subgroups for the roots $\pm\beta$ in the span of β, and we recall that $H = 1$ here. Indeed in effect, we could use $\{P, L\}$ as our indexing set Π for the parabolics, instead of $\{\alpha, \beta\}$—if we want to just keep our original stabilizer notation of G_P rather than G_α, etc.)

For a singleton $j \in J$, the subgroup G_j is a *maximal* (proper) parabolic; we obtain further parabolics of the the form $G_J = \cap_{j \in J} G_j$. In particular, $B = G_\Pi$; and we usually say that the $G_{\Pi \setminus \{j\}}$ are *minimal* parabolics. Furthermore $|\Pi|$ is equal to the Lie rank of G; so the rank also gives the number of non-conjugate maximal parabolics. The G_J are the parabolics above B; and so their conjugates are all the parabolic subgroups. (In our example of $G = SL_3(2)$, since the Lie rank is 2, the parabolics G_P and G_L are both maximal and minimal.) ♢

In a moment, we will explore some explicit examples of parabolic structure, using our classical matrix-group examples from the previous Section 2.1. But first, we quickly introduce several further related definitions.

[15] $O_p(X)$ denotes the largest normal p-subgroup of X.

[16] Some sources prefer the opposite indexing, determined by using J itself rather than $\hat{J}$ here; however, our choice of $\hat{J}$ is natural for the simplicial complex structure, in that G_J will stabilize a simplex on $|J|$ vertices.

DEFINITION 2.2.9 (The building as the poset of parabolic subgroups). For G of Lie type, we often informally[17] regard the *(Tits) building* of G as being given just by the poset $\mathcal{P}$ of parabolic subgroups of G.

But more formally, we usually want to consider the building as a simplicial complex. And so more precisely, our "internal" definition of the building is as the order complex $K(\mathcal{P})$ determined by the poset $\mathcal{P}$ of parabolics: that is, the simplices of the building are inclusion-chains of parabolic subgroups. $\diamondsuit$

This internal definition provides a context in which we can discuss the general case of a Lie type group G, and its building as $K(\mathcal{P})$. However, we emphasize from the outset that for the purpose of examples and practical computation, it is usually easier to work with a terser view of the building via some "external" complex K. (Such as the projective space in Example 2.2.5; but more generally, also compare with the terser complex of maximal parabolics in later Definition 2.2.36.)

Furthermore, in order to examine the parabolics in G, we will *not* in practice have to make the Chevalley construction with root subgroups as in Remark 2.2.7; instead, from such an external complex K we can recover the parabolics as the stabilizers of simplices in K. So to describe stabilizers, we will use the standard group-theoretic notation involving subscripts:

NOTATION 2.2.10 (stabilizers). If G acts on some set Ω, and $\omega \in \Omega$, the *stabilizer* (or *isotropy group*) in G of ω is the subset of elements:

$$G_\omega := \{g \in G : (\omega)g = \omega\}.$$

It is easily seen that G_ω is in fact a sub*group* of G. $\diamondsuit$

We can now state more formally our observation on "recovering" the parabolic subgroups:

LEMMA 2.2.11. *Assume that G is a finite group of Lie type.*
(1) A parabolic subgroup P of G is self-normalizing; i.e. $N_G(P) = P$.
(2) The stabilizer G_σ of a simplex σ of the building K is a parabolic subgroup.

PROOF. The property (1) is standard; for example: Write P in the Levi decomposition $G_J = U_J L_J$ of 2.2.7. Then [**Car89**, 8.5.2.iii] gives $G_J = N_G(U_J)$. But we saw that the unipotent radical U_J in fact gives $O_p(G_J)$; thus U_J is not just normal in G_J, but in fact *characteristic* (invariant even under outer automorphisms); so $N_G(G_J)$ must lie in $N_G(U_J) = G_J$, and hence equality holds.

For (2), a simplex $\sigma \in K$ is by Definition 2.2.9 a chain $G_1 < \cdots < G_m$ of parabolic subgroups; the action of G is by conjugation, so that the stabilizer G_σ of this chain is just the intersection of the normalizers of the terms. But each term is self-normalizing by (1); so this intersection is just the smallest term G_1—which is a parabolic, as required. $\square$

[17] In particular this explicit representation within G is in contrast to Tits's abstract axiomatization of buildings, which does not depend on groups of Lie type.

Some examples of parabolic subgroups. We are now ready to study some explicit parabolic subgroups. We will mainly follow the viewpoint of Lemma 2.2.11.2—exhibiting them as stabilizers of flags, in the projective and polar spaces of Section 2.1 (rather than using root subgroups from the Chevalley construction, as in our discussion in Remark 2.2.7).

EXAMPLE 2.2.12 (Parabolic subgroups in the linear group). We first study parabolics in the context of the "general" projective space $\mathbb{P}(V)$, introduced at Definition 2.1.3: where V is an n-dimensional space over $\mathbb{F}_q$. This time, we take G to be not the full linear group $GL(V)$, but instead the determinant-1 subgroup $SL(V)$: for this smaller G turns out to give the Lie type group $A_{n-1}(q)$.

Note that if we just followed the Chevalley construction of G as in Remarks 2.2.2–2.2.7, then the space "V" of that Remark would be the Lie algebra itself, i.e. the *adjoint* module of dimension n^2-1, and we could laboriously build up the parabolic subgroups there. But for our expository purposes, it will be easier and more natural to instead largely follow the stabilizer approach suggested by Lemma 2.2.11.2—since it is based on the natural module V of dimension n for G, which is smaller and more convenient than the adjoint module.

Recall that a maximal simplex of $K(\mathbb{P}(V))$ is given by a maximal flag in V, namely an inclusion chain $V_1 < V_2 < \cdots < V_{n-1}$, where V_i is a subspace of dimension i in V (generated by the first i standard basis vectors). Recall also from 2.1.1 that we assume vectors are written as rows, and that $G = SL(V)$ acts by multiplication on the right. With these conventions, we see that the Borel subgroup B stabilizing this maximal flag is the subgroup of lower triangular matrices (that is, the matrices whose entries are zero strictly above the main diagonal).

We can now exhibit the structure of parabolics G_J, as indicated in the *final* paragraph of Remark 2.2.7—that is, without going first via the root system and Chevalley construction: Namely we use the notation $\{1, 2, \ldots, n-1\}$ to denote the "simple system Π", used just as an *indexing set*. Then the maximal parabolic G_i for $i \in \Pi$ is the subgroup of G normalizing the subspace V_i; that is, G_i is G_{V_i}, using the stabilizer notation of 2.2.10. From the matrix viewpoint, if we partition the rows and columns at the i-th position, G_i consists of the matrices which are lower triangular in that block-partitioning (as we had seen in Example 2.2.5). And more generally, the parabolic G_J for $J \subseteq \Pi$ is the intersection $\cap_{i\in J} G_i$ of these subspace stabilizers (given by matrices which are lower triangular with respect to a more refined partition). Finally, the collection of *all* parabolic subgroups consists of the G-conjugates of such overgroups G_J of B.

We also give an indication of the relationship of the complex $K(\mathbb{P}(V))$ from projective space with the building complex $K(\mathcal{P})$ from the poset $\mathcal{P}$ of parabolics. (We will discuss this relationship somewhat more precisely later, at Lemma 2.2.20.) First consider just the posets: Given a flag like $F := V_1 < V_2 < V_3$ in $\mathbb{P}(V)$, the order complex $K(F)$ is a 2-simplex whose faces are the subflags—of lengths $0, 1, 2$. And then the barycentric subdivision in the sense of 1.6.8 of the flag-poset is the face poset $\mathcal{P}(K(F))$; this is the poset where we put those subflags at "levels" $0, 1, 2$. However, the elements of this poset correspond exactly with the poset of parabolics G_J, for nonempty $J \subseteq \{1, 2, 3\}$; only now with the inclusions in the former *reversed* from those in the latter. More generally, we obtain that the barycentric subdivision of $\mathbb{P}(V)$ is the poset *dual* (or *opposite*) to the poset $\mathcal{P}$ of parabolics. In view of the homeomorphism property of the subdivision in Lemma 1.6.4 (and the obvious homeomorphism of complexes from opposite posets), we conclude that the simplicial complexes from the corresponding posets are topologically equivalent—and so they both provide possible approaches to the building. (Of course for most purposes, the approach via $\mathbb{P}(V)$ is more elementary.)

Finally we mention that it is of course possible (as a lengthier exercise) to relate this polar-space viewpoint on the parabolics G_J to the full apparatus of the root system Φ in the Chevalley construction, as in Remarks 2.2.2–2.2.7. The reader can find a description of the root system of type A_{n-1} e.g. at [**Hum72**, p 64]. $\diamondsuit$

Having made these observations about projective spaces in general, we now turn to a more detailed consideration of the concepts, in the case of the smaller examples of projective spaces from the previous section. Observe that in the cases over the small field $\mathbb{F}_2$, the only nonzero scalar is 1—so that $SL_n(2)$ is all of $GL_n(2)$. Hence our restriction as above to SL_n in fact loses nothing.

EXAMPLE 2.2.13 (Parabolic subgroups of $SL_2(2)$). We take $G = SL_2(2)$ as in earlier Example 2.1.7, where the vector space V of dimension 2 affords the projective line $\mathbb{P}^1(2)$; here the Lie type is A_1. A maximal flag in V is given just by a 1-space V_1; and following the indexing-set convention of 2.2.12, we have $\Pi = \{1\}$, with the corresponding parabolic subgroup denoted G_1. But more usually we would regard V_1 as a projective point P, and so instead write $\{P\}$ for Π, and G_P for this stabilizer. Here G_P just coincides with the Borel subgroup B. The matrices in this subgoup are the lower triangular group:

$$B = \begin{pmatrix} 1 & 0 \\ * & 1 \end{pmatrix},$$

where $*$ represents a choice of any element of $\mathbb{F}_2$, so that the group has order 2. In this case, the diagonal entries are forced to be 1, since that is the only nozero scalar in $\mathbb{F}_2$. Thus in the decomposition $B = UH$ of 2.2.4, we have $H = 1$ and so $B = U$. (Here U is a single root group, corresponding to the unique positive root of the Lie algebra.)

We observe also that in this case, B is a maximal subgroup of G. So the proper parabolic subgroups are just the conjugates of B. (There are three such conjugates—since B of order 2 is self-normalizing in $G = SL_2(2) \cong S_3$ of order 6.) In particular, the poset $\mathcal{P}$ of parabolics in 2.2.9 consists of three elements, with no inclusions among them; and the building $K(\mathcal{P})$ consists of 3 isolated vertices. $\diamondsuit$

EXERCISE 2.2.14 (Parabolic subgroups of $SL_2(4)$). Mimic the previous example, but with V of dimension 2 now instead over $\mathbb{F}_4$. Obtain a similar lower triangular matrix form for the Borel subgroup $B = G_P$—but now there will be nonzero scalars other that 1 which can appear on the diagonal. Conclude from the determinant-1 property that H is the subgroup of diagonal matrices, of order 3; and that $B = UH$ has order $4 \cdot 3 = 12$. This time $\mathcal{P}$ has 5 elements, and $K(\mathcal{P})$ consists of 5 isolated vertices. $\diamondsuit$

Recall we previewed the parabolic subgroups of $SL_3(2)$ in earlier Example 2.2.5.

EXERCISE 2.2.15 (Parabolic subgroups of $SL_4(2)$). Mimic Example 2.2.5, but taking $G = SL_4(2)$, so that the underlying 4-space W affords the projective 3-space of Example 2.1.15; now the Lie type is A_3. A maximal flag is given by an incident triple $PL\pi$ consisting of a projective point, line, and plane; where we take $P = \langle 1 \rangle$, $L = \langle 1, 2 \rangle$, and $\pi = \langle 1, 2, 3 \rangle$. Exhibit the Borel subgroup $B = G_{PL\pi}$ as lower triangular matrices; and show that the proper parabolic subgroups above B are the $2^3 - 1 = 7$ stabilizers G_J of nonempty subflags J of $PL\pi$. Use partitioned matrices to give the Levi decomposition of each G_J; give the order of U_J, and describe L_J in terms of smaller matrix groups. ♢

In the projective-space examples considered above, the chosen Borel subgroup B is the *full* subgroup of lower triangular matrices. For general Lie types (in particular, those other than type A_n), we commented in Remark 2.2.4 that we can always choose a suitable basis so that B will at least be *contained in* the group of lower triangular matrics. And indeed we had made such a choice—for the particular hyperbolic pairs in our discussion of the polar space in Example 2.1.19; so we can now exhibit that resulting lower-triangular feature:

EXAMPLE 2.2.16 (Parabolic subgroups of $Sp_4(2)$). Take $G = Sp_4(2)$, so that V affords the polar space of 2.1.19; this time the Lie type is C_2.[18] A maximal flag is given by PL where $P = \langle 1 \rangle$ and $L = \langle 1, 2 \rangle$.

One computes directly that the full flag stabilizer G_{PL} is given by

$$B = \begin{pmatrix} 1 & 0 & 0 & 0 \\ a & 1 & 0 & 0 \\ b & d & 1 & 0 \\ c & b & a & 1 \end{pmatrix},$$

where a, b, c, d are taken from $\mathbb{F}_2$; so that B is a subgroup of order 2^4, in the full lower triangular group of order 2^6. Again we have $H = 1$, so that $B = UH = U$.

Furthermore the proper parabolic subgroups above B are given by

$$G_P = \left(\begin{array}{c|cc|c} 1 & 0 & 0 & 0 \\ \hline a & * & * & 0 \\ b & * & * & 0 \\ \hline c & b & a & 1 \end{array}\right) \text{ and } G_L = \left(\begin{array}{cc|cc} * & * & 0 & 0 \\ * & * & 0 & 0 \\ \hline b & d & *' & *' \\ c & b & *' & *' \end{array}\right),$$

where the final lower-right block in G_L involving $*'$ is the transpose, about the opposite of the usual diagonal, of the upper-left block involving $*$. Using the partitioning as in Example 2.2.5 to obtain the Levi decomposition $G_P = U_P L_P$, we find $|U_P| = 2^3$ and $L_P \cong SL_2(2)$. Similar statements hold for the Levi decomposition of G_L (where the Levi complement L_L is isomorphic to just a single copy of $SL_2(2)$ because of the above transpose property restricting the entries $*'$).

Again when we consult the poset diagram for the polar space (the analogue for $Sp_4(2)$ of the graph (2.1.14)), the conjugates of G_P and G_L correspond to vertices, the the conjugates of G_{PL} correspond to edges. That is, we must proceed to the barycentric subdivision, to obtain the poset $\mathcal{P}$ in which all parabolics including $G_{PL} = B$ are elements. Then the subdivision poset has two levels, and so it coincides with the building complex $K(\mathcal{P})$, i.e. the inclusion-chains of parabolics. ♢

[18]The group of type C_2 is isomorphic to the (5-dimensional orthogonal) group of type B_2; cf. [**GLS98**, p. 11] or [**Wil09**, 3.57]. So we often write BC_2 for this Lie type.

EXERCISE 2.2.17 (Parabolic subgroups of $Sp_4(2)$—regarded as S_6). Repeat the discussion in the previous Example 2.2.16; but using the alternative construction of V as $E/\langle 123456\rangle$ given in Remark 2.1.24. This amounts to taking $PL = \langle 1234\rangle < \langle 1234, 3456\rangle$. It takes just a little longer to work out the matrix forms of the parabolic subgroups, with respect to this particular basis. ♢

EXERCISE 2.2.18 (Parabolic subgroups of $Sp_6(2)$). Describe the parabolics of $G = Sp_6(2)$, where V affords the polar space of Example 2.1.26; now the Lie type is C_3. The initial analysis is similar to that in Exercise 2.2.15, in that a maximal flag has form $PL\pi$; and there will be $2^3 - 1 = 7$ proper parabolic subgroups over B corresponding to nonempty subflags of $PL\pi$. However, the parabolic subgroups will now also exhibit suitable symmetry about the anti-diagonal, extending the pattern indicated in Example 2.2.16. ♢

The mapping of projective space to the stabilizer poset. Having now discussed some individual examples of projective and polar spaces in the relation to buildings, we close the subsection with a general result describing that "subdivision" relationship for all those examples—in particular, this will make more precise the initial discussion we had given for projective space at the end of Remark 2.2.12. The point is not a crucial one for us at this stage; but it does help indicate the relationship of the viewpoints on a geometry as being either internal to the relevant group G (via stabilizing subgroups), or external to the group (here, via projective space).

So let $\mathcal{P}$ denote a projective or polar[19] space of the previous Section 2.1; and let $\mathcal{S}$ denote the poset of stabilizers in G of the *flags* in $\mathcal{P}$—that is, of the simplices in the order complex $K(\mathcal{P})$.

Observe first that the "stabilizer mapping" given by

$$f : \mathcal{P} \to \mathcal{S} \text{ defined by } f : x \mapsto G_x \tag{2.2.19}$$

is not a poset map when the Lie rank is greater than 1: for then in our examples, we have chains like $P < L$ in the spaces, whereas we see that $G_P \not\leq G_L$ for the corresponding stabilizers. This is one of the reasons why we need to pass to the barycentric subdivision poset $\mathrm{Sd}(\mathcal{P})$, in order to see more clearly the relationship with the stabilizer poset $\mathcal{S}$.

Thus we recall from Definition 1.6.8 that $\mathrm{Sd}(\mathcal{P}) := \mathcal{P}\big(K(\mathcal{P})\big)$ gives the barycentric subdivision of a poset. Using this, we now obtain the fundamental observation underlying our initial approach to buildings via classical groups and their classical projective geometries; namely the following explicit relation between the posets of this section, and of the previous section (where as mentioned earlier, we can expect the order relation to be *reversed* in the subdivision):

[19] We mention that for the building for an even-dimensional orthogonal group, the subdivision-relation is with the variant of the polar space given by the oriflamme geometry in Remark 2.1.27; here one must be a little more careful in describing the choice of flags to associate to parabolic subgroups. Cf. the discussion on pages 111–112 in [**Tit74**])

LEMMA 2.2.20. *For $\mathcal{P}$ and $\mathcal{S}$ as in the classical examples above, in analogy with* (2.2.19)*, define a stabilizer mapping* $\mathrm{Sd}(f)$ *(on* $\mathrm{Sd}(\mathcal{P})$*, namely* $\mathcal{P}(K(\mathcal{P}))$*) by* $\mathrm{Sd}(f) : \sigma \mapsto G_\sigma$*. Then*

$$\mathrm{Sd}(f) : (\mathcal{P}(K(\mathcal{P})), \subseteq) \to (\mathcal{S}, \geq)$$

is a poset isomorphism.

(Or: the map to $(\mathcal{S}, \leq)$ (using the usual direction of ordering) is a poset anti-isomorphism.)

PROOF. An inclusion in the poset $\mathrm{Sd}(\mathcal{P})$ just means an inclusion $\sigma \subseteq \tau$ of simplices in the order complex $K(\mathcal{P})$. Further σ and τ are inclusion-chains in $\mathcal{P}$, so that their vertices are sequences of elements of $\mathcal{P}$ related by $<$. Observe that the stabilizer of such a chain is by definition the intersection of the stabilizers of each member of the chain. So:

$$G_\sigma \overset{\text{bydef}}{=} \bigcap_{x \in \sigma} G_x \overset{(\sigma \subseteq \tau)}{\geq} \bigcap_{x \in \tau} G_x \overset{\text{bydef}}{=} G_\tau;$$

that is, $\subseteq_{\mathrm{Sd}(\mathcal{P})}$ is mapped to $\geq_{\mathcal{S}}$, so that $Sd(f)$ is a poset map. The map is onto by definition: For each member of $\mathcal{S}$ is the stabilizer of some simplex in $K(\mathcal{P})$, that is, of some element of the face poset $\mathrm{Sd}(\mathcal{P})$ of the order complex $K(\mathcal{P})$. Finally the map is 1:1—for it follows using the flag-transitivity of G on $\mathcal{P}$ (see Remark 2.1.4.2 for projective spaces, and the final observation in 2.1.16 for polar spaces) that the stabilizer G_σ of a flag σ in fact stabilizes only flags which are subflags of σ; and further that if σ is a proper subflag of τ, then $G_\sigma > G_\tau$. Hence distinct flags in $\mathcal{P}$ have distinct stabilizers in G. This completes the proof of the poset isomorphism. □

At this point, we can recall that Lemma 2.2.11.2 identifies the "internal" view of the building, more precisely the poset of parabolics in Definition 2.2.9, with the stabilizer poset $\mathcal{S}$. Then Lemma 2.2.20 in turn identifies $\mathcal{S}$ with the subdivision of the "external" version of the building arising from the projective or polar space $\mathcal{P}$.

A glimpse of apartments—as exhibited in generalized polygons. In the subsection after this one, we will use the above approach to buildings via parabolic subgroups, to see how our inductive viewpoint on geometries (introduced earlier starting at Example 2.1.12) leads naturally to the Dynkin diagram which is associated to the Lie type group G defining the building; and we will regard the diagram as encoding some axioms for the geometry.

In order to prepare for that discussion, we now mention that in the more standard original approach of Tits to buildings, that Dynkin diagram is primarily associated to the Weyl group W of G; and that W can be regarded as giving some axioms for the *sub*geometry of the building defined by an apartment. In this book, we will mostly be de-emphasizing the apartment viewpoint. However, in this subsection we will give just a quick overview of apartments—in the context of generalized polygons. This will useful for us at various later points, e.g. Example 3.4.13; but more immediately, we will be seeing shortly that in our inductive approach, the use of only generalized polygons as rank-2 links roughly[20] distinguishes buildings from more general geometries. Indeed this observation is the starting point for proceeding to further geometries which involve other possibilities for rank-2 links; this

[20]More precisely, it distinguishes the geometries of type M in later Definition 2.2.34.

leads to the sporadic geometries in the final Section 2.3 of the chapter. (However, we will no longer get apartment structures in those generalizations.)

Very roughly,[21] an *apartment* in a (finite) building is a certain natural subgeometry, which is a triangulation of a sphere (of the same dimension as the building complex itself); and the building is built up as a union of these spheres, with suitable identifications along their boundaries. We will give examples shortly, starting at (2.2.23).

In the language of Tits, the maximal simplices of the building are called *chambers*. The maximal faces of the chambers are called *panels*, and chambers sharing a panel are considered *incident*, leading to a notion of shortest paths (called *galleries*) between chambers. We focus here on just one important property of apartments:

(2.2.21) Any two chambers of a building lie in some apartment.

Indeed there is a maximal distance in the graph on chambers (given by $N = |\Phi^+|$);[22] and an apartment is determined by the set of galleries between two fixed chambers which are *opposite* in the sense of this maximal distance.

REMARK 2.2.22 (The BN-pair). We mention that the subgroup N of G which stabilizes an apartment-substructure has a normal subgroup given by a Cartan subgroup H (which fixes all the chambers of the apartment), extended by the Weyl group W (which permutes the chambers of the apartment in a regular orbit). This normalizer is the subgroup "N", which together with the Borel subgroup B, determines the *BN-pair* for G. (The pair gives one standard approach to the structure of G; cf. [**Car89**, Ch 8].) ◇

In this subsection, we concentrate on the case of G of Lie rank 2: As we saw earlier e.g. in(2.1.14), the building is then just a bipartite graph: So a chamber is just an edge, determined by two vertices, and a panel is a single vertex; hence an apartment will be a 1-sphere, i.e. a circuit (in fact, shortest-possible) in the graph. More precisely, an apartment in rank 2 will be a polygon—and the corresponding building will then be a *generalized polygon* of that same type. (See e.g. [**Ron89b**, p. 28] or [**Bro98**, p. 117] for fuller details.)

For example, a projective plane can be regarded as a generalized triangle, with apartments given by ordinary triangles; we now examine this terminology, for the plane $\mathbb{P}^2(2)$ of Example 2.1.12: We can perhaps indicate more clearly the motivation for the language of (generalized) triangles, if we first use the "terser" Fano-plane visualization given in (2.1.13). Thus a chamber like $1 \in \langle 1, 2\rangle$ is here not really an edge in a simplicial complex; but instead a point 1, lying on $\langle 1, 2\rangle$ which is drawn as a line. Then to illustrate property (2.2.21), we fix a second chamber, say $3 \in \langle 12, 3\rangle$: which we in fact choose at the maximal distance (namely 3) from $1 \in \langle 1, 2\rangle$ (so the chambers are opposite, in the sense mentioned after (2.2.21)). Below, we first draw the subdiagram of (2.1.13) determined by these initial points and lines. Then it is easy to see that we can embed these chambers in a subtriangle: The two initial

[21] See e.g. [**Car89**, Sec 15.5] for fuller details.

[22] In fact the distance of a second chamber, from some first chamber chosen to be fixed by our Borel subgroup B, is given by the *length* of a corresponding element of the Weyl group W—as a product of standard generators indexed by a simple system Π; and N turns out to give the maximum of this length.

lines intersect in the point 12, so we indicate its subsequent addition within a square bracket; and the two initial points are on the common line $\langle 1,3\rangle$, and we indicate its addition as a dotted line—completing our apartment-triangle. Note we have a total of 6 chambers on the final resulting 3 points and 3 lines.

(2.2.23)

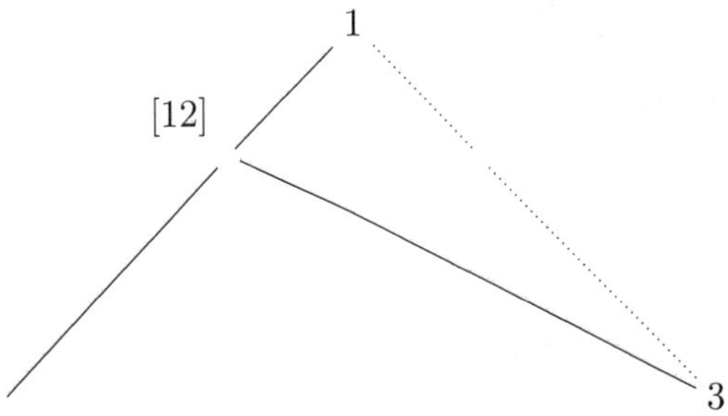

To recapitulate: our initial chambers $1 \in \langle 1,2\rangle$ and $3 \in \langle 12,3\rangle$ are opposite (i.e. at the maximal distance 3)—and the apartment is then given by the union of the galleries (shortest paths) between them: there are two, namely going either via $\langle 1,2\rangle \ni 12$ then $12 \in \langle 12,3\rangle$, or via $1 \in \langle 1,3\rangle$ then $\langle 1,3\rangle \ni 3$. (If the initial chambers had been chosen at less than maximal distance, they would in fact lie in more than one such apartment.)

Of course we can also use the less terse graph-as-simplicial-complex picture of (2.1.14): To make the corresponding analysis, this time we first draw, with solid edges, the subdiagram determined by our two initial (opposite) chambers. Now we add (within square brackets) the further point-vertex and line-vertex that we indicated above; and finally fill in (as dotted lines) the relevant edges among them which appear in (2.1.14). Now we see more explicitly the 6 total chambers needed for the apartment-"triangle"...

(2.2.24)

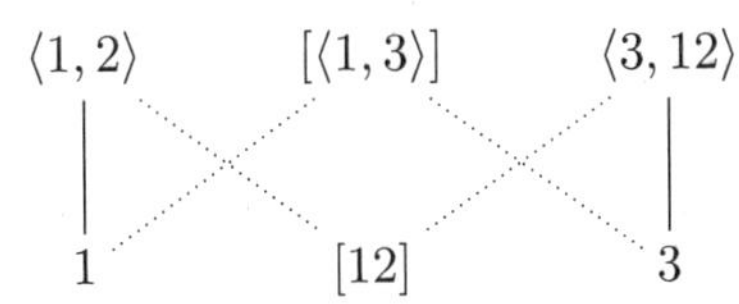

...but notice since the 3 lines are now vertices just like the 3 points, the previous triangle in (2.2.23) above has now been replaced by its barycentric subdivision—namely a hexagon. (Move the added vertices past each other to see the more usual orientation of the hexagon.) *The reader should expect this doubling-via-subdivision behavior in our remaining generalized polygon examples.*

In a similar way, the polar space of a symplectic 4-space provides an example of a *generalized quadrangle*. We sketch more briefly an analysis similar to the above:

EXAMPLE 2.2.25 (The generalized quadrangle for $Sp_4(2)$). Using the polar space for $Sp_4(2)$ in Example 2.1.19, we mimic the examination made in (2.2.24): Namely the chambers $1 \in \langle 1,2\rangle$ and $4 \in \langle 3,4\rangle$ are at distance 4 (which is maximal). So below we first draw these in using solid edges. And we will want to embed these in a "quadrangle"—which in this graph, as noted above, will actually be its barycentric subdivision, namely an octagon.

This could be done by trial and error. But we can roughly mimic the analysis in (2.2.24) above, this time using "nearness" as measured by orthogonality in the polar space: Note that 2 is the only point in the original line $\langle 1, 2\rangle$ which is perpendicular to the original point 4; and similarly 3 is unique in $\langle 3, 4\rangle$ perpendicular to 1. And new (isotropic) lines, involving these new points and the original points, are $\langle 1, 3\rangle$ and $\langle 2, 4\rangle$. Now we have the 4 points and 4 lines:

(2.2.26)

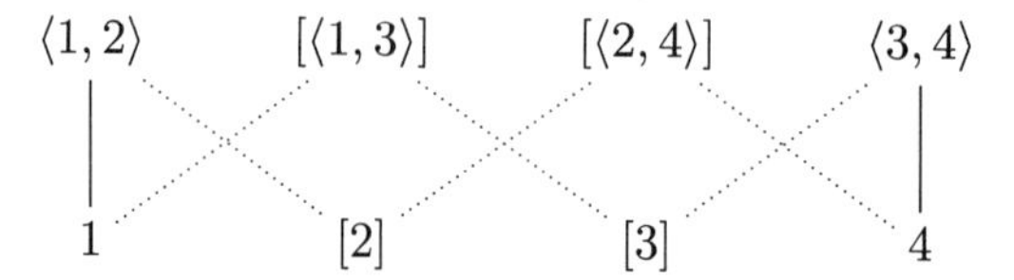

of the apartment-"quadrangle" (in its subdivided octagon-form). ◇

Next we simply mention that a building of type G_2 provides an example of a *generalized hexagon* (cf. [**Wil09**, 4.3.8]). Indeed in the case of $G_2(2)$ in Example 2.2.1, one could mimic the above calculations, using the explicit multiplication indicated there: to show (laboriously) that two chambers of form $P \in L$ are at distance at most 6, and can be embedded in a "hexagon"—i.e. the barycentric subdivision in the graph, namely a 12-gon.

We mention even more briefly that there are also examples of *generalized octagons*: these arise for Lie type groups of the "twisted" type ${}^2F_4$; they are too complicated to pursue here—the reader can consult e.g. Ronan [**Ron89b**, App. 5,6].

Finally we indicate the "trivial" example of a generalized polygon, namely the following notion: An ordinary *digon* means a "lens"-shaped graph, given by 2 vertices joined by 2 *distinct* edges.[23] This is the relevant notion of an apartment in a *generalized digon*—namely a complete bipartite graph. Of course, we have seen e.g. in Example 2.1.21 (and inductively e.g. in Example 2.1.15) that complete bipartite graphs do arise naturally in the geometries of interest to us.

An it turns out that the examples above in fact cover all the kinds of generalized polygons that we need: For the celebrated theorem of Feit and Higman [**FH64**] shows roughly that:

(2.2.27) a suitable generalized m-gon can only have value $m = 2, 3, 4, 6, 8$;

where the "suitable" restriction refers essentially to being defined over some field, e.g. as in the case of Lie type groups.

The generalized polygons are geometries of rank 2. We conclude the subsection by briefly suggesting some analogies for apartments in rank higher than 2: For example, in the projective space $\mathbb{P}^3(2)$ of Example 2.1.15, an apartment is a particular triangulation of a sphere, now in dimension 2—given by the barycentric subdivision of a tetrahedron: it has 24 chambers,[24] which are 2-simplices; and

[23]Normally we consider only graphs which do *not* have multiple edges. And indeed digons will for us just appear (as above) in a barycentric subdivision—given by a quadrangle, which does not have multiple edges. Notice that, within a complete bipartite graph, such a quadrangles does give a shortest (nontrivial) cycle, and hence defines the relevant apartment.

[24]Generically, this number is the order of the Weyl group W: here for Lie type A_3, W is isomorphic to the symmetric group S_4.

maximum distance 6. (Thus one might regard this projective space as a "generalized tetrahedron", and higher-dimensional examples as "generalized polytopes"; however, the literature uses Tits's language of a "geometry of type M", given in our later Definition 2.2.34.)

Associating the Dynkin diagram to the geometry of the building. In our projective-space examples in Section 2.1, starting with Example 2.1.12, we gave some prominence to what we called our "inductive" viewpoint: namely we observed that the links of vertices provided suitable but smaller subgeometries, still having roughly the same general type as that determined by the global geometry. We now make this viewpoint more precise, in the more general context of the building for a Lie type group G: by observing that these subgeometries correspond to subdiagrams, within the Dynkin diagram associated to G (via its Weyl group W). In effect, the diagram will now roughly represent certain (partial) axioms for the geometry.

In this procedure, the rank-2 subdiagrams will correspond to the generalized polygons discussed in the previous subsection. By contrast, in the final Section 2.3 of this chapter, we will generalize the procedure to obtain "diagram geometries" for sporadic simple groups—by allowing rank-2 links which are suitable geometries other than generalized polygons.

The inductive property of links of simplices in a building. So we begin this subsection with a discussion of the relevant inductive property of links in buildings, from the viewpoint of the Dynkin diagram. (We follow essentially the treatment in [**Tit81**], but the property had been indicated as 3.12 in Tits's apartment-based approach to buildings in [**Tit74**].)

Though we will be more precise later at Definition 2.3.1, our exposition here will continue to use just an informal notion of the "type" of a vertex in a complex K: For example, in projective space as in Example 2.1.3 (and in later special cases), the types are given by points, lines, planes, and so on; we may abbreviate the names these types e.g. by P, L, π, etc. In our examples such as buildings we have (and so more generally we will always assume):[25]

a simplex consists of vertices of *distinct* types.

And so a simplex also has a type—given by a subset of the possible vertex types, for example $P\pi$. (And since the action of G preserves the type of a vertex and hence of a simplex σ, we see that the stabilizer G_σ must in fact stabilize each vertex of σ—so that the action is *type-preserving* in the sense we indicated at earlier Remark 1.5.7.)

To indicate vertex types in the more general context of the building of a Lie type group G, we will refer fairly informally to a version of the building which provides an alternative to Definition 2.2.9; we indicate some further details in an appendix to the section, at Definition 2.2.36. In overview, we take the conjugacy classes of *maximal* parabolics to give the types of vertices; with a larger simplex defined by a set of such maximals (from distinct classes), which intersect in at least a Borel subgroup. (We mentioned in Example 2.2.12 that these maximal types can be identified with the roots in a simple system Π for G, viewed as an indexing set.)

In our exposition, a Dynkin diagram D will have a set of nodes (pictured as $\bullet$), along with suitable single or multiple bonds joining subsets of nodes. (We may

[25] Equivalently: two vertices of the *same* type are never on an edge of the complex.

informally use the name D of the diagram to also refer just to the *set* of nodes, without reference to the bonds.)

We will wish to label the nodes of D bijectively with the the types of vertices in K (writing for example $\overset{P}{\bullet}$ to label "points" P); so that the labeling satisfies:

The link of a simplex of of type $J \subseteq D$ is "axiomatized by" $D \setminus J$.

Then the geometry of K will be regarded as being built up inductively, using the subgeometries for links indicated in the diagram. The inductive content is clearest for a geometry of rank at least 3; but first we describe below the base step(s)—for links having rank just 1 or 2:

First, the axiomatization has very minimal content when $|D \setminus J| = 1$, i.e. when the simplex has all but one of the possible vertex types in J: For then by the distinct-types restriction above, the link in the complex consists of vertices which all have that same remaining type in $D \setminus J$, and so they cannot be incident. Correspondingly a single node $\bullet$ represents a link which is just *a set of disconnected vertices*. (In the case of a building, this link will be a projective line: with $(q+1)$ points, as in (2.1.5) using $n = 2$, when G is defined over $\mathbb{F}_q$ (and untwisted).)

Indeed the base step of the axiomatization procedure has its main content when $|D \setminus J| = 2$: For (essentially as the rank-2 subcase of the general building property we will indicate as later Remark 2.2.29) this link must be a building of rank 2—which in particular must be a generalized polygon. Indeed because G is defined over a field $\mathbb{F}_q$, the only possibilities are those from the Feit-Higman Theorem indicated in (2.2.27). So below we indicate each such class of polygons—along with the corresponding Lie type for the Weyl group W of G, and the Dynkin diagram for W:[26]

(2.2.28)

generalized-:	-digon	-triangle	-quadrangle	-hexagon	-octagon
for W of type:	$A_1 \times A_1$	A_2	$B_2\ (= C_2)$	G_2	$I_2(8)$
with diagram D:	$\bullet \quad \bullet$	$\bullet\!-\!\!\!-\!\bullet$	$\bullet\!=\!\!=\!\bullet$	$\bullet \equiv \bullet$	$\bullet \overset{8}{—} \bullet$

Typically the vertex types in a generalized polygon are points and lines; and we write e.g. $\overset{P}{\bullet} — \overset{L}{\bullet}$. To illustrate the above inductive requirement on the diagram: In $\mathbb{P}^2(2)$ in Example 2.1.12, we see that a line-vertex has type $J = \{L\}$; we remove that, so that on $D \setminus J$ we have $\overset{P}{\bullet}$ for the remaining subdiagram—and this represents the 3 discrete points in the link of that projective line $\mathbb{P}^1(2)$. Similarly, removing a point-vertex gives the subdiagram $\overset{L}{\bullet}$; representing the 3 discrete lines on that projective point (where these 3 lines also determine a $\mathbb{P}^1(2)$ configuration).

Now the axiomatization proceeds recursively: For a diagram of rank 3, namely with $|D| = 3$, we require rank-2 links to be generalized polygons from the above list. Then this condition should be satisfied by rank-3 links, when $|D| = 4$. And so on. We will explore some higher-rank examples in a moment; but first:

As we have suggested, the main motivation for considering this inductive requirement is that it is satisfied by buildings; we state this as:

---

[26]The diagram for the octagon is often drawn with a quadruple bond, rather than the single bond with label "8" given here. The Weyl group is the dihedral group D_{16} of order 16; and does not in fact arise from a Lie algebra as in the untwisted cases.

REMARK 2.2.29 (Inductive diagram property for links in buildings, cf. 3.12 in [**Tit74**]). For D the Dynkin diagram associated to G of Lie type with building K, and σ a simplex of type $J \subsetneq D$:

The link $\mathrm{Lk}_K(\sigma)$ is a building, whose type is given by the subdiagram $D \setminus J$. We recall here that when the subdiagram $D \setminus J$ is disconnected, the overall complex is the join of the buildings for the connected components (which are buildings for simple groups). $\diamondsuit$

We mention that Tits in [**Tit74**] uses certain conventions (e.g. for the term star, and the notation $\mathrm{St}(\sigma)$, cf. p. 2 there) somewhat different from the usage in the topological literature. In particular he uses the term *residue*, and writes $\mathrm{Res}(\sigma)$, essentially for the link $\mathrm{Lk}(\sigma)$. (The terminology of residues has correspondingly become fairly common in the literature on groups and geometries.) Thus from property 2.2.29, the building K can be said to satisfy the diagram D "residually". However, in this book we will normally adhere to the usual terminology of links.

The content of property 2.2.29 is primarily geometric; we also now indicate a variant of this inductive property which is expressed in terms of group structure—namely of the parabolic subgroups of the Lie type group G, which provide our view of the building in Definition 2.2.9. The property is phrased in the language of the Levi decomposition $G_J = U_J L_J$ of a parabolic in (2.2.8):

REMARK 2.2.30 (Inductive diagram property for parabolics). For G of Lie type given by D, with parabolic subgroup G_J for nonempty $J \subsetneq D$:

The Levi complement L_J is of the Lie type given by $D \setminus J$.

Here (for p the characteristic of G) it would be more precise to replace L_J by the subgroup $O^{p'}(L_J)$ generated by the p-elements of G;[27] this removes from L_J any "superfluous" elements from the Cartan subgroup H.

When the subdiagram on $D \setminus J$ is disconnected, this group is the commuting product of the individual groups (usually simple) whose types are given by the connected components.

We mention that this property follows from the description of G_J at 8.5.2 in [**Car89**], along with the Chevalley Commutator Formula 5.2.2 there. $\diamondsuit$

Examples of the inductive diagram properties. We now explore the properties in Remarks 2.2.29 and 2.2.30 in some of our earlier explicit examples. Recall for the classical matrix groups G in Section 2.1 that in view of Lemma 2.2.20, the building K of G is realized as the order complex $K(\mathcal{P})$, where $\mathcal{P}$ is the projective or polar space on the natural module V for G.

As we had mentioned above, 2.2.29 has minimal content for $G \cong SL_2(2)$ in Example 2.1.7, where G is of Lie rank 1 and of Lie type A_1: the diagram given by the single node $\overset{P}{\bullet}$ just says that the projective line $\mathbb{P}^1(2)$ is a set of 3 disconnected points.

We turn to our earlier cases of Lie rank 2: namely Example 2.2.5 of Lie type A_2 with diagram given by $\overset{P}{\bullet} \text{---} \overset{L}{\bullet}$; and Example 2.2.16 of Lie type C_2 with the diagram $\overset{P}{\bullet} = \overset{L}{\bullet}$. Here the content of property 2.2.29 is just that the link of any

[27] The p' in the notation refers to affording the maximal *quotient* which is a p'-group.

vertex x has subdiagram of form $\overset{y}{\bullet}$, namely of type A_1—indeed, it is a projective line $\mathbb{P}^1(2)$, as we had earlier observed in those examples. Similarly the content of property 2.2.30 is just that for the parabolic subgroup G_x stabilizing a vertex, we have L_x of Lie type A_1: and indeed, we had seen that $G_x/O_2(G_x) \cong SL_2(2)$—from the matrix forms for the parabolic subgroups given in the later continuation of those Examples at 2.2.5 and 2.2.16.

In fact the real strength of properties 2.2.29 and 2.2.30 only becomes apparent when we get to our earlier examples of Lie rank at least 3.

EXAMPLE 2.2.31 ($SL_4(2)$ has A_3-diagram geometry). Let $G = SL_4(2)$ as in Exercise 2.2.15, with building $K = K(\mathbb{P}(V))$ given by the order complex for the projective space $\mathbb{P}^3(2)$ of Example 2.1.15. The Lie type is A_3, and we label the corresponding Dynkin diagram D as $\overset{P}{\bullet} - \overset{L}{\bullet} - \overset{\pi}{\bullet}$.

We first verify property 2.2.29: For a plane π, we had seen in Exercise 2.2.15 that $\mathrm{Lk}(\pi)$ consists of the points and lines of a projective plane $\mathbb{P}^2(2)$—and this (generalized triangle) structure has the Lie type A_2, which indeed corresponds to the subdiagram on $D \setminus \{\pi\}$, namely $\overset{P}{\bullet} - \overset{L}{\bullet}$. Similarly we had seen that $\mathrm{Lk}(P)$ determines (with vertices of types L and π above P) a plane structure; giving type A_2 as in the subdiagram on $D \setminus \{P\}$, namely $\overset{L}{\bullet} - \overset{\pi}{\bullet}$. Finally for a line L, we had seen that $\mathrm{Lk}(L)$ (with vertices of types P and π) has the structure of a join of two projective lines, i.e. a generalized digon with type $A_1 \times A_1$—again corresponding to the subdiagram on $D \setminus \{L\}$, namely $\overset{P}{\bullet} \quad \overset{\pi}{\bullet}$. This completes the verification of property 2.2.29, for vertices of the building K. For edges, the links are just projective lines—corresponding to single nodes $\bullet$, so that the verification is essentially vacuous.

We turn to the verification of the group-theoretic property 2.2.30: We assume the solution to earlier Exercise 2.2.15, namely the partitioned-matrix forms for parabolics—analogous to those for $SL(3,2)$ in Example 2.2.5. For x of type P or π, we find that the stabilizer G_x has Levi complement $L_x \cong SL_3(2)$—of the type A_2 exhibited by the corresponding subdiagram. On the other hand for x of type L, we obtain $L_x \cong SL_2(2) \times SL_2(2)$, of the type $A_1 \times A_1$ of the relevant subdiagram. The verification of 2.2.30 for edges E follows from that for vertices—namely $L_E \cong SL_2(2)$ of type A_1. ♢

EXERCISE 2.2.32 ($SL_5(2)$ has A_4-diagram geometry). Mimic the previous example, but now with dimension $n = 5$; that is, explicitly verify that $SL_5(2)$ has diagram A_4 ($\bullet - \bullet - \bullet - \bullet$), in the sense of properties 2.2.29 and 2.2.30.

Indeed more generally, $SL_{n+1}(q)$ has diagram A_n: $\bullet - \bullet - \cdots - \bullet - \bullet$ with n nodes. ♢

EXAMPLE 2.2.33 ($Sp_6(2)$ has C_3-diagram geometry). Let $G = Sp_6(2)$ as in Exercise 2.2.18, with $K = K(\mathcal{P})$ given by the order complex for the polar space $\mathcal{P}$ of Example 2.1.26. The Lie type is C_3, and we label the corresponding Dynkin diagram D as $\overset{P}{\bullet} - \overset{L}{\bullet} = \overset{\pi}{\bullet}$.

Again we first verify property 2.2.29: For a plane π (much as in the previous Example), we had seen in Example 2.1.26 that $\mathrm{Lk}(\pi)$ is a projective plane—with type A_2 corresponding to the subdiagram $\overset{P}{\bullet} — \overset{L}{\bullet}$ on $D \setminus \{\pi\}$. For a line L (again much as in the previous Example), we had seen that $\mathrm{Lk}(L)$ is the join of two projective lines, with type $A_1 \times A_1$ as in the subdiagram on $D \setminus \{L\}$, namely $\overset{P}{\bullet} \quad \overset{\pi}{\bullet}$. However for a point P, this time we had seen that $\mathrm{Lk}(L)$ is instead a rank-2 polar space, of type C_2—but again we see that this corresponds to the diagram $\overset{L}{\bullet} = \overset{\pi}{\bullet}$ on the subdiagram $D \setminus \{P\}$. This verifies property 2.2.29 for vertices; and again the case of edges follows essentially vacuously.

Finally for the group-theoretic property 2.2.30: We assume the solution of Exercise 2.2.18, namely the partitioned-matrix form for parabolics—analogous to those for $Sp_4(2)$ in Example 2.2.16. We find that $L_P \cong Sp_4(2)$ of type C_2; and $L_L \cong SL_2(2) \times SL_2(2)$ of type $A_1 \times A_1$; and $L_\pi \cong SL_3(2)$ of type A_2—all corresponding to the subdiagrams indicated in the previous paragraph. Again the verification of property 2.2.30 for edges E amounts to $L_E \cong SL_2(2)$, which follows from the verification just made for the vertices. ♢

The inductive property for some geometries beyond buildings. The property in 2.2.29 suggests that we might obtain interesting classes of geometries—including, but not limited to, buildings—if we consider Coxeter diagrams D which are more general than the Dynkin diagrams associated to the complex simple Lie algebras.

As one step in this direction we indicate fairly informally one natural such class of geometries introduced by Tits. (We will further extend this class, in the final Section 2.3 of the chapter.)

DEFINITION 2.2.34 (Tits's geometries of type M). In [**Tit81**], Tits makes an inductive definition of the class of *geometries of type* M,[28] where M is a symmetric matrix—with entries given by nonnegative integers or ∞. One assumes the distinct-types restriction we had mentioned earlier: namely each simplex σ of the complex has vertices of distinct "types", taken from some type-set denoted I. (This set plays the role of the node-set, in our discussion above of Dynkin diagrams D.) The types index the rows and columns of M, and M has entries 1 on the diagonal; and if for $i \neq j$ the matrix entry $M_{i,j}$ has the (finite) value m, then the link in the geometry of a simplex with type given $I \setminus \{i, j\}$ is assumed to be a *generalized* m-*gon*. ♢

Notice that the possible m above are *not* restricted to those limited values (namely $2, 3, 4, 6, 8$ in (2.2.27)) which can occur for buildings of Lie type groups.

And to such a matrix M, we associate a corresponding Coxeter diagram: There is a node for each type in i; and when $M_{ij} = m$ (now including ∞), the nodes i and j determine a Coxeter subdiagram $\bullet \overset{m}{—} \bullet$, which corresponds to the dihedral Coxeter group D_{2m}. (As indicated earlier, in the cases $m = 2, 3, 4, 6, 8$ we may by convention draw $0, 1, 2, 3, 4$ bonds.)

EXAMPLE 2.2.35 (The matrices M for types A_3 and C_3). In Examples 2.2.31 and 2.2.33, the matrices M in Definition 2.2.34 are:

$$\begin{pmatrix} 1 & 3 & 2 \\ 3 & 1 & 3 \\ 2 & 3 & 1 \end{pmatrix} \quad \text{and} \quad \begin{pmatrix} 1 & 3 & 2 \\ 3 & 1 & 4 \\ 2 & 4 & 1 \end{pmatrix}$$

[28]Most authors now call these *Tits geometries*.

for types A_3 and C_3, respectively. $\diamond$

In the final Section 2.3 on diagram geometries, we will see diagram geometries which are defined in this inductive way; but which allow rank-2 links to be somewhat more general than the generalized polygons which arise in the geometries of type M (and in particular in buildings).

Some other viewpoints on buildings. This final subsection is essentially an appendix to the section, containing some details that we had omitted during the main exposition. The more impatient reader can just skip ahead directly to the final Section 2.3.

We had seen in 2.2.20 that for G a classical matrix group, the face poset of the simplicial complex $K(\mathcal{P})$ (determined by the natural geometry of the projective or polar space $\mathcal{P}$ for G) is isomorphic to the poset $(\mathcal{S}, \geq)$ of parabolic subgroups—where we are using the reverse $\geq$ of the usual inclusion $\leq$. (So $\mathcal{P}$ leads us to the building as given in Definition 2.2.9.)

However, for G a group of Lie type other than a classical matrix group (that is, for G of exceptional type or of a non-unitary twisted type), there is no immediate[29] analogue of such a "more elementary" poset $\mathcal{P}$ which leads naturally to the parabolics and the building.

Instead, the analysis of Tits (see 5.2 in [**Tit74**]) just *defined* the building for an arbitrary G of Lie type using the poset of parabolics—and we have followed this in our Definition 2.2.9. Indeed in Tits's language, the simplicial complex "is" the poset $(\mathcal{S}, \geq)$; whereas we will continue to distinguish between a complex and its face poset.

In that spirit, we will next indicate an equivalent description of the building—now as a certain abstract simplicial complex K (in the sense of Definition 1.2.5); in such a way that the face poset $\mathcal{P}(K)$ gives the parabolic poset $(\mathcal{S}, \geq)$—i.e., the building in the viewpoint of our original Definition 2.2.9). (We mention that in the language of later Remark 4.5.5, we are in effect constructing the *nerve* of the covering of the building poset defined by the cones under maximal parabolics.)

DEFINITION 2.2.36 (The building as a complex on *maximal* parabolics). We let the set K_0 of vertices of a complex K be given by the set of maximal parabolic subgroups of G; and a simplex σ of K is defined as any collection of maximal parabolics not conjugate in G, all containing some common Borel subgroup of G.

From this definition, it is clear that any subset of a simplex still contains that Borel in the subset-intersection. And this is the condition required for an abstract simplicial complex in Definition 1.2.5.

Indeed as the intersection of such a collection σ contains a Borel, that intersection is by definition (in Remark 2.2.4) itself a parabolic subgroup X of G; and since parabolic subgroups are self-normalizing (by Lemma 2.2.11.1), we see that

---

[29] However we mention one approach to a possible substitute for "$\mathcal{P}$": Using the viewpoint of [**RS85**, 2.1], for V any irreducible module for G in characteristic p, one could in general define a poset $\mathcal{P}$ via a suitable subset of the projective space $\mathbb{P}(V)$ of V, with elements corresponding to the parabolics $\mathcal{S}$. But typically the simplices in the building K (corresponding to the inclusions in the poset $(\mathcal{S}, \geq)$) would define more incidences than those given by the inclusion-chains in the order complex $K(\mathcal{P})$ of this $\mathcal{P}$.

the parabolic X is exactly the stabilizer G_σ, in the action of G on K determined by conjugation.[30]

Note that the type of a simplex is given by the set of conjugacy classes of the maximal parabolics which define it; and since G preserves these classes, the action is type-preserving in the sense we indicated at earlier Remark 1.5.7—in particular, the stabilizer in G of a simplex fixes all its vertices. ♢

And now, although we cannot for an arbitrary building find an analogue of the elementary poset $\mathcal{P}$ in Lemma 2.2.20, the above definition of K in effect provides an analogue at least of the complex "$K(\mathcal{P})$" there. Thus our analogue of that lemma for a general building shows that this complex K leads to the building viewed as the parabolic poset in Definition 2.2.9:

LEMMA 2.2.37. *For G a group of Lie type, with building K as constructed from maximal parabolics in Definition 2.2.36, define the stabilizer mapping $f : K \to \mathcal{S}$ by $f : \sigma \mapsto G_\sigma$. Define the induced map on $\mathcal{P}(K)$ by $\mathcal{P}(f) : \sigma \mapsto G_\sigma$. Then*

$$\mathcal{P}(f) : (\mathcal{P}(K), \subseteq) \to (\mathcal{S}, \geq)$$

is a poset isomorphism.

PROOF. For a simplex σ of maximal parabolics G_j for $j \in J$, the self-normalizing property shows that the stablizer of σ is the intersection $\cap_{j\in J} G_j$; and this is just the parabolic G_J (as we had already seen in Remark 2.2.7). From this it is essentially immediate that the map is a *set* isomorphism. To see that the map preserves the ordering, observe for $J \subseteq K$ that the intersection G_J of the smaller set J contains the intersection G_K for the larger set K. □

We conclude this appendix-subsection with some further background on several research directions in the area of buildings, which we will not pursue in this book.

A main theorem of Tits in [**Tit74**] (see e.g. page IV there, and our later discussion leading up to (9.3.5)) characterizes finite buildings of rank at least 3 (defined abstractly via the axioms on apartments) as arising only from finite groups G of Lie type. A corresponding result for finite buildings of rank 2, requiring the further hypothesis of a *split* BN-pair, was given by Fong and Seitz [**FS73**]. An alternative approach to chararacterizing buildings of rank 2 via generalized polygons satisfying the *Moufang* condition has been given recently by Tits and Weiss [**TW02**].

A finite building (and we usually consider only the finite case in this book) is often called a *spherical* building, since its (finite) Weyl group W acts on an apartment—which as a simplicial complex can be regarded as a triangulation of a sphere embedded in a finite-dimensional Euclidean space (cf. p 24 of [**Tit74**]). There is also a considerable literature on infinite buildings; especially the case of *affine* or *Euclidean* buildings, in which the *affine Weyl group* $\tilde{W}$ is infinite, and the apartment is now a triangulation of a finite-dimensional Euclidean space. The standard list of *extended* (or *affine*) Dynkin diagrams $\tilde{D}$ for affine Weyl groups $\tilde{W}$ is given e.g. on p 12 of [**GLS98**]. Affine buildings arise for example from p-adic Lie groups, and more generally from Kac-Moody groups. (See e.g. Ronan [**Ron89b**]

[30] Some readers may be more familiar with Tits's construction of the building as a *coset* complex: with simplices defined by the cosets of the parabolic subgroups over a fixed Borel subgroup B (rather than by the conjugates of those parabolics). But this construction by cosets is equivalent to the construction by conjugates, in view of the self-normalizing property above (e.g. using Theorem 3(2) and Proposition 6 of [**DF99**, Ch 4]).

for more details.) Infinite buildings are more difficult than finite buildings to classify, since they are "freer"—see e.g. [**RT87**]. However in the affine subcase, Tits in [**Tit86**] gave a classification of suitable buildings of rank at least 4 (cf. our later Remark 2.3.9, and our discussion leading up to Theorem 9.3.8). Furthermore Ronan and Tits (cf. [**Tit92**]) and others have pursued the related theory of *twin buildings*.

2.3. (Option S): Diagram geometries for sporadic simple groups

In the first subsection of this section, we will set up our general context of diagram geometries, as a generalization of the "inductive" property of buildings in Remark 2.2.29. Then in each of the four subsequent subsections, we introduce an example of a non-building diagram geometry for a simple group. In doing so, we mainly emphasize the diagram's inductive view of the links; but we also record numerical parameters as in earlier sections, since we will continue use these sporadic geometries as examples in later chapters.

A general setting for geometries with associated diagrams. Since Tits's "local approach" paper [**Tit81**], there has been considerable research on diagram geometries (especially for sporadic simple groups). There is some variation in the literature among the axiomatic setups used—and we will occasionally mention other approaches that we do not follow here.

Our abstract "internal" context for diagram geometries. In order to continue our theme of diagrams, we first define a wider geometric context in which to place it. This will be a group-theoretic variant of the general context in the work of Tits on geometries of type M (our Definition 2.2.34). In effect, it mimics certain aspects of the construction of the building via maximal parabolics, which we had introduced in Definition 2.2.36. (Cf. also the general context of Chapter 6 in [**BS08**]; e.g. Definition 6.3.2 there is a version of Definition 2.3.1 below.)

In particular, we mention that our abstract internal approach below was a later historical development: the original introduction of 2-local geometries in Ronan-Smith [**RS80**] (and hence also of the examples considered in [**RSY90**]) proceeded instead via the "external" approach we will indicate subsequently at Remark 2.3.2.

DEFINITION 2.3.1 (Intersection complex over a set I (of vertex types)). We work in some fixed group G. We choose a collection G_i (as i varies over some finite indexing set I) of represesentatives of distinct G-conjugacy classes of subgroups of G. We set $B := \bigcap_{i \in I} G_i$.

We form a simplicial complex K as follows: We define the vertex set K_0 to be the G-conjugates of the G_i; with simplices defined by any collection of conjugates of the G_i, at most one from each G-conjugacy class, containing a common conjugate of B. Just as in Definition 2.2.36, this condition on a collection is inherited by subcollections, as required for Definition 1.2.5 of an abstract simplicial complex.

The *rank* of the complex K is $|I| =: n$; while the topological *dimension* is $\dim K = (n-1)$. ◊

REMARK 2.3.2 (External complexes). In some cases, we may initially be given some more elementary "external" complex K', on which G acts; and then when we take the G_i to be the stabilizers of suitable vertices of various types in K', the group-theoretic complex K in Definition 2.3.1 will provide an alternative version of K' which is "internal" to G. ◊

Of course Definition 2.3.1 is a generalization of the motivating case of the building for a Lie type group G in Definition 2.2.36, where the role of "I" is played by the simple system Π, with G_i the maximal parabolics over a Borel subgroup B. Particular cases appear in Examples 2.2.31 and 2.2.33, where we identified Π with the type set $\{P, L, \pi\}$.

And as we had mentioned above, Definition 2.3.1 provides a group-theoretic context exhibiting the notion of a *geometry over* I in Section 1.2 of [**Tit81**]: our vertices are Tits's set "V"; his incidence relation "$*$" corresponds to the face relation in our simplex K; and his "type map τ" becomes, for a simplex σ in our situation, just the listing of the particular conjugacy classes $i \in I$ of the G_i which occur in the set σ. This context can be used to make precise the notion of the *type* of a vertex or simplex, which we had been using more informally in the earlier sections of this chapter.

In particular, K is an example of a *simplicial complex with type*;[31] and G has a natural action on K by conjugation—which is by construction *type-preserving* (in the sense we indicated at earlier Remark 1.5.7: notice a simplex stabilizer fixes all vertices of that simplex).

REMARK 2.3.3 (Some geometric features we typically expect). The above setup in 2.3.1 of an intersection complex K over I is reduced to the bare essentials; whereas in practice, our actual examples will usually share many further properties—which again are motivated by the model case where G is of Lie type, and K is its building. For example:

(1) Usually G will be a simple group; typically a sporadic group.

(2) Usually the G_i will be maximal subgroups of G; in particular, the G_i will typically be self-normalizing (in analogy with 2.2.11.1); and this will also hold for their various intersections containing B. And then (for reasons suggested in the proof of Lemma 2.2.37) we also get:

(a) The intersection X of the subgroups defining a simplex σ is the stabilizer G_σ; indeed typically we obtain an analogue of Lemma 2.2.37, in which the face poset of K is isomorphic to the stabilizer poset $(\mathcal{S}, \geq)$.

(b) The definition via conjugates provides a first step towards showing that G should be flag-transitive, that is, transitive on maximal simplices in K.

(c) If σ has type $J \subset I$, the above stabilizer G_σ is a conjugate of $G_J := \bigcap_{j \in J} G_i$. we may let U_J denote the kernel of G_J on $\mathrm{Lk}_K(G_J)$, as a generalization of the concept of the unipotent radical. And we may define a substitute for the concept of Levi complement by:

$$L_J := G_J/U_J.$$

In this notation, typically we will have analogues of the inductive group-theoretic properties in Remark 2.2.30 for buildings.

(3) Typically for a prime p, the groups G_i will be p-local subgroups; and often we take the intersection B of the G_i as the normalizer of a Sylow p-subgroup of G. (Of course when G is of Lie type in characteristic p, this is the case for a Borel

[31] One also sees the term *colored* simplicial complex, especially in the combinatorial literature.

subgroup B.) In this case, we may say that the geometry K is of *characteristic* p. $\diamondsuit$

We now turn to the property in Remark 2.2.29, namely the inductive viewpoint on links, as axiomatized by a diagram. So in the above context of a complex K over I, we discuss what it should mean for K to "residually satisfy" a diagram D (whose nodes are identified with the indexing set I):

First, just as at the start of our discussion of Dynkin diagrams for buildings in the previous section, a diagram of rank 1 is just a single node $\bullet$; and this represents a collection of disconnected vertices.

Next for rank 2: We recall that in the case of buildings, and indeed more generally for Tits's geometries of type M in Definition 2.2.34, the allowable diagrams of rank 2 are just the "$I_2(m)$", corresponding to generalized m-gons. Now instead our allowable class (call it $\mathcal{C}$) of rank-2 geometries will include, in addition to the generalized polygons, certain other "sporadic" rank-2 geometries—for which we will define suitable further diagrams. (We will begin to indicate some such sporadic rank-2 geometries to add to the class $\mathcal{C}$, in just a moment.)

And finally for I of rank at least 3, we will obtain a *diagram D on the indexing set I* by connecting the pairs of nodes in I using the bonding conventions of such assumed rank-2 diagrams; this should satisfy the analogue of the inductive property in Remark 2.2.29:

DEFINITION 2.3.4 (Diagram D for a complex K). Assume we are given a complex K for G, over an indexing set I. (In general we consider an intersection complex as in Definition 2.3.1, though when available we may use a more elementary external complex K as in Remark 2.3.2.) We say K *has diagram D* (where the nodes of D are indexed by I) if we have the recursive property (based on suitable class $\mathcal{C}$ of rank-2 possibilities as indicated above):

For each $i \in I$, the link $\mathrm{Lk}_K(G_i)$ of the vertex G_i has diagram $D \setminus \{i\}$.

We may refer to K less specifically as a *diagram geometry*. $\diamondsuit$

Indeed it should be clear that Definition 2.3.4 just extends the idea in Remark 2.2.29, by now allowing the rank-2 class $\mathcal{C}$ to contain geometries beyond just the generalized polygons.

And of course, the main purpose of this section is to discuss examples of diagram geometries which use a larger class $\mathcal{C}$; and which typically arise from sporadic groups, rather than from the buildings for Lie type groups. In particular, the more impatient reader may now wish to skip directly to the next subsection, which begins to introduce such examples; the remainder of this subsection provides primarily historical context on the development of this abstract viewpoint on diagram geometries.

Buekenhout's examples based on circle geometries. The work of Buekenhout in [**Bue79**] first drew widespread attention to the existence of diagram geometries for many of the sporadic simple groups.

Buekenhout proceeded by using, as an addition to the class $\mathcal{C}$ of allowable rank-2 geometries (beyond generalized polygons), what he called the *circle* geometries; by convention a c labels the bond in the corresponding diagram $\overset{P}{\bullet}\overset{c}{—}\overset{L}{\bullet}$. The geometry reflects a property of suitable permutation groups, using also the induced action on unordered pairs: The *points* are indexed by the left node, and each pair of points determines a unique *line* (indexed by the right node); and indeed each line consists of just *two* points. Thus lines may as well be identified with pairs of points. (Note that these two-point lines are in contrast with the projective lines over fields $\mathbb{F}_q$, with $q + 1 \geq 3$ points, which we have seen in our earlier examples.)

Using circle geometries as well as generalized polygons, Buekenhout was able to give diagram geometries for more than half of the sporadic groups; and his papers inspired a great deal of further investigation of various notions of geometries, by many authors.

Such diagrams with circle geometries seem to be particularly well suited to the description of interesting permutation representations of simple groups; for example, the multiply transitive representations of the Mathieu groups.

However, the diagrams do not seem so natural for our context of p-subgroup complexes in this book: since typically the maximal subgroups G_i for the nodes i in these diagrams are set stabilizers, which are not p-local subgroups for any prime p; and in particular, various desirable features in Remark 2.3.3 usually do not hold. For example, in the analogue of Remark 2.2.30 mentioned in 2.3.3.2, the subgroups U_J are often trivial—in contrast to unipotent radicals for Lie type groups, as well as suitable analogues that occur for many sporadic simple groups.

So in this book, we will instead follow the later approach given by:

The context of p-local geometries. In [**RS80**], Ronan and Smith introduced the (fairly informal) notion of a *p-local geometry*. The motivating observation in that work is that for most sporadic G, there is at least one prime p (most often $p = 2$, which is the case considered in [**RS80**]) for which a natural collection of p-local subgroups satisfies an analogue of the group-theoretic property in Remark 2.2.30; and in such a way that the diagrams D involved seem very close in spirit to the usual Dynkin diagrams for buildings (as compared say to the circle diagrams in Buekenhout's examples). In particular, one usually gets $U_J > 1$ in the language of 2.3.3.2. This group-theoretic behavior led in turn to an analysis of the "sporadic" rank-2 geometries to be added to the class $\mathcal{C}$, in the analogue of the geometric property in Remark 2.2.29. Below we briefly indicate a few aspects of these geometries.

There is no intrinsic definition of a p-local geometry: instead the examples were determined by ad-hoc observation, using experience with p-local structure in the groups. But to summarize very roughly, for G sporadic, and with order divisible by some reasonably high power of p, the groups G_i in Definition 2.3.1 can usually be taken to be the maximal p-locals M which are of "abstract characteristic p"—namely they satisfy $C_G(O_p(M)) \leq O_p(M)$ (and in particular are "p-constrained"). In certain cases, and indeed for certain particular purposes such as group cohomology (see e.g. the discussion of M_{23} in [**BS08**, Sec 7.4]), the "appropriate" p-local geometry may vary somewhat from this typical construction.

Historically the diagram D for a particular G was determined by group-theoretic inspection: Namely the groups L_i for vertices in 2.2.30 were often Lie type groups with Dynkin diagrams—and these could be overlappingly assembled into a "Dynkin-like" diagram D for the global geometry, satisfying the usual inductive requirement. This group-theoretic construction of the diagram D in turn suggested the appropriate rank-2 subgeometries needed. We now indicate this in overview for the case of $G = M_{24}$ (we will provide further detail later in the section at Example 2.3.11).

Here the global diagram D has the form $\overset{O}{\bullet} = \overset{T}{\bullet} - \overset{S}{\bullet} - \square$; this is similar to an ordinary Dynkin diagram of type C_4—except for replacing the right-hand node by a square $\square$. And indeed our convention here is that $\square$ is *not* a node, so that the geometry has rank 3 rather than 4. Now the subdiagram for a vertex of type O ("octad") has $L_O \cong GL_4(2)$, which corresponds to a diagram of type A_3. But here for the link of O, the subdiagram for $D \setminus \{O\}$ in fact represents the structure of a "truncated" rank-2 variant, with diagram $\overset{P}{\bullet} - \overset{L}{\bullet} - \square$, of the usual A_3 diagram of Example 2.2.31—in which we use only the points and lines (i.e., we ignore the planes). And similar considerations apply to the other links.

Some other directions. We close the subsection with some remarks on the more general literature in the area of p-local and more general diagram geometries.

An expository overview of some of the earlier work appears in [**Smi88b**]. Our approach above is based on a notion of "maximal parabolics"; a different approach based on "minimal parabolics" was introduced by Ronan and Stroth in [**RS84**]. The minimal-parabolic approach involves some rank-2 diagrams which differ from those appearing in [**RS80**] (and in this book): For example, geometries in [**RS80**] involving $\square$ are treated in [**RS84**] via the "tilde" geometry with diagram $\bullet \overset{\sim}{=} \bullet$, for the nonsplit triple cover $3A_6$ of the alternating group A_6. (This geometry can be regarded as a certain covering of the geometry for $A_6 = Sp_4(2)'$ in Example 2.1.19.) Similarly, geometries in [**RS80**] involving the "triples" geometry for S_5 (to be discussed shortly in Example 2.3.5) are treated in [**RS84**] using the "Petersen" geometry with diagram $\bullet \overset{P}{-} \bullet$; this uses the action of S_5 of the Petersen graph from the combinatorial literature. We emphasize that these "sporadic" rank-2 geometries are involved as links (in the sense of 2.3.4 above) in the larger diagrams for quite a number of sporadic groups (including the Monster M and the Baby Monster B). Much further work on these geometries has been done (from the minimal-parabolic viewpoint) by Ivanov and Shpectorov; see e.g. [**IS94a**], and the exposition in Ivanov's later book [**Iva99**]. A recent viewpoint on p-local geometries appears in Chapter 6 of [**BS08**].

Finally we mention that considerable research effort has also been devoted to results in the spirit of Tits's characterization theorems for buildings—namely results which aim to characterize more general geometries, essentially in terms of their diagrams: See for example Timmesfeld's [**Tim83**], and his more recent [**Tim01**].

In the rest of the section, we explore some examples of sporadic geometries and their diagrams.

The "triples" geometry for S_5. We begin by exhibiting a very easy case of a "sporadic" rank-2 geometry—to be included in our class $\mathcal{C}$, along with the generalized polygons.

In fact we are using the term "sporadic geometry" fairly loosely: e.g. in this case of the triples geometry for S_5, it is the unusual geometry which can be considered sporadic—rather than its group S_5, involving alternating A_5 which is not sporadic.

However, the geometry is involved as a link in larger geometries for various sporadic groups, such as the Mathieu group M_{22}, the Conway group Co_2, the Fischer group F_{22}, and the Baby Monster B. Indeed the geometry was introduced for just that purpose, in [**RS80**].

We remark at the outset that the more usual geometry for S_5 in characteristic 2 is of rank 1: Namely we regard A_5 as $SL_2(4)$ of Lie rank 1: and its building is the order complex of the projective line $\mathbb{P}^1(4)$ in Exercise 2.1.8. (This consists of 5 isolated vertices, corresponding to the 5 parabolic subgroups, each isomorphic to A_4). Hence one unusual feature of our triples geometry for S_5 in this subsection, is that it naturally has the larger rank 2.

So we consider $G = S_5$. Here it will be more elementary to first define our geometry K using an external complex in the sense of Remark 2.3.2. Let G act naturally as permutations on $\Omega := \{1, 2, 3, 4, 5\}$. As vertices of our first type, we define the *points* to be the 5 members of Ω, for example $P = 1$; so far we have just the rank-1 geometry mentioned above. To get a geometry of rank 2, we introduce a second vertex type: namely we define the *triples* to be the $\binom{5}{3} = 10$ triples from Ω, for example $T = 123$. The incidence of points with triples is the obvious inclusion, for example $1 \in 123$; such edges define a simplicial complex K, which here is just a bipartite graph.

Notice that we could also realize K as the order complex $K(\mathcal{P})$, where $\mathcal{P}$ is the subposet of the power set 2^{Ω} defined by the singletons and 3-subsets. Yet another approach to K arises as a certain extension of the orthogonal polar space for $S_5 \cong O_4^-(2)$ in Example 2.1.25: namely for points we have the singular vectors, and the triples are now replaced by the nonsingular vectors—with incidence given by orthogonality. (Thus an orthogonal pair defines a 2-subspace which is totally isotropic, but not totally singular.) Indeed in the set-theoretic viewpoint of 2.1.23 used in the final paragraph of Example 2.1.25, this reproduces just the complements of the points and triples: for a singular vector is a 4-set, orthogonal to a nonsingular vector when it contains that 2-set.

We also provide an internal version of K, via the intersection-complex viewpoint of Definition 2.3.1: Observe that G_P is the subgroup $G_1 \cong S_4$ acting on $\Omega \setminus \{1\}$; and $G_T \cong S_3 \times \langle (45) \rangle$ is the *set* stabilizer of T (as opposed to the stabilizer of each point of T). We record this view in:

EXAMPLE 2.3.5 (The triples geometry for S_5). In the notation above, we take $I = \{P, T\}$ as index set, and G_i for $i \in I$, in Definition 2.3.1—to obtain a complex K called the *triples geometry* for $G = S_5$.

One checks that $B = G_P \cap G_T = \langle (23), (45) \rangle \cong E_4$; and is of index 2 in a Sylow 2-subgroup of G. $\diamondsuit$

We find that many, but not all, of the desirable features in Remark 2.3.3 hold: The vertex stabilizers are maximal subgroups, and are self-normalizing—though this latter property fails for their intersection B; and B is only of index 2 in a Sylow 2-subgroup D_8. This has the further consequence that K is *not* obtained as the poset $(\mathcal{S}, \geq)$ of stabilizers of simplices: for notice that the stabilizer B of the edge PT is also the stabilizer of PT' where $T' = 145$. In particular, the correspondence between simplices and stabilizers is not 1:1.

The diagram invented for the triples geometry K in [**RS80**] is $\overset{P}{\bullet}\overset{t}{\text{—}}\overset{T}{\circ}$. The use of a "hollow node" $\circ$ at T (i.e., in the subdiagram corresponding to the link $\mathrm{Lk}(P)$), is a more general convention for the phenomenon called a "double cover" in [**RS80**]: For observe here that $\mathrm{Lk}(P)$ consists of the triples defined by $P = 1$ along with one of the $\binom{4}{2} = 6$ pairs of points in $\Omega \setminus \{1\}$. These 6 choices are determined by 3 complementary pairs; and these 3 complementary pairs can be considered a double cover of the behavior we might have expected from our earlier examples defined over $\mathbb{F}_2$, where rank-1 links were projective lines $\mathbb{P}^1(2)$ consisting of just 3 points. By contrast, Lk(T) does consist just of the 3 points of T; and so the filled node • at P (i.e. in the subdiagram for Lk(T)) says that we do *not* have any proper covering of the expected projective line. (Group-theoretically, the hollow node at T warns us that G_T does not contain a Sylow 2-subgroup of G.)

REMARK 2.3.6. In the spirit of our earlier examples, it is easy to compute parameters for this geometry: Indeed we got $\#P = 5$, $\#T = 10$, $\#\ P/T = 3$, and $\#\ T/P = 6$; and hence we have $\#PT = 30$. ◊

Our three examples in the remaining subsections will be of rank 3; and so they will more clearly exhibit the inductive content of Definition 2.3.4, in terms of rank-2 geometries from a specified allowable class $\mathcal{C}$.

Indeed the first two of these examples require only generalized polygons, and so define geometries of type M in the sense of Definition 2.2.34.

Neumaier's C_3 geometry for A_7. We turn to one of the smallest examples (having rank at least 3) of a diagram geometry which is not a building; because of its small size (and elementary definition), it is one of the easiest sporadic geometries in which to make sample computations—and indeed we will work with it frequently, in exemplifying various later developments.

The group of the geometry is A_7, which is alternating rather than sporadic; so again, it is the geometry rather than the group that can be considered sporadic. But as before, the geometry it is also involved as a link (in the sense of Definition 2.3.4) in a number of larger geometries for sporadic simple groups, e.g. that for the McLaughlin group McL—cf. p. 70 in [**RS84**].

Neumaier's geometry K has diagram C_3—which also arises for the polar space of $Sp_6(2)$ in Example 2.2.18.[32] In particular, K satisfies Definition 2.3.4, but without requiring any further assumed rank-2 geometries beyond the generalized polygons of Definition 2.2.34; so it is in fact a geometry of type M, for the matrix M for type C_3 mentioned in Example 2.2.35.

Indeed I believe that K is the only known non-building geometry with a Dynkin diagram for a *finite* (i.e. spherical) Weyl group. Later (see e.g. the $\tilde{C}_2$-geometry for $U_4(3)$ in Example 2.3.10 in the next subsection) we will see cases of non-building geometries having the Coxeter diagram for an affine (hence infinite) Weyl group.

The geometry was discovered by Neumaier in [**Neu84**] (and independently noticed by Aschbacher around that time). It is strikingly exceptional in having a spherical diagram (of type C_3) without being a building. For example, this fact demonstrates one reason why links of type C_3 must be *assumed* to afford buildings, in Tits's Theorem 1 in [**Tit81**]; and correspondingly in [**Tit86**, Cor, p 175], which characterizes affine buildings of rank at least 4. (Cf. Remark 2.3.9, and our later discussion from Theorem 9.3.6 through Theorem 9.3.8.)

[32]But the similarity between the two geometries is strictly "local", in the sense of isomorphic links of vertices; rather than "global", since K is not a building. (See the later discussion leading up to Theorem 9.3.12.)

An external view of the complex K via a poset $\mathcal{P}$. As in the previous subsection, again it is convenient to first describe K via an external complex in the sense of Remark 2.3.2. Indeed this time we can even realize K as the order complex $K(\mathcal{P})$ for a suitable poset $\mathcal{P}$. (This is in the spirit of Lemma 2.2.20, since the face poset of K then gives the poset $(\mathcal{S}, \geq)$ of stabilizers in G of chains in $\mathcal{P}$.)

We consider $G = A_7$. We let $\Omega := \{1,2,3,4,5,6,7\}$ denote the 7 letters on which G acts naturally as permutations.

For the members of our poset $\mathcal{P}$, it will be natural to use the language of points, lines, and planes—as in our earlier examples of projective spaces. (However, we will see at later Remark 2.3.8 that $\mathcal{P}$ cannot actually be embedded—with the expected projective dimensions $0, 1, 2$—in any projective space; by contrast, the building of type C_3 over $\mathbb{F}_2$ for $Sp_6(2)$ is naturally embedded as the polar space of Example 2.1.26.)

We first define the *points* of $\mathcal{P}$ to be the 7 letters of Ω; for example $P = 1$. So letting P also denote this vertex type for our eventual complex K, we have the parameter $\#P = 7$.

We next define the *lines* of $\mathcal{P}$ to be all the $\binom{7}{3} = 35$ possible 3-subsets of Ω. As usual, for notational brevity we will write such a line as $L = 123$, rather than $\{1,2,3\}$. (So as above we get $\#L = 35$, $\#\ P/L = 3$, and $\#\ L/P = \binom{6}{2} = 15$, so that $\#PL = 105$. We could define our complex K without the poset $\mathcal{P}$; but it is natural to regard the lines as a level in $\mathcal{P}$ "above" the points: with order relation defined by the obvious inclusion, such as $1 \in 123$.

Finally we define the *planes*, as certain projective-plane structures $\mathbb{P}^2(2)$ on the 7 points: namely a plane is a choice of a 7-subset π of the 35 lines, in which any two lines in π intersect in a unique point, and any two points determine a unique line in π. It is an elementary but tedious combinatorial exercise to check that there are 30 possible choices of planes—in two conjugacy classes of size 15 under A_7, interchanged by any element of $S_7 \setminus A_7$. We use *one* such A_7-class.[33] In particular, we have $\#\pi = 15$, with $\#P/\pi = 7 = \#L/\pi$; so that $\#\ P\pi = 105 = \#\ L\pi$, and $\#\ PL\pi = 315$. Hence $\#\ \pi/P = 15$ and $\#\ \pi/L = 3$ (using the alternative method demonstrated in Example 2.1.15). We regard the planes as a level in our poset $\mathcal{P}$ above the lines, with order relation given by the natural inclusion of a line in a plane.

Notice each point is contained in each plane (indeed we saw $\#\ \pi/P = 15$)—this guarantees the transitivity needed for the order relation (namely, $P \in L$ and $L \in \pi$ implies $P \in \pi$) so that our 3-level structure really is a poset. (However, this completeness of the point-plane bipartite graph is *very* unusual among our examples of complexes for simple groups.)

We can realize a plane π by assigning the 7 points to the positions on the Fano-plane triangle pictured in Example 2.1.12, and then read off the 7 lines of π. However, we emphasize that π is determined by its 7 lines; thus any two different such pictures with the same line-set determine the same plane π. Rather than repeatedly drawing the picture in 2.1.12, we will typographically abbreviate a plane in the form $\pi = \begin{smallmatrix} & 7 & \\ & 456 & \\ 1 & 2 & 3 \end{smallmatrix}$. Ideally this makes clear the 3 peripheral lines and the 3 diagonal lines; with the seventh "circular" line being given by 246.

[33]Hence $S_7 \setminus A_7$ interchanges two isomorphic copies of our poset $\mathcal{P}$, and hence interchanges two copies of our complex K.

Fortunately we will not often need to calculate directly with planes. However, to support our initial definition of the geometry, we will record the following facts—which again are a straightforward (if tedious) combinatorial exercise to verify:

We can distinguish the two A_7-orbits of size 15 on planes by the condition that a pair of distinct planes from the same orbit has just one line in common; while a pair from different orbits has zero or three lines in common. For example, the plane $\rho = \begin{smallmatrix} 6 \\ 457 \\ 1\,2\,3 \end{smallmatrix}$ has the three lines 123, 345, and 367 in common with the plane π from the previous paragraph; and so ρ lies in the G-orbit other than that of π (and hence ρ is *not* a plane in our poset $\mathcal{P}$, which we assume contains π). On the other hand, $\pi' = \begin{smallmatrix} 5 \\ 467 \\ 1\,2\,3 \end{smallmatrix}$ has only the one line 123 in common with π; thus π' lies in the same G-orbit as π, and hence π' *is* a plane in our poset $\mathcal{P}$ containing π. Indeed, repeated application of this criterion to candidate planes is one way to generate the G-class of all 15 planes in the orbit containing π.

We mention also that $\mathcal{P}$ contains exactly 3 planes on each line: for example, the planes on 123 are π, π', and $\pi'' = \begin{smallmatrix} 6 \\ 475 \\ 1\,2\,3 \end{smallmatrix}$. (We had already observed that $\#\ \pi/L = 3$.)

Thus we have now constructed our poset $\mathcal{P}$ with three "levels", given by points below lines below planes.

Finally the order complex $K(\mathcal{P})$ is our external representation of the desired geometry K.

The internal definition of K via subgroups of G. We next obtain the internal representation of K as an intersection complex in Definition 2.3.1, by using the stabilizer mapping as in Lemma 2.2.20: First fix the point $P = 1$. Then G_P is the natural subgroup A_6 acting on $\{2,3,4,5,6,7\}$. Next fix $L = 123$. Then G_L is the stabilizer in A_7 of the 3-set L, of structure $\mathbb{Z}_2/(A_3 \times A_4)$.[34] Finally fix $\pi = \begin{smallmatrix} 7 \\ 456 \\ 1\,2\,3 \end{smallmatrix}$; then G_π is a subgroup of A_7 isomorphic to $GL_3(2)$: namely those elements of A_7 permuting the points and lines of π (and hence stabilizing the projective plane structure $\mathbb{P}^2(2)$ that they define). We summarize this in:

EXAMPLE 2.3.7 (The Neumaier C_3-geometry for A_7). In the notation above, we take index set $I = \{P, L, \pi\}$, and G_i for $i \in I$, in Definition 2.3.1 to obtain a complex K.

One checks that $B = G_P \cap G_L \cap G_\pi \cong D_8$ is a Sylow 2-subgroup of G. $\diamondsuit$

We can verify many of the other expected features of Remark 2.3.3: In particular the G_i for $i \in I$ are maximal subgroups of G, and they and their intersections containing B are self-normalizing. Indeed many of those intersections are 2-local subgroups; so that we will view the geometry as having characteristic 2. The stabilizers which are not 2-local subgroups are G_P and G_π—which we saw are simple, and in particular we have $U_P = 1 = U_\pi$. Furthermore as in Lemma 2.2.20 we obtain that the face poset of K is the poset $(\mathcal{S}, \geq)$ of the stabilizers of the chains in $\mathcal{P}$; and K is isomorphic to $K(\mathcal{P})$.

---

[34] Notational convention: The expression A/B (for abstract groups A,B) denotes some extension of a normal subgroup isomorphic to B by a group isomorphic to A.

It remains to check that K (equivalently $K(\mathcal{P})$) has the diagram $\overset{P}{\bullet} - \overset{L}{\bullet} = \overset{\pi}{\bullet}$ of type C_3, in the sense of Definition 2.3.4:

First by explicit construction, the 7 points and 7 lines of $\mathrm{Lk}(\pi)$ give the structure of the projective plane $\mathbb{P}^2(2)$; this agrees with the A_2 subdiagram $\overset{P}{\bullet} - \overset{L}{\bullet}$.

Next consider $\mathrm{Lk}(L)$: We had indicated earlier that $\#\ P/L = 3 = \#\ \pi/L$, and that all points are incident with all planes. Thus $\mathrm{Lk}(L)$ is the join of those two projective lines, agreeing with the $A_1 \times A_1$ subdiagram $\overset{P}{\bullet} \quad \overset{\pi}{\bullet}$.

Finally consider $\mathrm{Lk}(P)$: We had indicated earlier that $\#\ L/P = 15 = \#\ \pi/P$. It is not difficult to see that these lines and planes have the incidence relations of the 4-sets and $2, 2, 2$-partitions of the 6-set $\Omega \setminus \{1\}$, which give the alternative view in Remark 2.1.24 of the symplectic polar space for $G_P \cong A_6 \cong Sp_4(2)'$. This agrees with the C_2-subdiagram $\overset{L}{\bullet} = \overset{\pi}{\bullet}$.

This completes, for vertices, the verification of Definition 2.3.4. It also follows for each edge σ in K that $\mathrm{Lk}(\sigma)$ is a projective line $\mathbb{P}^1(2)$, of type A_1.

We also examine the analogues of the inductive group-theoretic relations in Remark 2.2.30: Observe that in the notation for "L_J" of Remark 2.3.3.3, the group $G_\pi = L_\pi \cong L_3(2)$ is of Lie type A_2; while $G_P = L_P \cong A_6$, which has index 2 in $Sp_4(2) \cong S_6$ of type C_2. Similarly $L_L = G_L/U_L \cong \mathbb{Z}_2/(\mathbb{Z}_3 \times \mathbb{Z}_3)$, which is also of index 2 in $GL_2(2) \times GL_2(2)$, of type $A_1 \times A_1$. For σ any edge of K, we obtain $L_\sigma \cong GL_2(2)$ of Lie type A_1.

Final remarks. We mention that the Neumaier geometry has been studied by various authors, especially Pasini—see e.g. [**LP89**].

We emphasize that the above analogies of the Neumaier geometry K with buildings are very limited—for example, in K we do not obtain any analogy with apartments and their properties such as those demonstrated at (2.2.24).

(There has also been some investigation in the literature of possible analogues of apartments in sporadic geometries; but no really convincing theory has yet emerged.)

REMARK 2.3.8 ("non-embeddabilility" of the A_7-geometry). Finally, as we promised earlier, we show that the objects in our poset $\mathcal{P}$, which we have called points, lines, and planes, cannot actually be embedded as such in a corresponding projective (or polar) space over $\mathbb{F}_2$:

For suppose V is some vector space over $\mathbb{F}_2$, and $\phi : \mathcal{P} \to \mathbb{P}(V)$ is a poset map—which we require to take points to projective points, and lines to projective lines. (We will obtain a contradiction, even without further requiring that planes should be taken to projective planes.)

The image $\phi(1)$ of the point 1 is a projective point, which we may identify with the nonzero vector of a linear 1-subspace of V. Since $L = 123$ is a line, $\phi(L)$ is a projective line, arising from a linear 2-subspace of V—in which the three nonzero vectors sum to zero. Hence $\phi(1) + \phi(2) + \phi(3) = 0$.

Now as it is standard that A_7 is 5-transitive on Ω, we conclude that the images of the vectors of *any* line sum to zero. Applying this to the line 124, and summing with the relation above for 123, we obtain $\phi(3) + \phi(4) = 0$. Summing this in turn with the relation for 134, we obtain $\phi(1) = 0$—contradicting $\phi(1) \neq 0$.

We remark however that if we remove the restriction of preserving projective dimensions, then $\mathcal{P}$ *does* admit nonzero poset maps into projective spaces $\mathbb{P}(V)$ for certain V. ◇

Kantor's $\tilde{C}_2$ geometry for $U_4(3)$. Our next example is a geometry K having a Dynkin diagram which is now of *affine* type: namely the diagram •══•══•, with type $\tilde{C}_2$; this diagram is the standard "affine" extension (by one node) of the spherical diagram •══• of type C_2. (The various possible affine diagrams are given e.g. on p. 12 of [**GLS98**].) In particular, just as in our C_3-geometry for A_7 in Example 2.3.7 in the previous subsection, K will satisfy Definition 2.3.4 without making use of any further assumed rank-2 geometries beyond the generalized polygons of Definition 2.2.34; and so K is in fact a geometry of type M. (Indeed note that in the earlier diagram •—•══• of type C_3, we may replace the left-hand A_2-subdiagram with a C_2-subdiagram, in order to obtain the diagram •══•══• of type $\tilde{C}_2$ for K; and similarly we replace the corresponding entries of 3 in the matrix for type C_3 in Example 2.2.35 by a value of 4, we obtain the matrix M for our K of type $\tilde{C}_2$.)

The geometry K was introduced by Kantor in [**Kan81**]; it was also independently "guessed" by Ronan-Smith in the preparation of [**RS80**]—since it is the natural 2-local geometry for $U_4(3)$, which in particular is needed as a link, in the sense of Definition 2.3.4, in the larger 2-local geometry for the sporadic Fischer group F'_{24} in that paper.

This geometry arises for the classical unitary group $U_4(3)$, which is not a sporadic group; so again, it is the geometry itself which can be considered sporadic: For example, this geometry has characteristic 2 (in the sense of Remark 2.3.3.3), in contrast to the characteristic 3 of the group.

REMARK 2.3.9 (Infinite affine buildings and finite quotients). Furthermore, the geometry has the diagram for an infinite Weyl group, though its group $U_4(3)$ is finite. Indeed the geometry is one of a number of examples of finite-group geometries with affine diagrams, which were discovered around 1980—cf. the references mentioned at the start of [**Kan85**].

Tits in [**Tit81**] and [**Tit86**] gave results showing that for a suitable geometry with an affine diagram,[35] the *universal cover* (in essentially the usual sense of topology, applied to simplicial complexes; cf. [**Spa81**, Sec. 2.5, p. 80]) should usually be an infinite building, based on the full infinite affine Weyl group $\tilde{W}$. (Cf. our discussion of later Theorem 9.3.8.) And indeed, for the $U_4(3)$ example in this subsection, and for many others that had been found for finite groups, Kantor in [**Kan85**] explicitly determined that universal cover: namely he used the relevant spherical-type group over the 2-adics $\mathbb{Q}_2$, and passed to the corresponding affine (Bruhat-Tits) building (in which an apartment is not a sphere, but an affine space) over the residue field $\mathbb{F}_2$.

We will return to this theme later, when we give a somewhat fuller discussion leading up to Kantor's result (Theorem 9.3.10). ◇

[35] Tits's hypotheses in fact require rank at least 4—and also the exclusion of rank-3 residues of types C_3 and H_3. And indeed Neumaier's C_3 of Example 2.3.7 (though discovered independently) exhibits a corresponding complication: for it is simply connected (i.e. equal to its own universal cover), but is not a building.

So we consider $G = U_4(3)$. For further reference on this group, see e.g. p 52 of the Atlas [**CCN$^+$85**], or [**Wil09**, 3.12.2]. Below we will quote various structural details (often without specific reference).

In the next few pages, we indicate several different approaches to the geometry K for G—first via external complexes based on the group structure in characteristic 2, and then characteristic 3; and finally via an internal complex as usual.

An external complex K defined using characteristic-2 structure. Just as in the examples in the two previous subsections, we will begin with an external complex K; indeed we will indicate two possible external approaches.

Our description of the first external complex will be somewhat briefer: for this first viewpoint may seem a bit more indirect—and it is largely motivated by group structure, so that our later intersection-complex viewpoint is perhaps more natural in the long run.

Our starting point is the group-theoretic inclusion $U_4(3) < U_6(2)$. This allows us to roughly mimic the general procedure for Example 2.3.7 in the previous subsection: Namely we sketch how to find an appropriate poset $\mathcal{P}$ for G, as a subposet of the unitary polar space for $GU_6(2)$; its face poset will correspond as before to the poset $(\mathcal{S}, \geq)$ of stablizers in G of the simplices of K.

The unitary polar space is defined by a unitary form, in a fashion analogous to the symplectic polar space from a symplectic form. The natural module V for $GU_6(2)$ is of dimension 6—but it is defined over the larger field $\mathbb{F}_4$, since here $\mathbb{F}_2$ plays the role of the subfield fixed by the "conjugation" field automorphism. For our unitary 6-space, the polar space is determined by the totally isotropic linear subspaces—of dimensions $1, 2, 3$ over $\mathbb{F}_4$. As usual we refer to these as points, lines, and planes; and abbreviate their types by P, L, and π. Now the group $U_6(2)$ has twisted Lie type ${}^2A_5$, which determines its building and twisted diagram of type C_3—cf. [**Car89**, p. 221]; and we label the diagram as $\overset{P}{\bullet} - \overset{L}{\bullet} = \overset{\pi}{\bullet}$. This of course looks like the diagram for $Sp_6(2)$ in Example 2.2.33; but one difference not visible in the notation is that the quadrangle in $\mathrm{Lk}(P)$ is the unitary polar space for $U_4(2)$ in our present case, rather than the ("smaller") symplectic polar space for $Sp_4(2)$ in 2.1.19.

Now our G "is" the proper subgroup $SU_4(3)$ of $GU_6(2)$; and the subgroup G_π in G that stabilizes a plane π has quotient $G_\pi / U_\pi \cong 3A_6$. From the standard inclusion $3A_6 < SL_3(4)$, this quotient induces on π the action of the corresponding proper subgroup of the full group of a plane. Next as A_6 is of index 2 in the group $S_6 \cong Sp_4(2)$, it is now natural to replace a plane π with a "quad" Q, namely the symplectic polar space which is the generalized quadrangle for $Sp_4(2)$. This in turn involves replacing the original 21 points and 21 lines of π (defined over $\mathbb{F}_4$) with an appropriate subset of 15 points and 15 lines (now defined instead over $\mathbb{F}_2$). And once this procedure is applied throughout the geometry, we also find that the link $\mathrm{Lk}(P)$ is no longer a full unitary quadrangle—but instead a subcomplex giving an $Sp_4(2)$ quadrangle (so that $\mathrm{Lk}(P)$ also has the quad-structure).

To summarize our sketch: for our subgroup $U_4(3)$, these replacements define a subposet $\mathcal{P}$ of the unitary polar space, which leads to a complex described by the $\tilde{C}_2$ diagram, which we (temporarily) label as $\overset{P}{\bullet} = \overset{L}{\bullet} = \overset{Q}{\bullet}$.

An external complex defined using characteristic-3 structure. We now indicate our second approach to the geometry K, again leading to the same diagram; this viewpoint on K is again an external complex. But this time, we will use more of the natural group-theoretic structure of G, as a group defined in characteristic 3; and our treatment can now be somewhat more precise—for example, about the structure of the quad-group denoted by G_Q above. In particular, we can explicitly realize vertices of our complex K as certain 2-subgroups in G. (Although K will not precisely arise from a *poset* of 2-subgroups.)

We mention one other feature at the outset: We saw above that the vertices called points P and quads Q in fact had the same quadrangle structure in their links. Indeed we have $G_P \cong G_Q$, with these groups exchanged by an outer automorphism of G; so there is no a-priori reason to not also call P a quad. For this reason, in our discussion below we will replace P and Q by *left quads* L and *right quads* R. (And we will replace the previous notation of L for lines, with involutions z).

We will regard $G = U_4(3)$ via a standard isomorphism as the group $\Omega_6^-(3)$; that is, as $P S\Omega_6^-(3)$, namely the quotient of $\hat{G} = S\Omega_6^-(3)$ modulo its center Z (spanned by the negative $-I$ of the identity matrix), where $S\Omega_6^-(3)$ is the commutator subgroup of the group of 6×6 matrices of determinant 1 over $\mathbb{F}_3$ preserving an orthogonal form of minus-type. (Cf. Section 1.4 in [**Car89**].) As in Remark 2.1.2, it is often more convenient to work with non-simple $\hat{G}$, so that we can directly consider matrices acting on a vector space V of dimension 6 over $\mathbb{F}_3$. In particular, with respect to our form of minus-type, we may choose an orthonormal basis $v_1, \ldots, v_6$ of V. We will define the desired 2-subgroups of G in terms of matrix action on this basis. Set $V_i := \langle v_i \rangle$ for each i.

Corresponding to the vertices previously denoted by L, we now instead define vertices of our first type to be the involutions of G, namely the elements of order 2; there is just one conjugacy class of these in the simple group G. Using the above basis, we may choose a representative z to be diagonal, with -1 in the first two positions, and 1 in the remaining four positions. (This element of $\hat{G}$ is of course equivalent—modulo Z, that is, as an element of G—to the involution with four nontrivial eigenvalues of -1 in the last 4 positions). The stabilizer $G_z = C_G(z)$ is a maximal subgroup of G, and is an extension (see [**CCN**$^+$**85**, p. 52] for details) of an extraspecial 2-group 2_+^{1+4} by $\mathbb{Z}_4/(\mathbb{Z}_3 \times \mathbb{Z}_3)$. There are $|G : G_z| = 2835$ involutions in G (that is, our complex has parameter $\#z = 2835$); so that our geometry will be substantially larger than our earlier examples such as 2.3.7. However, it will not be significantly more difficult to compute in this larger geometry.

We next define vertices of our second type to be the *left quads*; we construct a particular representative L as follows: Consider, in the full orthogonal group on V, the diagonal subgroup—with each element acting as ± 1 on the members of the above basis. This is an *elementary* 2-group (i.e. an abelian 2-group where every nontrivial element has order 2) of order 2^6; we denote such a group by E_{2^6}. Its subgroup of determinant 1 (where elements have an even number of eigenvalues -1) is a subgroup $\hat{L} \cong E_{2^5}$ in $\hat{G}$; then in the quotient G modulo Z, we obtain our group $L \cong E_{2^4}$. The stabilizer G_L of L under the group action by conjugation is just the usual normalizer $N_G(L)$ of L, which in this case is a maximal subgroup of G; it is the "monomial" subgroup corresponding to our basis, namely an extension of L by a group A_6, acting as permutation matrices on the basis. We define the set of all left quads to be the G-conjugates of L; there are $|G : G_L| = 567$ such conjugates, so we have $\#L = 567$.

For vertices of our third type, we define *right quads* as conjugates of the left quads under a suitable outer automorphism α of order 2 on G: We may then define $\alpha : v_i \mapsto w_i$; where $w_1 := v_1 + v_2$ and $w_2 := v_1 - v_2$, and similarly for the other pairs 3, 4 and 5, 6 in place of 1, 2. Notice that α preserves orthogonality of basis elements, but interchanges vectors of squared lengths 1 and 2, so α is not itself an orthogonal transformation. Hence α induces an outer automorphism; and $\alpha^2 = -I$, so that α is of order 2 as an automorphism. We define $R := (L)\alpha$ as a representative of the right quads, which are then given by the G-conjugacy class of R. By this conjugacy, remarks analogous to those for L in the previous paragraph also hold for R; for example, we have $\#R = 567$.

We now observe that the permutation group A_6 in the quad stabilizers G_L and G_R is in fact isomorphic to $Sp_4(2)'$; this is essentially why we obtain the double-bond C_2-subdiagram for the link of each of L and R—and thus we again obtain the $\tilde{C}_2$-diagram for K.

It remains to indicate the higher-dimensional simplices in our complex K; we do this by defining incidence among the above vertices. Notice by our construction that $z \in L \cap R$. In fact $L \cap R$ is given by the matrices which act as ± 1 on the subspaces $V_1 + V_2$, $V_3 + V_4$, and $V_5 + V_6$; since G is the quotient of $\hat{G}$ by $Z = \langle -I \rangle$, we see that $L \cap R \cong E_{2^2}$. Notice furthermore that the subgroup A_6 in G_L (resp. G_R) preserves a symplectic form on L (resp. R); and the linear 2-subspace $L \cap R$ is an isotropic line in L (resp. R) since it is normal in a Sylow 2-subgroup of G. We extend the above remarks by G-conjugacy to define incidence of all types of vertices: Containment defines incidence of an involution with a left or right quad. And a left and right quad are incident, when they intersect in a subgroup E_4 which is an isotropic line in each quad (equivalently, which is normal in a Sylow 2-subgroup of G). Finally a 2-simplex is defined when the involution z lies in the intersection $L \cap R$ making L and R incident. It should be clear that this incidence relation is inherited by subsets, as required by Definition 1.2.5 for a complex.

Notice, as we had mentioned earlier, that our complex K defined via 2-subgroups does not however arise as the order complex of a subgroup poset: for we have incidence of left and right quads—but since they are distinct and of the same order, they could not satisfy proper containment relations in a subgroup poset.

Finally we indicate the computation of the remaining values for our usual geometric parameters. Our observation above, that L has a symplectic structure preserved by G_L, illuminates the significance of the group-theoretic structure of $G_L \cong A_6/L$, where $A_6 \cong Sp_4(2)'$, and in particular motivates the term "quad" for L; with similar remarks for R. In particular from our study of that quadrangle in Example 2.1.19, we obtain the parameter values $\#\ L/R = 15 = \#\ z/R$ and $\#\ z/L = 15 = \#\ R/L$; which we can then use to verify $\#\ L/z = 3 = \#\ R/z$. We also obtain $\#Lz = \#LR = \#zR = 567 \cdot 15$ and $\#LzR = 567 \cdot 15 \cdot 3$.

The internal definition of K via subgroups of G. We now indicate our third (and final) approach to the geometry K for G. We recall the representatives z, L, and R of vertices from above, and the structure of their stabilizers G_z, G_L, and G_R.

EXAMPLE 2.3.10 (Kantor's $\tilde{C}_2$-geometry for $U_4(3)$). In the notation above, we take $I = \{L, z, R\}$ as index set, and G_i for $i \in I$, in Definition 2.3.1 to obtain a complex K.

One checks that $B = G_L \cap G_z \cap G_R \cong D_8/L$ is a Sylow 2-subgroup of G. $\diamondsuit$

As in the previous subsections, we can investigate the expected features of Remark 2.3.3; indeed this time all the expected features do hold. In particular the G_i for $i \in I$ are maximal subgroups of G, and they and their intersections containing B are self-normalizing; and now all those intersections are 2-local subgroups, while the geometry as indicated earlier has characteristic 2. Again (this time cf. Lemma 2.2.37) the face poset of K is isomorphic to the poset $(\mathcal{S}, \geq)$ of the stabilizers of the simplices in K.

Although we had already examined the diagram, we now re-consider it in the form $\overset{L}{\bullet} = \overset{z}{\bullet} = \overset{R}{\bullet}$ determined by our newer notation.

First consider the link $\mathrm{Lk}(R)$ of the right quad R; we review its generalized-quadrangle structure in more detail: Each of the 15 involutions in the subgroup R gives an isotropic 1-space in the polar space of R. The left quad Q incident with R determines, as we saw earlier, the isotropic 2-subspace $L \cap R$ of R. Furthermore L is the only subgroup of the product group LR, other than R itself, isomorphic to E_{2^4}: this is because LR is isomorphic to a Sylow 2-subgroup of $GL_3(4)$, where this fact is easily checked from explicit matrix forms. This means that the 2-subspace $L \cap R$ determines the left quad L uniquely. Similar remarks hold for conjugates of L incident with R. We conclude that the involutions and left quads of $\mathrm{Lk}(R)$ afford the points and lines of the polar space for $Sp_4(2)$. This is consistent with the subdiagram $\overset{L}{\bullet} = \overset{z}{\bullet}$. (Alternatively, one could argue much as in the link of P in Example 2.3.7 of the previous subsection: namely an involution (resp. left quad) incident to R determines by its eigenvalues -1 a pair (resp. partition into three pairs) of our basis $\{v_1, \ldots, v_6\}$; so we obtain the polar space in the alternative viewpoint of $\tilde{E}$ in Remark 2.1.24.)

By conjugacy under α, similar remarks hold for L; in particular the involutions and right quads of $\mathrm{Lk}(L)$ afford the points and lines of the polar space for $Sp_4(2)$ of Example 2.1.19—which is consistent with the subdiagram $\overset{z}{\bullet} = \overset{R}{\bullet}$.

Finally consider $\mathrm{Lk}(z)$. The incident left and right quads determine $L \cap R \cong E_{2^2}$ containing z, and $L \cap R$ is normal in the Sylow 2-subgroup B of G. Furthermore we saw that $L \cap R$ determines L and R uniquely. Now L has 3 conjugates under an element of order 3 in $G_z \cap G_R$, so these conjugates are all incident with R. Similarly (using α-conjugacy) R has 3 conjugates under an element of order 3 in $G_z \cap G_L$. We obtain $3 \cdot 3 = 9$ conjugates of $L \cap R$; these demonstrate the incidences showing that $\mathrm{Lk}(z)$ is the join of the projective line $\mathbb{P}^1(2)$ given by the conjugates of L in $\mathrm{Lk}(z)$, and the similar projective line given by the conjugates of R in $\mathrm{Lk}(z)$. This is consistent with the subdiagram $\overset{L}{\bullet} \quad \overset{R}{\bullet}$.

This completes, for vertices, the verification that K has diagram $\tilde{C}_2$ in the sense of Definition 2.3.4. In particular it follows for each edge σ in K that $\mathrm{Lk}(\sigma)$ is a projective line $\mathbb{P}^1(2)$, of type A_1.

We also indicate the analogues of the group-theoretic relations in Remark 2.2.30: Observe that in the notation of "L_J" in Remark 2.3.3.3, we have $L_L \cong A_6 \cong L_R$, where A_6 is of index 2 in the full Lie type group $Sp_4(2)$ of type C_2. Similarly $L_z = G_z/U_z$ contains a subgroup $\mathbb{Z}_2/(\mathbb{Z}_3 \times \mathbb{Z}_3)$, which in turn is of index 2 in the full Lie type group $GL_2(2) \times GL_2(2)$ of type $A_1 \times A_1$. For σ any edge of K, we obtain $L_\sigma \cong GL_2(2)$ of Lie type A_1.

The geometry is sometimes called a *near-hexagon*, since its point-line distance behavior resembles some aspects of a generalized hexagon.

The 2-local geometry for M_{24}. Our final example in this section, of a diagram geometry not a building, was historically the first 2-local geometry to be observed; it then led to the larger collection of such geometries given in Ronan-Smith [**RS80**]. The discovery flowed from simple experimentation with possible sporadic analogues for the properties of buildings in Remarks 2.2.29 and 2.2.30: namely trying to assemble the diagrams for suitable local subgroups G_i of G into a coherent "global" diagram for a geometry for G. Unlike the Examples 2.3.7 and 2.3.10 in the previous two subsections, this time we do require in Definition 2.3.4 a rank-2 geometry in the allowed class $\mathcal{C}$ beyond the generalized polygons of Definition 2.2.34—namely a truncation of a projective space identified via a "square node", as previewed earlier.

As in Example 2.3.10 of the previous subsection, this geometry is not small—but it is still reasonably easy to compute in, especially in view of the availability of many standard results and techniques on the group M_{24} and its associated combinatorial structures. The geometry is also involved as a link, in the sense of Definition 2.3.4, in the 2-local geometries for larger sporadic groups such as the Conway group Co_1, the Fischer group F'_{24}, the Janko group J_4, and the Monster M—see [**RS80**].

We consider $G = M_{24}$. For further reference on this group, see e.g. p 97 in the Atlas [**CCN$^+$85**], and especially the material starting at p 216 in Conway [**Con71**], as well as [**Wil09**, Sec 5.2]. Below we will quote various facts from these sources, often without specific reference.

Let $\Omega = \{1, 2, \ldots, 24\}$ be the set of 24 letters on which $G = M_{24}$ acts naturally via permutations.

An external view of the complex K via a poset $\mathcal{P}$. Much as in the previous subsection, we indicate two separate (but now more clearly related) approaches to an external complex K. On the one hand, we construct K as a simplicial complex based on structures in the Steiner system $\mathcal{S}(5, 8, 24)$ (for reference see the citations indicated above). On the other hand, we will also obtain K as isomorphic with the order complex $K(\mathcal{P})$, where the poset $\mathcal{P}$ is a part of the projective space $\mathbb{P}(V)$ for the 11-dimensional irreducible "code" module V for G (which is determined by the extended binary Golay code), in which those Steiner system structures are naturally embedded. This gives us some flexibility in choosing one or the other viewpoint in developing various properties of the geometry.

As vertices of our first type (denoted O), we choose the *octads*: these are certain 8-subsets of Ω (see p. 225 of [**Con71**] or [**Wil09**, 5.2.2] for fuller detail), defining the *Steiner system* $\mathcal{S}(5, 8, 24)$; in particular, each 5-subset of Ω lies in just one octad. There are 759 octads in the Steiner system, so we will write $\#O = 759$ in our usual language of parameters.

Further if we regard the power set 2^{Ω} as a vector space over $\mathbb{F}_2$ (where addition is as usual given by symmetric difference), the octads span a 12-dimensional subspace called the *extended binary Golay code*, and denoted by $\mathcal{C}$ on p 224 of [**Con71**]. The 11-dimensional quotient given by $\mathcal{C}$ modulo the vector defined by Ω is the *irreducible code module V* for M_{24}. Thus we may identify octads with (certain of) the projective points, in the projective space $\mathbb{P}(V)$; or else with linear 1-subspaces of the vector space V—we may write V_O for the 1-space corresponding to the octad O.

As vertices of our second type (denoted T), we take the *trios*; these are partitions of Ω into three octads (see pp. 228ff of [**Con71**] for fuller detail). There are $\#T = 3795$ trios. We define incidence of an octad with a trio by the obvious inclusion of the octad as one part of the trio partition.

Observe that for a trio T, the 3 vectors in the corresponding octad spaces V_O sum to the vector for the set Ω—which is 0 in our quotient V; that is, T determines via these 3 vectors a linear 2-subspace V_T of V; or just a projective line of $\mathbb{P}(V)$. Thus incidence of an octad O with a trio T corresponds to the inclusion $V_O < V_T$ in $\mathbb{P}(V)$.

As vertices of our third (and final) type (denoted S), we take the *sextets*: these are partitions of Ω into 6 subsets of size 4 (called *tetrads*), such that any pair of tetrads is an octad; see pp 228ff of [**Con71**] or [**Wil09**, p 187] for fuller detail. There are in fact $\#T = 1771$ sextets. We define incidence of octads and trios with sextets by the obvious relations: when an octad is a union of two tetrads in the sextet; or when the partition for the trio is refined by that for the sextet. A 2-simplex is defined when the sextet is incident with an incident octad-trio pair; this relation is inherited by subsets, as required by Definition 1.2.5 for the face relation of a complex.

Since the octads V_O (resp. trios V_T) on a sextet S determine pairs (resp. partitions into three pairs) of the set of 6 tetrads, we conclude (as in the discussion of $\tilde{E}$ in Remark 2.1.24) that the V_O (resp. V_T) afford the isotropic points (resp. lines) of that alternative viewpoint on the polar space for $Sp_4(2)$. In particular, these span a linear 4-subspace V_S of V. Furthermore the incidences of O,T with S are reflected by subspace containments $V_O < V_S$ and $V_T < V_S$—that is, by the order relation in the projective space $\mathbb{P}(V)$.

Our incidence relations have now defined K as a simplicial complex. We have also obtained K as the order complex $K(\mathcal{P})$ as the subposet of the projective space $\mathbb{P}(V)$ defined by the corresponding subspaces V_O, V_T, and V_S of V.

The internal definition of K via subgroups of G. As in Example 2.3.7, we will obtain in analogy with Lemma 2.2.20 an isomorphism of the face poset of K with the poset $(\mathcal{S}, \geq)$ of stabilizers in G of the chains in $\mathcal{P}$. We first quote (e.g. from [**Con71**] or [**Wil09**, 5.2.6]) the structure of the vertex stabilizers; we will use the notation of U_J and L_J in Remark 2.3.3:

For an octad O, G_O is an extension of $U_O \cong E_{2^4}$ by $L_O \cong GL_4(2)$. For a trio T, G_T is an extension of $U_T \cong E_{2^6}$ by $L_T \cong GL_2(2) \times GL_3(2)$. For a sextet S, G_S is an extension of $U_S \cong \mathbb{Z}_3/E_{2^6}$ by $L_S \cong Sp_4(2)$.

EXAMPLE 2.3.11 (The 2-local geometry for M_{24}). In the notation above, we take $I = \{O, T, S\}$ as index set, and G_i for $i \in I$, in Definition 2.3.1 to obtain a complex which is isomorphic to (and hence which we may identify with) K constructed earlier.

One checks that $B = G_O \cap G_T \cap G_S$ is an extension of a Sylow 2-subgroup of G by a subgroup of order 3. $\diamondsuit$

Furthermore as in the previous subsections, we can verify most of the other expected features of Remark 2.3.3: in particular the G_i for $i \in I$ are maximal subgroups of G, and they and their intersections containing B are self-normalizing. Indeed those intersections are 2-local subgroups, so that the geometry has characteristic 2. In particular as in Lemma 2.2.20 we obtain that the face poset of K is the poset $(\mathcal{S}, \geq)$ of the stabilizers of the chains in $\mathcal{P}$.

We now indicate how K satisfies the diagram $\overset{O}{\bullet} = \overset{T}{\bullet} — \overset{S}{\bullet} — \square$; interpreting the square node in terms of suitable truncations of projective geometries. The necessary facts are perhaps more evident when we use the viewpoint of the subposet $\mathcal{P}$ of the projective space $\mathbb{P}(V)$ of the Golay code module.

For example, we have already seen that the octads and trios of $\mathrm{Lk}(S)$ afford the points and lines of the polar space for $Sp_4(2)$. This is consistent with the subdiagram $\overset{O}{\bullet} = \overset{T}{\bullet} \quad \square$: For the truncation corresponding to a single node simply ignores all points of a projective line, leaving nothing; so that in effect we are left just with the C_2 diagram for that generalized quadrangle. At this point, we have parameters $\#\ O/S = 15 = \#\ T/S$ and $\#OS = 759 \cdot 15 = \#TS$; we use these to obtain $\#OTS = 759 \cdot 15 \cdot 3$, along with $\#\ S/O = 35$ and $\#\ S/T = 7$.

Next consider $\mathrm{Lk}(O)$ for O an octad. The linear 2-subspaces V_T of V for trios T on O determine linear 1-subspaces $\tilde{V}_T$ in the quotient $\tilde{V} := V/V_O$; and it is a standard fact (cf. H.16.2 in [**AS04a**]) that these 2-subspaces generate a subspace W_O of dimension 5 in V. Then $\tilde{W}_O$ is a 4-subspace of $\tilde{V}$ in which the $\tilde{V}_T$ provide the 15 nonzero vectors. Further for sextets S on O, $W_S := V_S \cap W_O$ is the 3-subspace $V_0^{\perp}$ in the form on V_S; and the $\tilde{W}_S$ afford the 35 linear 2-subspaces of $\tilde{W}_O$. Thus the trios and sextets on O afford the points and lines of the truncation, by ignoring planes, of the projective space $\mathbb{P}^3(2)$. This is consistent with the truncated diagram $\overset{T}{\bullet} — \overset{S}{\bullet} — \square$; which as we mentioned earlier, indicates that this truncation is one of the new members in our class $\mathcal{C}$ of allowable rank-2 geometries beyond the generalized polygons. At this point, we get the further parameters $\#\ T/O = 15$ and $\#TO = 759 \cdot 15$.

Finally the linear 2-subspace V_T for a trio T has three linear 1-subspaces V_O for the 3 octads O on T. Furthermore the linear 4-subspace V_S for sextets S on T determine 2-subspaces in the quotient $\bar{V} := V/V_T$. Again it is a standard fact (with proof similar to that for the fact quoted for $\tilde{V}$ in the previous paragraph) that these 4-subspaces span an 8-subspace W_T of V, so that $\overline{W_T}$[36] is 6-dimensional; and further this 6-dimensional quotient is a tensor product of natural modules for the factors of $L_T \cong GL_2(2) \times GL_3(2)$. In particular the 7 linear 2-subspaces $\bar{W}_T$ each decompose into a sum of 2 linear 1-subspaces taken from two natural submodules for $L_3(2)$; and then the 7 linear 1-subspaces from a fixed natural submodule afford the points of the truncation, by removing lines, of the projective space $\mathbb{P}^2(2)$. In view of our construction of the poset $\mathcal{P}$, we conclude (either checking directly, or using Lemma 1.6.13) that $\mathrm{Lk}(T)$ is the join of the projective line $\mathbb{P}^1(2)$ afforded by the 3 octads on T with the truncation of the projective plane determined by the 7 sextets on T. This is consistent with the subdiagram $\overset{O}{\bullet} \quad \overset{S}{\bullet} — \square$. (We have a

[36]Here via our definition of $\bar{V}$ we also assume the usual "bar convention" that a bar is also used to denote *images* in $\bar{V}$ of other subspaces, such as W_T here.

complete bipartite graph $K_{3,7}$, which in particular is a generalized digon. So our subdiagram is essentially just • •.) We also have the parameter $\# \; O/T = 3$.

As for the group-theoretic relations in Remark 2.2.30, we have already indicated earlier the structures of L_i for vertices of type i, of Lie types consistent with the corresponding subdiagrams (including truncated type A_3). For edges, as before we get $L_{OS} \cong SL_2(2) \cong L_{TS}$; but this time because of the truncation involved, for the edge OT we instead get $L_{OT} \cong L_3(2)$ (corresponding to truncated type A_2), transitively permuting the 7 sextets in $\mathrm{Lk}(OT)$.

Final remarks. For the reader interested in further explicit examples, there is a substantial literature on sporadic geometries. Certain references already mentioned contain systematic collections of examples, e.g. Ronan-Smith [**RS80**], Ronan-Stroth [**RS84**], Ivanov-Shpectorov [**IS94a**][**Iva99**], and Benson-Smith [**BS08**, Ch 6]. Many other authors have also contributed to the area; for instance we mention a series of papers by Yoshiara on geometries for various individual sporadic groups, such as [**Yos89**].

Part 2

Fundamental Techniques

The next two chapters present various results and techniques concerned with homotopy equivalence of subgroup posets, and in particular with contractibility.

These results and techniques are heavily used in the later chapters on applications to various other topological properties.

CHAPTER 3

Contractibility

The most elementary topological spaces are those which are contractible to a point. This chapter develops some fundamental techniques for establishing contractibility, within our context of subgroups posets and complexes. (Typically we will use such contractibility proofs as an initial step toward establishing further properties which are developed in our later chapters.)

Indeed in our examples corresponding to simple groups, the full geometry will not itself usually be contractible; instead, there will often be certain natural *sub*geometries for which we will wish to exhibit contractibility.

There is particularly significant overlap of this chapter on contractibility with the subsequent Chapter 4 on homotopy equivalence: For on the one hand, contractibility is a special case of homotopy equivalence (namely with a point); while on the other hand, some important general homotopy equivalence methods (such as Quillen's Fiber Theorem 4.2.1) require as hypothesis the contractibility of certain subspaces. So in particular, the present chapter on contractibility must also introduce at least some of the basics of homotopy equivalence.

The reader may be interested in some further references which treat contractibility material much like that in this chapter, such as Benson [**Ben91**, Sec 6.5] and Adem-Milgram [**AM04**, Sec V.2].

The chapter is organized as follows:

First an introductory section briefly indicates some significant ways in which contractibility arises within the context of subgroup posets.

Then Section 3.1 reviews the relevant fundamental notions from topology—namely homotopy, homotopy equivalence, and contractibility; while indicating some special features for our case of subgroups posets and complexes.

The sections thereafter are then ordered according to the number of homotopy-equivalence "steps" needed—in effect, shrinking to smaller and smaller spaces, with the final step shrinking to a point. From this viewpoint, a cone requires just one step; and we expand on this concept in Section 3.2. The important case of "conical contractibility" (developed in Section 3.3) requires two steps: first shrinking to a suitable cone, and then to a point. However, in some cases our contractibility proofs will require many more steps; in Section 3.4 we will indicate various examples and techniques, including some related to the special case of contractibility given by collapsibility.

The final Section 3.5 briefly introduces a further optional track that the reader can follow, in this and subsequent chapters—namely Option G, on G-homotopy equivalence: Our posets and complexes normally admit the action of a group G; and here the idea is roughly that our various proofs of contractibility and homotopy equivalence can usually be very easily adapted to establish the stronger notions of G-contractibility and G-homotopy equivalence.

Preview: Cones and contractibility in subgroup posets

We will try to roughly indicate some of the situations where we apply these topological concepts to our poset context:

Probably the most obvious kind of contractible space (larger than just a single point) is provided by a cone. We will discuss this situation in Section 3.2; but first, just intuitively, for a cone we would have some picture like:

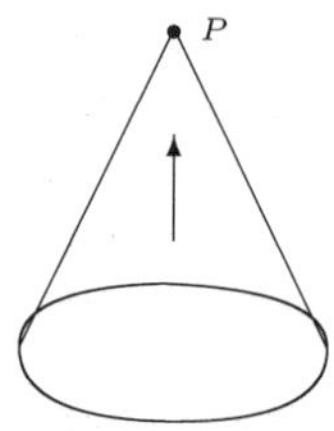

Here in order to contract to the point P, we continuously shrink upward (as the arrow is meant to suggest) along the "radial" lines emanating from P.

Now a poset $\mathcal{P}$ will involve some natural cone substructures: notably subposets of the form $\mathcal{P}_{\leq x}$ and $\mathcal{P}_{\geq x}$ (recall Definition 1.1.7). We will indicate some more formal details later at Lemma 3.2.3, though naively their contractibility is perhaps already obvious.

A notion almost as elementary as that of a cone is given by *conical contractibility*—meaning roughly that we first contract to a cone, and then to the "cone point". In our standard context of the p-subgroups poset $\mathcal{S}_p(G)$, conical contractibility was developed by Brown; and the technique is particularly emphasized in Quillen's paper [**Qui78**]. Indeed an important application of the approach is Quillen's result [**Qui78**, 2.4]: that if the largest normal p-subroup $O_p(G)$ is nontrivial, then $\mathcal{S}_p(G)$ is contractible. And in fact the proof first observes that set-multiplication by $O_p(G)$ defines a contraction of $\mathcal{S}_p(G)$ into $\mathcal{S}_p(G)_{\geq O_p(G)}$; and then in that cone, as suggested in the previous paragraph, we just contract down to $O_p(G)$ itself. We will treat this stiking result as Lemma 3.3.5; and indeed the result will motivate[1] our development in the earlier sections leading up to that initial goal: We will see that a more detailed viewpoint on these two easy steps will in fact provide natural examples of various basic concepts in our overall treatment of contractibility.

3.1. Topological background: Homotopy of maps, and homotopy equivalence of spaces

In our later chapters, we will often want to compare geometries which are not isomorphic, but which nonetheless can be regarded as interchangeable, in terms of the properties that we will study—including various topological invariants.

For topological spaces, the relevant equivalence relation is that of *homotopy equivalence*. So this section briefly reviews some of the basic concepts at the level of spaces, while also further exploring them at the level of posets or complexes.

[1]Thanks to Chris Cashen for suggesting this approach.

The experienced reader can of course skip over this background; while the reader interested in fuller detail can consult for example Section 19 of [**Mun84**]. In this chapter we will also sometimes refer to [**Coh73**]. The specialization of the topological notions to posets is summarized in Section 1 of Quillen's paper [**Qui78**].

A homotopy equivalence between spaces is roughly a "continuous deformation" from one to the other. However, it may not be obvious why we might also need a similar notion of deformation *between mappings*. We can approach this question by comparing with the stronger notion of homeomorphism—where now we use a "long form" to express the invertibitility of the map f in Remark 1.4.1; that is, we now require continuous maps $f : X \to Y$ and $g : Y \to X$, such that the compositions give the relevant identity mappings: namely $fg = \mathrm{Id}_Y$ and $gf = \mathrm{Id}_X$. This formulation suggests that in order to define the weaker relation of homotopy equivalence, we might wish to instead require that those composition maps should be just "deformable" to the corresponding identities.

Homotopy of maps. So our discussion will actually begin with the corresponding equivalence relation needed for mappings.

The topological definition of homotopy. We recall (essentially from [**Mun84**, p 103]) the notion of homotopy. In the background, we are regarding a deformation from f to g as taking place over "time" t in the interval $[0,1]$; so that all the intermediate stages between f and g can be incorporated into a single function F, defined on the "spacetime" given by the Cartesian product $[0,1] \times X$:

REMARK 3.1.1 (Homotopy of maps (for topological spaces)). A pair of continuous maps $f, g : X \to Y$ are *homotopic* (denoted $f \simeq g$) if there exists a *homotopy* of f with g—namely a continuous map:

$F : [0,1] \times X \to Y$, with $F(0,x) = f(x)$ and $F(1,x) = g(x)$ for all $x \in X$.

(We may abbreviate this requirement in the form $F(0,-) = f$ and $F(1,-) = g$.)[2]

It is elementary to check that homotopy is an equivalence relation; and also that restriction preserves homotopy—that is, for a subspace $X' \subseteq X$ in the role of "X", we get $f|_{X'} \simeq g|_{X'}$. ◇

REMARK 3.1.2 (Homotopy for maps of posets and complexes). We will continue the practice of applying topological concepts to posets and complexes, which we had introduced in Section 1.3 via the notion of the underlying topological space. Thus if f and g above are instead defined as maps of posets (or of complexes), then we can also refer to them as homotopic, whenever we in fact have homotopy of the induced topological maps $|f|$ and $|g|$ (recall these were indicated in Section 1.4, especially Remark 1.4.4).

Typically we give definitions such as those in Remark 3.1.1 above just in the language of topology—but we will implicitly assume the relevant extensions, and so might correspondingly refer informally to "the poset version of Remark 3.1.1". ◇

[2] We mention that when X and Y are G-spaces, we could further require that the maps f and g here should be G-equivariant—obtaining the stronger notion of a G-*homotopy*. And although most of the results in the chapter can be extended in this way fairly easily, classroom experience has suggested that it is instead better to simplify the exposition here, by initially presenting just the classical material—and postponing the G-equivariant extensions to the final Section 3.5 of the chapter, as our Option G.

In fact, since later we will mainly restrict to posets and complexes, we will not ordinarily need to explore in detail the actual topological content of Remark 3.1.1. However, we do provide an example below—and also follow it through the later development in this section. (The topological demonstration here is probably not really required for our overall exposition; but it will be relevant for our later discussion of collapsibility starting at Definition 3.4.7.)

EXAMPLE 3.1.3 (Collapsing a 2-simplex: homotopy with the identity). We continue the theme begun in our earlier Example 1.4.3, on collapsing a 2-simplex onto its base: There we saw, for the abstract 2-simplex K on vertices $\{1,2,3\}$, that the simplicial map f—which is induced by the set-map $(1,2,3) \mapsto (2,2,3)$—maps K down to a single edge $\{2,3\}$. We express this pictorially in the following triangle:

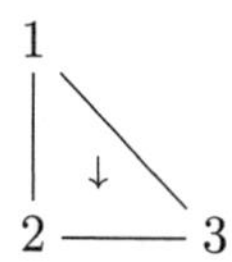

Here the arrow just informally suggests the "effect" on K of the endomorphism f. Now our earlier discussion of equivalence indicated that we are particularly interested in homotopy with the identity; so our goal will be to show that f is homotopic to Id_K.

As noted in Remark 3.1.2, this means that we need to examine the corresponding topological content in Remark 3.1.1, when we pass to the geometric realization $|K|$ and the induced mapping $|f|$. (Of course as f is simplicial, $|f|$ is continuous by 1.4.5.) So as suggested by the picture above, we realize $|K|$ in $\mathbb{R}^2$, as a filled triangle with the right angle on the axes, and in particular with the vertex 2 at the origin. Here the role of "points x of X" in 3.1.1 will be played by suitable co-ordinate pairs (x,y) in $|K|$. Now $|\mathrm{Id}_K|$ is just $\mathrm{Id}_{|K|}$, which of course satisfies $\mathrm{Id}_{|K|}\big((x,y)\big) = (x,y)$. And we see that $|f|\big((x,y)\big) = (x,0)$. Hence the above arrow suggests how to do the continuous homotopy process from $\mathrm{Id}_{|K|}$ to $|f|$: by shrinking down vertical lines, namely parallel to the edge $1-2$ on the y-axis. Thus the t-th stage of F should just shrink the y-coordinate by the fraction $(1-t)$; that is, $F\big(t,(x,y)\big) := (x,(1-t)y)$. Notice F is continuous (indeed even linear); and we have $F(0,-) = \mathrm{Id}_{|K|}$ and $F(1,-) = |f|$—as required for the homotopy in Remark 3.1.1. $\diamondsuit$

Homotopy for poset maps. In order to pursue the theme of poset homotopy indicated in Remark 3.1.2, we introduce:

DEFINITION 3.1.4 (Some poset constructions related to homotopy). We can exhibit, at the level of posets, some analogues of certain of the topological ingredients in Remark 3.1.1:

We write $\{0 < 1\}$ for the two-element poset whose unique nontrivial chain is $0 < 1$. Note that its order complex $K(\{0 < 1\})$ is just a 1-simplex—defined by those two vertices. And then the abstract complex given by this discrete set has a continuous geometric realization $|\{0 < 1\}|$—which may be taken to be the full unit interval $[0,1]$.

Given posets $\mathcal{P}$ and $\mathcal{Q}$, we define the *(Cartesian) product poset* on the Cartesian product $\mathcal{P} \times \mathcal{Q}$ using the order relation:

$$(x,y) \leq_{\mathcal{P}\times\mathcal{Q}} (x',y') \text{ whenever } x \leq_{\mathcal{P}} x' \text{ and } y \leq_{\mathcal{Q}} y'.$$

For the realization of the product we have (e.g. [**Qui78**, (1.2)]):

$$|\mathcal{P} \times \mathcal{Q}| \cong |\mathcal{P}| \times |\mathcal{Q}|. \tag{3.1.5}$$

We mention that the left-hand term is factored in a natural way via the product of the topological maps $|\mathrm{proj}_{\mathcal{P}}|$ and $|\mathrm{proj}_{\mathcal{Q}}|$—which are determined by the corresponding projections at the level of posets, from $\mathcal{P} \times \mathcal{Q}$ to the factors. (A full proof in fact involves some further technical details, which are not really important for our development. But the interested reader could pursue for example the material on products at [**Mun84**, 57.1] or [**Spa81**, Sec 5.3].) ♢

The most frequent situation where we obtain homotopy is when our poset maps satisfy the relation:

NOTATION 3.1.6 (Poset map dominance). If poset maps $f, g : \mathcal{P} \to \mathcal{Q}$ satisfy:

$$f(x) \leq_{\mathcal{Q}} g(x) \text{ for all } x \in \mathcal{P},$$

then we abbreviate this relation by writing either $f(-) \leq g(-)$—or usually even more briefly $f \leq g$.

In the literature, one sometimes sees the terminology of g *dominating* f. ♢

LEMMA 3.1.7 (poset "Homotopy Property", 1.3 in [**Qui78**]).
(1)Assume that $f, g : \mathcal{P} \to \mathcal{Q}$ satisfy:

(a) f and g are poset maps; and

(b) $f \leq g$.

Then f and g are homotopic. (That is, $|f|$, $|g| : |\mathcal{P}| \to |\mathcal{Q}|$ are homotopic.)

(2) In particular for a poset endomorphism with $f \leq \mathrm{Id}_{\mathcal{P}}$, we have $f \simeq \mathrm{Id}_{\mathcal{P}}$.

PROOF. (Sketch:) We expand slightly on the briefer discussion in [**Qui78**].[3]

We work with the product poset $\mathcal{R} := \{0 < 1\} \times \mathcal{P}$. Motivated by the definition of a homotopy, we set $F(0,\cdot) := f$ and $F(1,\cdot) := g$; and we claim that F then defines a poset map on $\mathcal{R}$. Thus whenever $(t,x) \leq_{\mathcal{R}} (t',x')$, we must show that $F\big((t,x)\big) \leq_{\mathcal{Q}} F\big((t',x')\big)$. When $t = 0 = t'$, we have:

$$F\big((0,x)\big) \overset{\text{bydef}}{=} f(x) \overset{(a)}{\leq} f(x') \overset{\text{bydef}}{=} F\big((0,x')\big);$$

and similarly with g when $t = 1 = t'$. In the remaining case we have $t = 0$ and $t' = 1$, and we see:

$$F\big((0,x)\big) \overset{\text{bydef}}{=} f(x) \overset{(a)}{\leq} f(x') \overset{(b)}{\leq} g(x') \overset{\text{bydef}}{=} F\big((1,x')\big),$$

completing the proof that F is a poset map.

Hence by 1.4.7.2, $|F| : |\mathcal{R}| \to |\mathcal{Q}|$ is continuous. Now we quote (3.1.5), which in effect conceals the only real technical details in the proof: namely we use the projections defining the homemomorphism there, to replace the realization $|\mathcal{R}|$ of the product with the product of the realizations $|\{0 < 1\}|$ and $|\mathcal{P}|$. And since we saw in Definition 3.1.4 that $|\{0 < 1\}| \cong [0,1]$, this factored form of $|F|$ gives the map on $[0,1] \times |P|$ needed in Remark 3.1.1, to exhibit the homotopy of $|f|$ and $|g|$.

Of course (2) is a special case of (1) (indeed it is the form that we will use most often). □

[3]We will also indicate an alternative proof in our later Remark 4.1.11.

In the spirit of earlier Remark 1.4.4, we note that the use of the dominance hypothesis (b), in the second displayed formula in the proof above, is what in effect allows the induced map $|F|$ to continuously "fill in" the stages of the homotopy, in the convex hull of the discrete vertices 0 and 1—that is, for values of t in the interior of the interval $[0, 1]$.

Poset homotopy using the set-product of subgroups. We now apply the homotopy in Lemma 3.1.7, as a first step toward the study of contractibility for $\mathcal{S}_p(G)$ (culminating eventually in Lemma 3.3.5). The relevant type of poset map is based on the set-product of subgroups; so to that end, we first quickly review some relevant standard facts from elementary group theory:

REMARK 3.1.8 (Some properties of the set-product of groups). Recall that for subgroups H, K of G, the (set-)product is just $HK = \{hk : h \in H, k \in K\}$—which is not in general a subgroup of G.

But this is not a problem when for H we take a normal subgroup N of G (or even just when K normalizes N); for we have (see e.g. [**DF99**, Ch 3, Prop 13]):

(1) If N is normal in G, then the set-product NK is also a subgroup of G.

Since we often focus on p-subgroup complexes, notice that in that context, it follows from (1) and the product-order formula $|NK| = \frac{|N||K|}{|N \cap K|}$ (e.g. [**DF99**, Ch 3, Cor 15]) that:

(2) If further N and K are p-subgroups, then NK is also a p-subgroup.

Notice that it follows from (2) that the product of all normal p-subgroups is $O_p(G)$ (the *largest* normal p-subgroup of G). ♢

These group properties have some consequences for subgroup posets:

LEMMA 3.1.9. *Assume that N is normal[4] in G.*
(1) Define the product map $\text{prod}_N(K) := NK$ *for K in the subgroup lattice $L(G)$. Then:*

(a) prod_N *is a poset endomorphism of $L(G)$, the identity on its image $L(G)_{\geq N}$.*
(b) $\text{prod}_N \geq \text{Id}_{L(G)}$, *so that* prod_N *is homotopic to* $\text{Id}_{L(G)}$.

(2) The special case $\text{prod}_{O_p(G)}$ *preserves the subposet $\mathcal{S}_p(G)$—on which we get the analogue of (1):*

(a) $\text{prod}_{O_p(G)}$ *is a poset endomorphism of $\mathcal{S}_p(G)$, the identity on $\mathcal{S}_p(G)_{\geq O_p(G)}$.*
(b) $\text{prod}_{O_p(G)} \geq \text{Id}_{\mathcal{S}_p(G)}$, *so that* $\text{prod}_{O_p(G)}$ *is homotopic to* $\text{Id}_{\mathcal{S}_p(G)}$.

PROOF. The single image $\text{prod}_N(K) = NK$ is a subgroup by Remark 3.1.8.1, hence a member of $L(G)$; so prod_N is at least a set endomorphism of $L(G)$. Indeed since $NK \geq N$, we see considering all such images that $\text{prod}_N\big(L(G)\big) \subseteq L(G)_{\geq N}$. When $K \leq K'$, we have $NK \leq NK'$; this shows that prod_N is a poset map. When $K \geq N$, we get $NK = K$, so that prod_N induces the identity on $L(G)_{\geq N}$. In particular $L(G)_{\geq N} \subseteq \text{prod}_N\big(L(G)\big)$, so equality holds since we saw the reverse inclusion earlier. This completes (a). For (b), from $NK \geq K$ we also obtain that $\text{prod}_N \geq \text{Id}_{L(G)}$; this gives $\text{prod}_N \simeq \text{Id}_{L(G)}$, by Lemma 3.1.7.2 (in the "dual" form where $\leq$ is replaced by $\geq$). Thus (1) holds.

[4]We may as well assume also that N is nontrivial: for when $N = 1$, prod_N is the identity map, so that the lemma has no real content.

For (2), notice when K is a nontrivial p-subgroup that so is $O_p(G)K$ by Remark 3.1.8.2. Thus $\mathrm{prod}_{O_p(G)}$ restricts to an endomorphism of $\mathcal{S}_p(G)$; and the remainder of the proof just follows that of (1). $\square$

We mention that homotopy of a poset map via a dominance relation with the identity, as exhibited for example in Lemma 3.1.9, will lead in later Proposition 3.1.12 to homotopy equivalence with the image of the map.

As an explicit instance of Lemma 3.1.9.2, and indeed of the beginning of our investigation of contractibility for $\mathcal{S}_p(G)$, we return to an example introduced in Chapter 1—which we will also continue to follow in our subsequent development in this section:

EXAMPLE 3.1.10 (The product map $\mathrm{prod}_{O_2(G)}$ in $\mathcal{S}_2(D_{12})$). We recall $\mathcal{S}_2(G)$ for the dihedral group $G = D_{12}$, given in earlier Example 1.1.19. Here $O_2(G)$ is given by the center $Z(G) = \langle b^3 \rangle$. By 3.1.9.2, we have a homotopy—of the identity map $\mathrm{Id}_{\mathcal{S}_2(D_{12})}$, with the product map $\mathrm{prod}_{O_2(G)}$ which takes each nontrivial 2-subgroup K to $O_2(G)K$. The "effect" of this product map is to take each of the left-hand order-2 subgroups (which involve a) to the unique order-4 subgroup above it; the other subgroups are unchanged. $\diamondsuit$

Homotopy equivalence. We now use the concept of homotopy of maps in recalling the standard definition (cf. [**Mun84**, p 108]):

REMARK 3.1.11 (Homotopy equivalence). Topological spaces X and Y are *homotopy equivalent* if there are continuous maps $f : X \to Y$ and $g : Y \to X$ whose compositions satisfy the homotopies $gf \simeq \mathrm{Id}_X$ and $fg \simeq \mathrm{Id}_Y$.

We use the notation $X \simeq Y$ also for homotopy equivalence. (The context of spaces should normally distinguish this from the same notation $f \simeq g$ used for homotopy of maps.)

Homotopy equivalence is easily seen to be an equivalence relation (cf. [**Mun84**, p 108]). The equivalence class of a space in the relation is usually called its *homotopy type*.

We define homotopy equivalence for posets or complexes as usual, by passing to induced maps on the underlying topological spaces. $\diamondsuit$

Notice that the homotopy $gf \simeq \mathrm{Id}_X$ required above in fact uses only the special case of homotopy for an *endomorphism* of X—and then just with the identity map. (That is, the roles of "X, Y, f, g" in the general definition of homotopy in Remark 3.1.1 are played by X, X, gf, Id_X for homotopy equivalence in 3.1.11.) Of course a similar remark holds for $gf \simeq \mathrm{Id}_Y$.

Indeed we see next that just a single homotopy of an endomorphism with the identity suffices to exhibit a homotopy equivalence; and this special case will in fact provide a great many of our later equivalences. The poset formulation of this standard fact is essentially implicit in 1.5 of [**Qui78**]. It can be applied in particular in the set-product context of our earlier Lemma 3.1.9:

PROPOSITION 3.1.12.

(1) Assume f is a continuous endomorphism of a topological space X, satisfying $f \simeq \mathrm{Id}_X$. Then X is homotopy equivalent to $f(X)$ (indeed to any intermediate

subspace $f(X) \subseteq Y \subseteq X$). In this situation we say that f induces *a homotopy equivalence.*

(2) Similarly for a poset endomorphism with $f \simeq \mathrm{Id}_{\mathcal{P}}$ (e.g. if $f \leq \mathrm{Id}_P$), we get $\mathcal{P} \simeq f(\mathcal{P})$.

(3) We get the conclusion of (2) for the product maps in Lemma 3.1.9*; in particular, $\mathcal{S}_p(G) \simeq \mathcal{S}_p(G)_{\geq O_p(G)}$.*

PROOF. We take $X, f(X), f, \mathrm{Id}_{f(X)}$ in the roles of "X, Y, f, g" in the definition of homotopy equivalence in Remark 3.1.11. We see that "gf" is $\mathrm{Id}_{f(X)}f = f$, which is homotopic to Id_X by hypothesis. Furthermore "fg" is $f\mathrm{Id}_{f(X)} = f_{f(X)}$, which is homotopic to $\mathrm{Id}_{f(X)}$—by restriction (cf. 3.1.1) to $f(X)$ of the hypothesis $f \simeq \mathrm{Id}_X$. This establishes the homotopy equivalence in (1); for notice we could just as well replace X by Y in the proof.

For (2), as usual the induced map $|f|$ is continuous by Lemma 1.4.7. Hence the homotopy equivalence follows from (1). (Of course the homotopy when $f \leq \mathrm{Id}_{\mathcal{P}}$ follows from the poset Homotopy Property 3.1.7.)

Finally (3) follows from (the $\geq$-dual form of) (2), using product maps which dominate the identity in 3.1.9. □

REMARK 3.1.13 (deformation retraction). We recall that a product map in Lemma 3.1.9 induces the identity on its image, and is homotopic with the identity. These conditions in fact show that Proposition 3.1.12.3 above provides an example of the special case of a homotopy equivalence which is called a *(strong) deformation retraction* in the topological literature. (See e.g. [**Mun84**, p. 108], and also [**Coh73**, p. 1].) ◊

We can use Proposition 3.1.12 to further explore some earlier examples:

EXAMPLE 3.1.14 (The product map in $\mathcal{S}_2(D_{12})$, continued). Recall from Example 3.1.10 that for the dihedral group $G = D_{12}$ with $O_2(G) = \langle b^3 \rangle$, we had seen that $\mathrm{prod}_{\langle b^3 \rangle} \geq \mathrm{Id}_{\mathcal{S}_2(G)}$. Now applying Proposition 3.1.12.3, we see that $\mathrm{prod}_{\langle b^3 \rangle}$ induces a homotopy equivalence of $\mathcal{S}_2(G)$ with $\mathcal{S}_2(G)_{\geq \langle b^3 \rangle}$. This moves us another step along, in our study of contractibility for $\mathcal{S}_2(G)$. ◊

EXAMPLE 3.1.15 (Collapsing a 2-simplex to an edge: homotopy equivalence). Recall in Example 3.1.3 that we had a continuous map $|f|$, which mapped the realization $|K|$ of a 2-simplex down to one of its edges (call it $|E|$); and we also had $|f| \simeq \mathrm{Id}_{|K|}$. So now by Proposition 3.1.12.1, $|K|$ is homotopy equivalent to $|E|$. (Indeed since $|f|$ induces the identity on its image, we see $|f|$ is a retraction in the sense of Remark 3.1.13.)

Of course the collapsing in Example 3.1.3 actually begins at the level of the abstract complex K, and then passes to the topological space $|K|$; so the above argument also shows that the 2-simplex K is homotopy equivalent to its edge E.

We mention that this example emphasizes that homotopy equivalence is substantially weaker than homeomorphism—here, the equivalence is certainly not injective; the dimension of simplices is not preserved, etc. ◊

EXERCISE 3.1.16 (Collapsing a simplex to a point). Extend the method in Example 3.1.15 above, further collapsing the edge E to a vertex v. More generally, obtain contractibility of any simplex—that is, a homotopy equivalence of a simplex with a point, for example any of its vertices, or its barycenter, etc. $\diamondsuit$

REMARK 3.1.17 (Dominance for simplicial maps). We briefly indicate a variant approach to the equivalence argument made in Example 3.1.15 above: Rather than passing from the complex K to the space $|K|$, we could instead pass to the face poset $\mathcal{P}(K)$. And then for the induced map we have the poset dominance relation that $\mathcal{P}(f) \leq \mathcal{P}(\mathrm{Id}_K) = \mathrm{Id}_{\mathcal{P}(K)}$—and we can apply (2) rather than (1) of 3.1.12 for the equivalence.

In a more general application of this idea of passing to the face poset, we could define a dominance relation "$f \subseteq g$" for simplicial maps, by $f(\sigma) \subseteq g(\sigma)$ for all $\sigma \in K$—to obtain homotopy in an analogue of Lemma 3.1.7 for complexes. And we obtain equivalence in an analogue of Proposition 3.1.12 for complexes, when we have $f \subseteq \mathrm{Id}_K$. (Our later development will consider equivalence of complexes in much more general circumstances—but this special case is still of some interest.) $\diamondsuit$

Our subsequent Chapter 4 will develop more advanced techniques for homotopy equivalences; the remainder of this section, and indeed of this chapter, will now focus primarily on the special case of homotopy equivalence given by contractibility.

Contractibility. We recall our final standard definition (cf. [**Mun84**, p 108]):

REMARK 3.1.18 (Contractibility). A topological space is *contractible* if it is homotopy equivalent to a single point.

We define contractibility of a poset or simplicial complex as usual by passing to the underlying topological space. $\diamondsuit$

EXAMPLE 3.1.19 (Contractibility of a simplex). Collapsing a simplex to any of its vertices, as indicated in Exercise 3.1.16, shows that a simplex is contractible. $\diamondsuit$

The remaining sections of the chapter will now explore various general techniques for establishing contractibility, mainly for posets or complexes.

3.2. Cones (one-step contractibility)

We now begin the main work of our chapter on contractibility, with a short section on cones; our approach will regard cones as the spaces which can be contracted to a point "in one step".[5] The material is mostly elementary—but the viewpoint of cones will be implicit in many later arguments related to contractibility.

[5] Of course a space given just by a single point is contractible (using *zero* steps in our approach); but this case is usually not very interesting structurally. For example, $\mathcal{S}_p(G)$ is a point only when G has a normal Sylow p-subgroup, of order exactly p.

Definition and basic examples. For our particular focus, it will be convenient to introduce cones at the level of complexes—where we take as our model the closed star $\overline{\mathrm{St}}(v)$ of a vertex v, in the language of Definition 1.6.10. Of course the cone-complex as defined below will then have a geometric realization which is a "cone on vertex v" in the usual topological sense of p. 44 of [**Mun84**]:

DEFINITION 3.2.1 (Cone). A simplicial complex K is a *cone* if it is a join of form $v * L$—for a subcomplex L, and a vertex v (which we may call a *cone point* for K).

Here L is in fact just the link of v, so in the viewpoint of Lemma 1.6.12, we can express the condition in equivalent ways via $K = v * \mathrm{Lk}_K(v) = \overline{\mathrm{St}}_K(v)$.

We say a poset $\mathcal{P}$ is a cone if its order complex $K(\mathcal{P})$ is a cone. ♢

For instance, a simplex is a cone, on any of its vertices.

And now our notion of "one-step" contractibility is expressed via:

LEMMA 3.2.2. *A cone is contractible (to any cone point).*

PROOF. (Sketch:) We assume a cone $K = v*L$ as in Definition 3.2.1. To implement formally the intuitive notion of "shrinking K down the rays to v", we need to construct a homotopy equivalence of K with v. In the context of Proposition 3.1.12, we use the "constant" map

$$k_v(\sigma) := v \text{ for all } \sigma \in K$$

for the contracting endomorphism; and we need a homotopy F of k_v with the identify Id_K.

At the level of the geometric realization $|K|$, a homotopy $|F|$ for $|k_v|$ is given by the standard classical (linear) map indicated in [**Mun84**, Cor. 20.6]. Of course that homotopy is just the obvious generalization from the collapsing map for a simplex indicated in our Exercise 3.1.16—which in turn is based on the idea of the linear homotopy F constructed for the collapsing of the 2-simplex to an edge at earlier Example 3.1.3). □

Here is the version of cones that we usually work with in our poset context:

LEMMA 3.2.3.

(1) Assume $\mathcal{P}$ has a unique maximal element x. Then $K(\mathcal{P}) = x * K(\mathcal{P}_{<x})$, and $\mathcal{P}$ is a cone.

(2) Now assume just $x \in \mathcal{P}$. Then $\mathcal{P}_{\leq x}$ and $\mathcal{P}_{\geq x}$ are cones (and hence contractible).

PROOF. By the unique-maximal hypothesis, any chain in $\mathcal{P}$ can be extended to include x; so it follows using Definition 1.6.10 that $K(\mathcal{P}) = \overline{\mathrm{St}}_{K(\mathcal{P})}(x)$. Hence from Definition 3.2.1, we see that $\mathcal{P}$ is indeed a cone. Furthermore $\mathcal{P}_{>x}$ is empty by the maximality of x, so that

$$\overline{\mathrm{St}}_{K(\mathcal{P})}(x) \overset{(1.6.12)}{=} x * \mathrm{Lk}_{K(\mathcal{P})}(x) \overset{(1.6.13)}{=} x * K(\mathcal{P}_{<x}).$$

This establishes (1).

And now (2) follows, with $\mathcal{P}_{\leq x}$ in the role of "$\mathcal{P}$" in (1). Of course for contractibility of the cone, we can just quote the general fact in Lemma 3.2.2.

But we can also demonstrate the contractibility more directly, at the level of posets: For the constant map k_x with image x is trivially a poset map; while the unique maximal hypothesis, in the form $x \geq y$ for all $y \in \mathcal{P}$, shows that we have $k_x \geq \mathrm{Id}_{\mathcal{P}}$—and then we get homotopy equivalence of $\mathcal{P}$ with x by the $\geq$-dual form of Proposition 3.1.12.2. □

There are of course obvious examples of posets satisfying the condition in (1) of Lemma 3.2.3: e.g. the full subgroup lattice $L(G)$, with unique maximal element G and minimal element 1 (cf. Example 1.1.12); or $\mathcal{S}_p(P)$ for a p-group P, which has unique maximal element P. But these cases are usually not so interesting for our approach in this book—at least partly because the contractibility is so obvious from the outset. Indeed (especially when we focus on examples from simple groups) the "main" poset $\mathcal{P}$ we study will usually not itself be contractible. However, in the course of our analysis of $\mathcal{P}$, subcones of the form $\mathcal{P}_{\leq x}$ given in (2) of Lemma 3.2.3 will come up frequently.

So we examine the concept within our running example:

EXAMPLE 3.2.4 (Still more $\mathcal{S}_2(D_{12})$). At Example 3.1.14, where $G = D_{12}$ and $O_2(G) = \langle b^3 \rangle$, we had obtained homotopy equivalence of $\mathcal{S}_2(G)$ with $\mathcal{S}_2(G)_{\geq \langle b^3 \rangle}$. Referring back to the original poset diagram in Example 1.1.19, we see that the latter subposet has the subdiagram:

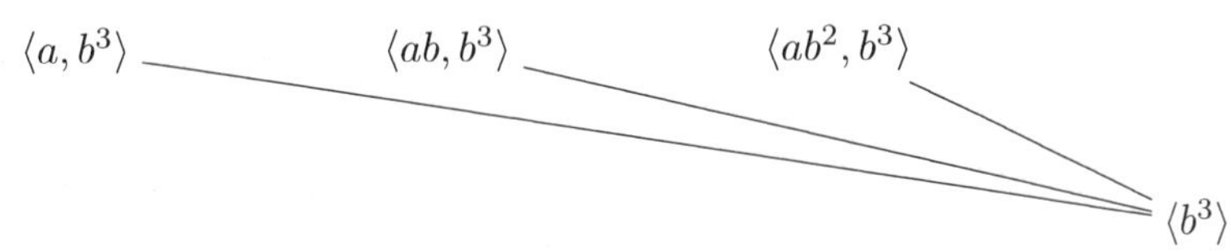

The cone structure is evident enough just from the complex viewed as a graph (here, a "3-claw", or complete bipartite graph $K_{3,1}$). ♢

REMARK 3.2.5. We warn the reader that the assumption $x \in \mathcal{P}$ in (2) of Lemma 3.2.3 is not as innocuous at it might appear: For if $\mathcal{P} \subseteq \mathcal{Q}$ with $x \in \mathcal{Q} \setminus \mathcal{P}$, then the subposet $\mathcal{P}_{\leq x}$ still makes sense in $\mathcal{P}$—but it need not be a cone, since it does not contain the element x which would play the unique-maximal role; that is, we see that $\mathcal{P}_{\leq x}$ is then really just $\mathcal{P}_{<x}$.

Hence for a subsposet $\mathcal{P}_{\leq x}$ in this situation, we would need to find other ways of checking whether contractibility does indeed hold. And later in fact, we will need to make such verifications: For instance, to apply Quillen's Fiber Theorem 4.2.1 for an inclusion $i : \mathcal{P} \to \mathcal{Q}$, when $x \in \mathcal{Q} \setminus \mathcal{P}$ we would need to show that the preimage $i^{-1}(\mathcal{Q}_{\leq x}) = \mathcal{P}_{\leq x} = \mathcal{P}_{<x}$ should be contractible. ♢

Digression: contractibility for some G-orbit structures. In this book, we usually focus on posets or complexes with a G-action; and we are not usually so interested in the corresponding structures determined by the *orbits* of G under those actions. (We will further discuss the relevance of orbit structures in our final Chapter 11.)

But since orbit structures do arise at least occasionally in our later material, we will quickly sketch here a few basic observations.

The orbit poset $\mathcal{S}_p(G)/G$. We mention for example that the conjugacy of Sylow p-subgroups in Sylow's Theorem means that the corresponding conjugacy class class is the unique maximal element in the orbit poset $\mathcal{S}_p(G)/G$—giving a cone structure.

So to review orbit structures, we begin at the level of posets:

DEFINITION 3.2.6 (Orbit poset). Let $\mathcal{P}$ be a G-poset. Then the *orbit poset* is the set $\mathcal{P}/G$ of G-orbits, under the induced ordering: $(x)G \leq_{\mathcal{P}/G} (y)G$ when there is $g \in G$ with $x \leq_{\mathcal{P}} (y)g$.

(To check e.g. that transitivity holds: Assuming $(x)G \leq (y)G \leq (z)G$ in $\mathcal{P}/G$ gives the relations $x \leq (y)g$ and $y \leq (z)h$ in $\mathcal{P}$; so $x \leq (y)g \leq \big((z)h\big)g = (z)(hg)$. Hence $x \leq (z)(hg)$ by transitivity of $\leq_{\mathcal{P}}$, so $(x)G \leq (z)G$ in $\mathcal{P}/G$.) ◊

EXAMPLE 3.2.7 (The orbit poset $L(S_3)/S_3$). In the subgroup lattice $L(S_3)$ for S_3 in Example 1.1.12, the three subgroups of order 2 are conjugate: thus we obtain the diagram for the orbit poset $L(S_3)/S_3$ by identifying those members (and the corresponding edges).

We now re-state more formally our earlier observation using conjugacy in Sylow's Theorem:

LEMMA 3.2.8. *$\mathcal{S}_p(G)/G$ is a cone, with unique maximal element given by the orbit $(P)G$ for $P \in \mathrm{Syl}_p(G)$.*

PROOF. We observed after the Definition 1.1.17 of $\mathcal{S}_p(G)$ that its maximal elements are provided by the Sylow p-subgroups $P \in \mathrm{Syl}_p(G)$. And we recall by Sylow's Theorem (cf. 1.1.16) that these form a single orbit $(P)G$ under G-conjugacy. Thus $(P)G$ is the unique maximal member of $\mathcal{S}_p(G)/G$, giving a cone as in Lemma 3.2.3.1. □

The orbit CW-complex $K\big(\mathcal{S}_p(G)\big)/G$. The elementary fact about the orbit poset $\mathcal{S}_p(G)/G$ in Lemma 3.2.8 above is perhaps not so interesting for its own sake. However, viewing it as a result about the order complex $K(\ \mathcal{S}_p(G)/G\ )$ (for the orbit poset) does suggest the considerably subtler question of whether an analogue should hold for $K\big(\mathcal{S}_p(G)\big)\ /\ G$—that is, for the orbit space of the complex for the original poset. In particular, this question about orbits exhibits one situation where our usual passage between posets and complexes is not in fact entirely routine. So here we will briefly discuss some of the issues involved.

One immediate complication is that for a G-complex K, the *orbit complex K/G* is not necessarily a *simplicial* complex: in general it has only the topological structure of a CW-complex (which we had mentioned in earlier Remark 1.2.17). The main reason for this failure is that the orbit complex K/G can lose the property in Definition 1.2.5 that a simplex is *determined* by its vertex set. For instance:

EXAMPLE 3.2.9 (A non-simplicial cell complex). Take $G = S_4$, and $\mathcal{P} = \mathcal{S}_2(G)$ as in Exercise 1.1.20; and let D denote the Sylow 2-subgroup $\langle(13),(1234)\rangle \cong D_8$.

Note first that the elements $(13)(24)$ and $(12)(34)$ of D are conjugate in G—e.g. by (23); thus they determine the same point in the orbit space $K(\mathcal{P})/G$.

Next consider the corresponding edges in $K(\mathcal{P})$, given by $\langle(13)(24)\rangle < D$ and $\langle(12)(34)\rangle < D$. These must instead lie in different G-orbits: for any conjugating element must normalize the common second term D, and hence also must normalize $Z(D)$ and its generator $(13)(24)$—rather than conjugating it to $(12)(34)$ as we would need for conjugacy of these edges. Thus these edges have different images in $K(\mathcal{P})/G$.

In particular, these images cannot be "edges" of a simplicial complex: since they would have the same two vertices, and so neither is uniquely determined by that vertex-set. Hence $K(\mathcal{P})/G$ is not a simplicial complex, just a CW-complex. ◊

Indeed Example 3.2.9 above demonstrates a further complication: For $K(\mathcal{P})$ is defined by chains in $\mathcal{P}$; and if we insert the intermediate group $\langle(13)(24),(12)(34)\rangle$ into both the chains on D indicated there, we obtain non-conjugate maximal chains in $\mathcal{P}$. Hence we cannot apply the unique-maximal argument used in Lemma 3.2.8, to get an obvious cone structure for $K(\mathcal{P})/G$.

Thus we would only hope to have an *exact* complex-analogue of the poset-proof in Lemma 3.2.8, when the G-action on $\mathcal{S}_p(G)$ is flag-transitive in the sense of earlier Remark 2.1.4. In fact flag-transitivity usually holds in our natural "local" geometries for simple groups G—but it is much less common for the full p-subgroups complex $\mathcal{S}_p(G)$, and for more general groups G.

However, it might still be plausible to hope for general G that $K\big(\mathcal{S}_p(G)\big)/G$ should be at least contractible. Indeed, this assertion appears as Conjecture 4.2 in Webb's paper [**Web87b**]—and it led to a certain amount of further research activity.

We first mention a number of early partial results toward the Conjecture: Webb had in fact established for $K\big(\mathcal{S}_p(G)\big)/G$ certain properties which are somewhat weaker than contractibility (as indicated in our later Lemma 5.2.3)—such as vanishing of the reduced Euler characteristic, in [**Web87a**, Prop. 8.2]; and even $\mathbb{F}_p$-acyclicity, in [**Web91**, Cor. 2.6.1]. And for certain special classes of groups G, the full conclusion of contractibility was obtained by Webb in [**Web87a**, Cor 8.3] and by Thévenaz in [**Thé92b**, Thms. A,B]. (Note that the Conjecture holds for the case of G of Lie type in characteristic p, using Quillen's equivalence in Theorem 4.4.8 of $\mathcal{S}_p(G)$ with the building—in view of the flag-transitive action of G on the building in earlier Remark 2.1.4.2, which of course makes the orbit complex just a single simplex, and in particular a cone.)

A few years later, the Conjecture for general G was established by Symonds:

THEOREM 3.2.10 (Symonds [**Sym98**]). *The orbit CW-complex $K\big(\mathcal{S}_p(G)\big)/G$ is contractible.* □

Symonds in reduces the proof to certain elementary properties of p-groups; however, the argument requires some more detailed topological concepts, which we have not yet introduced. Consequently we will postpone a sketch of his argument to later Remark 9.3.13.

We mention that a different proof, using Morse-theoretic methods, was subsequently given by Bux [**Bux99**]. Also recently Linckelmann in [**Lin09**] showed contractibility of the orbit space of the "fusion system" for a finite p-group P (we will indicate fusion systems as a topological generalization of $\mathcal{S}_p(G)$ in later Remark 7.1.3).

3.3. Conical (two-step) contractibility

This section presents *conical contractibility*, which is one of the contractibility techniques most frequently available in our standard poset contexts. Because of its importance, after indicating the method we present a lengthier treatment of initial examples of various representative types of applications.

From our overall viewpoint in this chapter, conical contractibility requires *two* steps: namely a first step mapping the original space into a cone; and then a second step contracting that cone to a point—as indicated in the previous section.

The definition of conical contractibility. The technique of conical contractibility was introduced by Brown in [**Bro75**], and was more explicitly emphasized in Quillen [**Qui78**]; we will follow essentially Quillen's treatment below. But we mention that analogous concepts also appear in the combinatorial literature—e.g. *join-contractibility* and *meet-contractibility* (cf. 10.13 in [**Bjö95**]); and the language of a *meet-semilattice* or a *conjunctive element* (cf. Lemma 2.4 and Definition 2.6 in [**HIÖ89**]).

DEFINITION 3.3.1 (Conical contractibility). A poset $\mathcal{P}$ is *conically contractible* if we have an element x_0 and a poset endomorphism f of $\mathcal{P}$ satisfying

$$x \leq f(x) \geq x_0 \tag{3.3.2}$$

for all $x \in \mathcal{P}$. ◊

That this really does suffice for contractibility was observed as Lemma 3 in [**Bro75**], and as 1.5 in the formulation in Quillen [**Qui78**]. Our more extended exposition up to this point has been tailored so that the assertion now follows easily from earlier statements:

LEMMA 3.3.3. *If a poset $\mathcal{P}$ is conically contractible, then it is contractible.*

PROOF. The first inequality in (3.3.2) above shows that $f \geq \mathrm{Id}_{\mathcal{P}}$, and the second inequality shows $f(\mathcal{P}) \subseteq \mathcal{P}_{\geq x_0}$; hence by (1) and (2) of 3.1.12, we have homotopy equivalence among the various terms in $f(\mathcal{P}) \subseteq \mathcal{P}_{\geq x_0} \subseteq \mathcal{P}$. And the cone $\mathcal{P}_{\geq x_0}$ is in turn homotopy equivalent to its cone point x_0, by Lemma 3.2.3.2. (Of course the latter argument is also based essentially on 3.1.12—but this time we would use the second inequality in (3.3.2) above, to get $\mathrm{Id}_{\mathcal{P}_{\geq x_0}}$ dominating the constant map k_{x_0}.) □

WARNING 3.3.4. Following the usage of Quillen in [**Qui78**], it has become customary in the literature to abbreviate proofs of conical contractibility by simply writing out the inequalities required in (3.3.2); and we may also follow this usage.

However, the reader is warned that when there are several different posets in the picture, we need to be explicitly careful about the remaining requirements of Definition 3.3.1: namely that all the elements in (3.3.2) (especially x_0) must be shown to actually lie in the particular poset $\mathcal{P}$ in question; and of course that f must also be checked to be a poset map. ◊

In the remainder of the section, we present a number of applications of conical contractibility—emphasizing some of the standard contexts where we need contractibility in our later development.

Quillen's result on conical contractibility for $\mathcal{S}_p(G)$. In particular in this subsection and the next, we will apply (3.3.2) in the situation where the poset endomorphism f is given by a set-product map; this continues a theme that we had introduced in general at (1) of Lemma 3.1.9.

Indeed for the special case of $\mathcal{S}_p(G)$, in (2) of Lemma 3.1.9 we had considered the product map $\text{prod}_{O_p(G)}$. In doing so, we were in effect analyzing the main components of Quillen's approach to $\mathcal{S}_p(G)$; and so that analysis now culminates in Quillen's result, which we had indicated as our intial goal in the chapter:

LEMMA 3.3.5 (2.4 in [**Qui78**]). *If $O_p(G) > 1$, then $\mathcal{S}_p(G)$ is conically contractible (to $O_p(G)$).*

PROOF. We use $O_p(G)$ and $\text{prod}_{O_p(G)}$ in the roles of "x_0, f"; so that in the shorthand notation of (3.3.2), what we mainly need is just the easy observation that for $X \in \mathcal{S}_p(G)$, we have:

$$X \leq XO_p(G) \geq O_p(G).$$

However (in accordance with Warning 3.3.4), for actual completeness we need to check that the terms in this expression are indeed in $\mathcal{S}_p(G)$: In particular we now very crucially use the nontriviality hypothesis $O_p(G) > 1$—needed for $O_p(G)$ to lie in the set $\mathcal{S}_p(G)$ defined as *nontrivial* p-subgroups of G. Nontriviality holds also for $XO_p(G)$ since $X \in \mathcal{S}_p(G)$; while we saw that $XO_p(G)$ is a p-subgroup (and hence lies in $\mathcal{S}_p(G)$) using the group-theoretic fact in Remark 3.1.8. (We had quoted this during the proof of Lemma 3.1.9.2a.) Finally we saw during the proof of Lemma 3.1.9.1a that a set-product map such as $\text{prod}_{O_p(G)}$ is automatically a poset map. This completes the checking of the needed details. □

This proof has in effect summarized more tersely the individual steps in our more detailed analysis, which had led up to this point in the chapter: in particular we had shown $\mathcal{S}_p(G)$ is homotopy equivalent to $\mathcal{S}_p(G)_{\geq O_p(G)}$ using (3) of Proposition 3.1.12; and followed this by homotopy equivalence of that cone with the cone point $O_p(G)$, using Lemma 3.2.3.2.

Of course Lemma 3.3.5 above applies in particular to our running example:

EXAMPLE 3.3.6 (Conical contractibility of $\mathcal{S}_2(D_{12})$). Recall we had seen in Example 1.1.19 that for $G = D_{12}$ we have $O_2(G) = \langle b^3 \rangle > 1$; hence $\mathcal{S}_2(G)$ is conically contractible to $\langle b^3 \rangle$ by Lemma 3.3.5.

We had in fact been been checking the various intermediate steps of the contraction for this particular group, in the sequence leading through Examples 3.1.10, 3.1.14, and 3.2.4. ◊

EXERCISE 3.3.7 (Conical contractibility of $\mathcal{S}_2(S_4)$). Mimic the analysis for Example 3.3.6 above—now for $\mathcal{S}_2(S_4)$ given in Exercise 1.1.20. Here $O_2(S_4)$ is given by $\langle (12)(34), (13)(24) \rangle$. ◊

Before proceeding with further results which establish (conical) contractibility, we briefly mention proceeding in the other direction: For Quillen also made the remarkable conjecture [**Qui78**, 2.9] that the converse of Lemma 3.3.5 should hold, namely:

CONJECTURE 3.3.8 (Quillen's Conjecture).
If $\mathcal{S}_p(G)$ is contractible, then $O_p(G) > 1$.

Quillen's paper established some important special cases, and further results were subsequently established by a number of authors—a topic we will develop in later Chapter 8. (That material is not so natural for this chapter on contractibility proofs—since the conjecture is usually approached via the contrapositive: namely assuming that $O_p(G) = 1$, and trying then to show that $\mathcal{S}_p(G)$ is *not* contractible; and indeed we will later re-state 3.3.8 in that form, as Conjecture 8.4.1.)

In pursuing examples in the next two subsections, it will be natural to introduce two important subposets of $\mathcal{S}_p(G)$, which are denoted by $\mathcal{A}_p(G)$ and $\mathcal{B}_p(G)$. These can be regarded as smaller and more convenient substitutes for $\mathcal{S}_p(G)$—since we will see in the following chapter (at Theorems 4.2.4 and 4.3.4) that they are homotopy equivalent to $\mathcal{S}_p(G)$. And in fact the contractibility in our examples here will be needed as a part of those later equivalence proofs.

Some conical contractibility results related to $\mathcal{A}_p(G)$. Other work of Quillen (for example on nilpotency in group cohomology) had emphasized the role of p-subgroups which are abelian, and indeed *elementary* abelian—that is, all non-trivial elements have order exactly p. Thus it was natural for Quillen to define the corresponding subposet of $\mathcal{S}_p(G)$:

DEFINITION 3.3.9 (The elementary p-subgroups poset $\mathcal{A}_p(G)$). Set
$\mathcal{A}_p(G) :=$ the poset of non-trivial elementary abelian p-subgroups of G.
In the literature this is sometimes referred to informally as the *Quillen poset*.

We recall the standard notation $m_p(G)$ for the p-rank of G—namely the rank of the largest elementary abelian p-subgroup of G. Notice then that $\mathcal{A}_p(G)$ has dimension $m_p(G) - 1$. ◊

EXAMPLE 3.3.10 (The elementary poset $\mathcal{A}_2(D_8)$). Since $G = D_8$ is a 2-group, $\mathcal{S}_2(G)$ is obtained from the full subgroup lattice $L(G)$ (given in Example 1.1.13) by removing just the identity 1_G. And then the elementary subposet $\mathcal{A}_2(G)$ is obtained by further removing G itself (as D_8 is not abelian) and $\langle(1234)\rangle$ (which is cyclic of order 4, rather than elementary); thus $\mathcal{A}_2(G)$ consists of the five subgroups of order 2, along with the two subgroups isomorphic to $\mathbb{Z}_2 \times \mathbb{Z}_2$. Thus we have the following suposet of 1.1.13:

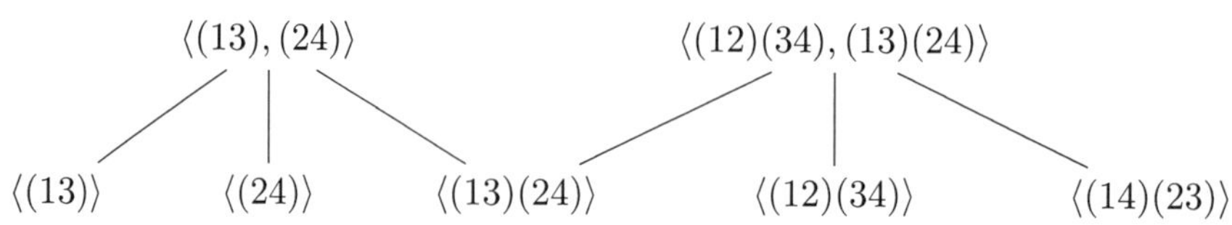

◊

The elementary poset $\mathcal{A}_p(G)$ (especially regarded as a more convenient substitute for $\mathcal{S}_p(G)$) has proved to be important for a wide variety of applications. In fact the substitution of $\mathcal{A}_p(G)$ for $\mathcal{S}_p(G)$ depends on Quillen's fundamental observation [**Qui78**, 2.1] that these two posets are homotopy equivalent: we will discuss Quillen's equivalence at Theorem 4.2.4—as a main application of equivalence methods in the next chapter. And correspondingly our next few examples will demonstrate contractibility results which can be used in several different approaches to the proof of that equivalence.

A contractibility result for equivalence via "removal". Our first technical application below might at first sight appear to deal with $\mathcal{S}_p(G)$ rather than $\mathcal{A}_p(G)$: but in fact it is part of the process of shrinking $\mathcal{S}_p(G)$ down to its subposet $\mathcal{A}_p(G)$, by "removing" those p-subgroups P which are *not* elementary abelian.

This approach via removal gives Bouc's alternative proof of Quillen's equivalence Theorem 4.2.4; we indicate that alternative in the next chapter, at Remark 4.3.3—though in effect it just iterates an equivalence method which we introduce later in this chapter, at Lemma 3.4.10. That method shows in particular that the removal from $\mathcal{S}_p(G)$ of a single element P will be a homotopy equivalence, when we can show that $\mathcal{S}_p(G)_{<P}$ is contractible. So we want to establish this property for those P which are not elementary abelian—in order to shrink $\mathcal{S}_p(G)$ down to $\mathcal{A}_p(G)$. (Below we will be writing such a subposet $\mathcal{S}_p(G)_{<P}$ in the simpler form $\mathcal{S}_p(P)_{<P}$.)

Now just as the poset properties in earlier Lemma 3.1.9 were consequences of elementary group theory facts in Remark 3.1.8, we will again derive our poset consequences from some corresponding facts in the theory of p-groups:

REMARK 3.3.11 (Properties of the Frattini subgroup of a p-group). For G a finite group, the *Frattini subgroup* is the normal (indeed characteristic) subgroup of G given by:

$$\Phi(G) := \text{the intersection of all maximal subgroups of } G.$$

We now specialize to G given by a nontrivial p-group P; and we single out some special properties of this case:

(1) $P/\Phi(P)$ is the maximal elementary abelian quotient of P; so $\Phi(P) < P$.

(2) If $Q < P$, then $\Phi(P)Q < P$.

For (1), see e.g. Exer 26.b after Section 6.1 in [**DF99**], or 23.2 in [**Asc00**]. For (2), see e.g. Exer 24 of Chapter 6 in [**DF99**], or 23.1.2 in [**Asc00**]. (Notice (1) means that $\Phi(P)$ is generated by the p-th powers of the elements of P.) ◇

For instance in the dihedral group D_8 of Example 1.1.13, $\Phi(D_8)$ is $\langle(13)(24)\rangle$.

We are now ready to state the promised result on $\mathcal{S}_p(P)_{<P}$. For comparison, we first recall by Lemma 3.2.3.2 that the related poset $\mathcal{S}_p(P)_{\leq P}$ is a cone on P; thus Bouc's approach (essentially via 3.4.10.2) focuses on situations where we can remove such a cone point, and still preserve contractibility:

LEMMA 3.3.12. *If the p-group P is not elementary abelian, then $\mathcal{S}_p(P)_{<P}$ is conically contractible (to $\Phi(P)$). On the other hand, if* P *is elementary abelian (and nontrivial), then $\mathcal{S}_p(P)_{<P}$ is not contractible.*

PROOF. The proof of the first statement parallels that of Lemma 3.3.5—now using the product map $\text{prod}_{\Phi(P)}$. As usual this is a poset map by Lemma 3.1.9.1. And the requirement of (3.3.2) is given by the easy fact that for $X \in \mathcal{S}_p(P)_{<P}$ we have:

$$X \leq \Phi(P)X \geq \Phi(P).$$

But again we must be careful (as in Warning 3.3.4) about where these terms lie: We get $\Phi(P)X < P$ by Remark 3.3.11.2, so that $\Phi(P)X \in \mathcal{S}_p(P)_{<P}$ as required; while since by hypothesis P is not elementary abelian, we get $1 < \Phi(P) < P$ from 3.3.11.1, so that also $\Phi(P) \in \mathcal{S}_p(P)_{<P}$.

For the converse given in the second statement: When P *is* elementary abelian, it can be regarded as a vector space over $\mathbb{F}_p$; and then $\mathcal{S}_p(P)_{<P}$ is just the projective space $\mathbb{P}(P)$—which is not contractible. (E.g. it is a building, and so it has a nonzero homology group giving the corresponding Steinberg module—cf. later Theorem 6.4.2). □

For instance, take P to be the dihedral group D_8—so that $\mathcal{S}_2(P)_{<P}$ is obtained by removing, from the full subgroup lattice $L(P)$ in Example 1.1.13, P itself as well as the identity 1_P. Now we had indicated that $\Phi(P) = \langle(13)(24)\rangle$ after Remark 3.3.11; so in the conical contractibility in Lemma 3.3.12, the relevant cone over $\Phi(P)$ consists of $\Phi(P)$ itself of order 2, below the three maximal subgroups, which have order 4.

A result for use in equivalence via contractible fibers. Our second technical application is used later in Quillen's original proof of Theorem 4.2.4—which proceeds via his Fiber Theorem 4.2.1: In that approach, we consider the mapping given by inclusion, namely $i : \mathcal{A}_p(G) \to \mathcal{S}_p(G)$. Given $P \in \mathcal{S}_p(G)$, the subposet $\mathcal{S}_p(G)_{\leq P}$ (i.e. just $\mathcal{S}_p(P)$) is of course a cone by Lemma 3.2.3.2; and we must show that $\mathcal{A}_p(G)_{\leq P}$ (i.e. $\mathcal{A}_p(P)$), which is the preimage (or *fiber*) of that cone under the inclusion i, is also contractible. Of course here we cannot simply apply Lemma 3.2.3.2, since the cone point P may not be elementary, i.e. might not lie in $\mathcal{A}_p(G)$; so this latter contractibility will require proof—which we indicate below.

As usual our poset consequences will be derived from some standard facts from group theory:

REMARK 3.3.13 (Some elementary abelian subgroups of a p-group). Let P denote a p-group. We use the standard notation of $\Omega_1(P)$ to denote the subgroup generated by all the elements of order p. Note that if A is an abelian subgroup of P, then $\Omega_1(A)$ is elementary abelian.

In particular, $Z := \Omega_1\big(Z(P)\big)$ is elementary abelian. Standard arguments such as Theorem 1 in Section 6.1 of [**DF99**] show in fact that:

(1) If N is normal in P, then $Z \cap N > 1$. In particular, $Z > 1$.

Indeed if P normalizes a p-group Q, applying (1) with Q, PQ in the roles of "N, P" shows that $C_Q(P) > 1$. Furthermore we get:

(2) If E is elementary abelian in P, then so is EZ (since Z centralizes E). ◊

We can now turn to $\mathcal{A}_p(P)$:

LEMMA 3.3.14 (2.2 in [**Qui78**]). *Assume P is a nontrivial p-group. Then $\mathcal{A}_p(P)$ is conically contractible (to $\Omega_1\big(Z(P)\big)$).*

PROOF. The proof parallels that of Lemmas 3.3.5 and 3.3.12—but we use the product map prod_Z for $Z := \Omega_1\big(Z(P)\big)$. As usual this is a poset map by Lemma 3.1.9.1. And the requirement of (3.3.2) is given by the easy observation that for $X \in \mathcal{A}_p(P)$ we have:

$$X \leq XZ \geq Z.$$

Once again being careful (as in Warning 3.3.4) about where these terms lie: We get XZ elementary by Remark 3.3.13.2, so that $XZ \in \mathcal{A}_p(P)$; while we have $Z > 1$ by 3.3.13.1 so that $Z \in \mathcal{A}_p(P)$. □

For instance, when P is the group D_8, with the poset diagram for $\mathcal{A}_2(D_8)$ given in Example 3.3.10, we have $Z = Z(P) = \langle (13)(24) \rangle$; and we see that the product map prod_Z maps each of the four noncentral subgroups of order 2 up to the unique subgroup of order 4 above it—so the image of the map prod_Z is the cone above Z.

A contractibility result related to $\mathcal{B}_p(G)$. We saw above that Bouc's approach (via removals) to the equivalence of $\mathcal{S}_p(G)$ with $\mathcal{A}_p(G)$ involved contractibility of subposets which have the form $\mathcal{S}_p(G)_{<P}$. It was then also natural for Bouc to consider a "dual" approach, namely removal via contractibility of subposets of the form $\mathcal{S}_p(G)_{>P}$; and this in turn led him to the following poset roughly dual to $\mathcal{A}_p(G)$. (The relevance of the technical condition on the normalizer in the definition below will appear more clearly after later Lemma 3.3.19.)

DEFINITION 3.3.15 (The poset $\mathcal{B}_p(G)$ of nontrivial p-radical subgroups). Set:
$\mathcal{B}_p(G) :=$ the poset of nontrivial p-subgroups P of G with $P = O_p\big(N_G(P)\big)$.
In the literature, this is sometimes referred to informally as the *Bouc poset*. Its members are usually called the *p-radical* subgroups. (Or *p-stubborn*, in some of the topological literature.) ◇

EXAMPLE 3.3.16 (The 2-radical poset $\mathcal{B}_2(D_{12})$). Take $G = D_{12}$ as in Example 1.1.19: Here each subgroup X of order 4 is self-normalizing, so that we have $X = N_G(X) = O_2\big(N_G(X)\big)$, and hence $X \in \mathcal{B}_2(G)$. However each Y of order 2, other than $\langle b^3 \rangle$, has normalizer $N_G(Y)$ given by the unique X of order 4 above it, which a 2-group; so that $X = N_G(Y) = O_2\big(N_G(Y)\big) > Y$, and hence $Y \notin \mathcal{B}_2(G)$. Finally $Z := \langle b^3 \rangle = O_2(G)$ is normal in G, so that $N_G(Z) = G$; thus $Z = O_2(G) = O_2\big(N_G(Z)\big)$, and hence $Z \in \mathcal{B}_2(G)$. So $\mathcal{B}_2(G)$ consists of Z of order 2, and the three subgroups X of order 4. That is, for $\mathcal{B}_2(D_{12})$ we have the subdiagram of 1.1.19 given by:

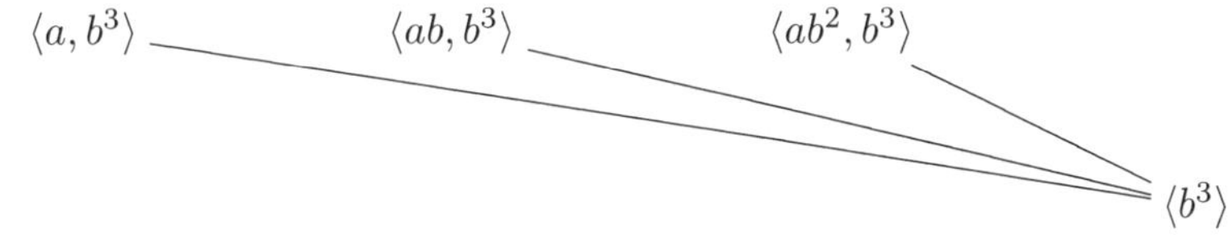

◇

The p-radical poset $\mathcal{B}_p(G)$ has also proved to be important in many applications: for example in approaches to Alperin's Conjecture in p-modular representation theory (which we discuss in Section 11.2); and in the work on decompositions of group cohomology (indicated in our Theorem 7.4.9) of Jackowski-McClure-Oliver [**JMO92**]—where the terminology of p-stubborn subgroups was instead used. (Such applications motivated an effort—carried out by An, Conder, Sawabe, Yoshiara, and others—to systematically determine $\mathcal{B}_p(G)$ for all the sporadic simple groups G—and all primes p, especially $p = 2$. A summary of results for sporadic groups is given in Yoshiara [**Yos05b**].)

So we now turn to investigating conical contractibility for $\mathcal{S}_p(G)_{>P}$.

For comparison, we recall that when in Lemma 3.3.12 we considered the subposet of the dual form $\mathcal{S}_p(G)_{<P}$, we could write it as $\mathcal{S}_p(P)_{<P}$—entirely within the p-group P; and we were then able to use the normal subgroup $\Phi(P)$ of P, implicitly applying a set-product map in the approach of (3) in Proposition 3.1.12.

By contrast, for $\mathcal{S}_p(G)_{>P}$ we cannot express members of the subposet within P; and there will be no set-product map for a normal subgroup of P which we could then use to define the poset endomorphism "f" in (3.3.2).

Indeed rather than obtaining conical contractibility directly via (3.3.2), we can instead first establish a homotopy equivalence that converts $\mathcal{S}_p(G)_{>P}$ to a more convenient form—namely one to which we can later just apply Quillen's result in Lemma 3.3.5 (on $\mathcal{S}_p(G)$ when $O_p(G) > 1$) to deduce the conical contractibility. And for that preliminary equivalence, we will instead use the method of (2) of 3.1.12, with a poset endormorphism defined via a normalizer construction. And once again, the poset properties we need will be derived from a fact in the theory of p-groups:

(3.3.17) If $Y < X$ for a p-group X, then also $Y < N_X(Y)$.

(For this, again see e.g. Theorem 1 in Section 6.1 of [**DF99**].) We obtain:

LEMMA 3.3.18 (See proof of 6.1 of [**Qui78**]). *If P is a nontrivial p-subgroup, then $\mathcal{S}_p(G)_{>P}$ is homotopy equivalent to $\mathcal{S}_p\big(N_G(P)\big)_{>P}$, which in turn is isomorphic to $\mathcal{S}_p(\ N_G(P)/P\ )$.*

PROOF. We will begin with the second assertion: Set $N := N_G(P)$, and then $\bar{N} := N/P$. From elementary group theory (e.g., Theorem 20 of Chapter 3 in [**DF99**]), the quotient homomorphism $N \to \bar{N}$ gives a bijection—between subgroups of N containing P, and subgroups of $\bar{N}$. This bijection preserves inclusions, so the poset isomorphism $\mathcal{S}_p(N)_{>P} \cong \mathcal{S}_p(\bar{N})$ follows.

We turn to the first assertion: Let $X \in \mathcal{S}_p(G)_{>P}$, so that $X > P$. We define a "normalizer map" on $\mathcal{S}_p(G)_{>P}$ by $n := n_-(P)$: that is, $n : X \mapsto N_X(P)$. The image $N_X(P)$ is indeed in $\mathcal{S}_p(G)_{>P}$, since $P < N_X(P)$ by (3.3.17); so n is at least a set endomorphism of $\mathcal{S}_p(G)_{>P}$. When $X \leq Y$, we have $N_X(P) \leq N_Y(P)$; so n is a poset map. Finally $N_X(P) \leq X$ shows that $n \leq \mathrm{Id}_{\mathcal{S}_p(G)_{>P}}$; and also $N_X(P) \leq N_G(P) = N$ shows that $n\big(\mathcal{S}_p(G)_{>P}\big) \subseteq \mathcal{S}_p(N)_{>P} \subseteq \mathcal{S}_p(G)_{>P}$. Now we can conclude from (1) and (2) of Proposition 3.1.12 that $\mathcal{S}_p(G)_{>P}$ is homotopy equivalent to $\mathcal{S}_p(N)_{>P}$. This completes the proof. □

As a corollary, we now obtain the desired contractibility—for those elements which we will want to remove, in the proof of later Theorem 4.3.4 (which gives the equivalence of $\mathcal{S}_p(G)$ with $\mathcal{B}_p(G)$):

LEMMA 3.3.19 (6.1 in [**Qui78**]). *Assume that $P \in \mathcal{S}_p(G) \setminus \mathcal{B}_p(G)$. Then $\mathcal{S}_p(G)_{>P}$ is contractible (to $O_p\big(N_G(P)\big)$).*

PROOF. In the notation of Lemma 3.3.18 and its proof, $\mathcal{S}_p(G)_{>P}$ is homotopy equivalent to $\mathcal{S}_p(\bar{N})$; while by hypothesis $P < O_p(N)$, so that $O_p(\bar{N}) > 1$—and so $\mathcal{S}_p(\bar{N})$ is conically contractible to $O_p(\bar{N})$ by Lemma 3.3.5. □

For instance, we saw in Example 3.3.16 for $G = D_{12}$ that the subgroups Y of order 2 lying outside $\mathcal{B}_2(G)$ have $\mathcal{S}_2(G)_{>Y}$ given by just a single point X; which in particular provides the cone point $O_2\big(N_G(Y)\big)$ here.

Of course the application of Lemma 3.3.5 in this proof is equivalent to applying, within N, the product map determined by the preimage Q of $O_p(\bar{N})$: then the requirements of (3.3.2) are given by $X \leq XQ \geq Q$.

Conical contractibility for some fixed-point subposets. For comparison with examples to be developed in this subsection, we first recall certain aspects of the previous examples: So far we have mainly been considering posets $\mathcal{P}$ fairly closely related to $\mathcal{S}_p(G)$. We have studied contractibility for subposets of forms such as $\mathcal{P}_{<x}$—typically using group structure, e.g. the product map determined by a suitable normal subgroup. And the contractibility will often be applied later in homotopy equivalence proofs using the method of contractible fibers as in Theorem 4.2.1, or of removal as in later Lemma 3.4.10.

By contrast, in this subsection: We will study more general posets $\mathcal{P}$ and complexes, including geometries for simple groups. Now the emphasis will be not on group structure, but instead on the G-action—and correspondingly our contractibility examples will be for subposets of the form $\mathcal{P}^P$, namely the points fixed under the action of a p-subgroup P of G. And the contractibility will be applied later in various ways—including homotopy equivalence via Quillen's method of closed sets (see e.g. the discussion in Remark 4.4.7), but also to obtain various properties studied in Chapters 5–7. (See e.g. the discussion in Remarks 6.3.1 and 6.3.3.)

Contractibility of fixed points in $\mathcal{S}_p(G)$. Again our basic example is $\mathcal{S}_p(G)$. For fixed points, an analysis (including suitable infinite-group structures for G) appears in the proof of Theorem 2 in Brown [**Bro75**]; but below we state it in the more familiar form of Quillen, for finite groups. Again we use a set-product map—though this time rather than using a normal subgroup, we instead use the "just normalizes" variant of Remark 3.1.8 (and also of results depending on that Remark).

LEMMA 3.3.20 (see the proof of Prop. 4.1 of [**Qui78**]). *For a nontrivial p-subgroup P, the fixed subposet $\mathcal{S}_p(G)^P$ is conically contractible (to P).*

PROOF. We take P, prod_P in the roles of "x_0, f" in (3.3.2). Since P normalizes all members of $\mathcal{S}_p(G)^P$ by definition, the product map prod_P is a poset map using the abovementioned "normalizes" variant of Lemma 3.1.9.2. So what we mainly need is the usual easy observation that:

$$X \leq XP \geq P,$$

where the crucial further facts are that P normalizes itself, and also normalizes XP since we have $X \in \mathcal{S}_p(G)^P$—so that we indeed get $P, XP \in Sp(G)^P$ as needed. □

EXAMPLE 3.3.21 (Fixed points in $\mathcal{S}_2(D_{12})$). Take $\mathcal{P}$ to be our running example of $\mathcal{S}_2(D_{12})$, with poset diagram given in Example 1.1.19; and let P given by $\langle a \rangle$. We can check that the subposet $\mathcal{P}^a$ of subgroups normalized by a consists of:

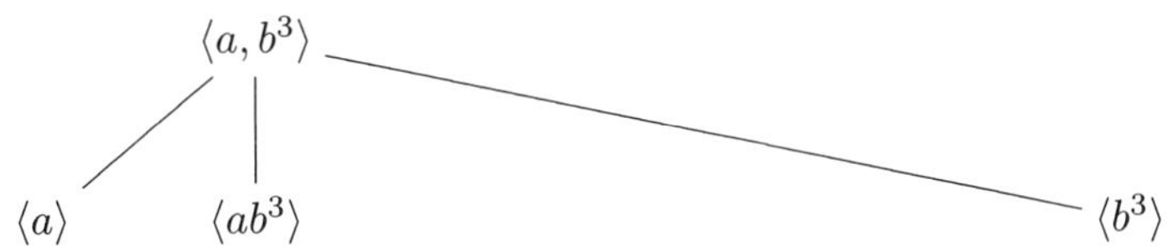

Thus here $\mathcal{P}^a$ turns out to be just the cone subposet $\mathcal{P}_{\leq \langle a,b^3 \rangle}$—which is already visibly contractible. But to follow the viewpoint of Lemma 3.3.20: The set product with $\langle a \rangle$ moves the groups $\langle ab^3 \rangle$ and $\langle b^3 \rangle$ of order 2 up to $\langle a, b^3 \rangle$ of order 4, giving $\mathcal{P}_{\geq \langle a \rangle}$; and this is followed by contraction down to $\langle a \rangle$. ◇

One can of course ask if the analogue of Lemma 3.3.20 above also holds for fixed points in the related posets $\mathcal{A}_p(G)$ and $\mathcal{B}_p(G)$. And indeed we will present the analogue for $\mathcal{A}_p(G)$ in the next section as Lemma 3.4.1—since our proof there will provide a good example of three-step contractibility, in contrast to the two-step contractibility of this section. (Later at Remark 4.2.5, we will discuss an alternative approach, which gives the analogue of 3.3.20 for both posets.)

Contractibility of fixed points in projective spaces. We now turn to analogues of Lemma 3.3.20 for group geometries which are not necessarily defined via posets of p-subgroups such as $\mathcal{S}_p(G)$; but instead are natural geometries for simple groups, such as the geometries exhibited in Chapter 2. So it will be natural to start with the projective spaces of Section 2.1.

For projective space in general, we essentially follow the treatment of Quillen, at the end of Section 3 of [**Qui78**]: As in Notation 2.1.1, we consider the case of $G := GL_n(q) = GL(V)$ for V of dimension n over $\mathbb{F}_q$; and as our poset $\mathcal{P}$ we take the projective space $\mathbb{P}(V)$ (as in Definition 2.1.3), given by the nonzero proper subspaces of V under inclusion. Letting P denote some p-subgroup of G, we see that $\mathbb{P}(V)^P$ is the set of such subspaces which are fixed under the conjugation action—that is, normalized by P.

LEMMA 3.3.22 (see after 3.1 in Quillen [**Qui78**]). *For a nontrivial p-subgroup P of $GL(V)$, $\mathbb{P}(V)^P$ is conically contractible (to $C_V(P)$).*[6]

PROOF. We define a "centralizer map" c on $\mathbb{P}(V)^P$ by $c := c_-(P)$—that is, for $W \in \mathbb{P}(V)^P$ (so that P normalizes W), we map $c : W \mapsto C_W(P)$, i.e. to the subspace *centralized* by P. First observe that the image $C_W(P)$ is indeed in $\mathbb{P}(V)^P$: For we have $C_W(P) \leq W < V$, while $C_W(P) > 0$ by 3.3.13.1, so $C_W(P) \in \mathbb{P}(V)$; and P centralizes $C_W(V)$, so that $C_W(P) \in \mathbb{P}(V)^P$. Thus c is at least a set endomorphism of $\mathbb{P}(V)^P$. And when $W \leq X \in \mathbb{P}(V)^P$, we have $C_W(P) \leq C_X(P)$, so that c is a poset map.

Now we will obtain conical contractibility via the *dual* form[7] of (3.3.2)—namely we have the obvious relations:

$$W \geq C_W(P) \leq C_V(P).$$

For the remaining details (in the viewpoint of 3.3.4): We have $C_V(P) > 0$ using Remark 3.3.13.1, while $C_V(P) < V$—here crucially using our hypothesis that P is nontrivial. Thus $C_V(P) \in \mathbb{P}(V)$; and P centralizes $C_V(P)$, so $C_V(P) \in \mathbb{P}(V)^P$. □

EXAMPLE 3.3.23 (Contractibility of fixed points in $\mathbb{P}^2(2)$). To explore an explicit example, we take $G = GL_3(2)$, with $\mathcal{P}$ the projective plane $\mathbb{P}^2(2)$ whose poset diagram we gave at (2.1.14). From within the Borel subgroup $B = U$ which we had

[6]Here we have written $C_V(P)$, where in the usual topological convention of Definition 1.5.8, we could instead write V^P—to denote the fixed points of the action of P on *vectors* of V. (Indeed Quillen in [**Qui78**] uses the notation V^P.) However, we are already using $\mathcal{P}^P$ for fixed points in the action permuting *subspaces* in $\mathcal{P}$; so to avoid confusion between the two actions, we have chosen to use the group theory centralizer notation $C_V(P)$ for fixed points in the action on vectors.

[7]Notice in particular that the first inequality, from the poset map c, contracts into the *downward* cone $\mathbb{P}(V)_{\leq C_V(P)}$.

indicated in Example 2.2.5, we choose $P = \langle z \rangle$, where z of order 2 has matrix form:

$$\begin{pmatrix} 1 & 0 & 0 \\ 0 & 1 & 0 \\ 1 & 0 & 1 \end{pmatrix}.$$

One checks that $C_V(z)$ is the 2-space $\langle 1, 2 \rangle$; while z normalizes two more 2-spaces given by $\langle 1, 3 \rangle$ and $\langle 1, 23 \rangle$ (but fixes only the vector 1 in each). Thus the fixed suposet $\mathcal{P}^z$ has the subdiagram:

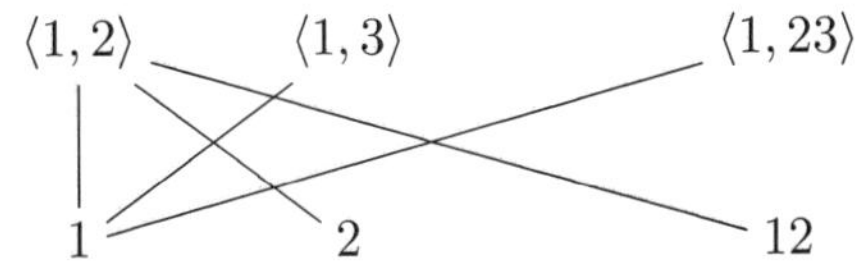

Now the first step in Lemma 3.3.22, via the centralizer map c, is to contract each of the 2-spaces $\langle 1, 3 \rangle$ and $\langle 1, 23 \rangle$ down to the unique vector 1 which is centralized by z inside each—resulting in the cone $(\mathcal{P}^z)_{\leq \langle 1,2 \rangle}$.

Since $G = L_3(2)$ has just one conjugacy class of involutions, we get an isomorphic structure for $\mathcal{P}^P$, for any other choice of P of order 2.

We leave as an exercise the determination of $\mathcal{P}^P$ for P of order 4 or 8: One can take $\langle z \rangle < P$ in the Sylow group U, and then $\mathcal{P}^P$ will be a (contractible) substructure of that determined above—in each case containing at least the edge $1 - \langle 1, 2 \rangle$. (And we also mention that in our later development—see e.g. Remark 6.3.1 and especially Lemma 6.3.3—it will often suffice for various applications to check contractibility of $\mathcal{P}^P$ just for P of order exactly p.) ♢

EXERCISE 3.3.24 (Contractibility of fixed points for $Sp_4(2)$). Lemma 3.3.22 has an analogue for the subposet of projective space given by a polar space $\mathcal{P}$ on V. But the proof above does not go over directly—since $C_V(P)$ need not be isotropic (i.e. might not lie in $\mathcal{P}$). Instead the analogue follows using e.g. the later result Lemma 3.4.16 applying to buildings in general.

So: mimic Example 3.3.23 above, but using $G = Sp_4(2)$ (with matrix forms given in Example 2.2.16), acting on the polar space $\mathcal{P}$ given in Example 2.1.19 (or its alternative version at Remark 2.1.24). That is, verify directly the contractibility of $\mathcal{P}^P$.

This time to analyze $\mathcal{P}^P$ for P of order 2, there are three conjugacy classes of involutions. In the matrix form for G_P in 2.2.16, these are given by z defined by $a = b = 0, c = 1$; and by t defined by $a = c = 0, b = 1$; and the product tz. ♢

Fixed points in buildings and sporadic geometries. We continue the theme above, for readers who are following our optional tracks indicated in Sections 2.2 and 2.3.

(Option B:) We had seen that the projective spaces of Section 2.1 in particular provide examples of the buildings for Lie type groups in Section 2.2. Hence by proceeding from Lemma 3.3.20 for $\mathcal{S}_p(G)$ to the analogue Lemma 3.3.22 for projective space, we have now seen an instance of what we called the Building Principle at Remark 2.0.3.

In fact we will discuss the analogue of Lemma 3.3.20 for general buildings, mentioned just above, as later Lemma 3.4.16—since its usual proof involves multi-step contractibility, rather than conical contractibility.

(Option S:) Similarly, in accordance with our Vague Sporadic Principle 2.0.4, we will find that there are often analogues of Lemma 3.3.20, where $\mathcal{S}_p(G)$ is replaced by a "sporadic" diagram geometry K for a suitable group G (usually also sporadic). Again, most proofs of such results require multi-step contractibility, rather than conical contractibility; e.g. we consider one instance later as Example 3.4.11. But we can exhibit here a very small example, in which fixed-point subposets are contractible in at most two steps:

EXAMPLE 3.3.25 (Contractibility of fixed points for the triples geometry of S_5). Let $G := S_5$, acting on its triples geometry in characteristic 2; we use the "external" version of the geometry indicated before Example 2.3.5, namely the poset $\mathcal{P}$ given by the points (i.e. singleton-sets) and the triples (i.e. 3-sets of points).

We will determine $\mathcal{P}^P$, for P of order 2. (For P of order greater than 2, see later Example 4.4.11.) So we take representatives $z = (12)(34)$ and $x = (12)$ of the two conjugacy classes of involutions in S_5 (namely inside and outside A_5). Recall that this poset is neither the building for G, nor $\mathcal{S}_2(G)$; so that there is no general result forcing contractibility of fixed point subposets. Instead we will simply check directly that contractibility in at most two steps does occur.

First, $\mathcal{P}^x$ is the subposet:

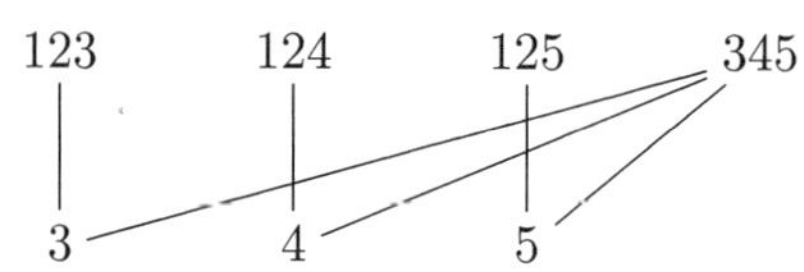

And in the spirit of the centralizer map c of Lemma 3.3.20, we see (now just geometrically) that we can contract each of the triples 123, 124, 125 which are not pointwise fixed down to the unique point in it which *is* fixed by x; resulting in the cone $(\mathcal{P}^x)_{\leq 345}$ (contractible to 345).

Next, $\mathcal{P}^z$ is the "right-hand half" of $\mathcal{P}^x$ above: it consists just of the point 5, below the triples 125 and 345. This is a cone on its unique minimal element 5, hence contractible in only one step. ♢

Contractibility of fixed points in sporadic geometries is investigated methodically in Ryba-Smith-Yoshiara [**RSY90**]; and we will be returning to this topic at various points in our later chapters. However, we emphasize that in doing so, we will also encounter some cases of sporadic geometries $\mathcal{P}$ where the analogy with Lemma 3.3.20 fails: that is, $\mathcal{P}^P$ may *not* be contractible, for at least some p-subgroups P. (E.g. take P generated by the involution "x" in our later Example 6.6.5.)

Some further directions related to fixed points. We mention that the proof of Lemma 3.3.20 generalizes easily to give the extension below, which combines two of our two earlier themes, namely normal p-subgroups and fixed points. The extension is then relevant for results proved using such tools as Conlon's induction theorem in representation theory; see e.g. [**Web87a**], and also [**Ben91**, 6.7.3]—these involve the special case where $H/O_p(H)$ is cyclic, as indicated in our proof of later Theorem 6.3.4.

LEMMA 3.3.26 (Cf. (a) of Thm 1 in [**Web87a**]). *Assume H is a subgroup of G with $O_p(H) > 1$. Then $\mathcal{S}_p(G)^H$ is conically contractible (to $O_p(H)$).*

PROOF. (Sketch:) Roughly, apply (3.3.2) with $X \leq XO_p(H) \geq O_p(H)$. □

For instance in our running Example 1.1.19 for $\mathcal{S}_2(D_{12})$, take $H := \langle b \rangle \cong \mathbb{Z}_6$, with $O_2(H) = \langle b^3 \rangle$. In this case $\mathcal{S}_2(G)^{\langle b^3 \rangle}$ consists *only* of the single point $O_2(H)$.

We mention that the contractibility in Lemma 3.3.20, of fixed points under a *single* p-subgroup P, is used by Quillen to obtain a further result on contractibility of a union of fixed-point subsets over *various* p-subgroups: namely of the "singular set" $\cup_{Q \in \mathcal{S}_p(P)} \mathcal{S}_p(G)^Q$. We present this later as Proposition 4.4.15, since its proof requires the more advanced techniques on homotopy equivalences developed in Chapter 4.

We end the section by emphasizing that many further instances of conical contractibility can be found in the literature; typically they are used as part of proofs of other results, such as homotopy equivalences. We mention for example the proofs of 2.5, 2.7bcd, 4.1, and 4.2 in [**KT85**], and of 2.4 and 6.1 in [**Thé87**].

3.4. Multi-step contractibility and collapsibility

In addition to the two-step procedure in conical contractibility, there will also be various situations where contractibility proofs require three or more steps—a theme we will start to explore in this section. Of course the steps prior to the final contraction to a point are "more general" homotopy equivalences; and indeed as we proceed we will also be introducing some further important equivalence methods: these will be variants on the theme of removal mentioned earlier—namely via *elementary collapses*, and especially via contractible links.

Multi-step contractibility. But first we will demonstrate such a multi-step procedure which uses only an elementary method for homotopy equivalence which we have already introduced—via poset map dominance in Proposition 3.1.12.2.

To indicate the overall approach, we observe that (3.3.2) concatenates two inequalities that determine equivalences—so we may as well concatenate three or more such inequalities.

We exhibit this by establishing, as promised earlier, the analogue for $\mathcal{A}_p(G)$ of Lemma 3.3.20 for fixed points in $\mathcal{S}_p(G)$. Notice first that the proof of that earlier lemma does not just go over—since our p-subgroup P may not be elementary abelian (and so need not itself lie in $Ap(G)$). To compensate for this, we will extend certain earlier ideas, such as the use of $Z = \Omega_1(Z(P))$ in Lemma 3.3.14, and the centralizer map in Lemma 3.3.22; and we obtain:

LEMMA 3.4.1 (See 4.4 in [**Qui78**]). *Assume P is a nontrivial p-subgroup of G. Then $\mathcal{A}_p(G)^P$ is contractible (to $\Omega_1(Z(P))$).*

PROOF. Set $Z := \Omega_1(Z(P))$. Much as in (3.3.2), note for $X \in \mathcal{A}_p(G)^P$ that:
$$X \geq C_X(P) \leq C_X(P)Z \geq Z,$$

and we will now examine the consequences of this sequence[8] of relations in further detail:

The first containment $X \geq C_X(P)$ is another instance of the centralizer map of Lemma 3.3.22; and shows via Proposition 3.1.12.2 that $\mathcal{A}_p(G)^P$ is homotopy equivalent to the image of the map: namely the subposet of elementary subgroups which are *centralized* by P—for which we will use the fairly obvious notation $C_{\mathcal{A}_p(G)}(P)$.

And now we may restrict the final two containments above to $Y \in C_{\mathcal{A}_p(G)}(P)$—where they take the form $Y \leq YZ \geq Z$, giving conical contractibility for $C_{\mathcal{A}_p(G)}(P)$ as usual by (3.3.2).

So composing the equivalences above, we see $\mathcal{A}_p(G)^P$ is contractible. □

EXAMPLE 3.4.2. Consider G given by the dihedral group D_8, with $\mathcal{A}_2(G)$ given in Example 3.3.10.

First consider P outside $\mathcal{A}_2(G)$, say $P = \langle(1234)\rangle$. This element interchanges in pairs the subgroups of order 2 other than $Z = \langle(13)(24)\rangle$, so that $\mathcal{A}_2(D_8)^{(1234)}$ consists of Z and the two elementary groups X above it. Now the centralizer map in 3.4.1 above takes those X to $C_X\big((1234)\big) = Z$; and in this case the contraction to the point Z is already complete, just after this first step.

So now consider instead some choice inside $\mathcal{A}_2(G)$, say $P = \langle(13)\rangle$; here we have $P = \Omega_1\big(Z(P)\big) = Z$. To obtain $\mathcal{A}_2(D_8)^{(13)}$, we exclude only the subgroups $\langle(12)(34)\rangle$ and $\langle(14)(23)\rangle$ which are interchanged by (13). The centralizer map now takes $X = \langle(12)(34), (13)(24)\rangle$ to $C_X\big((13)\big) = \langle(13)(24)\rangle$; the other members are unchanged. Next the product with $Z = \langle(13)\rangle$ takes $\langle(24)\rangle$ and $\langle(13)(24)\rangle$ to $\langle(13), (24)\rangle$. And finally the latter is mapped down to the cone point $\langle(13)\rangle$. So this P exhibits all three steps in the proof of Lemma 3.4.1. ◊

For an alternative proof of Lemma 3.4.1, see our later discussion at Remark 4.2.5.

Topological background: collapses and collapsibility. As mentioned at the start of the section, our multi-step contractibility proofs will frequently involve homotopy equivalence methods which are somewhat more specialized than our basic method of poset map dominance in Proposition 3.1.12.2, as used in Lemma 3.4.1 above. We now begin to introduce some of those further methods; and in the remainder of the section, we will focus on several aspects of what we had earlier called the removal method.

Some motivation from posets. To suggest part of the context for our development of elementary collapses, we first examine what is involved in removing an element x from a poset $\mathcal{P}$:

Notice that in the corresponding order complex $K(\mathcal{P})$, we must then remove all chains that contain x. In the language of Lemmas 1.6.12 and 1.6.13, we want to remove simplices which are in the closed star $\overline{\mathrm{St}}(x)$—but not in the link $\mathrm{Lk}(x)$. At the level of topological spaces, namely of the geometric realization $|K(\mathcal{P})|$ of the complex, this amounts to removing the open star $\mathrm{St}_{|K(\mathcal{P})|}(x)$ as in Definition 1.6.10. And at the level of abstract simplicial complexes, this involves removing the discrete set $\mathrm{St}_{K(\mathcal{P})}(x)$ corresponding to the open star in 1.6.10.

Of course for our purposes, we will want to exploit situations where such a removal is actually "harmless"—that is, where the removal induces a homotopy

[8] Cf. the terminology of a *zigzag* of equivalences in algebraic topology.

equivalence. Thus we will want to actually implement the removal by a suitably continuous process: in which the closed star is smoothly "collapsed" down onto the link.

If we further restrict attention to a single simplex σ in the closed star $\overline{\mathrm{St}}(x)$, this process involves collapsing σ down to $\mathrm{Lk}_\sigma(x)$, namely the maximal subface of σ given by the vertices other than x. Thus we are in particular returning to our earlier theme of collapsing a simplex, started at Remark 1.4.3 and continued through Example 3.1.3; the new feature is that our simplex is now embedded in a larger complex, and we want the collapsing to be an equivalence also on that larger complex.

Definitions for an elementary collapse in a complex. The preceding discussion for a poset and its order complex suggests that we might generalize by considering an arbitrary complex K, and asking when removal of some maximal simplex σ might be an equivalence. Here, rather than collapsing σ as above to a single maximal subface, we could just attempt to collapse less stringently—down to the union of *all except one* of the faces. This corresponds to the standard topological notion of an *elementary collapse*.

We now briefly sketch some of the relevant topological background. Except perhaps for the main statements, the details of this material will not really be required for our later development.

For the more motivated reader, we mention that the basic ideas go back to J. H. C. Whitehead around 1940; they are developed in the subsequent classical topological literature—and the theory seems to be more or less assumed in more modern topological references. One fairly recent source is Section 11 of Björner [**Bjö95**]; more detail is provided in Section 2 of Cohen [**Coh73**] (and a briefer treatment is given in Section 11.1 of Davis-Kirk [**DK01**]).

We start by indicating the standard condition which will suffice to guarantee that removal affords an equivalence: we use essentially the language of 11.1 of [**Bjö95**]; but we will see in a moment that this leads to the version of the condition for collapse given on page 3 of [**Coh73**]:

Definition 3.4.3 (Freeness and elementary collapse). Assume that σ is a maximal simplex in a complex K, and that τ is one of the maximal faces of σ.

If σ is the *only* simplex of K above τ, then we say that σ is *free over* τ. In that case we may also say that τ is a *free face* of σ.

And then the removal of $\{\sigma, \tau\}$ from K is called an *elementary collapse*. $\diamondsuit$

Example 3.4.4. In our earlier Example 1.2.13, the various 2-simplices F_i are *not* free over the edge E which they have in common. However, each F_i is free over any of its other two edges.

Note that collapsing such an F_i would reduce it to the union of a remaining edge D_i with E. (But then D_i is free over its vertex not on E; and so D_i would collapse down to E.) $\diamondsuit$

Next we further explore certain aspects of Definition 3.4.3:

An elementary collapse reduces K to the subcomplex $L := K \setminus \{\sigma, \tau\}$. We might roughly suggest the situation via some diagram like (cf. [**Coh73**, p 3]):

(3.4.5)

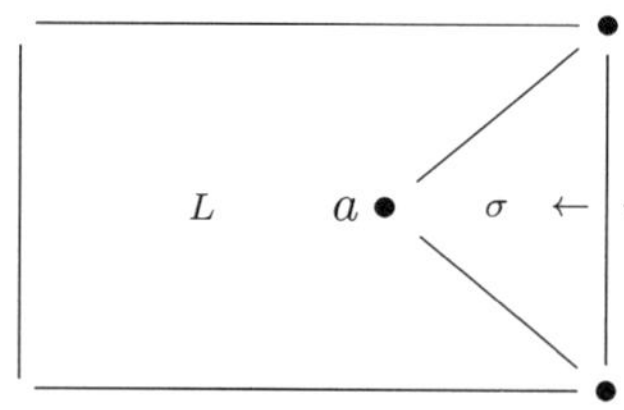

where L just stands for whatever the remaining subcomplex might be. The idea of the collapse is roughly to push "through" the free edge τ: that is, to take the points of τ continuously towards points in the remaining edges of σ, until the "interior" of σ (more precisely, that of $|\sigma|$ and $|\tau|$) has also been removed. (This is of course similar to the collapsing procedure in earlier Example 3.1.3.)

Note that the freeness condition in Definition 3.4.3 says, at the level of the topological space $|K|$, that the open star $\mathrm{St}_{|K|}(\tau)$ to be removed consists just of the interiors of $|\sigma|$ and $|\tau|$, as mentioned above; or more simply at the level of the abstract complex K, that the corresponding discrete star $\mathrm{St}_K(\tau)$ to be removed is just $\{\sigma, \tau\}$. Thus the simplicial map which removes σ and τ from K (to get L) induces at the level of spaces a continuous map taking $|K|$ to $|L|$ by removing the open star $\mathrm{St}_{|K|}(|\tau|)$ of τ.

As promised we also indicate how freeness gives the version of the hypotheses for elementary collapse on page 3 of [**Coh73**]: For let τ' denote the remainder of the boundary $\partial\sigma$, namely the union of the maximal subfaces other than τ; then we have the required conditions:

$$K = L \cup \sigma \text{ with } L \cap \sigma = \tau'.$$

(We can in fact write $\sigma = a * \tau$ where a is the unique vertex of σ not in τ; and then $\tau' = a * \partial\tau$.)

The equivalence property for collapses. As indicated earlier, collapses will be important for us since they provide a useful special case of homotopy equivalence. (Indeed even of retraction; cf. line 4 of p 3 in [**Coh73**].)

LEMMA 3.4.6 (standard). *Assume that σ is free over τ in a complex K, and set $L := K \setminus \{\sigma, \tau\}$. Then the simplicial collapsing map $r : K \to L$ is homotopic to Id_K; as is the inclusion $i : L \to K$.*

Hence either map induces a homotopy equivalence of complexes. (Namely by passing to the underlying topological spaces, and there applying Proposition 3.1.12.*1.)*

We will usually abbreviate by saying that an elementary collapse "is" a homotopy equivalence.

PROOF. (Sketch) The more impatient reader can just assume this standard result, since the methods in our discussion here will not be required in our later development.

Actually it does not seem easy to locate a detailed proof of the result, since the modern topological literature seems to take it entirely for granted. Here I am grateful to Pete Bousfield: first for searching through his substantial library, and also for then providing the following sketch—the argument is primarily combinatorial in flavor, and so fits in with the overall approach in this book:

Let a denote the vertex of σ which is not in τ.

We first consider the complex K_b obtained from K by adding a new vertex b, corresponding to the barycenter of τ in $|K|$—with edges to all vertices of σ, i.e. including a; and with the obvious corresponding higher-dimensional simplices. (In our picture (3.4.5), b would be placed at the base of the arrow, dividing the edge τ into two new edges; and we get a further new edge from b to a.) The resulting K_b is in the spirit of a "partial" barycentric subdivision of K, as described in Remark 2.1.27; and of course, using the corresponding extension of Lemma 1.6.4, $|K_b|$ is homeomorphic to $|K|$.

So we replace our original K with K_b; that is, we may as well assume that K *is* K_b. And in this version of K, "σ" is now no longer a single simplex, but a union of simplices: it has the form $\sigma = \tau' * b$, where as before $\tau' := \sigma \cap L$ is the union of the maximal faces of σ other than τ.

Now consider the simplicial collapsing map $r : K \to L$, defined by $r(v) := v$ for vertices $v \neq b$, but $r(b) := a$.

Notice that for each maximal face ρ in τ', r just applies to the simplex $\rho * b$ a collapsing set map—in the spirit of our Example 1.4.3; while r is simply the identity on "the rest of K", namely on L. Thus we might expect to be able to suitably generalize the corresponding argument made in Example 3.1.3: to show that shrinking (now down lines parallel to the edge $a*b$) should lead to a homotopy of r with the identity Id_K; and hence homotopy equivalence via Proposition 3.1.12.1.

And indeed that is possible; we summarize the argument, in the language of Section 3.5 of Spanier [**Spa81**]—to which the reader can refer for fuller details: It is immediate that the continuous maps $|r|$ and $|\mathrm{Id}_K| = \mathrm{Id}_{|K|}$ are *contiguous* for the pair (K, L) in the sense of [**Spa81**, p. 130, before 3.5.1]—that is, the union of the images under the two maps of a simplex is always a simplex (in either $|K|$ or $|L|$). Hence by [**Spa81**, 3.5.2], the maps are *homotopic relative to* L (cf. [**Spa81**, Secs. 1.3, 1.4]). This gives the homotopy of r (or i) with Id_K. So we get homotopy equivalence of K with L via Proposition 3.1.12.1. □

Collapsibility and contractibility. Since an elementary collapse defines a homotopy equivalence by Lemma 3.4.6 above, we can (and often will) use sequences of such collapses in various later proofs of homotopy equivalence.

And indeed very frequently, we use collapses to establish a strong form of contractibility, namely:

DEFINITION 3.4.7 (Collapsibility). We say a complex K is *collapsible* if it can be shrunk to a point (necessarily some vertex of K)—using a sequence of elementary collapses.

In particular by Lemma 3.4.6, collapsibility implies contractibility.

And we say a poset $\mathcal{P}$ is collapsible if the order complex $K(\mathcal{P})$ is collapsible. ◇

We conclude the subsection by mentioning a characterization of contractibility phrased in the language of simple homotopy theory and collapses (again this material is not really essential for our later development):

REMARK 3.4.8 (Anti-collapses and contractibility). An *elementary anti-collapse* (or *elementary expansion*; (see e.g. [**Coh73**, p 3]) is roughly a map as in Definition 3.4.3, but read in the opposite direction—in effect blowing up from potential boundary faces to a new higher dimensional simplex (which will be free over the new part of its boundary). And in analogy with Lemma 3.4.6, an anti-collapse is also a homotopy equivalence.

And then we have the standard characterization (here restricted to our context of a complex K):

(3.4.9)

K is contractible iff it is equivalent to a point via collapses and anti-collapses.

(The reader curious about why anti-collapses might be needed can refer to the standard topological example of the "house with two rooms", discussed e.g. on pp 2–3 of [**Coh73**].)

The statement (3.4.9) is in fact just a special case of the characterization of homotopy equivalence, in the context of simply connected CW-complexes, as being given by a sequence of collapses and anti-collpases. (See e.g. [**Coh73**, 24.2].) That is, for simply connected complexes, homotopy type is the same as "simple-homotopy type". $\diamondsuit$

Our contractibility arguments in this book will most often establish the stronger form of collapsibility. And indeed in the remainder of the section, we will present some representative applications of collapsibility techniques.

Equivalence via contractible links. We start with a special case of equivalence via collapses, which has proved particularly useful in many applications in the literature on group geometries, especially diagram geometries: namely removal of vertices whose links are contractible. Indeed such equivalence seem to be most frequently used in multi-step contractibility (or collapsibility) proofs.

The method. This particular removal technique seems to have been well known in the literature on combinatorial geometry for a long time. Below we present the version stated as 2.1 in [**RSY90**],[9] and used heavily there and in subsequent related work. We will continue to use the notation from Definition 1.6.10 of $\mathrm{St}_K(x)$ for the "open star in K", i.e. the set of all simplices containing x.

LEMMA 3.4.10 (standard; e.g. special case of 5.3′ in [**Coh73**]).

(1) Assume for a vertex v of a complex K that the link $\mathrm{Lk}_K(v)$ is contractible. Then the inclusion $K \setminus \mathrm{St}_K(v) \subseteq K$ induces a homotopy equivalence. In this situation, we will say informally that "we can remove v".

(In the statement we could replace the vertex v by a larger simplex σ.)

(2) Assume for an element x in a poset $\mathcal{P}$ that either $\mathcal{P}_{>x}$ or $\mathcal{P}_{<x}$ is contractible. Then the inclusion $\mathcal{P} \setminus \{x\} \subseteq \mathcal{P}$ induces a homotopy equivalence. In this situation, we will say informally that "we can remove x".

PROOF. (Sketch:) We will assume that $\mathrm{Lk}_K(v)$ is in fact collapsible; in view of (3.4.9), a proof for contractibility can be obtained by adjusting the argument, as given below for collapses, also to the case of anti-collapses.

So suppose that a sequence collapsing $\mathrm{Lk}_K(v)$ to a point w begins with the simplex σ, which is free over its maximal subface τ in $\mathrm{Lk}_K(v)$. Now any ρ which is a maximal simplex in K over $v*\tau$ must in fact lie in $\mathrm{St}_K(v)$—and $\overline{\mathrm{St}}_K(v) = v*\mathrm{Lk}_K(v)$ by Lemma 1.6.12; hence $\rho = v * \sigma'$ for some σ' in $\mathrm{Lk}_K(v)$ over τ. But then by the

[9] We mention: that paper tends to use the term "homotopic" somewhat loosely, where "homotopy equivalent" would actually be appropriate.

freeness of σ in $\mathrm{Lk}_K(v)$, $\sigma' = \sigma$—so that $\rho = v{*}\sigma$. We conclude then that $v{*}\sigma$ is free over $v * \tau$ in K. Thus collapsing $v * \sigma$ is a homotopy equivalence by Lemma 3.4.6. And indeed continuing in this way through the rest of the collapsing sequence for $\mathrm{Lk}_K(v)$, we reduce $\overline{\mathrm{St}}_K(v)$ to the edge $v * w$; but again this is free over v, so we can similarly collapse it to $w \in \mathrm{Lk}_K(v)$. This establishes (1) (subject to our remark above about adjusting for anti-collapses).

Since by 1.6.13 we have $\mathrm{Lk}_{K(\mathcal{P})}(x) = \mathrm{Lk}_{K(\mathcal{P})}(\mathcal{P}_{<x}) * \mathrm{Lk}_{K(\mathcal{P})}(\mathcal{P}_{>x})$, we see that (2) follows from (1). □

(Option S) An application to fixed points in a sporadic geometry. We had begun to explore this general theme in a small sporadic geometry, in earlier Example 3.3.25; here, as a demonstration of how Lemma 3.4.10 is often used in practice, we present a three-step collapsibility proof (which is among those appearing in [**RSY90**]):

EXAMPLE 3.4.11 (Collapsibility of fixed points for the C_3-geometry for A_7). We take $G = A_7$, and consider the Neumaier C_3-diagram geometry of Example 2.3.7. It is perhaps easiest to use the external viewpoint given there of the geometry as a three-level poset $\mathcal{P}$, with order relations given by suitable set-inclusions.

As indicated in the previous section (notably at the end of Example 3.3.23), we focus on fixed points under a subgroup P of order exactly 2; and as a representative of the unique conjugacy class of involutions in G, we will take P generated by (45)(67). This choice is made so that the fixed subposet $\mathcal{Q} := \mathcal{P}^{(45)(67)}$ will be contractible to the obvious (pointwise) fixed line 123; and in particular, the three planes in $\mathcal{Q}$ are those over 123 indicated in the preliminary discussion of 2.3.7, namely $\pi = \overset{\overset{7}{456}}{123}$, $\pi' = \overset{\overset{5}{467}}{123}$, and $\pi'' = \overset{\overset{6}{475}}{123}$. Clearly the points which are fixed by P are 1, 2, and 3; while the other lines which are fixed (only setwise) by (45)(67) are 146, 157, 246, 257, 346, and 357. And then the fixed subposet $\mathcal{Q}$ has the diagram:

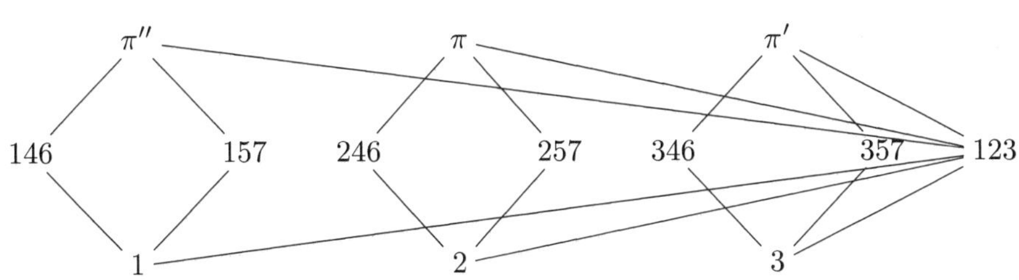

Next to envision the 2-dimensional order complex $K(\mathcal{Q})$: We first need to add further edges—such as $1 - \pi'$, which is implied by the chain $1 < 146 < \pi''$ of length 2; and also $2 - \pi'$ implied by $2 < 123 < \pi''$; and so on. (Indeed recall that all points are incident to all planes.) And thereafater we need to "fill in" all triangles, to get the 2-simplices. With this fairly complicated resulting picture, it may or may not be clear that the complex $K(\mathcal{Q})$ is contractible.

However, we claim that the subposet $\mathcal{R} := \mathcal{Q}_{<\pi''}$ is collapsible: For $\mathcal{R}$ has the diagram:

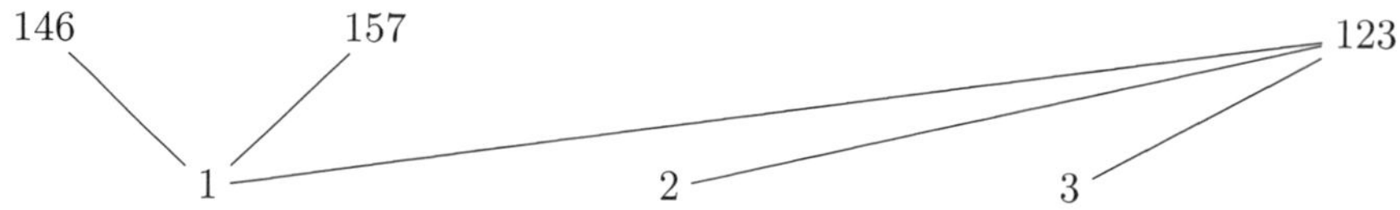

And of course visibly this graph is contractible (indeed we established conical contractibility[10] of an isomorphic poset in Example 3.3.23). However, let us pursue relentlessly (if perhaps somewhat pedantically) our theme of applying Lemma 3.4.10: Note that $\mathcal{R}_{<146}$ is the single point 1, and indeed so is $\mathcal{R}_{<157}$; thus by (2) of the Lemma we can remove the lines 146 and 157 from $\mathcal{R}$, obtaining a subposet $\mathcal{R}'$. Then a similar argument in $\mathcal{R}'$ allows us to remove the points 1, 2, and 3—leaving just a single vertex given by the line 123, and so establishing our claim that $\mathcal{R} = \mathcal{Q}_{<\pi''}$ is collapsible.

The claim now allows us to apply Lemma 3.4.10.2 in $\mathcal{Q}$, to remove the plane π''; and similarly we can remove π and π': We conclude that $\mathcal{Q}$ is homotopy equivalent to the subposet $\mathcal{Q}'$ given by only its points and lines, that is, the bottom two levels of $\mathcal{Q}$. In particular this is now a 1-dimensional complex, i.e. a graph. But then we can collapse $\mathcal{Q}'$ to a point, with just a slight adaptation of the argument made above for $\mathcal{R}$; so we conclude that the fixed subposet $\mathcal{Q} = \mathcal{P}^P$ is collapsible. $\diamondsuit$

(Option B) Collapsibility in the Solomon-Tits argument. We will most frequently pursue collapsibility via arguments for removal using contractible links, as in Lemma 3.4.10 of the previous subsection. Thus the Solomon-Tits argument presented in this final subsection is more optional reading (even within our Option B on buildings)—in that it more directly applies collapses just via the definition, a direction that we will not usually follow in our later development.

On the other hand, the argument does have considerable theoretical importance: In particular, it is the heart of the proof of the Solomon-Tits Theorem [**Sol69**, Thms. 1,2]—which we treat later as Theorem 9.4.6—showing that the fundamental topology of the building is that of a *bouquet*[11] of spheres (given by apartments). And indeed later in this subsection at Lemma 3.4.16, we will see that essentially this argument is used by Quillen, at (b) in the proof of 3.1 in [**Qui78**], to establish the contractibility of fixed points in a building.

(Finally: I confess that I have always regarded the Solomon-Tits argument as exceptionally beautiful—ever since Mark Ronan first explained it to me, in about 1979.)

To give a rough initial summary of the argument, we use the language introduced our brief discussion of paths etc. in a building (of rank ≥ 2) around (2.2.21): Fix some particular chamber (maximal simplex) σ in a building K, and arrange the remaining chambers in order of increasing distance from σ. In particular let $K^{\leq N-1}$ denote the subcomplex obtained by removing all chambers "opposite" to σ (i.e. by

---

[10] And more generally, z is nontrivial in the action of $G_{\pi''} \cong L_3(2)$ on the building π'' (the kernel $K_{\pi''}$ is trivial since here $G_{\pi''}$ is simple); so contractibility also follows from later Lemma 3.4.16.

[11] a union of spheres (including a point in common), with suitable identifications along their boundaries

removing those at the maximal distance N). Then the subcomplex $K^{\leq N-1}$ of chambers at less than maximal distance is collapsible: namely we can first remove the chambers at distance exactly $N-1$ via elementary collapses; then remove those at distance $N-2$, and so on; eventually reducing (at distance 0) to the original simplex τ itself, which is of course collapsible.

Some further background on paths. To indicate the Solomon-Tits argument in fuller detail, we need to extend a little beyond that earlier discussion around (2.2.21).

As above we fix a chamber σ; and now let $d(\sigma')$ denote the distance from σ of a chamber σ'. And extending the notation indicated above let $K^{\leq i}$ denote the the subcomplex of chambers σ' with distance $d(\sigma') \leq i$.

Furthermore for a panel τ, we write $d(\tau) = i$, where i is the minimum of the values of $d(\sigma')$, over chambers σ' of which τ is a maximal face. We can use this language to express the property (cf. [**AB08**, 3.105]) which will be crucial for the collapsibility argument; essentially it extends to the whole building a property of opposites in a single apartment-sphere on σ—namely:

(3.4.12) If $d(\tau) = i$, then exactly one of the chambers σ' on τ has $d(\sigma') = i$.

(That is, all the remaining chambers σ'' on τ have $d(\sigma'') = i+1$.) In Tits's original treatment [**Tit74**], this property can be deduced from his more general theory of "projections": Namely we see using [**Tit74**, 3.19] with σ, τ in the roles of "B, A", that our distinguished chamber σ' arises as the projection $\mathrm{proj}_\tau(\sigma)$.

We mention that (3.4.12) is sometimes called the "gate" property; presumably the idea underlying the terminology is that the unique closest chamber σ' on τ is the "gate", through which all shortest paths to σ using any chamber σ'' on τ must pass. In particular, Dress and Scharlau in [**DS87**] show that this gate property alone comes very close to characterizing buildings among more general geometries in metric spaces. (Indeed we remark for Option S that this is presumably why there seems to be no real analogue of this property in sporadic geometries—and so exhibits one situation where our Sporadic Principle 2.0.4 fails.)

EXAMPLE 3.4.13 (The gate property for $L_3(2)$). We return to $G = L_3(2)$, whose building K is the projective plane with poset diagram given in (2.1.14).

We take the leftmost edge $1 - \langle 1, 2\rangle$ in (2.1.14) as our fixed initial chamber σ. We have maximum distance $N = 3$ in the graph K, as in the discussion of paths leading up to (2.2.24).

Indeed we now choose an apartment like that in (2.2.24); but this time instead of the opposite edge $3 - \langle 12, 3\rangle$ used there, we instead use the opposite edge σ'' given by $3 - \langle 2, 3\rangle$. This has the convenient property that the apartment-hexagon, determined by σ and σ'' as in (2.2.24), is now given just by the left-hand three "columns" of (2.1.14).

Now the panel $\tau := 3$ in σ'' is at distance 2: Indeed the gate σ' corresponding to τ just in the apartment is given by the chamber $3 - \langle 1, 3\rangle$ at distance 2—since the other chamber on 3 in the apartment is $3 - \langle 2, 3\rangle$, which is at the larger distance 3.

Now return to the full poset diagram of (2.1.14): the above statement, about σ' being the gate for τ in the apartment, is in fact still valid in this full building graph: for there is just one further chamber on τ, namely $3 - \langle 3, 12\rangle$; and it is also at the larger distance 3. ◊

EXERCISE 3.4.14 (The gate property for $Sp_4(2)$). Mimic the example above, using $G = Sp_4(2)$, with building given by the polar space in Example 2.1.19. For instance, fix σ as the chamber $1-\langle 1, 2\rangle$. First consider the apartment given in (2.2.26), determined by the opposite chamber σ'' given by $4 - \langle 3, 4\rangle$ at distance 4; for τ use the panel 4 at distance 3, and determine the gate in the apartment. Then in the full building, compute the remaining chamber (beyond σ'') on τ, and verify that it also is at the larger distance 4, so that σ' is also the gate in the full building graph. (The computations should also be easy in the set-theoretic alternative version 2.1.24 of the polar space.) $\diamondsuit$

The collapsibility argument. The main result of the subsection is:

THEOREM 3.4.15 (Solomon-Tits argument; cf. proof of Thms 1,2 in [**Sol69**]). *Let G be a finite group of Lie type, with building K having maximal distance N.*

Then the subcomplex $K^{\leq N-1}$ (of chambers not opposite to some fixed chamber σ) is collapsible.

PROOF. If G is of rank 1, then in the language of Remark 2.2.7 there is just one root in the simple system Π, and hence just one in the positive system Φ^+; so for the maximal distance we have $N = |\Phi^+| = 1$. That is, all chambers distinct from σ are in fact opposite to it. So $K^{\leq N-1}$ is just $K^{\leq 0}$, namely the single chamber σ—a simplex, which is certainly collapsible.

Thus in the remainder of the proof, we may assume that G has rank at least 2, and in particular that $N \geq 2$; so that our earlier discussion of paths in the building K has nontrivial content.

Now consider some panel τ with $d(\tau) = N - 1 \geq 1$. Since $K^{\leq N-1}$ contains no chambers at distance N from σ, the gate property 3.4.12 shows that in $K^{\leq N-1}$, there is a unique chamber σ' over τ; that is, σ' is free over τ in $K^{\leq N-1}$.

Of course this means by Lemma 3.4.6 that we can remove σ'; however, since we want to remove all such σ' independently, we must be a little more careful: Notice that the argument shows that σ' is free over *any* of its panels at distance $N - 1$; while at least some panel is at the next smaller distance, namely $N - 2$ (which is non-negative since $N \geq 2$). So we apply a slight extension of Lemma 3.4.6: namely we collapse σ' down, not to *all* panels other than our original τ, but instead to the nonempty subset of panels at distance $N - 2$. This gives us the freedom we wanted, to simultaneously remove all panels σ' at distance $N-1$. In particular, we collapse $K^{\leq N-1}$ down to $K^{\leq N-2}$.

Now we just apply this same argument to collapse $K^{\leq N-2}$ down, and so on: More precisely, we want to replace $N-1$ above with any $1 \leq i \leq N-1$; for given τ at distance i with gate σ', we need $i-1 \geq 0$ to guarantee that we have a nonempty set of panels of τ at the smaller distance $i - 1$.

Thus $K^{\leq N-1}$ collapses down to $K^{\leq 0}$, namely just the single simplex σ. And of course the simplex σ itself is collapsible to a point. $\square$

As mentioned earlier, essentially the above argument was used by Quillen in obtaining the analogue of Lemma 3.3.20 for the building K. However, Quillen's proof summarizes this argument just by mentioning contractibility via "uniqueness of geodesic": this amounts to using the uniqueness of the gate for τ in (3.4.12) roughly to reconstruct a shortest path from τ all the way back to σ. But for our purposes, it will be more instructive below to indicate the collapsibility in a bit more detail, as another example of a multi-step contractibility proof:

LEMMA 3.4.16 ((b) in the proof of 3.1 in [**Qui78**]). *Let G be of Lie type in characteristic p, and K its building. Assume P is a nontrivial p-subgroup of G.*

Then the fixed-point subcomplex K^P is collapsible (hence contractible).

PROOF. Set $L := K^P$. We can embed the p-subgroup P in a Sylow p-subgroup U of G, and we saw in 2.2.4 that U lies in a Borel subgroup B of G; while in Definition 2.2.9, B defines a simplex σ of the building K—and of course B normalizes itself, and hence fixes σ. Thus $\sigma \in K^P = L$. Similarly notice for any simplex $\tau \in L$ that P lies in the stabilizer G_τ, which is a parabolic and so contains some Borel subgroup (again see 2.2.4); so the Sylow argument above also shows that P normalizes some chamber on τ. Thus we conclude that L is in fact a union of chambers.

Next, the action of G on K is simplicial, and so takes paths to paths, preserving length. Since P fixes σ, it must in fact permute the members of each $K^{\leq i}$.

We wish to adapt the Solomon-Tits collapsibility argument in 3.4.15, by applying it to the relevant subcomplexes $L^{\leq i} := L \cap K^{\leq i}$; and we must just be careful about a few details:

First $L \leq K^{\leq N-1}$ for the maximal distance N in K: This is because σ and any opposite σ'' determine some apartment A; and we had mentioned after (2.2.21) that the stabilizer in G of the pair (A, σ) is a Cartan subgroup H, which we saw (again 2.2.4) is a p'-group. (This is where we use the hypothesis that P is a *nontrivial* p-group.)

Now let i be the maximum distance (of any simplex) in L, so that $L = L^{\leq i}$, where we just saw $i \leq N - 1$. If $i = 0$, then $L = L^{\leq 0} = K^{\leq 0}$ is just the simplex σ, which is indeed collapsible; so we may assume that $i \geq 1$.

Our argument above that L is a union of chambers, along with the maximality of i, shows that any simplex in L at distance i is in fact in a chamber σ'' of L, which is also at distance i. As $i \leq 1$, some panel τ of σ'' is at distance $i - 1$; and then the gate σ' for τ in (3.4.12) is also at distance $i-1$. And now as P fixes τ, it permutes the set of chambers on τ; and since we saw that P also permutes $K^{\leq i-1}$, the uniqueness of the gate in $K^{\leq i-1}$ shows that P fixes σ'; that is, we have $\sigma' \in L$. This gives the crucial nonemptiness of the set of chambers at that smaller distance, which is the "target" for the collapsing procedure. Now the rest of the argument for 3.4.15, in particular using $i \leq N - 1$ for freeness over τ, goes through: collapsing $L^{\leq i}$ down to $L^{\leq i-1}$, and so on, down to $L^{\leq 0}$ given by the simplex σ, and then to a point. □

The collapsing procedure can be examined explicitly for instance in Example 3.4.13.

For an alternative proof of Lemma 3.4.16, see the end of Remark 4.4.9.

3.5. (Option G): G-homotopy equivalence and G-contractibility

Of course in this book, most of our posets, complexes, and spaces will admit the action of a group G; and then it is sometimes important (cf. our later Remark 7.3.1) to require that the various relevant homotopies should be G-equivariant.

So in this final appendix-section of the chapter, we give a brief overview of the necessary background material. Fortunately, many of our arguments on "ordinary" homotopy equivalence and contractibility (in this chapter and also in our later development) will extend without real difficulty to the stronger versions defined for a G-action.

The applications of such strengthened results will mainly be to group cohomology, in our later Sections 7.3 and 7.4; and we will indicate more about the role of G-equivalence e.g. at Remark 7.3.1.

For the reader interested in a more detailed treatment: In the particular context of subgroup complexes, many of the standard homotopy equivalences in the literature were shown to extend to G-homotopy equivalences by Thévenaz and Webb in [**TW91**]; this is essentially our default reference for such extended equivalences.

DIGRESSION 3.5.1. We mention that these extended equivalences are particularly relevant for later application to *G-equivariant cohomology*, for example in our later Section 7.3. For just as an ordinary homotopy equivalence will preserve the homology of a space X, a G-equivalence for a G-space X will preserve the G-equivariant homology of X (or equivalently, will preserve the Borel construction of X). We will discuss these concepts briefly at later Remark 7.3.1. But for more general background on G-equivariant cohomology, one standard reference is Bredon [**Bre67**]. A quicker overview of the backround material, adapted for our present context, is provided in Section 2.5 of Benson–Smith [**BS08**]. And see also Section 6.4 in Benson's [**Ben91**]. $\diamondsuit$

Definitions. First we need to make the appropriate (and fairly obvious) extensions of the definitions given in Section 3.1.

REMARK 3.5.2 (G-equivalence). Assume the spaces X and Y admit a G-action.

Maps $f, g : X \to Y$ are *G-homotopic* if there is a homotopy F as in Remark 3.1.1 which is G-equivariant; and then we may write $f \simeq_G g$.

The spaces X and Y are *G-homotopy equivalent* (or just G-equivalent) if there are homotopies of fg and gf with the relevant identity maps as in Remark 3.1.11, which are in fact G-homotopies (i.e., G-equivariant); then we may write $X \simeq_G Y$.

Finally we say X is *G-contractible* if there is a homotopy equivalence of X with a point (as in Remark 3.1.18) which is in fact a G-homotopy equivalence.

We extend these definitions from spaces to posets and complexes as usual. $\diamondsuit$

These definitions are straightforward, and indeed will usually be fairly easy to check; for this reason we adopt:

REMARK 3.5.3 (Our convention of "G-equivariant versions"). Whenever our ordinary equivalence arguments (say for a result numbered x.y) can be easily extended to the G-equivariant form, *we will usually omit the further proof, and just refer to "the G-equivariant form of $x.y$".*

Thus normally we will only discuss G-equivariance arguments explicitly, when there is some special subtlety to deal with. $\diamondsuit$

First as an instance of the easily-extended case, we indicate (in a little more detail than we would usually give):

EXAMPLE 3.5.4 (The G-equivariant form of Proposition 3.1.12). Recall that for an ordinary homotopy equivalence, it was sufficient in 3.1.12 to have just one map f which is homotopic to the identity.

Now for a G-homotopy equivalence, *it suffices to assume also just that the single map f (homotopic to the identity) is in fact G-equivariant.* (At the level of posets, as in (2) of 3.1.12, this extension appears as [**TW91**, 1.2].)

Here is the mild technicality needed, which we would normally omit: The equivariance says $\big(f(x)\big)g = f(xg)$ for $x \in X$ and $g \in G$, so $\big(f(X)\big)G \subseteq f(X)$—hence the image space $f(X)$ also admits a G-action. And now that we know that $f(X)$ is a G-subspace of X, we see that the homotopies in the proof of Proposition 3.1.12 are in fact G-equivariant, since f is. ♢

Next, by contrast with Example 3.5.4, we indicate a situation using a map f (now *not* homotopic to the identity) which is G-equivariant—but which does not induce a G-equivalence. This theme is fairly standard in the literature, and appears e.g. in Warning 2.5.2 and Example 2.5.3 in [**BS08**]; below we will briefly interpret that idea in our present context:

WARNING 3.5.5. The smoothness of the above G-equivariant version of Proposition 3.1.12 might lull us into a false sense of security: we might think similarly that for a contractible G-space X, a map c contracting X to a point p should in fact induce a G-homotopy equivalence: for G acts trivially on p, so that the map c is automatically G-equivariant. However, note that here the original contraction c involves a homotopy of Id_X with the constant map k_p with image p; whereas in Example 3.5.4, we had a homotopy of the original map f with Id_X.

And the analogy does break down: For since we are not assuming that our p lies in X, the proof of Proposition 3.1.12 fails to go through: Indeed when we try to mimic the argument in Example 3.5.4, where the image "$f(X)$" was a G-subspace of X, we see that for a homotopy inverse $r : p \to X$ to be G-equivariant, we would need for $r(p)$ to be fixed by all $g \in G$; whereas X might not contain such a G-fixed point.

Hence for a map, such as c here, which just "induces" a homotopy equivalence, going on to establish G-equivariance is *not* sufficient to guarantee G-equivalence. ♢

A standard method for G-equivalence. We have just seen in Warning 3.5.5 that for a G-homotopy equivalence, in general we will need more than just an ordinary homotopy equivalence induced by a G-equivariant map $f : X \to Y$. One standard additional condition is the requirement that ordinary homotopy equivalence should also hold for the fixed-point subcomplexes X^H and Y^H under all subgroups $H \leq G$. This characterization of G-homotopy equivalence is due to James and Segal; see for example [**Bre67**, Sect. II]. We state it below in the context of CW-complexes (recall Remark 1.2.17), so in particular it applies in our usual context of simplicial complexes and simplicial maps:

THEOREM 3.5.6. *Let $f : X \to Y$ be a G-equivariant cellular map of CW-complexes. Then f induces a G-homotopy equivalence $X \simeq_G Y$ if and only if for each $H \leq G$, $f|_H$ induces an ordinary homotopy equivalence $X^H \simeq Y^H$.* □

Notice that the case $H = 1$ of the condition gives an ordinary homotopy equivalence $X \simeq Y$.

The method in Theorem 3.5.6 above was used by Thévenaz and Webb [**TW91**] to extend many known equivalences to G-equivalences; and correspondingly it applies to extend many of the equivalence proofs in this book. For instance:

EXAMPLE 3.5.7 (The G-equivariant form of Lemma 3.3.5). In the proof of 3.3.5, the condition (3.3.2) was provided by the containments $X \leq XO_p(G) \geq O_p(G)$—for general $X \in \mathcal{S}_p(G)$.

Now given some $H \leq G$, we observe that if H normalizes X (i.e. $X \in \mathcal{S}_p(G)^H$), then it also normalizes $O_p(G)$ and $XO_p(G)$, so that they also lie in $\mathcal{S}_p(G)^H$. Thus the same containments as above, restricted to $\mathcal{S}_p(G)^H$, establish via (3.3.2) the ordinary contractibility of $\mathcal{S}_p(G)^H$.

So using Theorem 3.5.6, we see that when $O_p(G) > 1$ holds, the contractibility of $\mathcal{S}_p(G)$ extends to G-contractibility. $\diamondsuit$

We also recall that within a group G, we may make some construction with regard to a proper subgroup P; and in that case, we might expect that an ordinary equivalence argument would extend only to say a P-equivalence. For instance, the homotopy equivalence in Lemma 3.3.18 for $\mathcal{S}_p(G)_{>P}$ extends to a P-homotopy equivalence; indeed it extends to an $N_G(P)$-homotopy equivalence—this observation appears as [**TW91**, 2.1.i]. Further if we can simultaneously remove a G-orbit of such P, where each individual removal is an $N_G(P)$-equivalence, then the overall removal is a G-equivalence; cf. [**TW91**, 2.2, Thm 1].

We conclude the section with a more methodical listing of results in this chapter which were given G-equivariant extensions in [**TW91**]:

Result	G-version in [**TW91**]
Lemma 3.1.9	1.1
Proposition 3.1.12	1.2
Lemma 3.3.12	2.1.i
Lemma 3.3.19	2.1.ii
Lemma 3.4.10	1.7

CHAPTER 4

Homotopy equivalence

In this chapter, we assume the introductory material on homotopy equivalence developed in earlier Section 3.1, and at various other points in the previous chapter; and we go on to develop a number of more advanced techniques and results on homotopy equivalence for posets and complexes.

And then in our chapters after this one, the main applications of homotopy equivalence will be to replace some initially-given poset by an equivalent one which will be more convenient to work with. Of course because of the equivalence, many basic topological constructions and properties will still hold for the more convenient version. And especially the theme of simplifying computation via a more convenient equivalent poset will be prominent in the next few chapters.

The chapter is organized as follows:

Section 4.1 quickly reviews the classical topological procedure for homotopy via a *contractible carrier*: The purpose of this background is to provide some combinatorial insight into the relevance of posets for the homotopy equivalence method in Quillen's Fiber Theorem.

That result is our initial goal in the chapter, and it will be developed in the subsequent Section 4.2—after which we provide some direct applications of the result. In the sections thereafter, it also leads to some more specialized equivalence methods: namely Bouc's methodical approach to the removal method, in Section 4.3; and Quillen's method of closed sets in a product, in Section 4.4. We also indicate equivalences via the Nerve Theorem, in Section 4.5.

We conclude the chapter with a summary in Section 4.6 of the "standard" homotopy type determined by the p-subgroups poset $\mathcal{S}_p(G)$. (We will frequently focus on its members in the later chapters.)

Here are some historical remarks on the material: A number of the methods and results presented in this chapter were introduced by Quillen in [**Qui78**]. We will also be indicating some further developments made by Kratzer and Thévenaz [**KT85**], Bouc [**Bou84a**], and Walker [**Wal81b**]. (Under Option G in earlier Section 3.5, we had already introduced the theme of Thévenaz and Webb in [**TW91**] of extending equivalence proofs to G-homotopy equivalence—and that thread also will be continued as we proceed through this chapter.) Another reference for some of the equivalence material in this chapter is Benson [**Ben91**, Sec 6.6].

4.1. Topological background: Homotopy via a contractible carrier

In the section following this one, we will discuss (as Theorem 4.2.1) the homotopy equivalence technique in Quillen's Fiber Theorem [**Qui78**, 1.6]. But since the proofs of that result indicated by Quillen in [**Qui78**] are phrased in more modern topological language involving category theory, we have preferred instead to present a proof which is phrased in the more classical topological language of simplicial complexes—since complexes provide our basic context in this book.

And this approach will require a little further topological background—on contractible carriers for complexes, which we briefly indicate in this section. (So the reader not really interested in the details of our proof of Quillen's Fiber Theorem can just skip over the section.)

Our approach primarily follows Walker in [**Wal81b**, Sec. 2]—who emphasizes those parts of the arguments which are essentially combinatorial in flavor. (More classical topological material on carriers can be found e.g. in [**Mun84**, Sec 13] or [**Spa81**, p 112]. A further combinatorial discussion appears in [**Bjö95**, Sec 10].)

Summary. Here is a quick overview of the material in the section: For a simplicial complex K, we wish to "automate" the process of establishing homotopy among continuous maps c from the associated space $|K|$ into another space Y.

For this purpose, we will use a *carrier* C, which will describe the image of c only "up to homotopy": that is, for each simplex σ it will specify a possible "range" $C(\sigma) \subseteq Y$ into which the image $c(|\sigma|)$ should fit.

Now a simplex σ is contractible; so it should seem at least plausible that we also want each target space $C(\sigma)$ to be contractible—in which case, we say that C is a *contractible* carrier. One technical reason for this assumption is the possibility of "building up" our map c, from lower dimensions to higher—using the standard extension property for contractible spaces in Lemma 4.1.4 below. And a further motivation is that in our usual poset context, the cones of form $\mathcal{P}_{\geq x}$ (recall Lemma 3.2.3) will lead very naturally to corresponding constructions of contractible carriers.

Finally, standard arguments using induction on the dimension of simplices will show at Theorem 4.1.7 that all maps c, which are *carried* by C in this way, are indeed homotopic.

An informal introduction to contractible carriers. As initial motivation for the ideas mentioned above, we now quote an analogue in the topological context of the poset equivalence in Proposition 3.1.12.2:

LEMMA 4.1.1 (standard, e.g. from 13.3 in [**Mun84**]). *For a complex K, assume f is a continuous endomorphism of $|K|$, which satisfies:*

$$f(|\sigma|) \subseteq |\sigma| \text{ for all simplices } \sigma \text{ of } K.$$

Then f is homotopic to $\mathrm{Id}_{|K|}$, so f induces a homotopy equivalence $|K| \simeq f(|K|)$.

Our discussion below will just assume this result. (But we will in fact indicate a proof later, in the final subsection of this section. And in the spirit of similar earlier results, we observe: The condition that $f(|\sigma|) \subseteq |\sigma|$ is the obvious analogue of the poset condition "$f \leq \mathrm{Id}_{\mathcal{P}}$" in 3.1.12.2; indeed it is a slight extension (in fact to topological spaces) of the notion of the notion of dominance "$f \subseteq \mathrm{Id}_K$" for simplicial maps, which we had indicated at Remark 3.1.17.)

From the viewpoint of our quick overview above, $|K|$ plays the role of the image space "Y", and the geometric realization $| - |$ (i.e. the convex-hull map $\sigma \mapsto |\sigma|$ on simplices)[1] plays the role of the carrier "C". Thus the above containment hypothesis $f(\sigma) \subseteq |\sigma|$ (where $|\sigma|$ plays the role of the range "$C(\sigma)$") will just say that that f is indeed "carried" by $| - |$ (in the sense of later Definition 4.1.6).

[1] Indeed see e.g. [**Spa81**, p. 112, before 3.1.18] for the original usage of the term "carrier" as a simplex, rather than as a functor defined on simplices.

Definitions related to carriers. As usual, we will focus our exposition on the context of simplicial complexes; though at times we will need to refer to the context of CW-complexes which is more usual in topology.

DEFINITION 4.1.2 (Carriers). A *carrier* from a complex K to a space Y is a mapping:

$$C : \{ \text{ simplices of } K \ \} \to \{ \text{ subspaces of } Y \ \},$$

satisfying the obvious functorial requirement: (for categories in the spirit of earlier Remark 1.2.16):

$$\text{If } \sigma \subseteq \tau \text{ in } K, \text{ then } C(\sigma) \subseteq C(\tau) \text{ in } Y.$$

(We could of course express this in the language of a poset map: from the face poset $\mathcal{P}(K)$ to the analogous "subspace poset $\mathcal{P}(Y)$".) ♢

In particular, the induced inclusion $|\sigma| \subseteq |\tau|$ for convex hulls shows that $|-|$ is indeed a carrier from K to $|K|$; this is of interest for the above context of Lemma 4.1.1.

DEFINITION 4.1.3 (Contractible carrier). A carrier C as in Definition 4.1.2 is *contractible* if:

$$C(\sigma) \text{ is contractible}$$

for each simplex σ of K. ♢

In particular, the contractibility of the convex hull $|\sigma|$ shows that $| - |$ is a contractible carrier; again this is of interest for the context of Lemma 4.1.1.

The importance of the contractibility assumption is that we can then use the following elementary lemma to inductively build up, from one dimension to the next higher dimension, continuous functions such as homotopies which are needed in the Contractible Carrier Theorem 4.1.7.

LEMMA 4.1.4 (Standard, e.g. Cor 13 on p 27 in [**Spa81**]). *Let σ be a simplex, and assume c is a continuous map from the boundary $\partial|\sigma|$ to a contractible space Y.*

Then c can be extended to a continuous map defined on all of $|\sigma|$.

PROOF. (Sketch:) Intuitively, given a point p in the interior of $|\sigma|$, we can locate p on a path between two points on the boundary $\partial|\sigma|$—by following a "line" of a contraction of Y (i.e. the track followed by the images of a boundary point during the contraction); and then we can extend c to p essentially via the corresponding linear combination of the values at those boundary endpoints. □

There are various more general statements, such as extending from a subspace A to the cone on A—cf. [**Rot88**, p 23, Ex 1.7]; in our special case, the simplex σ arises as the cone on $\partial\sigma$ with cone point given by the barycenter of σ.

DIGRESSION 4.1.5 (The viewpoint of homotopy groups). For the reader interested in a bit more detail, we also sketch the more general topological approach—roughly previewing some material which we develop at later Remark 9.1.1: Namely for σ of dimension $n \geq 1$, the obstructions to such an extension from the boundary $\partial\sigma$ (which is a triangulation of a sphere of dimension $n-1$) determine the corresponding *homotopy group* $\pi_{n-1}(Y)$; see our brief review, at Lemma 9.1.2, of this standard result (or cf. [**Spa81**, p. 43]). But since Y is contractible, these homotopy groups are trivial (cf. 9.1.5). And then Lemma 4.1.4 follows from 9.1.2.

Of course, this approach really uses not contractibility itself, but rather the triviality of the homotopy groups. Indeed a consequence of this triviality (cf. later Lemma 9.1.6) is the vanishing of the homology groups; so that $C(\sigma)$ is *acyclic*, in the language of later Definition 5.2.1. Correspondingly, this material is often developed instead in the more general context of homology: For example, our Contractible Carrier Theorem 4.1.7 has a natural analogue given by the Acyclic Carrier Theorem (e.g. [**Mun84**, 13.3])—in which the conclusion is *homology* equivalence rather than homotopy equivalence. ♢

The final definition we need is:

DEFINITION 4.1.6 (Maps carried by a carrier). A carrier C as in Definition 4.1.2 is said to *carry* the continuous map $c : |K| \to Y$ if:

$$c(|\sigma|) \subseteq C(\sigma)$$

for all simplices σ of K. ♢

For instance, the containment hypothesis $f(|\sigma|) \subseteq |\sigma|$ in Lemma 4.1.1 just says that f is carried by the carrier $| - |$. Notice that it is immediate that the identity $\mathrm{Id}_{|K|}$ is also carried by $| - |$.

The Contractible Carrier Theorem. Below we state the Contractible Carrier Theorem essentially in the form given as Lemma 2.1 in Walker [**Wal81b**], who in turn references Lemma 4 of Lakser [**Lak72**].

We mention that later we will only use conclusion (1) (the existence of carried functions) in a fairly technical way (namely for construction of a homotopy inverse during the proof of 4.2.1). Instead conclusion (2) is the part we will quote more frequently, to obtain homotopy—indeed typically even a homotopy equivalence, via a homotopy with the identity as in Proposition 3.1.12.

THEOREM 4.1.7 (Contractible Carrier Theorem). *Assume that C is a contractible carrier from the complex K to the space Y (as in Definition 4.1.3). Then:*

(1) We can construct a continuous map $c : |K| \to Y$ which is carried by C.

(2) Any two continuous maps $c, d : |K| \to Y$ carried by C are homotopic.

PROOF. (Sketch) For (1), we must define c on $|\sigma|$ for each simplex σ of K. In summary, we will start at vertices, and proceed inductively, using Lemma 4.1.4, to simplices of higher dimension.

The base step of our induction occurs when $\dim \sigma = 0$, so that our initial approximation c_0 to c will be defined on vertices v. We can then freely choose $c_0(|v|)$ to be any point in the subspace $C(v) \subseteq Y$—this is required since we want c_0 to be carried by C in the sense of Definition 4.1.6.

Next of course we must extend c_0 to a map c_1 defined on edges E of K. Assume the boundary ∂E is given by vertices v and w. Then by the hypothesis that C is a carrier, Definition 4.1.2 shows that the points $c_0(|v|)$ and $c_0(|w|)$ lie in $C(E)$. So we must define c_1 on the interior of $|E|$ by a "path" in $C(E)$ between these points. And this is possible because of the hypothesis of contractibility of C, so that $C(E)$ is contractible from Definition 4.1.3—for then we can apply Lemma 4.1.4 for the extension. (Cf. our remarks in the sketch-argument for that result, about using the "lines" of a contraction.)

"And so on": For the general inductive step for dimension $r > 0$, assume we have so far constructed c_{r-1} satisfying the "carried" condition 4.1.6 on simplices τ of dimension up to $r-1$, namely $c(|\tau|) \subseteq C(\tau)$. So for σ of dimension r, c_{r-1} is defined on the boundary $\partial|\sigma|$, and in fact

$$(4.1.8) \qquad c_{r-1}(\partial|\sigma|) \stackrel{\text{bydef}}{=} \bigcup_{\tau\in\partial\sigma} c_{r-1}(|\tau|) \stackrel{4.1.6}{\subseteq} \bigcup_{\tau\in\partial\sigma} C(\tau) \stackrel{4.1.2}{\subseteq} C(\sigma).$$

Now $C(\sigma)$ is contractible by hypothesis (again using Definition 4.1.3), so we may apply Lemma 4.1.4 to extend c_{r-1} to c_r defined on all of $|\sigma|$ (i.e. now including the interior). That construction satisfies $c_r(|\sigma|) \subseteq C(\sigma)$; so we see on applying this procedure to all σ of dimension r that we have extended c_{r-1} to c_r which satisfies 4.1.6 up to dimension r.

Continuing by induction, we extend to c defined on all of $|K|$, with c carried by C in the sense of 4.1.6.

For (2), we must construct a homotopy F; that is, we need to define a continuous map $F : [0,1] \times |K| \to Y$ with $F(0,-) = c$ and $F(1,-) = d$. In summary, the construction of F proceeds via induction on dimension, very much as in the proof of (1) above. The main difference is that this time we need the CW-complex versions of the carrier definitions (and of Lemma 4.1.4)—since the Cartesian product $[0,1] \times |K|$ is built up from cells of form $[0,1] \times \sigma$, rather than just from simplices σ.

The boundary of the cell $[0,1] \times \sigma$ will require only a little more discussion than the simplex boundary $\partial\sigma$ above: For in addition to the obvious terms of form $[0,1] \times \tau$ for $\tau \in \partial\sigma$, we also get the terms $0 \times \sigma$ and $1 \times \sigma$. To handle these latter terms, we combine c and d into an auxiliary map H defined just on $\{0,1\} \times |K|$: namely $H(0,-) := c$ and $H(1,-) := d$. Our definition of F will extend from H, adding further cells of the more obvious form $[0,1] \times \sigma$.

Again the base step of the induction has $\dim\sigma = 0$, so that σ is a vertex v of K. Here for the 1-cell $[0,1] \times |v|$, since $\partial|v|$ is empty, the boundary $\partial([0,1] \times |v|)$ consists *only* of the terms $0 \times |v|$ and $1 \times |v|$ of the special form indicated above. So at this step, we only need the auxiliary map H above; we set:

$$F_0(0 \times |v|) := H(0 \times |v|) = c(|v|) \text{ and } F_0(1 \times |v|) := H(0 \times |v|) = d(|v|).$$

By the hypothesis that c and d are both carried by C, these two points lie in $C(v)$. By the contractibility hypothesis on C, $C(v)$ is contractible. And then we can use the CW-complex version of Lemma 4.1.4, in order to extend F_0 to a continuous map $F_1 : [0,1] \times |v| \to C(v)$. Doing this for all v, we obtain a definition of F_1 on all 1-cells of the form $[0,1] \times v$.

For the general inductive step, consider σ with $\dim\sigma = r > 0$. Our inductive assumption is that F_{r-1} has been defined so as to extend H on cells of form $0 \times \sigma$ and $1 \times \sigma$; and on cells of form $[0,1] \times \tau$ for τ of dimension up to $r-1$, so as to satisfy:

$$(4.1.9) \qquad F_{r-1}([0,1] \times \tau) \subseteq C(\tau).$$

Thus our earlier boundary discussion in (4.1.8) is replaced by the slightly more complicated form:

$$\begin{aligned} F_{r-1}\big(\partial([0,1] \times \sigma)\big) &\stackrel{\text{bydef}}{=} H(0 \times |\sigma|) \cup H(1 \times |\sigma|) \cup \textstyle\bigcup_{\tau\in\partial\sigma} F_{r-1}([0,1] \times |\tau|) \\ &\stackrel{4.1.9}{\subseteq} c(|\sigma|) \cup d(|\sigma|) \cup \textstyle\bigcup_{\tau\in\partial\sigma} C(\tau) \stackrel{4.1.6,4.1.2}{\subseteq} C(\sigma). \end{aligned}$$

Now we finish just as before: Namely $C(\sigma)$ is contractible using the CW-complex version of 4.1.3; hence we may apply the CW-complex version of Lemma 4.1.4 to extend F_{r-1} to a continuous map $F_r : [0,1] \times |\sigma| \to C(\sigma)$.

Continuing by induction, we get a continuous map $F : [0,1] \times |K| \to Y$, which (by our construction extending H) satisfies $F(0,-) = c$ and $F(1,-) = d$. So we have obtained the required homotopy. □

Some initial applications of the Contractible Carrier Theorem. Recall that in the context of Lemma 4.1.1 where $f(|\sigma|) \subseteq |\sigma|$, we had observed after Definition 4.1.6 that both f and $\mathrm{Id}_{|K|}$ are carried by contractible $|-|$. Hence by the Contractible Carrier Theorem 4.1.7.2, f is homotopic to $\mathrm{Id}_{|K|}$; and so by Proposition 3.1.12.1, we obtain a proof of the homotopy equivalence in Lemma 4.1.1.

We end the section with one further application—essentially giving a quick preview of a theme from the subsequent Section 4.2: namely that when we focus our initial attention on a poset $\mathcal{P}$ (as opposed to starting with a complex K), the cone subposets of the form $\mathcal{P}_{\geq x}$ will lead us naturally to some useful contractible carriers for the order complex $K(\mathcal{P})$. To this end, we will adopt the following convention:

NOTATION 4.1.10 (The initial element in a poset chain). Given a simplex x in the order complex $K(\mathcal{P})$ of a poset $\mathcal{P}$, we will let x_0 denote the element of $\mathcal{P}$ which is the initial element of the ordered chain x of elements from $\mathcal{P}$.

(That is, we may just write x_0, even when we do not write out in full some expression such as $x = x_0 < x_1 < \cdots < x_n$ with $x_i \in \mathcal{P}$.) ◇

REMARK 4.1.11 (An alternative proof of the poset Homotopy Property 3.1.7). Our earlier discussion of the Homotopy Property 3.1.7 had sketched Quillen's proof given at [**Qui78**, 1.3]. Below we present a more combinatorial proof, essentially following 5.3 in the 1981 M.I.T. thesis [**Wal81a**] of Walker (who in turn references a private communication from Björner).

Assume the hypotheses of Lemma 3.1.7: namely that f and g are poset maps from $\mathcal{P}$ to $\mathcal{Q}$ satisfying $f \leq g$.

Recall from Notation 4.1.10 that for a simplex x in the order complex $K(\mathcal{P})$, $x_0 \in \mathcal{P}$ denotes the initial element of the chain x. We exploit the natural cone subposet $\mathcal{Q}_{\geq f(x_0)}$ to define:

$$C(x) := |\mathcal{Q}_{\geq f(x_0)}|;$$

and we claim that C gives contractible carrier from the complex $K(\mathcal{P})$ to the space $|\mathcal{Q}|$. For suppose that $x \subseteq_{K(\mathcal{P})} y$, so that x is a sub-chain of y, and hence we have $y_0 \leq_{\mathcal{P}} x_0$; then since by hypothesis f is a poset map, we obtain that $f(y_0) \leq_{\mathcal{Q}} f(x_0)$, and so $\mathcal{Q}_{\geq f(x_0)} \subseteq \mathcal{Q}_{\geq f(y_0)}$—which implies, on passing to the geometric realizations, the inclusion $C(x) \subseteq C(y)$ as required for a carrier in 4.1.2. Furthermore the subposet $\mathcal{Q}_{\geq f(x_0)}$ is contractible (a cone, in 3.2.3.2), which gives the contractibility of $C(x)$ required in 4.1.3.

Next $|f|$ and $|g|$ are continuous by 1.4.7.2; and we claim both are carried by C: Namely write out x in full as $x_0 < x_1 < \cdots < x_n$. By hypothesis $f \leq g$, so that $g(x_0) \geq f(x_0)$; and f and g are poset maps, so that $f(x_i) \geq f(x_0)$ and also $g(x_i) \geq g(x_0) \geq f(x_0)$. That is, we have $f(x_i), g(x_i) \in \mathcal{Q}_{\geq f(x_0)}$, so that we get $|f|(|x_i|), |g|(|x_i|) \in |\mathcal{Q}_{\geq f(x_0)}|$; and it follows from the construction of the geometric simplex $|x|$ as essentially the convex hull of the points $|x_i|$ that:

$$|f|(|x|), |g|(|x|) \subseteq |\mathcal{Q}_{\geq f(x_0)}| = C(x),$$

establishing the requirement for being carried by C in 4.1.6.

So now it follows from the Contractible Carrier Theorem 4.1.7.2 that $|f|$ and $|g|$ are homotopic; and of course this is what we mean at the level of posets when we say that f and g are homotopic. ♢

4.2. Equivalences via Quillen's Fiber Theorem

In this section, we will begin by stating Quillen's Fiber Theorem on homotopy equivalences; and we provide a combinatorial proof of the result, based on contractible carriers. Then we present some initial sample applications; and we will continue to use the method at various points in our later development.

Here we are following the fairly common practice of referring to the influential result [**Qui78**, Prop. 1.6] as *Quillen's Fiber Theorem*. It provides in the context of posets a result which is in a tradition of fiber theorems in various other settings in the topological literature.

We digress briefly to discuss the terminology. Classically a *fiber* of a map f is just the preimage $f^{-1}(x)$ of a point x; and for the injectivity condition in an isomorphism, we would want each such fiber to also be a point. But in the more general viewpoint of homotopy theory, a point is replaced by a "homotopy point"—that is, a contractible space, which is homotopy equivalent to a point. And changing our focus from isomorphism to homotopy equivalence, we might now want fibers to be contractible; or indeed more generally, we might want the preimages $f^{-1}(Y)$ of suitable contractible subspaces Y to also be contractible. This is roughly the background for the preimages of the form $f^{-1}(\mathcal{Q}_{\geq y})$ (for elements y of a poset $\mathcal{Q}$) appearing in Quillen's Fiber Theorem 4.2.1.

(In fact in the categorical language of Quillen's "Theorem A" used in his proof of the Fiber Theorem (see [**Qui78**, p 103]), the preimage $f^{-1}(\mathcal{Q}_{\geq y})$ appears as the "fiber of a functor"—namely all $x \in \mathcal{P}$ such that $f(x)$ is connected to y by a morphism of the poset $\mathcal{Q}$ considered as a category (in the spirit of Remark 1.2.16); that is, connected via a poset relation $f(x) \geq y$. However we will instead adhere more closely to our poset viewpoint; and so we will refer to this preimage of the cone $\mathcal{Q}_{\geq y}$ as a *cone fiber*. In particular this will distinguish it from the usual notion of fiber as the preimage of a point.)

Quillen's Fiber Theorem (with a proof via contractible carriers). Here is Quillen's fundamental observation:

THEOREM 4.2.1 (Quillen's Fiber Theorem, 1.6 in [**Qui78**]). *Assume that $\mathcal{P}$ and $\mathcal{Q}$ are posets, and that $f : \mathcal{P} \to \mathcal{Q}$ satisfies*

(a) f is a poset map, and
(b) $f^{-1}(\mathcal{Q}_{\geq z})$ is contractible, for all $z \in \mathcal{Q}$.

Then f induces a homotopy equivalence of $\mathcal{P}$ with $\mathcal{Q}$. (A dual form holds using $\mathcal{Q}_{\leq z}$.)

(Option G:) The G-equivariant form requires G_z-contractibility in (b), to obtain the conclusion of G-homotopy equivalence—this is Theorem 1 in [**TW91**].

PROOF. As discussed at the start of the previous Section 4.1, we are not presenting the original proof of Quillen in [**Qui78**] (since it requires more advanced techniques); instead we are giving a more combinatorial proof, phrased in the language of carriers which we sketched in that Section. Indeed below we essentially follow 2.2 in [**Wal81b**]; the proof is almost entirely formal, and certain details are very similar to parts of the argument of Walker which we gave in our "preview" in Remark 4.1.11 above. During the proof we will construct several contractible carriers.

The first carrier will be used to obtain a continuous map $c : |\mathcal{Q}| \to |\mathcal{P}|$—which will give our homotopy inverse to $|f|$. Recall from 4.1.10 that for a chain y in the order complex $K(\mathcal{Q})$, $y_0 \in \mathcal{Q}$ denotes the initial element of y. As partly suggested by the form of the terms in hypothesis (b), we now exploit the cone subposet $\mathcal{Q}_{\geq y_0}$, and also preimages under f, to define:

$$C(y) := |f^{-1}(\mathcal{Q}_{\geq y_0})|;$$

and we claim that C is a contractible carrier from the complex $K(\mathcal{Q})$ to the space $|\mathcal{P}|$. For suppose $x \subseteq_{K(\mathcal{Q})} y$; this leads, just as in the argument in Remark 4.1.11, to $\mathcal{Q}_{\geq x_0} \subseteq \mathcal{Q}_{\geq y_0}$—so that on applying $|f^{-1}(-)|$, we get the inclusion $C(x) \subseteq C(y)$ required for a carrier in 4.1.2. Furthermore hypothesis (b) with y_0 in the role of "z" gives the contractibility of $C(y)$ required in 4.1.3, establishing the claim. Now (for the only time) we apply the "existence" conclusion (1) in the Contractible Carrier Theorem 4.1.7, to obtain a continuous map $c : |\mathcal{Q}| \to |\mathcal{P}|$—which is carried by C in the sense of 4.1.6, namely:

(4.2.2) $$\text{For } y \text{ a simplex of } K(\mathcal{Q}),\ c(|y|) \subseteq C(y) \overset{\text{bydef}}{=} |f^{-1}(\mathcal{Q}_{\geq y_0})|.$$

To see that any such carried c is a homotopy inverse for $|f|$, we must show that the compositions $|f|c$ and $c|f|$ are homotopic to the corresponding identities. And we will construct further carriers D and E for this purpose.

For $|f|c$, we work with $\mathcal{Q}$. Here we just remove the preimage f^{-1} in our construction of C above—that is, we use only the cone subposet $\mathcal{Q}_{\geq y_0}$, to define:

$$D(y) := |\mathcal{Q}_{\geq y_0}|.$$

And then D is a contractible carrier from $\mathcal{Q}$ to $|\mathcal{Q}|$—again just remove f^{-1} in the proof for C above. We next show that the relevant maps are carried by D: This time we remove the "f" from the argument in Remark 4.1.11: if y is of form $(y_0 < y_1 < \cdots < y_n)$, we get $|y_i| \in |\mathcal{Q}_{\geq y_0}|$, and then $|y| \subseteq |\mathcal{Q}_{\geq y_0}| \overset{\text{bydef}}{=} D(y)$; this is the requirement in 4.1.6 showing that $\mathrm{Id}_{|\mathcal{Q}|}$ is carried by D. And similarly (finally starting to make significant use of the preimage f^{-1} in the definition of C)

we have:

$$\begin{array}{ccccc} |f|(c(|y|)) & \overset{4.2.2}{\subseteq} & |f|(C(y)) & \overset{\text{bydef}}{=} & |f|(|f^{-1}(\mathcal{Q}_{\geq y_0})|) = |f|(|f|^{-1}(|\mathcal{Q}_{\geq y_0}|)) \\ & = & |\mathcal{Q}_{\geq y_0}| & \overset{\text{bydef}}{=} & D(y), \end{array}$$

so we see that $|f|c$ is also carried by D. Hence by the more usual conclusion (2) in the Contractible Carrier Theorem 4.1.7, we obtain $|f|c \simeq \mathrm{Id}_{|\mathcal{Q}|}$.

For $c|f|$, we will work instead with $\mathcal{P}$. Consider a chain x in $K(\mathcal{P})$, having the usual form $(x_0 < x_1 < \cdots < x_n)$. We modify the earlier definition of C by letting $f(x_0)$ play the role of "y_0" there, defining:

$$E(x) := |f^{-1}(\mathcal{Q}_{\geq f(x_0)})|.$$

And then E is a contractible carrier from $\mathcal{P}$ to $|\mathcal{P}|$: Again the proof essentially follows that for C above—but this time, when $x \subseteq_{K(\mathcal{P})} y$ so that $y_0 \leq_{\mathcal{P}} x_0$, we finally use hypothesis (a) that f is a poset map, to obtain $f(y_0) \leq_{\mathcal{Q}} f(x_0)$; this gives the inclusion $\mathcal{Q}_{\geq f(x_0)} \subseteq \mathcal{Q}_{\geq f(y_0)}$, so that as before on applying $|f^{-1}(-)|$ we satisfy the carrier condition 4.1.2 (with hypothesis (b) again giving the contractibility in 4.1.3). So we turn to showing that the relevant maps are carried by E, much as for D above: Again using the poset-map hypothesis (a), we have $f(x_i) \geq f(x_0)$; hence $|f|(|x_i|) \in |\mathcal{Q}_{\geq f(x_0)}|$, and then:

$$|f|(|x|) \subseteq |\mathcal{Q}_{\geq f(x_0)}|;$$

so taking preimages $|f|^{-1}$ we see:

$$|x| \subseteq |f|^{-1}(|f|(|x|)) \subseteq |f|^{-1}(|\mathcal{Q}_{\geq f(x_0)}|) = |f^{-1}(\mathcal{Q}_{\geq f(x_0)})| \overset{\text{bydef}}{=} E(x),$$

as required in 4.1.6 to show that $\mathrm{Id}_{|\mathcal{P}|}$ is carried by E. Similarly since $f(x_0)$ is the minimal element $f(x)_0$ of the chain $f(x)$ in $K(\mathcal{Q})$,[2] we have:

$$\begin{array}{ccccc} c(|f|(|x|)) & = & c(|f(x)|) \overset{4.2.2}{\subseteq} C(f(x)) & \overset{\text{bydef}}{=} & |f^{-1}(\mathcal{Q}_{\geq f(x)_0})| \\ & \underset{f(x)_0 = f(x_0)}{=} & |f^{-1}(\mathcal{Q}_{\geq f(x_0)})| & \overset{\text{bydef}}{=} & E(|x|) \end{array},$$

so that $c|f|$ is also carried by E. Thus again using the Contractible Carrier Theorem 4.1.7.2, we get $c|f| \simeq \mathrm{Id}_{|\mathcal{P}|}$.

This completes the proof that f induces a homotopy equivalence of $\mathcal{P}$ with $\mathcal{Q}$. (Of course for the dual form with $\mathcal{Q}_{\leq z}$, the proof uses the *final* element of a chain y, rather than the initial element y_0 used above.) □

(Option S:) We mention that in their program developing methods for establishing uniqueness of sporadic groups via their geometries, Aschbacher and Segev in [**AS92a**] establish an analogue of the equivalence in the Quillen's Fiber Theorem 4.2.1, in the context of "clique complexes". (In particular Section 4 of that work uses the approach via contractible carriers, as we have done in the first two sections of this Chapter.)

REMARK 4.2.3 (Fiber theorems with other properties as conclusions). We also mention, digressing somewhat from our main theme in this chapter, that Quillen in [**Qui78**] also uses the viewpoint of fibers (i.e. the preimages $f^{-1}(\mathcal{Q}_{\leq y})$) in other contexts: where hypotheses on the fibers weaker than contractibility then lead to a conclusion not of homotopy equivalence, but rather to the transfer from $\mathcal{Q}$ back to $\mathcal{P}$ of other corresponding weaker structural properties.

---

[2] More precisely we should write $K(f)(x)$ for this chain, using the map $K(f)$ induced by f at the level of the order complexes; but we use the abbreviation $f(x)$ here to simplify notation.

For instance, 7.6 in [**Qui78**] shows that "locally" assuming that pre-images are *n-connected* (i.e. have trivial homotopy groups in dimensions through n; see later Remark 9.1.3) suffices to show "globally" that n-connectedness for $\mathcal{Q}$ is equivalent to that for $\mathcal{P}$.

Similarly [**Qui78**, 9.1] is an extension of 7.6 above, in which the n-connectedness condition is roughly replaced by *n-sphericality* (i.e. $(n-1)$-connectedness, for a poset of dimension n).

Furthermore since Cohen-Macaulayness (see later Remark 9.1.9) is roughly n-sphericality defined inductively over all links, Quillen at [**Qui78**, 9.7] shows that assuming appropriate-dimensional Cohen-Macaulayness in fibers suffices to transfer that structure from $\mathcal{Q}$ back to $\mathcal{P}$. (This technical result is used in the proof of Theorem 9.4.5.)

These fiber-theorem methods have been used and extended in many other ways in the subsequent literature. They are relevant for example in Aschbacher's work on simple connectivity, e.g. as mentioned in the discussion of the solvable case after the statement of Theorem 9.4.5; and in the work of Lehrer-Rylands and of Vogtmann, mentioned at the end of the subsection on Cohen-Macaulayness of buildings in Section 9.4. $\diamondsuit$

Application of the Fiber Theorem to equivalence with $\mathcal{S}_p(G)$. We can now start to establish the promised equivalences with $\mathcal{S}_p(G)$ of certain subposets introduced earlier; we begin with $\mathcal{A}_p(G)$:

THEOREM 4.2.4 (2.1 in [**Qui78**]). *$\mathcal{A}_p(G)$ is homotopy equivalent to $\mathcal{S}_p(G)$. (Indeed equivalence holds for any intermediate poset $\mathcal{A}_p(G) \subseteq \mathcal{P} \subseteq \mathcal{S}_p(G)$.)*

(Option G:) The conclusion of G-homotopy equivalence is [**TW91**, Thm 2.i].

PROOF. Let i denote the inclusion of the subposet $\mathcal{A}_p(G) \subseteq \mathcal{S}_p(G)$, and take $P \in \mathcal{S}_p(G)$; then $i, \mathcal{A}_p(G), \mathcal{S}_p(G), P$ will play the roles of "$f, \mathcal{P}, \mathcal{Q}, z$" in the Fiber Theorem 4.2.1: Notice that $i^{-1}\left(\mathcal{S}_p(G)_{\leq P}\right)$ is just $\mathcal{A}_p(G)_{\leq P} = \mathcal{A}_p(P)$. And we had established contractibility of $\mathcal{A}_p(P)$ in the previous chapter at Lemma 3.3.14. Consequently i induces a homotopy equivalence by 4.2.1. (We could replace $\mathcal{S}_p(G)$ by any intermediate $\mathcal{P}$.) $\square$

We will see a different approach to the proof, via the removal method of Bouc, in the next section at Remark 4.3.3.

REMARK 4.2.5 (Using the equivalence in 4.2.4 for alternate proofs). Let P be a nontrivial p-group.

We consider P in the role of "G" in the Quillen Fiber Theorem 4.2.1 (ignoring for a moment how we proved that result). Then $\mathcal{A}_p(P)$ is homotopy equivalent to $\mathcal{S}_p(P)$. And of course $\mathcal{S}_p(P)$ has unique maximal element P, and so is contractible by Lemma 3.2.3.2. So the equivalence gives contractibility also of $\mathcal{A}_p(P)$—and we get an alternative proof of Lemma 3.3.14. Of course, to avoid circularity in this deduction, we could not then use our proof above of 4.2.4, which quotes 3.3.14 along the way; we would need a different proof of 4.2.4, e.g. that given in later Remark 4.3.3.

(Option G:) Now we recall that the G-homotopy equivalence $\mathcal{A}_p(G) \simeq_G \mathcal{S}_p(G)$ (i.e. the G-equivariant form of Theorem 4.2.4) gives via Theorem 3.5.6 an ordinary homotopy equivalence of fixed points $\mathcal{A}_p(G)^P \simeq \mathcal{S}_p(G)^P$. We saw in Lemma 3.3.20 that $\mathcal{S}_p(G)^P$ is contractible; so the equivalence gives contractibility also of $\mathcal{A}_p(G)^P$—and we get an alternative proof of Lemma 3.4.1. (Or, arguing in the other direction, we could use 3.4.1 to prove 3.3.20.) ◇

4.3. Equivalences via simultaneous removal

In our earlier discussion which led up to $\mathcal{B}_p(G)$ in Definition 3.3.15, we had mentioned that Bouc analyzed the equivalence of $\mathcal{S}_p(G)$ with $\mathcal{A}_p(G)$ roughly as an extension of the fiber approach in Quillen's Theorem 4.2.1—by in fact considering essentially repeated applications of the method of removals (which we began to introduce at Lemma 3.4.10). In this section, we indicate certain aspects of that analysis.

As background, we first examine how the fiber and removal methods in fact essentially co-incide, in the case where we want to remove just a *single* element x from a poset $\mathcal{P}$:

In the fiber viewpoint of Quillen's Theorem 4.2.1, we take $\mathcal{P} \setminus \{x\}$, $\mathcal{P}$ in the roles of "$\mathcal{P}, \mathcal{Q}$", with the inclusion i in the role of "f". Then for the fiber we have $i^{-1}(\mathcal{P}_{\geq z}) = (\mathcal{P} \setminus \{x\})_{\geq z}$. For $z \neq x$, this subposet has unique minimal element z, and so is contractible by Lemma 3.2.3.2. Thus for Quillen's hypothesis, we only need to verify contractibility of $(\mathcal{P} \setminus \{x\})_{\geq x} = \mathcal{P}_{>x}$.

And of course this condition is also sufficient for the removal of x, in the removal method via Lemma 3.4.10.2.

So we can regard the multiple-removal method as iterating a special case of Quillen's Fiber Theorem. But in our exposition in this section, it will be more natural to emphasize the viewpoint of iterating the single-removal method given in Lemma 3.4.10.2.

The method of simultaneous removals. In fact Bouc as well as several other authors asked essentially the question about iteration indicated above: or, viewed in terms of overall outcome, whether one can *simultaneously* remove all such x satisfying the above contractibility condition.

This is not necessarily obvious: since the removal of one x might conceivably alter the necessary property for the next y; that is, we must then check for contractibility of $(\mathcal{P} \setminus \{x\})_{>y}$, whereas we had originally assumed it instead for $\mathcal{P}_{>y}$. However, the key will be, in the viewpoint of Lemma 3.4.10.2, to assume the contractibility of the subposets of the form $\mathcal{P}_{>x}$ for *all* x (rather than mixing in some cases of $\mathcal{P}_{<y}$, as would also be allowed in that Lemma). For then we will see in the proof of Theorem 4.3.2 below that if we order our removals by always taking away a minimal possible element, we can continue—without altering the needed contractibility in the process.

Versions of Theorem 4.3.2 appeared in Kratzer-Thévenaz [**KT85**, Prop 1.6] and Bouc [**Bou84a**, Prop 4]. A later exposition of this material appears in [**Ben91**, Sec 6.6].

To describe the result of such simultaneous removals, we introduce a (not necessarily standard) shorthand notation:

DEFINITION 4.3.1. For a poset $\mathcal{P}$, set
$$\mathcal{P}^{>} := \{x \in \mathcal{P} : \mathcal{P}_{>x} \text{ is not contractible }\}.$$
Define $\mathcal{P}^{<}$ dually using $\mathcal{P}_{<x}$. ◊

THEOREM 4.3.2 (Kratzer-Thévenaz/Bouc). *Assume that $\mathcal{P}$ is a finite poset. Then for $\mathcal{P}^{<} \subseteq \mathcal{Q} \subseteq \mathcal{P}$, the inclusions induce homotopy equivalences. (And dually for $\mathcal{P}^{>}$).*

(Option G:) The G-equivariant version appears as [**TW91**, Prop 1.7]; it removes (G-orbits of) elements z such that $\mathcal{P}_{>z}$ is in fact G_z-contractible, and the conclusion is G-homotopy equivalence.

PROOF. Of course it suffices to consider the general inclusion $\mathcal{Q} \subseteq \mathcal{P}$.

Let x denote an element of $\mathcal{P} \setminus \mathcal{Q}$, chosen minimal in the restriction of the ordering $<$ to $\mathcal{P} \setminus \mathcal{Q}$. Now $\mathcal{P}_{>x}$ is contractible by our hypothesis $\mathcal{P}^{>} \subseteq \mathcal{Q}$. So by Lemma 3.4.10.2, $\mathcal{P}_1 := \mathcal{P} \setminus \{x\}$ is homotopy equivalent to $\mathcal{P}$.

So we next examine what is involved in repeating the argument: We choose y minimal in the restriction of $<$ to $\mathcal{P}_1 \setminus \mathcal{Q}$. The earlier minimal choice of x guarantees that $x \notin (\mathcal{P}_1)_{>y}$; it follows that $(\mathcal{P}_1)_{>y} \overset{\text{bydef}}{=} (\mathcal{P}\setminus\{x\})_{>y} = \mathcal{P}_{>y}$ is again contractible by hypothesis. And then similarly we get $\mathcal{P}_2 := \mathcal{P}_1 \setminus \{y\} = \mathcal{P} \setminus \{x, y\}$ homotopy equivalent to $\mathcal{P}_1$ and hence to $\mathcal{P}$.

Continuing in this way through further $\mathcal{P}_i$, we shrink $\mathcal{P}$ homotopically down to $\mathcal{Q}$. □

We mention that with a little more care, the result extends to a poset $\mathcal{P}$ which is infinite, but has all chains of finite length; see [**Bou84a**, Prop 4].

Applications of simultaneous removal to equivalence with $\mathcal{S}_p(G)$. Here is one application already promised in the previous chapter:

REMARK 4.3.3 (An alternative proof of Quillen's equivalence 4.2.4). Recall for $P \in \mathcal{S}_p(G)$ that Lemma 3.3.12 shows that $\mathcal{S}_p(G)_{<P} = \mathcal{S}_p(P)_{<P}$ is contractible if and only if P is *not* elementary abelian; that is,
$$\mathcal{S}_p(G)^{<} = \mathcal{A}_p(G).$$
So as a consequence of Theorem 4.3.2, we get $\mathcal{A}_p(G) \simeq \mathcal{S}_p(G)$—that is, we obtain another proof of Quillen's equivalence Theorem 4.2.4.

(Option G: We mention that the G-equivariant version of 4.2.4 (at [**TW91**, Thm 2.i]) is proved using simultaneous removal—namely via the G-equivariant version of 4.3.2 which is given at [**TW91**, 1.7].)

Notice also that $\mathcal{A}_p(G)$ is contained in the poset of nontrivial abelian (but not necessarily elementary) p-subgroups of G—call this poset say $\mathcal{A}$. Then Theorem 4.3.2 also shows that $\mathcal{A}$ is homotopy equivalent to $\mathcal{S}_p(G)$. This abelian poset also sometimes appears in the literature. ◊

A further innovation of Bouc was to "dually" discuss the poset $\mathcal{S}_p(G)^{>}$: Namely we consider some $P \in \mathcal{S}_p(G)\setminus\mathcal{B}_p(G)$; then by Quillen's observation in Lemma 3.3.19, $\mathcal{S}_p(G)_{>P}$ is contractible—so that contrapositively:
$$\mathcal{S}_p(G)^{>} \subseteq \mathcal{B}_p(G).$$
(Indeed Quillen's Conjecture 3.3.8 would give equality here.) So as another consequence of Theorem 4.3.2 we get:

THEOREM 4.3.4 (Cor, p 50, [**Bou84a**]). *The posets $\mathcal{B}_p(G)$ and $\mathcal{S}_p(G)$ are homotopy equivalent.* □

(Option G:) The conclusion of G-homotopy equivalence is [**TW91**, Thm 2.i].

We conclude the section by indicating one further relative of $\mathcal{S}_p(G)$, which has been used at some points in the literature:

REMARK 4.3.5 (Benson's poset $\mathcal{Z}_p(G)$). Benson continued the removal theme of Bouc, by introducing at [**Ben91**, Exercise, p 231] the following subposet of $\mathcal{A}_p(G)$:

$$\mathcal{Z}_p(G) := \{ E \in \mathcal{A}_p(G) : E = \Omega_1 O_p Z\big(C_G(E)\big) \}.$$

For assume that $E \in \mathcal{A}_p(G) \setminus \mathcal{Z}_p(G)$, so that $E < \Omega_1 O_p Z\big(C_G(E)\big) =: Z$, and hence $Z \in \mathcal{A}_p(G)_{>E}$. Now for any $X \in \mathcal{A}_p(G)_{>E}$, we see X lies in $C_G(E)$ and hence centralizes Z, so that $XZ \in \mathcal{A}_p(G)_{>E}$ also. Thus we may apply (3.3.2) with $X \leq XZ \geq Z$, to conclude that $\mathcal{A}_p(G)_{>E}$ is conically contractible (to Z). It follows that

$$\mathcal{A}_p(G)^{>} \subseteq \mathcal{Z}_p(G);\text{ and hence } \mathcal{Z}_p(G) \text{ is homotopy equivalent to } \mathcal{A}_p(G).$$

(This uses Theorem 4.3.2—and of course we also get equivalence with $\mathcal{S}_p(G)$. Indeed in Option G, we get G-homotopy equivalence via the usual extension methods.)

For an explicit example, recall we computed $\mathcal{A}_2(D_8)$ in earlier Example 3.3.10. Here we have $Z := \langle (13)(24) \rangle = \Omega_1 Z(D_8)$; it follows that $Z \in \mathcal{Z}_2(D_8)$—and indeed that $Z \leq \Omega_1 O_2 Z\big(C_{D_8}(E)\big)$ for any $E \in \mathcal{A}_2(D_8)$, so that $\mathcal{Z}_2(D_8) \subseteq \mathcal{A}_2(D_8)_{\geq E}$. In fact equality holds in this case: for the two members E of $\mathcal{A}_2(D_8)_{\geq E}$ other than Z (these E have order 4) are self-centralizing in D_8.

We saw in Remark 4.3.3 that $\mathcal{A}_p(G) = \mathcal{S}_p(G)^{<}$, so $(\mathcal{S}_p(G)^{<})^{>} \subseteq \mathcal{Z}_p(G)$. (I am not aware of any other work in the literature considering iterations of the $<$- and $>$-removal constructions.) ◇

4.4. Equivalences via closed sets in products

In this section, we indicate another equivalence method which essentially extends from the Fiber Theorem 4.2.1. We will see that this more specialized method of Quillen, namely closed sets in products, can be particularly useful in comparing a poset of *subgroups* of G with a more general G-poset (or G-complex).

The method of closed sets in products. We will expand slightly on the treatment starting before 1.7 in Quillen [**Qui78**].

As an initial overview, the idea will be to establish the homotopy equivalence of posets $\mathcal{P}$ and $\mathcal{Q}$, proceeding via equivalence of each with a suitable "intermediate" subposet $\mathcal{R}$ of the Cartesian-product poset $\mathcal{P} \times \mathcal{Q}$ (recall Definition 3.1.4). The fundamental property of the intermediate subposet will be:

DEFINITION 4.4.1 (Closed subset of a poset). A subset $\mathcal{R}$ of a poset $\mathcal{P}$ is *(downward-)closed* if

$$\text{whenever } x \in \mathcal{R} \text{ and } x' \leq_{\mathcal{P}} x, \text{ then also } x' \in \mathcal{R}.$$

One can define upward-closed sets in the obvious dual way. (The term *order ideal* is also used in the combinatorial literature, e.g. [**Sta86**, p 100].) ◇

For instance, a cone-subposet $\mathcal{P}_{\leq x}$ is obviously downward-closed. More generally a closed set $\mathcal{R}$ containing x must contain $\mathcal{P}_{\leq x}$, and so $\mathcal{R}$ is the union of the downward cones determined by its maximal elements.

We now restrict attention to the case where our $\mathcal{R}$ is in fact closed in a product poset $\mathcal{P} \times \mathcal{Q}$. We will wish to exploit the preimages appearing in Quillen's Fiber Theorem 4.2.1, determined by the projection maps from $\mathcal{R}$ to the two factors (for example, $\text{proj}_{\mathcal{P}}^{-1}(\mathcal{P}_{\geq x})$ for $x \in \mathcal{P}$). And we will see shortly in Lemma 4.4.3 that these cone fibers in the subset $\mathcal{R}$ of the product can be conveniently replaced by certain closed sets in the factors—defined as follows using just the ordinary "point" fibers (such as $\text{proj}_{\mathcal{P}}^{-1}(x)$):

DEFINITION 4.4.2 (The sets R_x and R_y). For projection maps proj_P and proj_Q defined on a subset $\mathcal{R}$ of the product poset $\mathcal{P} \times \mathcal{Q}$, for $x \in \mathcal{P}$ and $y \in \mathcal{Q}$ define

$$\mathcal{R}_x := \text{proj}_{\mathcal{Q}}\big(\text{proj}_{\mathcal{P}}^{-1}(x)\big) = \{y \in \mathcal{Q} : (x,y) \in \mathcal{R}\},$$
$$\mathcal{R}_y := \text{proj}_{\mathcal{P}}\big(\text{proj}_{\mathcal{Q}}^{-1}(y)\big) = \{x \in \mathcal{P} : (x,y) \in \mathcal{R}\}.$$

If $\mathcal{R}$ is closed in $\mathcal{P} \times \mathcal{Q}$, then one checks directly that $\mathcal{R}_x$ is closed in $\mathcal{Q}$, and $\mathcal{R}_y$ is closed in $\mathcal{P}$. ◊

The desired relation of cone fibers to point fibers is obtained using just the basic Homotopy Property 3.1.7:

LEMMA 4.4.3 (implicit in the proof of Prop 1.7 of [**Qui78**]). *Define projection maps on a subset $\mathcal{R}$ which is downward-closed in the poset product $\mathcal{P} \times \mathcal{Q}$.*

For $x \in \mathcal{P}$, $\text{proj}_{\mathcal{Q}}$ induces a homotopy equivalence of the preimage $\text{proj}_{\mathcal{P}}^{-1}(\mathcal{P}_{\geq x})$ with $\mathcal{R}_x = \text{proj}_{\mathcal{Q}}\big(\text{proj}_{\mathcal{P}}^{-1}(x)\big)$. (And analogously for $y \in \mathcal{Q}$ and $\text{proj}_{\mathcal{P}}$.)

PROOF. Set $\mathcal{S}_x := \text{proj}_{\mathcal{P}}^{-1}(\mathcal{P}_{\geq x})$. Then an element of $\mathcal{S}_x$ has form $(x', y) \in \mathcal{R}$ with $x' \geq x$. So $(x,y) \leq_{\mathcal{P}\times\mathcal{Q}} (x',y)$. And now we make crucial use of the hypothesis that the set $\mathcal{R}$ is closed: Namely we get that $(x,y) \in \mathcal{R}$ also; and hence $y \in \mathcal{R}_x$. Thus proj_Q does indeed map (x',y) into $\mathcal{R}_x$—while $(x,-) : y \mapsto (x,y)$ maps $\mathcal{R}_x$ back into $\mathcal{S}_x$; and both are visibly poset maps. Furthermore

$$(x,-) \circ \text{proj}_Q : (x',y) \mapsto (x,-)(y) = (x,y) \leq (x',y),$$

so that $(x,-)\circ\text{proj}_Q \leq \text{Id}_{\mathcal{S}_x}$, and hence these maps are homotopic by Lemma 3.1.7.1. Finally the reverse composition $\text{proj}_Q \circ (x,-)$ just equals $\text{Id}_{\mathcal{R}_x}$, completing the homotopy equivalence. □

Now the Fiber Theorem gives Quillen's equivalence method via "closed sets in products":

THEOREM 4.4.4 (Prop 1.7,Cor 1.8 in [**Qui78**]). *Define projection maps on a subset $\mathcal{R}$ which is downward-closed in the poset product $\mathcal{P} \times \mathcal{Q}$.*

(1) If $\mathcal{R}_x$ is contractible for all $x \in \mathcal{P}$, then $\mathcal{P}$ is homotopy equivalent to $\mathcal{R}$.

(2) If also $\mathcal{R}_y$ is contractible for all $y \in \mathcal{Q}$, then $\mathcal{P}$ is homotopy equivalent to $\mathcal{Q}$.

(Option G: I am not sure if the conclusion of G-homotopy equivalence is explicitly stated in the literature—but it follows from the usual extension method in [**TW91**, 1.7] with the hypotheses of G_x- and G_y-contractibility.)

PROOF. For (1), the contractibility of $\mathcal{R}_x$ in the hypothesis implies contractibility of the homotopy fiber $\text{proj}_{\mathcal{P}}^{-1}(\mathcal{P}_{\geq x})$ by the equivalence in Lemma 4.4.3. So Quillen's Fiber Theorem 4.2.1 gives the equivalence of $\mathcal{P}$ with $\mathcal{R}$.

For (2), apply (1) to get equivalence also of $\mathcal{Q}$ with $\mathcal{R}$; and compose with the original result of (1), to get equivalence of $\mathcal{P}$ with $\mathcal{Q}$. □

The closed set of stabilizing pairs. Quillen applies the closed-set method in a number of rather similar situations, which are also essentially visible in the earlier work of Brown. Some of these common features are emphasized in the recent book of Adem-Milgram [**AM04**, p 174], developing the notion of the "singular set" of a G-complex. We will discuss that development later, at Definition 4.4.13—within a more general context for G-complexes, which we now introduce.

Namely we will explicitly axiomatize a concept which seems only implicit in Quillen's work and the subsequent literature: namely a certain very naturally-defined closed set—which turns out to be particularly useful for comparing a poset of *subgroups* of G with a G-complex.

Recall from our earlier Remarks 1.5.7 and 2.3.3 that we usually expect our G-action on a complex K to be *type-preserving*, in that the stabilizer of a simplex in fact fixes all vertices of that simplex. And indeed that property will be important in the situation below, so that we in fact explicitly assume it in:

EXAMPLE 4.4.5 (Stabilizing pairs). Assume that

(1) $\mathcal{Q}$ is a poset of nontrivial subgroups of G (i.e. a subposet of the subgroup lattice $L(G)$); and

(2) G has a type-preserving action on a complex K (so the face poset $\mathcal{P}(K)$ is a G-poset).

In particular we recall from Remark 1.5.7 that the type-preserving condition in (2) holds when K is the order complex $K(\mathcal{P})$ of a G-poset $\mathcal{P}$. In that case, the face poset $\mathcal{P}\big(K(\mathcal{P})\big)$ is just the barycentric subdivision $\mathrm{Sd}(\mathcal{P})$ (Definition 1.6.8) of the original poset $\mathcal{P}$.

Now in the product $\mathcal{Q}\times\mathcal{P}(K)$ with the face poset, define the subset of *stabilizing pairs* as:

$$\mathcal{S} := \{(H,\sigma) : H \leq G_\sigma\}.$$

Equivalently of course we could re-phrase the above condition as "H stabilizes σ" or "$\sigma \in K^H$".

This set appears for example in the proof of [**Qui78**, 4.1]. Indeed Quillen indicates there that the set is closed; but the verification of this property is left implicit. So since we saw that closure is fundamental for Lemma 4.4.3, and hence for the results that rely on it, we exhibit this verification in detail: Namely assume that $(J,\tau) \leq (H,\sigma) \in \mathcal{S}$. We have $J \leq H$ from the definition of the product poset, while $H \leq G_\sigma$ from the definition of the stabilizing pairs $\mathcal{S}$. Similarly from the product poset, we get $\tau \subseteq \sigma$ in the face poset $\mathcal{P}(K)$; and we now have $G_\sigma \leq G_\tau$—here crucially using our type-preserving hypothesis (2). Thus:

$$J \leq H \leq G_\sigma \leq G_\tau,$$

so that $J \leq G_\tau$, and hence $(J,\tau) \in \mathcal{S}$ from its definition, as required for closure.

Now that we have seen that $\mathcal{S}$ is closed in the product, we get an especially natural form for the corresponding closed sets in the factors from Definition 4.4.2, namely:

$$\mathcal{S}_H = \mathcal{P}(K^H) \qquad \text{and} \qquad \mathcal{S}_\sigma = \mathcal{Q}_{\leq G_\sigma}. \tag{4.4.6}$$

Finally, to explain our assumption of *non*trivial subgbroups in (1) above, note that otherwise 1 would be the unique minimal element in $\mathcal{Q}$—so that the goal of homotopy equivalence would in fact be to show that K is also contractible. And then proceeding via the closed set $\mathcal{S}$ in Theorem 4.4.4.2 would in particular require establishing contractibility of $\mathcal{S}_1 = \mathcal{P}(K^1)$, namely of K itself—i.e. the method would in that case provide no real reduction. $\diamondsuit$

REMARK 4.4.7 (The relevance of fixed points to equivalence). Thus in the above setup of stabilizing pairs $\mathcal{S}$, to establish homotopy equivalence of $\mathcal{Q}$ and K via Quillen's closed-set method 4.4.4.2, it suffices to show contractibility of the structures K^H and $\mathcal{Q}_{\leq G_\sigma}$ appearing in (4.4.6).

Now often contractibility for $\mathcal{Q}_{\leq G_\sigma}$ will be fairly easy: For example if we have $G_\sigma \in \mathcal{Q}$, then $\mathcal{Q}_{\leq G_\sigma}$ is a cone on G_σ. Or more frequently, we will take $\mathcal{S}_p(G)$ as "$\mathcal{Q}$"—and then whenever $O_p(G_\sigma) > 1$, we can apply Lemma 3.3.5 for contractibility of $\mathcal{S}_p(G)_{\leq G_\sigma} = \mathcal{S}_p(G_\sigma)$. (This corresponds with observation (a) in Quillen's proof of his equivalence [**Qui78**, 3.1]—we will give this result below, as Theorem 4.4.8.)

Thus the emphasis typically shifts to establishing contractibility for the fixed-point subcomplex K^H. This finally begins to establish the connection between contractible fixed points and equivalence, which we had mentioned in our earlier discussion—leading up to the various fixed-point contractibility results starting at Lemma 3.3.20. (Indeed contractibility for K^H corresponds to observation (b) in Quillen's proof of [**Qui78**, 3.1].)

And in fact in Webb's later analysis—applying to more general G-complexes, and establishing various properties beyond homotopy equivalence—it will even suffice (e.g. in hypothesis (b) of Theorem 6.3.4; cf. Lemma 6.3.3) to obtain contractibility of fixed points K^H just for H of order exactly p; we had already mentioned this starting at Example 3.3.23, and we will return to the theme for example at later Remark 6.3.1 $\diamondsuit$

Applications to equivalence of simple-group geometries with $\mathcal{S}_p(G)$. In applications of equivalence methods in this chapter so far, we have focused on equivalence of various other posets with $\mathcal{S}_p(G)$. Now the closed-sets method, notably via the stabilizing-pairs closed set $\mathcal{S}$ of Example 4.4.5, will allow us to compare $\mathcal{S}_p(G)$ with further G-posets, and especially G-complexes K: we will concentrate here on geometries for simple groups G, which we began to introduce in Chapter 2.

(Option B:) Equivalence with the building. We begin with the "model case" of Section 2.2: namely when G is a group of Lie type over a field of characteristic p, we consider the Tits building K for G. (Elementary examples are of course provided by the projective spaces of Section 2.1.)

We have in fact already developed, at various earlier points, the main subcomponents of the proof of Quillen's fundamental observation:

THEOREM 4.4.8 (Thm 3.1 in [**Qui78**]). *Assume that G is of Lie type in characteristic p. Then the building of G is homotopy equivalent to $\mathcal{S}_p(G)$.*

(Option G:) For the conclusion of G-homotopy equivalence, see [**TW91**, p. 180]; we will expand a little on their observation in Remark 4.4.9 below.

PROOF. Let K denote the building for G. We take this K, and $\mathcal{S}_p(G)$ in the role of "$\mathcal{Q}$", in the stabilizing-pairs set $\mathcal{S}$ of Example 4.4.5. Recall we saw (e.g. in

Definition 2.2.36) that the G-action on K is type-preserving, as required for (2) there; while $\mathcal{S}_p(G)$ consists of nontrivial subgroups, as needed for (1) there. Hence the set $\mathcal{S}$ is closed in the product $\mathcal{S}_p(G) \times \mathcal{P}(K)$ with the face poset of K.

Now for a simplex σ in the building K, we saw at Lemma 2.2.11.2 that the stabilizer G_σ is a (proper) parabolic subgroup, say of form G_J; and at Remark 2.2.4 that $O_p(G_J) = U_J > 1$. Thus we may apply Quillen's result Lemma 3.3.5 to see that $\mathcal{S}_p(G_J) = \mathcal{S}_p(G_\sigma)$ is contractible—and $\mathcal{S}_p(G_\sigma)$ gives the set $\mathcal{S}_\sigma = \mathcal{S}_p(G)_{\leq G_\sigma}$ corresponding to a cone fiber in (4.4.6).

Next for $P \in \mathcal{S}_p(G)$, the relevant subposet in (4.4.6) is $\mathcal{S}_P = \mathcal{P}(K^P)$, i.e. the face poset of the fixed-point subcomplex K^P; and we established the contractibility of K^P at Lemma 3.4.16. (The elementary case of K given by a projective space had in fact been covered earlier in Lemma 3.3.22.)

Thus we have the hypotheses of the closed-set method in Theorem 4.4.4.2, from which we obtain the equivalence of $\mathcal{S}_p(G)$ with the face poset $\mathcal{P}(K)$, and hence with the building K. □

This result is of course an important instance of the theme in our Building Principle 2.0.3.

REMARK 4.4.9 (The characterization of $\mathcal{B}_p(G)$ in Lie type G via unipotent radicals). A different approach to the equivalence of $\mathcal{S}_p(G)$ with the building K in Theorem 4.4.8 above is provided by Theorem 2.3 in [**TW91**]:

For that result shows that:

(4.4.10) For Lie type G in characteristic p, the unipotent radicals give $\mathcal{B}_p(G)$.

That is, we get the conjugates of the subgroups U_J mentioned in the proof of 4.4.8 above.

Now it is standard that the 1:1 correspondence $U_J \leftrightarrows G_J$ in fact defines a poset anti-isomorphism: between the poset of unipotent radicals (giving $\mathcal{B}_p(G)$), and the poset of parabolic subgroups—which gives the building K in our original Definition 2.2.9. And then we can compose the resulting equivalence, with the homotopy equivalence of $\mathcal{B}_p(G)$ and $\mathcal{S}_p(G)$ from earlier Theorem 4.3.4.

(Option G:) We had already mentioned the extension to G-homotopy equivalence indicated at [**TW91**, p. 180]. Indeed since that proof of the G-equivalence does not quote Lemma 3.4.16, we can get an alternative proof of the contractibility of K^P in that Lemma, in the spirit of earlier Remark 4.2.5: For by the G-equivalence we can use Theorem 3.5.6 to get ordinary homotopy equivalence of K^P with $\mathcal{S}_p(G)^P$; and $\mathcal{S}_p(G)^P$ is contractible by Lemma 3.3.20. ◊

(Option S:) Equivalence with with a sporadic geometry. In the light of Theorem 4.4.8 for the building, according to our Vague Sporadic Principle 2.0.4 we should also try to establish analogues for sporadic geometries, such as those we introduced in Section 2.3.

And for the first and smallest case in that Section, once again we have already established at various earlier points most of the subcomponents of the desired equivalence proof:

EXAMPLE 4.4.11 (Equivalence of the triples geometry with $\mathcal{S}_2(S_5)$). We consider the triples geometry K for $G = S_5$ of earlier Example 2.3.5—using the external-complex form: where the vertices are defined by the points and triples from $\{1,2,3,4,5\}$, with the obvious inclusion. We take this K, and $\mathcal{S}_2(G)$ in the role of "$\mathcal{Q}$", in the setup of stabilizing pairs $\mathcal{S}$ in Example 4.4.5. The members of $\mathcal{S}_2(G)$ are nontrivial subgroups, as required for (1) there; and the G-action on K obviously preserves the two vertex types (points and lines), as required for (2) there. So the set $\mathcal{S}$ of stabilizing pairs is closed in the product $\mathcal{S}_2(G) \times \mathcal{P}(K)$ with the face poset $\mathcal{P}(K)$.

For a simplex σ, we must establish contractibility of $S_\sigma = \mathcal{S}_2(G)_{\leq G_\sigma} = \mathcal{S}_2(G_\sigma)$. The simplex types are represented up to conjugacy by $\sigma = 5$, 345, and $5 \in 345$; and we see respectively that $G_\sigma \cong S_4$, $S_2 \times S_3$, and $S_2 \times S_2$; in each case $O_2(G_\sigma) > 1$, so that $\mathcal{S}_2(G_\sigma)$ is contractible by Quillen's Lemma 3.3.5.

Next for $P \in \mathcal{S}_2(G)$, we must show that $\mathcal{S}_P = \mathcal{P}(K^P)$ is contractible, and of course we work directly with K^P. In fact in Example 3.3.25, we had verified this contractibility in the case of P of order exactly 2: namely for P generated by $z = (12)(34)$ and $x = (12)$. For larger 2-subgroups P, representatives up to conjugacy are given by $P = \langle z, x\rangle$, $\langle z, (13)(24)\rangle$, and $\langle (1324)\rangle$ of order 4, and a Sylow group $\langle x, (1324)\rangle \cong D_8$. The first of these has K^P equal to K^z, which we saw is contractible; while for the others, K^P is already just a single point, namely 5.

Thus we have the hypotheses for the closed-set method in Theorem 4.4.4.2, from which we obtain the equivalence of $\mathcal{S}_2(G)$ with the face poset $\mathcal{P}(K)$, and hence with the triples geometry K.

However, we conclude with a warning: If we replace $G = S_5$ by A_5 in the discussion above, then equivalence does *not* hold: for example $\mathcal{S}_2(S_5)$ is connected, whereas $\mathcal{S}_2(A_5)$ has 5 connected components. And correspondingly, in trying to mimic the above proof, we would still have K^P contractible for the subcollection of groups P which arise in the subgroup A_5; but the argument would break down for instance where the stabilizer in A_5 of the triple 345 is isomorphic just to S_3—where that group has $O_2(S_3) = 1$. ◇

REMARK 4.4.12 (Some other sporadic equivalences and nonequivalences). And indeed more generally we comment, with regard to our Vague Sporadic Principle 2.0.4, that the analogue of the building equivalence in Theorem 4.4.8 does sometimes fail for various sporadic geometries. One example is given by the C_3 geometry for A_7—which we had introduced at Example 2.3.7. Here, just as for A_5 above, attempting an equivalence proof via closed sets would fail: on one hand, we still get contractibility of K^P (as we saw for P of order 2 in Example 3.4.11); but on the other hand, we get certain vertex stabilizers G_v, namely given by A_6 or $L_3(2)$ when v is a point or plane, which are simple groups—and these in particular have $O_2(G_v) = 1$. (Indeed a proof that equivalence does *not* hold is indicated at later Example 6.3.7.)

Equivalence does still turn out to hold for the other geometries we indicated in Section 2.3, namely for $U_4(3)$ and M_{24} at Examples 2.3.10 and 2.3.11: In fact this was essentially established[3] by contractibility proofs for K^P in [**RSY90**]—since in those cases the G_σ are 2-local subgroups, so that $O_2(G_\sigma) > 1$. This observation about the work of [**RSY90**] is part of the development in [**SY97**], which made a more methodical study of equivalence of sporadic geometries with $\mathcal{S}_p(G)$. And 1.4

[3]The equivalence for M_{24} was first demonstrated by Mark Ronan; cf. our discussion following Theorem 9.3.12.

there is a specialized version of the closed-sets method in Theorem 4.4.4.2. ◇

However, the A_7 example above does show that the situation in Option S may not always proceed by the most direct analogies with buildings. In fact, what often still holds is that a sporadic geometry K will be equivalent to some useful standard poset associated to G, *other* than $\mathcal{S}_p(G)$ and its relatives. We will return to this theme later—notably in Section 4.6, and in Remark 7.4.11.

Application to the singular set of a G-action. We now indicate a contractibility result, namely Proposition 4.4.15 below—which we present here, rather than in the previous Chapter 3 on contractibility, since its standard proof at [**Qui78**, 4.1] uses the closed-set equivalence method introduced in this section.

In fact Quillen's argument at [**Qui78**, 4.1] applies to the order complex of $\mathcal{S}_p(G)$, along with the specific subgroup P of G given by a Sylow p-subgroup. In order to be able to use the method elsewhere, we first abstract a part of the argument which applies to a more general complex and subgroup. In summary, we will wish to extend contractibility from the individual fixed-point sets for certain subgroups to the union of those sets. Correspondingly we define (cf. [**AM04**, p 174]):

DEFINITION 4.4.13 (The singular set of a G-complex). Assume G is type-preserving on a complex K, and $H > 1$ is some subgroup of G. Then the *singular set* of H on K is the subcomplex given by the union of fixed-points of nontrivial subgroups:

$$K^{H\text{-sing}} := \bigcup_{1<J\leq H} K^J.$$

Notice that we could equivalently describe $K^{H\text{-sing}}$ as the union of the simplices σ of K such that $H_\sigma > 1$. And using this viewpoint, we emphasize that $K^{H\text{-sing}}$ is not necessarily an *induced* subcomplex: For example we might have vertices v, w with $H_v, H_w > 1$, where $v - w$ gives an edge in K; but if $H_v \cap H_w = 1$, v and w will *not* define an edge in the subcomplex $K^{H\text{-sing}}$.

Finally note that the nontrivial subgroups J of H are the members of the subposet $L(H)_{>1}$ of the subgroup lattice of H; of course this subposet is a cone on its unique maximal element H. ◇

We mention that a main purpose of collecting the fixed points into the singular set is that H will then act freely on the remainder of the complex K; this will be useful at various later points, beginning with (5.3.6). The result on extending contractibility is:

LEMMA 4.4.14 (implicit in the proof of 4.1 in [**Qui78**]). *Assume G is type-preserving on a complex K, and $H > 1$ is a subgroup of G whose action satisfies:*

K^J is contractible for all $J \in L(H)_{>1}$, i.e. for all $1 < J \leq H$.

Then the singular set $K^{H\text{-sing}}$ is homotopy equivalent to the cone $L(H)_{>1}$, and hence contractible.

PROOF. We take $L(H)_{>1}$ and $K^{H\text{-sing}}$ in the roles of "$\mathcal{Q}, K$" in the construction of the set $\mathcal{S}$ of stabilizing pairs in Example 4.4.5. We are using nontrivial subgroups of H, as required for (1) there; and our hypothesis of type-preserving action on K is inherited by $K^{H\text{-sing}}$, as needed for (2) there. Thus the set $\mathcal{S}$ is closed in the product $L(H)_{>1} \times \mathcal{P}(K^{H\text{-sing}})$ with the face poset $\mathcal{P}(K^{H\text{-sing}})$.

Now for $J \in L(H)_{>1}$, we have $\mathcal{S}_J = \mathcal{P}\big((K^{H\text{-sing}})^J\big) = \mathcal{P}(K^J)$, which is contractible by hypothesis. On the other hand, for $\sigma \in K^{H\text{-sing}}$, we saw $H_\sigma > 1$ in the Definition 4.4.13 of the singular set. Thus H_σ is itself a member of our poset $L(H)_{>1}$, giving the easy case for contractibility that we had mentioned in Remark 4.4.7: namely $\mathcal{S}_\sigma = \big(L(H)_{>1}\big)_{\leq H_\sigma}$ which is a cone on its unique maximal element H_σ, and hence contractible (indeed it is just $L(H_\sigma)_{>1}$). Thus we have the hypotheses of the closed-set method in Theorem 4.4.4, which then gives the equivalence of $L(H)_{>1}$ with the face poset $\mathcal{P}(K^{H\text{-sing}})$, and hence with $K^{H\text{-sing}}$. □

Now as a special case, we obtain as Proposition 4.4.15 below the result originally appearing as 4.1 in [**Qui78**] (in the case where "P" is in fact a Sylow p-subgroup of G). For earlier versions of the basic idea in Brown's work, cf. [**Bro75**, Lemma, Sec 2] and [**Bro74**, Proof of Thm 5, p 244].

PROPOSITION 4.4.15 (4.1 in [**Qui78**]). *Assume that P is a nontrivial p-subgroup of G. Then the P-singular subset* $\mathcal{S}_p(G)^{P\text{-sing}} = \bigcup_{Q \in \mathcal{S}p(P)} \mathcal{S}_p(G)^Q$ *is homotopy equivalent to the cone $\mathcal{S}_p(P)$, and hence contractible.*

PROOF. We take P in the role of "H" in Lemma 4.4.14, so in particular $\mathcal{S}_p(P)$ plays the role of "$L(H)_{>1}$". And we take the order complex[4] $K(\mathcal{S}_p(G))$ in the role of "K" there—giving the type-preserving property for the G-action, as we had observed in Remark 1.5.7. The hypothesis of contractibility of *individual* fixed-point sets $K\big(\mathcal{S}_p(G)\big)^Q = K\big(\mathcal{S}_p(G)^Q\big)$ is given by our earlier result Lemma 3.3.20 for $\mathcal{S}_p(G)^Q$; so Lemma 4.4.14 gives the desired equivalence with $\mathcal{S}_p(P)$. □

We close the section by briefly mentioning the combinatorial notion of a *Galois connection* between posets $\mathcal{P}$ and $\mathcal{Q}$: namely a pair of order-reversing maps in each direction, such that the composites dominate the respective identity maps. Of course this domination makes the two maps homotopy inverses, in our viewpoint in Lemma 3.1.7. However, Walker in [**Wal81b**, start of Sec 4] mentions a proof of this homotopy equivalence for a Galois connection, via the method of closed sets: where that method appears as Theorem 2.5 in his treatment. He also indicates a further application of Galois connections to Möbius functions in his Theorem 4.1. (The notion of a Galois connection goes back to work of Rota [**Rot64**].)

4.5. Equivalences via the Nerve Theorem

In this section, we present at 4.5.2 a poset version of the Nerve Theorem from classical topology; indeed we will regard this version as yet another equivalence method which could be derived via Quillen's Fiber Theorem 4.2.1 However, this material is not really essential reading for our later development—since equivalence

---

[4] The requirement of type-preserving action is the reason for working (e.g. as Quillen does in [**Qui78**, 4.1]) with the barycentric subdivision $\mathcal{P}(K(\mathcal{S}_p(G)))$, rather than trying to directly use the original poset $\mathcal{S}_p(G)$, in the closed-set method of Theorem 4.4.4. (I thank David Green for clarification here.)

via the Nerve Theorem does not seem to be heavily used in the literature on subgoup complexes.

Nonetheless, we have several reasons for introducing the material at this particular point. First, we will be able to extend our recurring theme in this chapter of equivalences with $\mathcal{S}_p(G)$: by presenting some equivalences (observed by Alperin using the nerve approach) with certain other complexes which had earlier arisen in finite group theory—independently of the development of subgroup complexes. Second, there are various points in the literature on subgroup complexes where the nerve viewpoint seems at least implicit: among other examples, we will indicate some remarks of Thévenaz in this direction.

Background: Contractible Covers and the Nerve Theorem. We now briefly sketch, for the context of posets, some of the basic concepts involved in the Nerve Theorem. In overview, the idea is that given a *contractible cover* C of a poset $\mathcal{P}$ by suitable subposets, $\mathcal{P}$ will then be homotopy equivalent with the *nerve* $\mathcal{N}(C)$—a complex determined by the nonempty intersections of members of C.

For the more classical development of the material in the context of topological spaces, see for example Section 73 of Munkres [**Mun84**] on Čech cohomology, particularly [**Mun84**, Thm 73.2]; or Spanier [**Spa81**, p 152]. A more combinatorial approach appears in Björner [**Bjö95**, 10.6,10.7]; there are historical references to the topological work of Leray and Weil.

DEFINITION 4.5.1 (The nerve of a cover). A *cover(ing)* C of a finite poset $\mathcal{P}$ is a finite collection $C_i (i \in I)$ of subsets of $\mathcal{P}$, such that $\mathcal{P} = \cup_{i\in I} C_i$: that is, each $x \in \mathcal{P}$ lies in some member C_i of C. In fact typically we want the cover C to be *closed*: namely each subset C_i should be (downward) closed[5] in the sense of Definition 4.4.1

A cover is *contractible* if, whenever for some $J \subseteq I$ we have $C_J := \cap_{i \in J} C_i$ nonempty, then in fact C_J is a contractible subposet of $\mathcal{P}$.

For example, if I denotes the set of elements of $\mathcal{P}$ which are maximal in the ordering $<$ on $\mathcal{P}$, then the collection $C := \{\mathcal{P}_{\leq x} : x \in I\}$ is a closed cover. Furthermore each individual $\mathcal{P}_{\leq x}$ is a cone—and indeed the cover C will be contractible if, whenever a subset $J \subseteq I$ has some lower bound in $\mathcal{P}$, it in fact has a (unique!) greatest lower bound x_J: for then $\cap_{x\in J} \mathcal{P}_{\leq x}$ is the cone $\mathcal{P}_{\leq x_J}$.

A cover C determines, via the intersections $C_J := \cap_{i\in J} C_i$ of its members, an abstract simplicial complex on the vertex set I, called the *nerve* of C:

$$\mathcal{N}(C) := \{J \subseteq I : C_J \neq \emptyset\}.$$

For the subset-closure condition on a simplex in Definition 1.2.5, notice that if we have $\emptyset \neq K \subseteq J$, then the intersection C_K over the smaller set K contains C_J, and so is also nonempty. ◊

In our context of posets, we can obtain our version of the Nerve Theorem as a corollary of Quillen's Fiber Theorem 4.2.1:

[5] This restriction guarantees that the union property for $\mathcal{P}$ extends to the analogous property for the order complex, namely $K(\mathcal{P})$ is the union of the $K(C_i)$; and then the $|C_i|$ are also closed sets in the topological space $|\mathcal{P}|$. Now classically for a cover of a topological space, the subsets C_i would be open. Our use of closed sets is one reason for requiring the collection C to be finite.

THEOREM 4.5.2 (Nerve Theorem). *Assume that C is a contractible closed cover of the poset $\mathcal{P}$. Then $\mathcal{P}$ is homotopy equivalent to the nerve $\mathcal{N}(C)$ of C.*

PROOF. We will abuse notation slightly by letting $\mathcal{N}(C)$ be an abbreviation also for the face poset $\mathcal{P}(\mathcal{N}(C))$ of the nerve. Then we define a map $I_* : \mathcal{P} \to \mathcal{N}(C)$ by $x \mapsto I_x := \{i \in I : x \in C_i\}$; of course I_x is just the smallest simplex J in the nerve such that $x \in C_J$.

If the cover C is in fact *downward*-closed, we can without loss replace $\mathcal{P}$ by its opposite poset—so in any case, we can work under the hypothesis that C is *upward*-closed. Now assume $x < y$. Then if $i \in I_x$ so that $x \in C_i$, since C is upward-closed we get also $y \in C_i$ so that $i \in I_y$. Thus $I_x \subseteq I_y$, and hence I_* is a poset map. (For downward-closed C we would get an order-*reversing* map.)

Now for some $J \in \mathcal{N}(C)$, so that $C_J \neq \emptyset$, we check that the homotopy fiber $I_*^{-1}(\mathcal{N}(C)_{\geq J})$ is just the intersection C_J: For if x lies in this preimage, then we have $I_x \supseteq J$; thus for $i \in J$ we have $i \in I_x$ and hence $x \in C_i$—so that in fact x lies in the intersection C_J. And these steps are reversible; so we conclude that the preimage is exactly C_J.

But C_J is contractible, by our hypothesis that the cover C is contractible; so the equivalence follows from the Fiber Theorem 4.2.1. □

Some applications of the Nerve Theorem. First we continue our running theme of equivalences with $\mathcal{S}_p(G)$.

REMARK 4.5.3 (Some covers of $\mathcal{S}_p(G)$ and $\mathcal{A}_p(G)$). Using the viewpoint of covers, we now indicate some complexes which had arisen independently in finite group theory. We define some notation just for convenience, since there seem to be no standard names in the literature. We begin with $\mathcal{S}_p(G)$ in the role of our poset "$\mathcal{P}$".

(1) For our first cover C, as the index set I we take the set $\mathrm{Syl}_p(G)$ of all the Sylow p-subgroups of G—namely the maximal elements of $\mathcal{S}_p(G)$; and given some $P \in \mathrm{Syl}_p(G)$, we take $C_P := \mathcal{S}_p(G)_{\leq P}$. Then C is a downward-closed cover of $\mathcal{S}_p(G)$. Furthermore for a subset $J \subseteq \mathrm{Syl}_p(G)$, the set-intersection C_J is nonempty precisely when the corresponding Sylow-group intersection $I_J := \cap_{P \in J} P$ is nontrivial—and then C_J is just the cone $\mathcal{S}_p(G)_{\leq I_J} = \mathcal{S}_p(I_J)$; hence the cover C is in fact contractible. We will denote the nerve of this covering by $I_p(G)$, and will refer to it as the *complex of Sylow intersections*. We mention also that the collection of intersections I_J determines the *poset* of Sylow intersections, as a subposet of $\mathcal{S}_p(G)$; it also has the covering C, leading to the same nerve complex $I_p(G)$.

Sylow intersections arise in various places in finite group theory: for example, for control of G-conjugacy within a Sylow p-subgroup via normalizers of Sylow intersections, see e.g. Alperin [**Alp67**] (and Quillen's discussion in [**Qui78**, Sec 6]); or for defect groups in modular representation theory as Sylow intersections, see Green's theorem [**Fei82**, Thm 8.14], or [**Tho67**].

(2) We consider more briefly the obvious dual of the construction in (1) above: Namely now for I we instead take the *minimal* elements of $\mathcal{S}_p(G)$: namely the subgroups P of order exactly p—for convenience we will just denote these by $\mathcal{S}_p(G)_{\min}$. And we take $C_P := \mathcal{S}_p(G)_{\geq P}$; so that this time C is instead an *upward*-closed cover of $\mathcal{S}_p(G)$. For a subset $J \subseteq \mathcal{S}_p(G)_{\min}$, the set intersection C_J is nonempty when the group-theoretic span $S_J = \langle P : P \in J \rangle$ is contained in some p-group—and hence

that span itself is a member of $\mathcal{S}_p(G)$, so that C_J is the cone $\mathcal{S}_p(G)_{\geq S_J}$, and hence the cover C is contractible. We will refer to the nerve of this covering as $J_p(G)$.

(3) However, there is considerable use of a variant of (2) above, applied instead when the poset $\mathcal{P}$ is $\mathcal{A}_p(G)$: Again for the index set I we take the elements of order p: that is, $\mathcal{A}_p(G)_{\min} = \mathcal{S}_p(G)_{\min}$; of course now for $P \in \mathcal{A}_p(G)_{\min}$ we take $C_P := \mathcal{A}_p(G)_{\geq P}$—to get an upward-closed cover of $\mathcal{A}_p(G)$. This time for a subset $J \subseteq \mathcal{A}_p(G)_{\min}$, the set-intersection C_J is nonempty when the span S_J lies in an *elementary abelian* p-group: This occurs precisely when all the $P \in J$ mutually commute; and then S_J itself is a member of $\mathcal{A}_p(G)$, so that C_J is the cone $\mathcal{A}_p(G)_{\geq S_J}$—and hence the cover C is contractible. We will denote the nerve of this covering by $C_p(G)$, and refer to it as the *commuting complex on subgroups of order p*.

Commuting complexes (generalized from the initial case of commuting graphs) arise in many places in finite group theory. In the classification of finite simple groups, they are relevant to the basic construction of *signalizer functors*; see e.g. [**GLS96**, p 126]. For other applications see e.g. Aschbacher [**Asc93**] on simple connectivity (which we discuss briefly in later Section 9.3), and [**Asc94**, Ch 12] on application to uniqueness proofs for simple groups; also see [**PY01**] on duality between commuting and noncommuting complexes. ◊

These coverings and their nerves are discussed by Alperin in his (unpublished) U. Chicago lecture notes [**Alp89**]; a summary appears in [**Alp90**] as:

THEOREM 4.5.4 (Thm 3 in Alperin [**Alp90**]). *The poset $\mathcal{S}_p(G)$ is homotopy equivalent with the Sylow intersection complex $I_p(G)$, the dual complex $J_p(G)$ on subgroups of order p, and the commuting complex $C_p(G)$ on subgroups of order p.*

(Option G: The conclusion of G-homotopy equivalence follows via the usual extension methods.)

PROOF. In Remark 4.5.3, we obtained each of these as the nerve of a contractible closed cover of $\mathcal{S}_p(G)$ (or of $\mathcal{A}_p(G)$—which is course is equivalent to $\mathcal{S}_p(G)$ by Theorem 4.2.4). Hence the desired equivalences follow by the Nerve Theorem 4.5.2. □

We close the section with remarks about some other situations where the Nerve Theorem is (or could be) relevant.

Jacques Thévenaz read an early draft of this book, and made numerous very useful suggestions. In particular, he commented that the contractibility of the singular set in Lemma 4.4.14, which we had proved using the closed-set method in Theorem 4.4.4.2, can also be established by a direct use of the Nerve Theorem: As our poset $\mathcal{P}$ we take the (face poset of the) singular set $K^{H\text{-sing}}$. As index set I, we take $L(H)_{>1}$; and for $J \in L(H)_{>1}$ we set $C_J := K^J$. Since $K^{H\text{-sing}}$ is just the union of these K^J, we see this defines a cover C of the singular set. The cover is downward-closed by the type-preserving action of G in the hypothesis of 4.4.14: that is, if $\sigma \subseteq \tau \in K^J$, then also $\sigma \in K^J$. Finally to check contractibility of C: Assume $X \subseteq L(H)_{>1}$—say X consists of $J_1, \ldots, J_n$, and set $J_X := \langle J_i : J_i \in X \rangle$. (In particular we include the case $n = 1$, i.e. X is a single member of $L(H)_{>1}$.) We see that

$$C_X = \cap_{J_i \in X} C_{J_i} = \cap_{J_i \in X} K^{J_i} = K^{J_X};$$

and by the hypothesis of 4.4.14, this set is contractible—so that the cover C is contractible. Thus by the Nerve Theorem 4.5.2, the singular set $K^{H\text{-sing}}$ is homotopy equivalent to the nerve $\mathcal{N}(C)$. But the vertices of the nerve are just the members of $L(H)_{>1}$. And we saw above that for *any* subset X of these vertices, the intersection C_X is assumed contractible—and so in particular is nonempty; so that any X defines a simplex of the nerve. For example, we can take X given by *all* the vertices in $L(H)_{>1}$; and we see that the nerve as a complex is just a simplex, so it is certainly contractible.

REMARK 4.5.5 (Connection with some earlier constructions). Next we mention (within our Option B) that our earlier alternative Definition 2.2.36 of the bulding of a group G of Lie type, as a "complex of maximal parabolics", was implicitly using the viewpoint of the Nerve Theorem:

For let $\mathcal{P}$ denote the poset of parabolic subgroups of G, giving our original view of the building in Definition 2.2.9. As our index set I, we take the subset $\mathcal{P}_{\max}$ of maximal elements, i.e. the maximal parabolics. For $P \in \mathcal{P}_{\max}$ we set $C_P := \mathcal{P}_{\leq P}$. Then this C gives a downward-closed cover of $\mathcal{P}$.

Furthermore for a subset $J \subseteq \mathcal{P}_{\max}$ such that $C_J = \cap_{P \in J} C_P$ is nonempty, this set-intersection consists of certain members P_i of $\mathcal{P}$. In particular a Borel subgroup of some P_i lies in all the $P \in J$—and from the description in earlier Remark 2.2.4 of the parabolics over a Borel, we see that the group-intersection $P_J := \cap_{P \in J} P$ is itself a parabolic. It follows that C_J is the cone $\mathcal{P}_{\leq P_J}$, so that the cover C is contractible.

Thus in fact the Nerve Theorem 4.5.2 gives a proof of homotopy equivalence for the two definitions above of the building—in effect, an alternative to earlier Lemma 2.2.37.

(Option S:) Much the same remark applies to our generalization of Definition 2.2.36, to the notion of an intersection complex in Definition 2.3.1. Here the index set I stands for *all* "maximal parabolics"; whereas there the type set I was just the smaller set of representatives of conjugacy classes of those maximals.

And this nerve viewpoint is also in the background of the similar constructions such as Definition 6.3.2 in [**BS08**]; indeed often that work carries out constructions in the explicit language of various ostensibly-different subgroup posets, which often can be checked to have the same nerve. $\diamondsuit$

Thévenaz also observed that the equivalence of $\mathcal{S}_p(G)$ with the Sylow intersections $I_p(G)$ can be estabished in alternative ways via methods from our earlier sections. For example, let us abuse notation slightly by letting $I_p(G)$ denote the *poset* of Sylow intersections. Define $f : \mathcal{S}_p(G) \to I_p(G)$ by $f(P) := \cap_{P \leq Q \in \mathrm{Syl}_p(G)} Q$, and let g denote the inclusion of $I_p(G)$ in $\mathcal{S}_p(G)$. In summary, f and g are poset maps, with $fg = \mathrm{Id}_{I_p(G)}$ and $gf \geq \mathrm{Id}_{\mathcal{S}_p(G)}$. Then use Lemma 3.1.7 to complete the proof of equivalence. Or–apply the Fiber Theorem 4.2.1; the cone fiber $g^{-1}(\mathcal{S}_p(G)_{\geq P})$ will have a unique minimal element given by $f(P)$ above.

Finally we mention that in the combinatorial literature, the *crosscut* construction for a poset $\mathcal{P}$ determines a covering, and an associated nerve equivalent to $\mathcal{P}$; see e.g. Björner [**Bjö95**, 10.8].

4.6. Summary: The "standard" homotopy type determined by $\mathcal{S}_p(G)$

In this chapter on homotopy equivalence methods, our applications have been mainly focused on equivalence with the poset $\mathcal{S}_p(G)$ of nontrivial p-subgroups. So in this final section, we will pause briefly to summarize what we have accomplished in this direction—and also we then briefly preview some other related directions that we will pursue in later chapters.

REMARK 4.6.1 (The homotopy type of $\mathcal{S}_p(G)$—and its G-homotopy type). By now we have seen that the homotopy type of $\mathcal{S}_p(G)$ contains (at least) the posets:

$\mathcal{S}_p(G)$, $\mathcal{A}_p(G)$, $\mathcal{B}_p(G)$, $\mathcal{Z}_p(G)$, $\mathcal{A}$ (abelian p-groups);

and also the complexes:

$I_p(G)$, $J_p(G)$, $C_p(G)$; and the building for G of Lie type in characteristic p.

(The relevant statements are at 4.2.4, 4.3.4, 4.3.5, 4.3.3, 4.5.4, and 4.4.8. Indeed we will indicate one further equivalent complex $N_p(G)$ in Remark 4.6.2 below.)

Of course this means that all these members of the homotopy type will share basic topological invariants such as Euler characteristic, homology and homotopy groups, etc; see e.g. later Remark 5.1.15.

(Option G:) We had also indicated the extension of those equivalence results to G-homotopy equivalences—so that all these posets and complexes in fact lie in the G-homotopy type of $\mathcal{S}_p(G)$. In view of Theorem 3.5.6, this further implies, for each X in the type and each subgroup $1 < H \leq G$, that we get an ordinary homotopy equivalence of fixed points $X^H \simeq \mathcal{S}_p(G)^H$. In particular we get the analogue for X of contractible fixed points under a p-group in Lemma 3.3.20. These facts are in turn important for later applications such as G-equivariant cohomology in Section 7.3. (We will return to this point in a moment.) $\diamondsuit$

The thrust of Remark 4.6.1 is that very many natural constructions of posets and complexes will end up just being (G-)equivalent to $\mathcal{S}_p(G)$. And in fact, Webb had observed that the analogue, for some poset X, of the contractible fixed-points property in 3.3.20 would *almost* be enough to guarantee equivalence of X with $\mathcal{S}_p(G)$.

However, we will see later that there are also some natural posets which are *not* in general homotopy equivalent to $\mathcal{S}_p(G)$—but are still important for various applications (notably the group cohomology of G in Chapter 7). So in later Remark 7.4.11, we will collect together $\mathcal{S}_p(G)$ along with these other useful homotopy types (due for example to Jackowski-McClure [**JM89**] and Jackowski-McClure-Oliver [**JMO92**]) into a standard "super-type". We will briefly preview here some of that later discussion.

We mention for example that the "strengths" of the various equivalences which do, and do not, in general hold among the posets in Remark 7.4.11 are methodically determined in Grodal-Smith [**GS06**].

(Option G, continued:) The use, for the group cohomology of G, of the posets X in Remark 7.4.11 typically involves the G-equivariant cohomology of X. This can be approached via the Borel construction of K—which we discuss briefly in Remark 7.3.1, and which in particular is preserved by G-homotopy equivalence.

(Option S:) For G a sporadic group, we had mentioned earlier that the work of Ryba-Smith-Yoshiara [**RSY90**] essentially determined which sporadic geometries fall into the standard homotopy type of Remark 4.6.1, with $\mathcal{S}_p(G)$. These equivalences were more methodically investigated, using a version of the closed-set method of Theorem 4.4.4, in Smith-Yoshiara [**SY97**]. But those papers also encountered some sporadic geometries which are *not* equivalent to $\mathcal{S}_p(G)$; so in those cases, it was desirable to establish equivalence with one of the other natural posets in the super-type given in Remark 7.4.11. Indeed this process had to wait while the theory of those posets was being developed; but finally the work of Benson-Smith [**BS08**] then showed for *any* sporadic group (and $p = 2$) that a suitable p-local geometry is homotopy equivalent to one of the posets in 7.4.11.

We also mention some extensions, of equivalence methods such as the above, made by Sawabe in [**Saw03**]; primarily for use with sporadic groups via the collection of centric radical subgroups of Definition 7.4.8. Equivalence methods are also developed by Maginnis and Onofrei in [**MO08**]—in particular (in the spirit of [**GS06**]) relating their collection of "distinguished" p-subgroups with the standard collections in the literature; further applications appear in [**MO10**].

Chapter appendix on normal p-chains. We close the chapter with mention of a member of the standard homotopy type in Remark 4.6.1, which did not fit so naturally into our overall organization of the material earlier in the chapter:

REMARK 4.6.2 (Normal chains of p-groups and the chain-pairing method). Robinson introduced (see e.g. [**KR89**, 2.3]) the subcomplex $N_p(G)$ of the order complex $K(\mathcal{S}_p(G))$ given by the "normal" chains: i.e. chains $P_0 < P_1 < \cdots < P_n$ with $P_i \in \mathcal{S}_p(G)$, where in fact each P_i is normal in the final term P_n.

Robinson in effect established equivalence of this complex with $\mathcal{S}_p(G)$—for the purposes of computations involving alternating sums from consecutive dimensions of chain groups or homology groups (we give a brief discussion of these concepts from homological algebra in Section 5.1)—by the method of "chain-pairing" (cf. [**KR89**, proof of 3.3]): roughly, to remove non-normal chains, show that they arise in natural pairs in adjacent dimensions, which then cancel out in the alternating sum. This method provides a very elementary combinatorial viewpoint, which in particular avoids much of the more topological material that we have sketched in our development in this book. (A recent application of the method, establishing contractibility for the orbit structure of the "fusion system" for a finite p-group, appears in [**Lin09**]; we briefly mention fusion systems in later Remark 7.1.3.)

(Option G:) The extension of Robinson's equivalence to G-homotopy equivalence appears as Theorem 2.ii of [**TW91**], with a proof based on the usual methods from this chapter; it requires several steps, so we give just a brief summary: The "target" for the main equivalence is the abelian p-groups poset $\mathcal{A}$—which is equivalent to $\mathcal{S}_p(G)$ in Remark 4.3.3. The cone fiber corresponding to a normal chain is shown to be conically contractible to the intersection of the centers of the elements of the chain—where nontriviality of the intersection is guaranteed by the normality hypothesis on the chain. (The argument could use $\mathcal{A}_p(G)$ in place of $\mathcal{A}$, by using $\Omega_1(Z(-))$ in place of just $Z(-)$.) Then the G-equivariant form of the Fiber Theorem 4.2.1 completes the proof. $\diamondsuit$

Part 3

Basic Applications

The next three chapters present results on the topics which seem to have attracted the greatest attention in the literature: namely Euler characteristics, Lefschetz modules, and group cohomology.

CHAPTER 5

The reduced Euler characteristic $\tilde{\chi}$ and variations on vanishing

Much of the modern literature on p-subgroup complexes was inspired by a single very striking result of K. S. Brown [**Bro75**, Cor 2] on the Euler characteristic of $\mathcal{S}_p(G)$ (which we had previewed at (0.0.1)). Correspondingly this initial chapter on "applications" will survey a number of viewpoints on that result, together with a number of other results which deal specifically with Euler characteristics of subgroup posets.

So we will be treating Brown's result in considerably more detail than we gave in our introduction at (0.0.1); we present the result as Theorem 5.3.1 below—it will in effect provide our first main goal for the chapter.

However before that treatment, we will first present some background material from homological algebra, in the initial Section 5.1; and then in Section 5.2, we indicate in particular that contractibility implies vanishing of homology and reduced Euler characteristic—a principle which we will invoke frequently, during many proofs in this chapter (and indeed in the remainder of the book).

Thus in Section 5.3, we can begin to discuss a number of different viewpoints on the proof of Brown's theorem. In particular, we will view Brown's result as divisibility by $|G|_p$ of the reduced Euler characteristic $\tilde{\chi}(\mathcal{S}_p(G))$ (see Definition 5.1.1 below)—hence, as vanishing modulo $|G|_p$.

And then the subsequent several sections, beginning with Section 5.4, will consider further variations on the notion of vanishing—still of the reduced Euler characteristic. The chapter concludes with Section 5.6, in our Option G—sketching Brown's use of the *G-equivariant* Euler characteristic.

5.1. Topological background: Chain complexes and homology

Probably most readers will have already seen the notion of Euler characteristic—and so could safely skip over the introductory material in this section, proceeding directly to the applications in Section 5.2 (or even Section 5.3).

An ad-hoc definition of the Euler characteristic. In much of our work in this chapter, the Euler characteristic will appear just as a simple numerical invariant of a poset or complex—so that our viewpoint will often be combinatorial rather than explicitly topological.

Hence we will begin with a subsection giving just a bare-bones definition of the Euler characteristic—which should allow the less experienced reader to skip over the deeper topological background in the remaining subsections of this section, and still obtain an overall understanding of most of the material thereafter, from Section 5.3 onward.

DEFINITION 5.1.1 (Euler characteristic). Assume that K is a simplicial complex of dimension d; and recall from Notation 1.2.6 that K_i denotes the set of simplices of dimension i. Then we define (cf. [**Mun84**, Sec 22]) the *Euler characteristic* of K as the alternating sum of the *numbers* of such simplices:

$$\chi(K) := \sum_{i=0}^{d} (-1)^i \ \#(K_i).$$

(Here we are using the notation of $\#(-)$ rather than $|-|$ for cardinality, to avoid confusion with the use of $|-|$ for the geometric realization of simplices and complexes in Definition 1.3.1.)

In fact we will more frequently use the *reduced* Euler characteristic (cf. [**Mun84**, 7.2; Sec 13]), which often has more convenient algebraic properties; it is obtained by just subtracting 1 from the usual Euler characteristic. (This subtraction can be viewed as a consequence of "augmenting" the complex, with a single formal simplex in dimension -1; this in effect compensates for connectedness.) That is:

$$\tilde{\chi}(K) \ := \ \chi(K) - 1 \ = \ \sum_{i=-1}^{d} (-1)^i \ \#(K_i).$$

For a poset $\mathcal{P}$, as usual we define $\tilde{\chi}(\mathcal{P})$ via the order complex, namely as $\tilde{\chi}\big(K(\mathcal{P})\big)$. For a topological space X, we can use any complex K which triangulates X: for it is standard (see e.g. [**Mun84**, 19.5], or Remark 5.1.15 below) that the Euler characteristic is an invariant of the homotopy type of K; so this gives a definition of $\tilde{\chi}(X)$ which is independent of the particular triangulation K chosen for X. ◇

We indicate a few initial computations, using some earlier examples.

EXAMPLE 5.1.2 (Simplices and contractible spaces). First consider K given by the 3-simplex (solid tetrahedron) of Definition 1.2.1; we have:

$$\tilde{\chi}(K) = -1 + \#(K_0) - \#(K_1) + \#(K_2) - \#(K_3) = -1 + 4 - 6 + 4 - 1 = 0.$$

(Exercise:) Indeed show that a simplex K of *any* dimension d has $\tilde{\chi}(K) = 0$. The calculation of $\tilde{\chi}(K)$ turns out to be just the binomial expansion of $-\big(1+(-1)\big)^{d+1}$.

More generally (cf. later 5.2.2), any contractible complex K has $\tilde{\chi}(K) = 0$. This exhibits a nice algebraic feature of the reduced Euler characteristic: for in this "trivial" case of a space equivalent to a point, the additive identity 0 is more convenient than the multiplicative identity 1—which is the value that would be provided by the ordinary Euler characteristic $\chi(K)$ here. ◇

EXAMPLE 5.1.3 (Boundaries of simplices—hence spheres). Next consider L given by the *surface* of the tetrahedron, as in Example 1.2.2: Here we simply omit the final term, for the full 3-simplex K_3, from the calculation for K in Example 5.1.2 above, to get:

$$\tilde{\chi}(L) = -1 + 4 - 6 + 4 = 1.$$

(Exercise:) Adjust the more general exercise given thereafter in Example 5.1.2, to show that for the boundary L of a simplex σ of dimension d, we get $\tilde{\chi}(L) = (-1)^{d+1}$.

In fact this boundary L is a triangulation of the $(d-1)$-sphere S^{d-1}; and so we conclude that $\tilde{\chi}(S^{d-1}) = (-1)^{d+1} = (-1)^{d-1}$, or more simply $\tilde{\chi}(S^d) = (-1)^d$. We will return to this feature in our later discussion of spheres, at (8.1.6). ◇

The reader interested just in this chapter's results on Euler characteristics could now proceed directly to the applications in the subsequent Section 5.2—skipping over the further background material below on chain complexes and homology.

Chain complexes. In the remainder of the section, we give a very abbreviated summary of some basic homological algebra for simplicial complexes: the goal is just to provide definitions and statements of some fundamental properties. The reader interested in fuller details can consult e.g. [**Mun84**, Sec 5] or [**Spa81**, Ch 4].

In overview, to the combinatorial structure of a simplicial complex K, we will associate the purely algebraic module structures of the chain complex and its homology.

REMARK 5.1.4 (Chain spaces for a simplicial complex). Assume we are given a finite simplicial complex K. For technical reasons, we begin by choosing[1] some linear ordering on the finite set K_0 of vertices of K. Then we obtain an induced linear ordering on the vertices of each simplex σ of K.

Now for each $i = 0, 1, \cdots, \dim K$, the *i-th chain group* is the free $\mathbb{Z}$-module on the members of K_i: namely formal linear combinations $\sum_{\sigma \in K_i} c_\sigma \sigma$ of i-simplices σ,[2] with coefficients $c_\sigma \in \mathbb{Z}$. We denote this space by $C_i(K; \mathbb{Z})$

Sometimes in the construction we will want to replace $\mathbb{Z}$ by some other coefficient ring R (for example, $\mathbb{F}_p$); and then we write $C_i(K; R)$ for the result. Indeed often the ring will either be clear from the context, will be assumed to be $\mathbb{Z}$ by default; and then we may write just $C_i(K)$. (We may even write just C_i, since usually the complex K will also be fixed.)

The topological structure of boundaries within K is then modeled algebraically using the *boundary maps* $\partial_i : C_i(K) \to C_{i-1}(K)$: If the vertices of σ in the above ordering are $v_0 < v_1 < \cdots < v_i$, let σ_j denote the face obtained by removing v_j from σ, and set $\partial_i(\sigma) := \sum_{j=0}^{i} (-1)^j \sigma_j$. An easy formal calculation shows that applying two successive boundary maps causes each term $(\sigma_j)_k$ in the image to arise in two different ways—with opposite sign; so that:

$$(5.1.5) \qquad \partial_{i-1} \circ \partial_i = 0; \text{ hence } \operatorname{Im}(\partial_i) \subseteq \operatorname{Ker}(\partial_{i-1}). \text{ (We may say “}\partial^2 = 0\text{”.)}$$

Finally the collection of chain spaces equipped with these boundary maps is called the *chain complex* for K, and denoted by $C_*(K)$.

Often it is technically convenient to add a further space $C_{-1}(K)$ spanned by a single formal simplex σ_{-1} in “dimension -1”, with boundary map ∂_{-1} mapping all vertices to σ_{-1}; this gives the *augmented* chain complex $\tilde{C}_*(K)$. (We will see later, e.g. at Remark 5.1.7, that the augmented complex has properties which can be more natural for algebraic purposes.)

More generally, one can define an abstract chain complex just from spaces C_i, equipped with boundary maps satisfying the condition $\partial^2 = 0$ in (5.1.5). But the particular chain complex $C_i(K)$ above, for a simplicial complex K, is the case that we will usually consider in this book. ♢

In fact we will rarely have to calculate explicitly with boundary maps; but for an instance, see e.g. the discussion of $H_0(K)$ in Remark 5.1.7 below.

However we do observe at this point that we get an alternative formulation of the Euler characteristic via an alternating sum defined on the corresponding chain

[1]Choice of a different ordering could only change the construction below by a sign—when the orderings differ by an odd permutation.

[2]Here by σ regarded as a “basis element”, we mean more precisely—cf. [**Mun84**, p 27]—the corresponding *oriented* simplex, determined by the above induced ordering on the vertices of σ.

complex: For the i-simplices in K_i give a $\mathbb{Z}$-basis of the i-th chain space $C_i(K;\mathbb{Z})$, so that the cardinality $\#(K_i)$ gives the $\mathbb{Z}$-rank of that space. We will typically call this rank the *dimension* of the $\mathbb{Z}$-chain space—this is for consistency with the vector-space case, namely where we are working with a coefficient ring R which is in fact a field. Thus we can re-phrase Definition 5.1.1 in the language of dimensions:

$$\tilde{\chi}(K) = \tilde{\chi}(C_*(K)) = \chi(\tilde{C}_*(K)) := \sum_{i=-1}^{d} (-1)^i \dim(C_i(K)). \tag{5.1.6}$$

Homology groups (and modules). The observation (5.1.6) uses only the dimensions in the chain complex, to produce the topological invariant given by the Euler characteristic. The more sophisticated invariants given by the homology groups use the further structure recorded in the boundary maps:

REMARK 5.1.7 (Homology of a chain complex). Assume a chain complex $C_*(K)$, with coefficient ring $\mathbb{Z}$. In (5.1.5) we saw the property $\partial^2 = 0$; namely the relation $\partial_{i-1} \circ \partial_i = 0$, which gives $\mathrm{Im}(\partial_i) \subseteq \mathrm{Ker}(\partial_{i-1})$. Thus if we set:
$Z_i(K) := \mathrm{Ker}(\partial_i)$ (the group of *i-cycles* of K), and
$B_i(K) := \mathrm{Im}(\partial_{i+1})$ (the group of *i-boundaries*; take $B_d(K) := 0$),
then we can define (for $i = 0, \ldots, d$) the quotient:
$H_i(K) := Z_i(K)/B_i(K)$ (the *i-th homology group* of K).
The abelian group $H_i(K)$ is the quotient of submodules of the free $\mathbb{Z}$-module $C_i(K)$, which are therefore also free—but the quotient is *not* necessarily $\mathbb{Z}$-free.

We can make the analogous constructions over an arbitrary coefficient ring R, and then we use notation such as $H_i(K;R)$; when R is a field, we usually refer to homology *spaces*. (The relationships of the groups $H_i(K;R)$, as the ring R varies but the complex K is fixed, are described via the Universal Coefficient Theorem, e.g. [**Mun84**, Sec 10; 55.1].)

Very roughly, the i-cycles $Z_i(K)$ record suitable appropriate closed subcomplexes of dimension i in K (i.e. subcomplexes with zero boundary, e.g. i-spheres). And then the topological boundaries of $(i+1)$-simplices determine the elements of the boundary space $B_i(K)$; these define a natural notion of equivalence among the cycles—which we factor out, to define inequivalent classes—giving the elements of the homology group $H_i(K)$. These groups are *invariants*, in the sense of being preserved by suitable transformations such as homotopy equivalences (see e.g. Remark 5.1.15 below).

Aside from the elementary observation in the following paragraph, we will postpone more representative sample computations of homology groups to later chapters, e.g. (8.1.7); the reader may also wish to refer to [**Mun84**, Sec 5].

For the moment, we will just mention one standard fact (see e.g. [**Mun84**, Theorem 7.1]): namely $H_0(K)$ is $\mathbb{Z}$-free, and of dimension equal to the number of connected components of K. So for example, a connected complex K has $\dim H_0(K) = 1$. (Notice this is because the boundary map takes an edge to the difference of its two vertices—so those two vertices are identified in $H_0(K)$, where such boundary images in $B_0(K)$ have been set to zero.) In particular this holds for K given by a point, or more generally for K any contractible complex.

However, in the abelian-group context provided by chain complexes, it would be algebraically more convenient for the "trivial" complex afforded by a point to have homology groups given by the trivial abelian group 0 (which in particular has dimension 0). And this observation begins to explain the usefulness of instead working with the *augmented* chain complex $\tilde{C}_*(K)$ (which we had mentioned in Remark 5.1.4): For if we make the analogous definitions of $\tilde{Z}$, $\tilde{B}$ and $\tilde{H}$ with the augmented complex (namely obtaining the *reduced* homology of K), then the only actual difference occurs at the lowest term $\tilde{H}_0(K)$, where the effect (see e.g. [**Mun84**, Theorem 7.2]) is to reduce the dimension by 1:

$$\dim \tilde{H}_0(K) = \dim H_0(K) - 1. \tag{5.1.8}$$

Thus in the case of K a point or a contractible space, we do get $\dim \tilde{H}_0(K) = 0$, so that $\tilde{H}_0(K)$ is indeed just 0 as an abelian group, as desired. (We will return to the vanishing of the other homology groups of a point later at Lemma 5.2.2; and also to an additional desirable algebraic property of reduced homology at (8.1.1).) ♢

Of course our complexes K will usually admit the action of a group G. And we then obtain an induced G-action on the cycle, boundary, and homology groups of K—so that we can regard these spaces as *G-modules*.

In particular these spaces determine the reduced Lefschetz module of K, which will be the main focus of our analysis in later Chapter 6—but we will introduce it here, since it makes a few appearances in the present chapter as well. Namely we consider the alternating sum of the spaces whose dimension terms arose in (5.1.6):

DEFINITION 5.1.9 (Reduced Lefschetz (virtual) module). For a G-complex K of dimension d, we define the *reduced Lefschetz (virtual) module* as a formal (we expand on this point in a moment) alternating sum of chain groups:

$$\tilde{L}(K) := \bigoplus_{i=-1}^{d} (-1)^i \, C_i(K).$$

As usual our default coefficient ring is $\mathbb{Z}$; but we may wish to work over some more general ring R, in which case we will write $\tilde{L}(K; R)$ for the construction.

Often we write the module in a different formulation—essentially as the alternating sum of terms indexed by the simplices σ of K, each with sign term given by $(-1)^{\dim \sigma}$. Furthermore it is often convenient to consider the subspaces given by the G-orbits on simplices; and we write the indexing in the form "$\sigma \in K/G$". Each of these permutation modules for G can be written as the module from the trivial module ("1") of the stabilizer; so we get the form:

$$\tilde{L}(K) = \bigoplus_{\sigma \in K/G} (-1)^{\dim \sigma} \, \mathrm{Ind}_{G_\sigma}^{G}(1) \tag{5.1.10}$$

(where we include the formal empty simplex $\emptyset$ of dimension -1, whose stabilizer is all of G).

For the *character* of the reduced Lefschetz module $\tilde{L}(K)$ (namely the set of values at the various $g \in G$ of the *trace* of the action of g), we will usually use the abbreviation $\tilde{\Lambda}(K)$.

(We will defer computation of *explicit* examples of the Lefschetz module and its character until later, e.g. Example 6.2.2.) ♢

We will also further explore the *meaning* of the values of the Lefschetz character $\tilde{\Lambda}(K)$, in a moment.

But before doing so, we mention that the formal $\mathbb{Z}$-linear combination above is well-defined, up to a suitable equivalence of modules; and the signs are why we refer to $\tilde{L}(K)$ as a *virtual* module for G. (See e.g. [**Lan65**, p 43,p 101] for this standard construction.) Indeed there are several possible contexts in which we can view $\tilde{L}(K)$—depending on how we wish to define that equivalence for our virtual modules:

DEFINITION 5.1.11 (Rings of representations). One useful choice for the equivalence, which is sufficient to preserve the most basic information about modules, is defined by letting any short exact[3] sequence $0 \to A \to B \to C \to 0$ force the identification of B with the sum $A \oplus C$. This gives the *Grothendieck group* for G; in which a member is equivalent just to the (net) sum of the irreducible modules it involves. For this reason, a module viewed in the Grothendieck group is determined just by its character. The Grothendieck group in fact becomes a commutative ring, when we consider the multiplication given by the tensor product of modules.

However, when we need information not just about irreducibles, but also about the deeper module structure involving extensions among them, we can instead use the natural equivalence defined just by isomorphism in the category of modules. That is, we only make the above identification when the sequence is *split*, that is, when we actually have $B \cong A \oplus C$ *as modules*. This choice gives the *representation ring* or *Green ring* of G—in which members are viewed as direct sums of *indecomposable* modules.

Some of the literature on subgroup complexes focuses instead on the *Burnside ring*. This is related to permutation modules in the representation ring; but the Burnside ring is "logically prior", in that it is generated by the *G-sets*: Namely the basis elements correspond to transitive G-actions, denoted in the form G/H, for $H \le G$; but we identify G/H and G/K only when H and K are conjugate in G—and *not* necessarily when the corresponding permutation modules are isomorphic in the sense of representation theory. Hence when we consider the natural mapping λ which takes the G-set G/H to the permutation module $\mathrm{Ind}_H^G(1)$, we get a map of the Burnside ring B to the subring induced by permutation modules in the representation ring R—and the map often has a large (and indeed interesting) kernel. (For the ring structure: The Cartesian product of G-sets is mapped to the tensor product of the corresponding permutation modules.)

Of course in this book, we often consider some G-complex K; and since G permutes the simplices defining a basis of the chain spaces $C_*(K)$, we see that many of our resulting concepts—such as Lefschetz modules and Euler characteristics—can be studied either in the context of the Burnside ring B, or in the corresponding permutation subring $\lambda(B)$ of the representation ring R. (For example, we may write $\tilde{\mathcal{L}}(K)$ for the *Lefschetz invariant*: namely the appropriate alternating sum of G-sets—whose image under λ is the Lefschetz module $\tilde{L}(K)$, in turn defined by (5.1.10) in the representation ring.) Now in fact reasonably often in informal situations, we will not really need to distinguish between these two contexts. However, at some other points (e.g. the proof of Theorem 6.2.1), we will also indicate

[3]That is, where the image of each map is equal to the kernel of the next. The *short* case here is equivalent to having $A \le B$, and $C \cong B/A$, as modules.

arguments do emphasize the distinction—for example, arguments which involve the kernel of λ.

Finally we emphasize that, although we will sometimes work with representations defined in characteristic p, in the above context of representation rings we are considering the representations primarily as formal objects—so that their linear combinations are given with coefficients in the integers $\mathbb{Z}$, rather than in $\mathbb{F}_p$; in particular, a sum of p copies of such a representation is not 0. $\diamondsuit$

As promised earlier, we now return to the Lefschetz character $\tilde{\Lambda}(K)$ determined by the module $\tilde{L}(K)$. First from the Definition 5.1.9 of the module we have:

$$\tilde{\Lambda}(K)(g) = \mathrm{Tr}\big(g, \textstyle\sum_{i=-1}^{d} (-1)^i\ C_i(K)\big) = \textstyle\sum_{i=-1}^{d} (-1)^i\ \mathrm{Tr}\big(g, C_i(K)\big)\ .$$

Now since g just permutes the i-simplices in K_i, which give a basis for $C_i(K)$, it is an elementary fact about permutation modules that the trace of g is just the number $\#(\ (K_i)^g\ )$ of such simplices which are in fact fixed by g. Equivalently this is the number $\#(\ (K^g)_i\ )$ of i-simplices of the fixed subcomplex K^g—which gives $\dim C_i(K^g)$. Hence the character value at g is in fact given by the reduced Euler characteristic of the g-fixed subcomplex:

(5.1.12)

$$\tilde{\Lambda}(K)(g) = \mathrm{Tr}\big(g, \tilde{L}(K)\big) = \sum_{i=-1}^{d} (-1)^i\ \#(K_i^g) = \sum_{i=-1}^{d} (-1)^i\ \dim C_i(K^g) = \tilde{\chi}(K^g).$$

This relation is sometimes called the Lefschetz Fixed-Point Formula.[4]

In particular when $g = 1$ in (5.1.12) above, we have $K^1 = K$; and the trace of 1 just gives the dimensions of the full chain spaces $C_i(K)$. So using earlier (5.1.6), we can exhibit the Euler characteristic as the dimension of the Lefschetz module:

$$\tilde{\chi}(K) = \dim\big(\tilde{L}(K)\big) = \tilde{\Lambda}(K)(1). \tag{5.1.13}$$

Now as we had mentioned in Definition 5.1.11, if we view the Lefschetz module $\tilde{L}(K)$ as a member of the Grothendieck group, it is determined by the Lefschetz character $\tilde{\Lambda}(K)$ in (5.1.12). In fact, we get the same character (and hence the same module in the Grothendieck group) if we compute the trace instead on the corresponding alternating sum of homology groups: this is the classical formula of Hopf (e.g. [**Mun84**, 22.1, 22.2]):

---

[4]More precisely, this trace (stated for the alternating sum of homology groups as in 5.1.14 below) is the classical *Lefschetz number* of g. If K^g is empty, then the value of the reduced Euler characteristic is -1; and so any value other than -1 implies contrapositively that g has at least one fixed point on K—which is the content of the Lefschetz Fixed-Point Theorem (see e.g. [**Mun84**, 22.3]). This suggests the terminology of *Lefschetz* module.

LEMMA 5.1.14 (Hopf Trace Formula). *For G on K of dimension d, and $g \in G$:*

$$\mathrm{Tr}(\ g\ ,\ \sum_{i=0}^{d} (-1)^i\ H_i(K)\ ) = \mathrm{Tr}(\ g\ ,\ \sum_{i=0}^{d} (-1)^i\ C_i(K)\ ),$$

where for $H_i(K)$ with $\mathbb{Z}$-coefficients we in fact consider just the quotient of $H_i(K)$ modulo its torsion subgroup.

In particular $g = 1$ gives dimension terms, so using (5.1.13) *we get:*

$$\sum_{i=0}^{d} (-1)^i\ \dim H_i(K) = \sum_{i=0}^{d} (-1)^i\ \dim C_i(K)\ = \Lambda(\tilde{K})(1) = \tilde{\chi}(K)\ .$$

That is, we can express the reduced Euler characteristic via the alternating sum of homology, as well as of chains. □

Some sources give this homological formulation as the primary definition of Euler characteristic.

We end the subsection by more formally recording our earlier mention that the homology groups are indeed suitable topological invariants, in the sense of being preserved by homotopy equivalences:

REMARK 5.1.15 (Homotopy equivalence preserves homology groups and Euler characteristic). See e.g. [**Mun84**, 19.5] for the standard fact that:

A homotopy equivalence induces an isomorphism in homology.

(This does *not* force an isomorphism of the corresponding chain spaces!)

In particular for the purposes of this chapter on Euler characteristic, from the homological formulation in Lemma 5.1.14 we obtain the corollary that:

A homotopy equivalence induces an equality of Euler characteristics.

In fact a homotopy equivalence has the stronger property of preserving (as the name suggests) a complex's homotopy groups—as we had mentioned in Digression 4.1.5, and see also later (9.1.4).

We mention that it is also possible for maps which are weaker than homotopy equivalences to still preserve homology groups—the reader will find references to such *homology equivalences* in the topological literature; and also to *weak (homotopy) equivalences*, preserving homotopy groups. ◊

The material in this book will make heavy use of the homotopy equivalence methods that we developed in Chapter 4; and in such applications we will very frequently then quote (or later on, even just implicitly assume) the invariances in Remark 5.1.15.

5.2. Contractibility and vanishing of homology and $\tilde{\chi}$

In the later sections of this chapter, our proofs will often make use of conditions involving the vanishing of the reduced Euler characteristic, typically for a suitable subcomplex L of our original complex K.

In particular in view of the homological formulation in Lemma 5.1.14, that vanishing will in turn follow if we can establish the stronger condition of vanishing of the reduced homology groups:

DEFINITION 5.2.1 (Acyclic complexes). A complex K is *acyclic* if $\tilde{H}_i(K;\mathbb{Z}) = 0$ for all i. (From our discussion of (5.1.8), this is equivalent to K being connected, along with vanishing of the ordinary homology groups $H_i(K;\mathbb{Z})$ for $i > 0$.)

If K is acyclic, then in particular $\tilde{\chi}(K) = 0$ (using Lemma 5.1.14).

We say K is *R-acyclic* if this vanishing holds for coefficients in some ring R. ◊

Furthermore our most frequently-arising acyclic complexes K will in fact be contractible—where typically we will have established contractiblity using results developed earlier, e.g. in Chapters 3 and 4. So in all those situations, we can then obtain as a corollary the vanishing of $\tilde{\chi}$ which had been suggested in earlier Example 5.1.2, namely (cf. [**Mun84**, 19.5]):

LEMMA 5.2.2. *If K is contractible, then K is acyclic, and $\tilde{\chi}(K) = 0$.*

PROOF. First consider the special case where K is a point. Then K is connected; and since K has no simplices in dimensions $i > 0$, we get $H_i(K) = 0$. Thus K is acyclic in the language of Definition 5.2.1; and in particular $\tilde{\chi}(K) = 0$.

Now let K be any contractible complex. Then by definition K is homotopy equivalent to a point; so this new K is also acyclic, with $\tilde{\chi}(K) = 0$—since by Remark 5.1.15, the equivalence preserves these properties obtained just above in the case of a point. □

As an example of the corollaries from earlier results as mentioned above, observe when $O_p(G) > 1$ that from the contractibility of $\mathcal{S}_p(G)$ in Lemma 3.3.5, it follows that $\tilde{\chi}(\mathcal{S}_p(G)) = 0$ by Lemma 5.2.2.

And indeed in this way, the statement in Lemma 5.2.2, obtaining vanishing of $\tilde{\chi}$ from contractibility, will be sufficient for many of our purposes later in this chapter.

However, to give a somewhat broader context, we will extend this result by indicating a sequence of logical implications, among a number of further conditions related to contractibility; for example, we had seen earlier at 3.4.7 that collapsibility implies contractibility. The list below is adapted from Björner [**Bjö95**, 11.1] (which the reader may wish to consult for fuller details).

LEMMA 5.2.3 (A hierarchy of conditions related to contractibility). *For a complex K, the following implications are strict (i.e. not reversible in general):*

K is a cone
$\Rightarrow$ K is collapsible
$\Rightarrow$ K is contractible
$\Rightarrow$ K is ($\mathbb{Z}$-)acyclic.
$\Rightarrow$ K is F-acyclic (for any field F)
$\Rightarrow$ K is $\mathbb{Q}$-acyclic
$\Rightarrow$ $\tilde{L}(K;\mathbb{Z}) = 0$
$\Rightarrow$ $\tilde{\Lambda}(K;\mathbb{Z}) = 0$
$\Rightarrow$ $\tilde{\chi}(K) = 0$.

As a partial converse: If K is $\mathbb{F}_p$-acyclic for all primes p, then it is ($\mathbb{Z}$-)acyclic. □

Finally, we mention that some of the literature replaces topological terminology as above with more combinatorial language. For example, Lemma 2.4 in [**HIÖ89**] (cf. their remark on p. 1030) in effect replaces conical contractibility with the notion of a meet-semilattice, which we had mentioned before Definition 3.3.1; and also replaces the consequent vanishing of $\tilde{\chi}$ by the vanishing of the corresponding expression in the Möbius function (which we will indicate later in Definition 5.3.15).

5.3. Vanishing of $\tilde{\chi}(\mathcal{S}_p(G))$ mod $|G|_p$: Brown's Theorem

Recall that early on at (0.0.1), we had for overall motivation introduced Brown's result on the Euler characteristic of the p-subgoups poset $\mathcal{S}_p(G)$. In fact there we had emphasized the ordinary Euler characteristic $\chi(\mathcal{S}_p(G))$, and the analogy of the result with the numerical part of Sylow's Theorem. But in our treatment of Brown's result in this chapter, it will instead usually be more natural to work with the reduced Euler characteristic $\tilde{\chi}(\mathcal{S}_p(G))$; and to re-state the result as an instance of a certain kind of vanishing:

THEOREM 5.3.1 (Brown's Theorem). *For a finite group G,*

$$\tilde{\chi}(\mathcal{S}_p(G)) \equiv 0 \pmod{|G|_p}.$$

In the rest of this section, we will explore Quillen's version of Brown's original proof, as well as some other approaches and generalizations developed by other authors during the subsequent years.

But before turning to the proof, we examine the result in some small examples.

EXAMPLE 5.3.2 ($\tilde{\chi}(\mathcal{S}_2(S_3 \times S_3))$). Take $G = S_3 \times S_3$ and $p = 2$; then we have $|G|_2 = 2^2 = 4$. The poset $\mathcal{S}_2(G)$ consists of the subgroups of orders 2 and 4; and the order complex $K(\mathcal{S}_2(G)) =: K$ is a certain bipartite graph on those subgroups as vertices. To describe these in more detail:

Let the "left" S_3 permute the symbols $\{1,2,3\}$, while the "right" S_3 permutes $\{a,b,c\}$. Then subgroups of order 2 are determined by cycle structures such as $(1,2)$ or (a,b) or $(1,2)(a,b)$; so there are $3,3,9$ of each of these types. Similarly subgroups of order 4 look like $\langle(1,2),(a,b)\rangle$—for which there are 9 choices. These subgroups give the vertices K_0 of the complex K. Furthermore each subgroup of order 4 contains 3 subgroups of order 2—one of each of the above cycle types; so we have $9 \cdot 3$ such containments. These containments give the edges K_1 of the complex.

Consequently we compute that:

$$\tilde{\chi}(K) = -1 + \#(K_0) - \#(K_1) = -1 + [(3+3+9)+9] - [9 \cdot 3] = -4,$$

and this is indeed congruent to 0 modulo $|G|_2 = 4$ (as in Brown's Theorem). ♢

EXERCISE 5.3.3 ($\tilde{\chi}(\mathcal{S}_2(A_5))$). Let $G = A_5$ and $p = 2$. Show that the order complex K of $\mathcal{S}_2(G)$ consists of 5 connected components, each a complete bipartite graph $K_{1,3}$. Deduce that $\#(K_0) = 5 \cdot (1+3)$ and $\#(K_1) = 5 \cdot 3$; leading to the computation of $\tilde{\chi}(K)$ as 4, which is indeed congruent to 0 modulo $|G|_2 = 4$. ♢

REMARK 5.3.4. In the small cases given by both Example 5.3.2 and Exercise 5.3.3, the value of $\tilde{\chi}(\mathcal{S}_p(G))$ turned out to be equal to $(\pm)|G|_p$ *exactly*. We warn the reader that this single multiple of $|G|_p$ should *not* be expected in any generality.

(Option B:) But here, it happens that these two particular groups G are in fact of Lie type in characteristic 2. We saw at Theorem 4.4.8 that $\mathcal{S}_2(G)$ is then homotopy equivalent to the building. And we will see at Theorem 6.4.2 that the Lefschetz module of the building is $(\pm)$ the Steinberg module (cf. Remark 6.4.1) of G—which has dimension equal to $|G|_p$. (And we have $|G|_2 = 2^2$ in these two groups.)

(Option S:) However, we emphasize that while there are analogues for sporadic groups of the building geometry, whose Lefschetz module might be considered a "Steinberg module", the dimension term given by the Euler characteristic will *not* necessarily be a single multiple of $|G|_p$. The values of $\tilde{\chi}(K)$ for a number of sporadic geometries K are in fact given in [**RSY90**]; and for instance, in the case of the C_3-geometry K for A_7 of Example 2.3.7, we compute later at Example 6.3.7 that $\tilde{\chi}(K) = 56 = 2^3 7$—compared with $|A_7|_2 = 2^3$. ♢

In the remainder of the section, we will explore several different approaches to the proof of Brown's theorem.

Proof 1: the Brown/Quillen proof via homological algebra. Brown's Theorem 5.3.1 is given as Corollary 4.2 in [**Qui78**]; and below we will sketch the proof given there by Quillen.

DIGRESSION 5.3.5. But we emphasize that Quillen's argument can be regarded as an adaptation[5] of Brown's original proof for his more general results in [**Bro75**], especially Theorem 2 there; indeed, following Quillen primarily allows us to avoid certain complications in Brown's more general analysis. Here we will just briefly mention some of those complications (postponing a longer discussion to the final Section 5.6 of the chapter): First, Quillen restricts to finite groups G; whereas Brown considers groups which may be infinite, but still have finite "homological type". Furthermore, Brown works not just with the usual Euler characteristic, but also espeically with the *G-equivariant* Euler characteristic; and this latter material falls within our Option G. So following our usual convention, we have separated out this optional-track material into Section 5.6. ♢

We will begin by giving, in a moment, a "two-line" summary of Quillen's proof. But for comparison, we first recall one standard proof (e.g. [**Hal59**, p 45]) of the numerical part of Sylow's Theorem. Let P denote a Sylow p-subgroup of G, so that $|P| = |G|_p$. To conclude that $|\mathrm{Syl}_p(G)| \equiv 1 \pmod{p}$, that argument examines orbits on $\mathrm{Syl}_p(G)$ under conjugation by P, observing that:

(α) There is exactly 1 point fixed by P (namely P itself).

(β) All other orbits have length greater thatn 1—indeed length divisible by p.

And together these observations imply the assertion.

Now consider the poset $\mathcal{S}_p(G)$, and its order complex $K(\mathcal{S}_p(G)) =: K$. Quillen's proof shows analogously that:

(a) We can collect together all the simplices fixed by any of the *nontrivial* elements of P into a subcomplex L (the P-singular set)—which is contractible, so that $\tilde{\chi}(L) = 0$. (This corresponds to a value of 1 for $\chi(L)$, in analogy with the unique fixed point in (α).)

(b) Next any orbits of P on K *outside* L consist of simplices which are fixed only by 1; and so these orbits have length $|P|$. And we will see that $|P|$ must correspondingly divide the contribution of the rest of K to the value of $\tilde{\chi}(K)$—yielding Brown's Theorem.

So we now turn to expanding out the details for (a) and (b).

[5] I thank Jacques Thévenaz for clarifying this aspect of the arguments.

Exploiting contractibility of the singular set. Most of the work for (a) has already been done in earlier chapters: First, we had seen in Definition 4.4.13 that the P-singular set $\mathcal{S}_p(G)^{P\text{-sing}}$ is the subposet which collects together all those members of $\mathcal{S}_p(G)$ which are normalized by any nontrivial subgroup of P. So the subcomplex "L" in (a) is just the order complex $K\big(\mathcal{S}_p(G)^{P\text{-sing}}\big)$ of the singular set. We saw in Proposition 4.4.15 that L is contractible. Hence:

$$\tilde{\chi}(L) = \tilde{\chi}\big(\mathcal{S}_p(G)^{P\text{-sing}}\big) = 0 \tag{5.3.6}$$

using Lemma 5.2.2.

Computing the remainder with the relative chain complex. For (b), we need to calculate the contribution to $\tilde{\chi}(K)$ from that part of K *not* in the singular set L.

So note that if Q is any member of $\mathcal{S}_p(G)$ *outside* the singular set $\mathcal{S}_p(G)^{P\text{-sing}}$, then the stabilizer $P_Q = N_P(Q)$ of Q in P is necessarily the trivial subgroup 1; so that the P-orbit containing Q has length equal to $|P : P_Q| = |P : 1| = |P|$. That is, P acts *freely* on such subgroups Q. Hence P also acts freely on inclusion-chains which contain such a subgroup.

To finish, we first indicate an elementary numerical argument: We recall from Definition 5.1.1 that $\tilde{\chi}(L) = \sum_{i=-1} (-1)^i \ \#(L_i)$. And since orbits of P on chains containing a subgroup outside L have length $|P|$, we see that

$$\#(L_i) \equiv \#(K_i) \pmod{|P|}. \tag{5.3.7}$$

Now applying Definition 5.1.1 to K, we get $\tilde{\chi}(L) \equiv \tilde{\chi}(K) \pmod{|P|}$; and using this in (5.3.6), we obtain Brown's Theorem.

For the reader going on to [**Qui78**], we note that Quillen finishes with a more general topological argument, via a further standard construction from homological algebra; we just sketch this briefly (see e.g. [**Mun84**, Sec 9] for further detail): The subcomplex L determines a chain subcomplex $C_*(L)$, and the quotient spaces $C_i(K)/C_i(L)$ are called *relative* chain spaces; these in turn define the *relative chain complex* $C_*(K, L)$. A standard calculation relates the corresponding Euler characteristics:

$$\tilde{\chi}(K) = \tilde{\chi}(L) + \tilde{\chi}(K, L). \tag{5.3.8}$$

Now for our particular L which is contractible, the term $\tilde{\chi}(L)$ on the right vanishes by (5.3.6); so we get $\tilde{\chi}(K) = \tilde{\chi}(K, L)$ in our present situation.

And we saw above that P permutes freely the simplices which afford a basis of each relative chain space $C_i(K, L)$. So by the analogue of (5.1.6), $|P|$ divides $\tilde{\chi}(K, L) = \tilde{\chi}(K)$—yielding Brown's Theorem.

Here is a further slight variant: We can in fact see the equality $\tilde{\chi}(K, L) = \tilde{\chi}(K)$ above from a somewhat more "structural" viewpoint. Namely the short exact sequence

$$0 \to C_*(L) \to C_*(K) \to C_*(K, L) \to 0 \tag{5.3.9}$$

leads via standard homological algebra to a *long* exact sequence (e.g. [**Mun84**, Thm 23.3]) of the corresponding homology groups:

$$\cdots \to \tilde{H}_i(L) \to \tilde{H}_i(K) \to \tilde{H}_i(K, L) \to \tilde{H}_{i-1}(L) \to \cdots \tag{5.3.10}$$

Now since we saw that L is contractible, it is in fact even acyclic by Lemma 5.2.3; so the outer terms $\tilde{H}_i(L)$ and $\tilde{H}_{i-1}(L)$ above in fact vanish. It then follows from the exactness of the sequence that we get $\tilde{H}_i(K) \cong \tilde{H}_i(K, L)$ for the inner terms. So we obtain $\tilde{\chi}(K) = \tilde{\chi}(K, L)$, using the homological formulation for $\tilde{\chi}$ in 5.1.14.

Proof 2: Surowski's argument via the Lefschetz character. An algebraic variant of Quillen's approach, using very elementary representation theory (and presumably inspired by Quillen's module-theoretic Corollary 4.3 in [**Qui78**]), was given somewhat later by Surowski [**Sur85**].

Restricting the Lefschetz character to a Sylow group. The argument begins not with the full union of fixed subcomplexes given by the singular set $\mathcal{S}_p(G)^{P\text{-sing}}$, but instead with just the fixed points $\mathcal{S}_p(G)^g$ for any particular nontrivial $g \in P$: We had given the more elementary proof of (conical) contractibility of $\mathcal{S}_p(G)^g$ at earlier Lemma 3.3.20. So again invoking Lemma 5.2.2, we get $\tilde{\chi}(\mathcal{S}_p(G)^g) = 0$. However we saw at (5.1.12) that this Euler characteristic gives the value at g of the character of the reduced Lefschetz module $\tilde{L}(\mathcal{S}_p(G))$:

$$(5.3.11) \qquad \Lambda(\mathcal{S}_p(G))(g) = \mathrm{Tr}(\ g\ ,\ \tilde{L}(\mathcal{S}_p(G))\ ) = \tilde{\chi}(\mathcal{S}_p(G)^g) = 0,$$

for $1 \neq g \in P \in \mathrm{Syl}_p(G)$. Thus the restriction to P of the Lefschetz character vanishes away from the identity; and this restriction is a generalized character of P, that is, a $\mathbb{Z}$-linear combination of ordinary (complex) irreducible characters. So by elementary representation theory (e.g. [**Isa06**, 2.8,2.10]), the restriction $\Lambda((\mathcal{S}_p(G))|_P$ must be a multiple of the *regular* character of P—which in turn has dimension $|P|$. Hence $|P|$ divides $\dim \tilde{L}(\mathcal{S}_p(G)) = \Lambda((\mathcal{S}_p(G))(1)$; and this is $\tilde{\chi}(\mathcal{S}_p(G))$ by 5.1.14, completing our second proof of Brown's Theorem.

The deeper viewpoint of projective Lefschetz modules. We now return to our earlier mention of Corollary 4.3 in Quillen [**Qui78**]—which again comes from an earlier observation of Brown, and which we treat later as Theorem 6.2.1. That result in fact gives the stronger conclusion that $\Lambda((\mathcal{S}_p(G))$ is the character of some $\mathbb{F}_p$-*projective* module, which is the main theme of the following Chapter 6. And by elementary modular representation theory (as we mention in Definition 6.1.3), the character of a projective module has the above property of vanishing on the non-trivial elements of a Sylow group—so one can again complete the proof of Brown's Theorem as above. (Cf. [**Ben91**, 6.7.4].) And indeed the various further projectivity results in Chapter 6 will similarly give corollaries on vanishing of $\tilde{\chi}$ modulo $|G|_p$.

Proof 3: The Gluck/Yoshida argument in the Burnside ring. Finally we present an approach to the proof of Brown's Theorem which might be regarded as intermediate between the viewpoints of topology and algebra, which were separately prominent in the two proofs given above: namely via the context of the Burnside ring—which roughly emphasizes the more combinatorial aspects of the algebra involved in permutation modules.

We mention that tom Dieck [**tD79**] was especially influential in demonstrating the relevance of the Burnside ring in algebraic topology. And the viewpoint of the Lefschetz invariant $\tilde{\mathcal{L}}(K)$ (whose image under λ in the representation ring is the Lefschetz module $\tilde{L}(K)$, as indicated in Remark 5.1.11) is developed notably in Thévenaz [**Thé87**].

In this book, we will be examining Burnside-ring arguments comparatively infrequently. But since the approach arises at various points in the literature on subgroup complexes, we will begin the subsection with a very informal introduction to some of its basic features. For fuller details, the reader can consult standard sources such as [**Sol67**], [**tD79**], and [**Thé87**].

A quick glimpse of the Burnside ring (and algebra). In Definition 5.1.11, we had introduced the Burnside ring B of G as the ring of G-sets—whose image under λ is the subring generated by permutation modules within the representation ring of G. Since below we will wish to consider formulas involving denominators, we let $B_{\mathbb{Q}}$ denote result of "extending scalars" to the rationals $\mathbb{Q}$—obtaining the Burnside *algebra*.

Next recall that a transitive permutation representation for G has a basis given by the members of the coset space G/H, where H is the stabilizer of a point in that orbit. Sometimes G/H is called the G-*set* determined by H; and in this set-viewpoint, the ring addition is given just by disjoint union, and the multiplication is given by the Cartesian product.

Hence a basis for the Burnside ring B is given by elements indexed by these G-sets; we will let the basis element for G/H be denoted by $u_{G/H}$—where H runs over representatives of the conjugacy classes of members of the subgroup lattice $L(G)$ of G. We can write the corresponding index set as the orbit space $L(G)/G$. (Of course the poset $\mathcal{S}_p(G)$ of interest to us here is embedded as a subposet of the full lattice $L(G)$.)

Now the commutative algebra $B_{\mathbb{Q}}$ is semisimple (see e.g. [**Sol67**, p 610]), with a "Wedderburn" decomposition via primitive idempotents—which we denote by e_H, and which similarly are indexed by the $H \in L(G)/G$. And of course we can express this basis of $B_{\mathbb{Q}}$ as a $\mathbb{Q}$-linear combination of the above basis of G-sets—that is:

$$e_H = \sum_{K \in L(G)/G} c_{H,K} \cdot u_{G/K}, \tag{5.3.12}$$

with coefficients $c_{H,K}$ lying in the rational numbers $\mathbb{Q}$.[6]

Furthermore Dress in [**Dre69**] was able to obtain additional restrictions on these coefficients, by studying suitable *sums* $\tilde{e}_H$ of the idempotents e_L. Indeed he showed that these sums determine the primitive idempotents of the related algebra of the related ring $B_{\mathbb{Z}_{(p)}}$—where the coefficients are now taken over the subring of $\mathbb{Q}$ given by the localization $\mathbb{Z}_{(p)}$ of $\mathbb{Z}$ over the prime ideal (p) (namely, in which all primes other than p are inverted). In more detail, Dress set:

$\tilde{e}_H := \sum_L e_L$—with the sum over $L \in L(G)/G$ with $O^p(L) = O^p(H)$;
recall that $O^p(L)$ denotes the largest normal subgroup of L which is a p-group (equivalently, $O^p(L)$ is generated by the p'-elements of L). Thus Dress's result includes the restriction on the relevant coefficient (see e.g. [**Glu81**, Lemma 2] for a more explicit proof) given by:

$$c_{\tilde{H},K} := \Big(\sum_{O^p(L)=O^p(H)} c_{L,K} \Big) \text{ lies in the ring } \mathbb{Z}_{(p)} \text{ of } p\text{-local integers.} \tag{5.3.13}$$

A connection with $\mathcal{S}_p(G)$ emerges when we specialize to $H = 1$: Here the condition that $O^p(L) = O^p(H) = O^p(1) = 1$ just means that L is a p-group—and in particular includes the case $L = 1$. Thus if we adjoin 1 to our p-subgroups poset, we can

---

[6] In addition (see e.g. [**Sol67**, Thm 4]) we have $c_{H,K} = 0$ unless $K \leq H$, in the usual sense of $\leq$ for the orbit poset $L(G)/G$.

write $\tilde{e}_1 = \sum_{L \in [1 \cup \mathcal{S}_p(G)]/G} e_L$, and hence:

$$c_{\tilde{1},K} \stackrel{\text{bydef}}{=} \sum_{L \in [1 \cup \mathcal{S}_p(G)]/G} c_{L,K}. \tag{5.3.14}$$

Lattice combinatorics and the Möbius function. The more purely combinatorial literature approaches many of the same issues by considering not the orbit poset $L(G)/G$, but instead the structure of essentially the lattice $L(G)$ itself.

Here a standard tool is provided by the Möbius function: The function was applied to subgroups by P. Hall [**Hal36**]; and developed in a more general framework of poset and lattice theory notably by Rota [**Rot64**]. The case of the subgroup lattice $L(G)$ was further investigated particularly by Kratzer–Thévenaz in [**KT84**] and Hawkes–Isaacs–Özaydin in [**HIÖ89**]. (The latter was aimed partly at replacing Burnside-ring arguments with more explicitly combinatorial analysis.) See also the further work in Pahlings [**Pah93**] [**Pah95**]. We refer the reader to these sources for fuller details, as we will given only a brief sketch below.

DEFINITION 5.3.15 (The Möbius function of $L(G)$). The *Möbius function* (in general for a poset—here we will use $L(G)$) is the function $\mu : L(G) \times L(G) \to \mathbb{Z}$ defined in the case $H = K$ by setting $\mu(H, H) := 1$ for $H \in L(G)$; and when $H < K$, by defining $\mu(H, K)$ inductively via the relation

$$\sum_{H \leq J \leq K} \mu(H, J) = 0. \tag{5.3.16}$$

For the indexing set above, typically the combinatorial literature uses interval-style notation, of the form "$J \in [H, K]$". In our standard poset notation, we could also denote that indexing set by $L(G)^{\leq K}_{\geq H}$. ♢

Indeed when $H < K$ as above, it is standard (see e.g. [**HIÖ89**, 9.1]) that the value of the Möbius function is just the reduced Euler characteristic of the corresponding *open* interval:

$$\text{If } H < K, \text{ then } \mu(H, K) = \tilde{\chi}\big(L(G)^{<K}_{>H}\big). \tag{5.3.17}$$

So we could use this Euler characteristic formulation in the definition of μ, as an alternative to the inductive relation (5.3.16). In fact this formulation also leads to an expression in terms of μ for the Euler characteristic of suitable subposets, such as $\mathcal{S}_p(G)$: For we can let $\overline{\mu}$ denote the extension of μ from $[1 \cup \mathcal{S}_p(G)]$ to a larger poset $[1 \cup \mathcal{S}_p(G)] \cup \infty$, where ∞ denotes a formal unique maximal element. Then we get

(5.3.18)

$$\tilde{\chi}\big(\mathcal{S}_p(G)\big) \stackrel{(5.3.17)}{=} \overline{\mu}(1, \infty) \stackrel{(5.3.16)}{=} - \sum_{L \in [1 \cup \mathcal{S}_p(G)]} \overline{\mu}(1, L) = - \sum_{L \in [1 \cup \mathcal{S}_p(G)]} \mu(1, L).$$

We emphasize that the sum[7] on the right is over the full poset $[1 \cup \mathcal{S}_p(G)]$, whereas the similar sum in earlier (5.3.14) is only over the orbit poset $[1 \cup \mathcal{S}_p(G)]/G$; we will adjust for this difference in the final calculation that follows below.

[7]We also mention that [**HIÖ89**] treats this sum as the primary object of interest; without necessarily interpreting it as an Euler characteristic.

Brown's Theorem via the Gluck/Yoshida formula. The proof below of Brown's Theorem appeared independently in both Gluck [**Glu81**] and Yoshida [**Yos83**]:

The main observation of these papers is a formula for the coefficients $c_{H,K}$ in (5.3.12) above, which is phrased in terms of the values of the Möbius function:

$$(5.3.19) \qquad c_{H,K} = \frac{|K|}{|N_G(H)|}\mu(K,H); \text{ recall } \mu(K,H) = 0 \text{ when } K \not\leq H.$$

We now apply this formula to calculate the coefficient $c_{\hat{1},1}$, that is, taking $K = 1$ in our earlier discussion of $c_{\hat{1},K}$; and we adjust the computation from the orbit poset to the full poset $\mathcal{S}_p(G)$, using the multiplicity $|G : N_G(L)|$ of L in the latter:

$$\begin{aligned} c_{\hat{1},1} &\overset{(5.3.14)}{=} \sum_{L\in[1\cup\mathcal{S}_p(G)]/G} c_{L,1} \overset{(5.3.19)}{=} \sum_{L\in[1\cup\mathcal{S}_p(G)]/G} \frac{1}{|N_G(L)|}\mu(1,L) \\ &= \sum_{L\in[1\cup\mathcal{S}_p(G)]/G} \frac{|G:N_G(L)|}{|G|}\mu(1,L) = \frac{1}{|G|}\sum_{L\in[1\cup\mathcal{S}_p(G)]}\mu(1,L) \\ &\overset{(5.3.18)}{=} -\frac{\tilde{\chi}(\mathcal{S}_p(G))}{|G|}. \end{aligned}$$

Now since $c_{\hat{1},1}$ is a p-local integer by (5.3.13), we see that the p-part $|G|_p$ of the denominator $|G|$ must in fact divide the numerator $\tilde{\chi}(\mathcal{S}_p(G))$; completing our third proof of Brown's Theorem.

We close the section by mentioning that at later Corollary 6.6.4, we will indicate a generalization of Brown's Theorem 5.3.1 by Thévenaz, developed in the context of representation theory, which considers divisibility of $\tilde{\chi}(K)$ by smaller powers of p (indeed typically only slightly smaller).

5.4. Vanishing of $\tilde{\chi}(K)$ for suitable K modulo other divisors of $|G|$

In this section, we discuss a generalization of Brown's Theorem 5.3.1, which in effect replaces $|G|_p$ by some divisor n of $|G|$—where now n is not necessarily a prime power. For such an n, we set

$$\mathcal{S}_n(G) := \{H \in L(G)_{>1} : |H| \text{ divides } n\}.$$

(Notice that our usual p-subgroups poset $\mathcal{S}_p(G)$ would be described in this new notation by replacing the subscript p with the particular value of $n = |G|_p$.)

And then the generalization of 5.3.1 for general n (established by several authors independently, as described below) is:

THEOREM 5.4.1. *For n dividing $|G|$,*

$$\tilde{\chi}(\mathcal{S}_n(G)) \equiv 0 \pmod{n}.$$

One proof of Theorem 5.4.1 above was given by Thévenaz in [**Thé86**]; and in the subsequent deeper theoretical development of Brown and Thévenaz [**BT88**], the result was also a corollary of their more general result (1.1)—see (1.4) there, taking $H = 1$. In fact the result (1.4) obtained in [**BT88**] was also established independently[8] by Hawkes, Isaacs, and Özaydin as 5.1 in [**HIÖ89**] (with Theorem 5.4.1 above indicated as Corollary 5.4 there).

These proofs extending Brown's Theorem are somewhat more complicated, and in particular are too lengthy for us to present in detail here. So instead, in this section we give a briefer sketch of the arguments—trying to emphasize how they build further on the methods which we had discussed in the previous Section 5.3.

However before going on to those proofs, we first consider some small examples of Theorem 5.4.1.

EXAMPLE 5.4.2 ($\tilde{\chi}(\mathcal{S}_6(A_5))$). Let G be A_5, and take $n := 6$. The subgroups of G of order 6 are isomorphic to S_3; and subgroups of the other possible divisors 3 and 2 are in fact contained in these S_3-subgroups. These inclusions are as in the (nontrivial part of the) poset diagram in earlier Example 1.1.12: namely 1 subgroup of order 3, and 3 subgroups of order 2. Furthermore there are 10, 10, 15 conjugate subgroups of the orders 6, 3, 2: represented for example by $\langle (12)(34), (345) \rangle$, $\langle (345) \rangle$, and $\langle (12)(34) \rangle$. Consequently we compute

$$\tilde{\chi}(\mathcal{S}_6(G)) = -1 + (10 + 10 + 15) - 10(3 + 1) = -1 + 35 - 40 = -6.$$

And this is indeed congruent to 0 mod 6, as required for Theorem 5.4.1. ♢

EXERCISE 5.4.3 ($\tilde{\chi}(\mathcal{S}_6(S_4))$). Take $G := S_4$ and $n := 6$. Here there are *no* subgroups of order 6; so our poset $\mathcal{S}_6(G)$ is disconnected. Check that there are 4 subgroups of order 3, and $(3 + 6)$ subgroups of order 2; so that $\tilde{\chi}(\mathcal{S}_6(G)) = 12$. which is indeed congruent to 0 mod 6, as required. ♢

We mention that one particular case of Theorem 5.4.1, namely where n is $|G|_\pi$ for a set of primes π, had been established by Bouc as [**Bou84b**, (ii) of Cor, p 11].

The proof of Thévenaz via Burnside ring idempotents. The approach of [**Thé86**] to Theorem 5.4.1 is perhaps the most direct extension of part of our discussion in the previous section, namely Proof 3 of Gluck/Yoshida via the Burnside ring:

Thévenaz now considers a "family" of subgroups, which is closed under taking subgroups and conjugates; we let $\mathcal{F}$ denote the nontrivial part, such as $\mathcal{S}_n(G)$, of a family. In analogy with $e_{\hat{1}}$, he studies the sum $e_{\mathcal{F}} := \sum_{L \in [1 \cup \mathcal{F}]/G} e_L$; and for the corresponding coefficient $c_{\mathcal{F},1}$ defined in analogy with (5.3.14), he applies the Gluck/Yoshida idempotent formula just as in the calculation after (5.3.19):

[8]Historical anecdote—seemingly I first bore the bad tidings of the overlap: I heard of the result informally from Thévenaz, at the June 1985 group theory program at ENSJF in Paris. And later I was in the audience when Hawkes lectured on the result at the U. of Chicago group theory seminar in March 1986. (As I understand, Gruenberg also mentioned the coincidence to Brown at Oberwolfach in June 1986.) Of course the authors communicated, and adjusted the papers to reference each other. Note that the "preprint [25]" thus mentioned in the first paragraph of [**HIÖ89**] was apparently intended to be [26] in the references there; this preprint was not published as such, but instead was subsumed into [**BT88**].

$$c_{\mathcal{F},1} \overset{(5.3.19)}{=} \sum_{L\in[1\cup\mathcal{F}]/G} \frac{1}{|N_G(L)|}\mu(1,L) = \sum_{L\in[1\cup\mathcal{F}]} \frac{1}{|G|}\mu(1,L) \overset{(5.3.18)}{=} -\frac{\tilde{\chi}(\mathcal{F})}{|G|}.$$

But at this point, Thévenaz replaces the earlier reference to Dress [**Dre69**] (for the p-local integers) with a notion of the *exponent* of the family, which describes the possible denominators in the coefficients for $e_{\mathcal{F}}$. Thévenaz in fact makes use of the standard *mark* homomorphisms on the Burnside algebra: Namely for $K \in L(G)/G$, the map ϕ_K takes the basis element $u_{G/H}$ to the size $\#\big((G/H)^K\big)$ of the fixed-point set under K. And indeed (e.g. [**Thé87**, after 1.1]), this value gives the coefficient at K, when we express u_H in terms of the basis of idempotents e_K; that is, the matrix of $\mathbb{Q}$-coefficients in earlier (5.3.12) arises as the inverse of the $\mathbb{Z}$-matrix of these sizes. If follows that:

(5.4.4) For x in the Burnside algebra, $x = \sum_{K\in L(G)/G} \phi_K(x)e_K.$

His Proposition 2.2 exploits this to show that the exponent is a multiple of the gcd of the indices in G of the subgroups in $\mathcal{F}$.

Thus in particular when $\mathcal{F} = \mathcal{S}_n(G)$, this gcd is $|G|/n$. And then since $c_{\mathcal{F},1}$ has "at most" this denominator, it follows that n must divide the numerator term $\tilde{\chi}\big(\mathcal{S}_n(G)\big)$ in expression above—completing the proof.

The Hawkes-Isaacs-Özaydin proof via the Möbius function. The main result of [**HIÖ89**] establishes the following divisibility relation for an arbitrary subgroup H of G (where taking $H = 1$ yields as a corollary their 5.4, which is our Theorem 5.4.1):

THEOREM 5.4.5 (5.1 in [**HIÖ89**]). *Assume H is a subgroup of G, and n divides $|G|$. Then*

$$\frac{|G|}{|N_G(H):H|}\ \tilde{\chi}\big(\mathcal{S}_n(G)_{>H}\big) \equiv 0 \pmod{n}.$$

Here we have used the language of $\tilde{\chi}$, where [**HIÖ89**] instead uses the language of the sum $\sum_{H\leq L\in[1\cup\mathcal{S}_n(G)]} \mu(H,L)$—these are equivalent just as in (5.3.18).

In summary, the proof of [**HIÖ89**, 5.1] obtains divisibility of the above expression by $n_p := |G|_p$ for any prime p, using induction on $|G|_p/n_p$.

The base step where $n_p = |G|_p$ (that is, when $|G|_p$ divides n) is handled by their earlier 3.9, which in turn is a corollary of an intermediate divisibility result 3.7. The proof of 3.7 uses essentially the same method in Surowksi's Proof 2 in the previous section: namely that vanishing of the character in (5.3.11) away from a normal subgroup B of some group A leads to a multiple of the regular character of A/B, and hence to divisibility by the order of that quotient. The relevant vanishing (see [**HIÖ89**, p 1013, using Lemma 2.4]) in effect uses a mild generalization of the contractibility of $\mathcal{S}_p(G)^g$, which we had quoted via Lemma 3.3.20 in describing Surowski's argument.

Some comments on the Brown-Thévenaz proof. Theorem 5.4.5 above also appeared (though phrased in somewhat different language) as 1.4 in the work of Brown and Thévenaz [**BT88**]—where it is deduced from the main result Theorem 1.1 there.

There are some similarities, in broad outline, with the proof of Hawkes-Isaacs-Özaydin indicated just above: Section 3 of [**BT88**] argues by induction—where the case of powers of a fixed p is handled in Section 2 there, in a manner somewhat like the base step in [**HIÖ89**, 3.9].

However, the details of the analysis in [**BT88**] largely follow the original method of Brown in [**Bro75**]—though now carried out in the language of the Möbius function. (Recall we had postponed indicating Brown's approach to our later Option G discussion in Section 5.6; though of course we had described Quillen's version of Brown's argument as Proof 1 in the previous section.)

We do mention here that the statement of [**BT88**, 1.1] involves a set D of divisors of $|G|$: where D is considered as a poset, with "divides" in the role of the order relation $\leq$. And then the power of p in divisiblity conditions is determined essentially by the number of p-powers which lead to conical contractibility of the poset D: namely powers p^a such that $\mathrm{lcm}(p^a, d)$ is in D for all $d \in D$, so that multiplication by p^a is a poset map taking D into the cone above p^a.

In particular we note, following [**BT88**, (A), p 415]: If D consists entirely of (nontrivial) powers of p, then for the subposet $\mathcal{P}$ of $\mathcal{S}_p(G)$ of subgroups of the corresponding orders, from their result 1.1 we get that $\tilde{\chi}(\mathcal{P})$ divisible by $p^{|D|}$. Thus for example when D consists just of the single value $|G|_p$, so that $\mathcal{P}$ is just $\mathrm{Syl}_p(G)$, we get Sylow's Theorem; while when D consists of all nontrivial powers[9] up to $|G|_p$, so that $\mathcal{P}$ is all of $\mathcal{S}_p(G)$, we get Brown's Theorem. That is: their 1.1 generalizes both the Sylow and Brown Theorems.

DIGRESSION 5.4.6 (The set of elements of order dividing n). The results described in this section, for the poset $\mathcal{S}_n(G)$ of subgroups of G of order dividing n, can also be used to study a classical theme of Frobenius, namely the *set*:

$$E_n(G) := \{ \text{ elements of } G \text{ of order dividing } n \ \}.$$

Indeed in all three of the papers indicated above—namely at [**Thé86**, Cor 3.3], [**HIÖ89**, Thm 6.3], and [**BT88**, Cor 1.6]—there are deductions of the well-known theorem of Frobenius [**Fro03**] that:

$$|E_n(G)| \equiv 0 \pmod{n}. \tag{5.4.7}$$

Furthermore one could hope (as I suspect those authors might have been hoping) that the methods should also be applicable to the famous related conjecture of Frobenius—namely that if the set $E_n(G)$ has size *exactly* n, then it must even be a sub*group* of G. However, as far as I know, no such approach has succeeded.

Instead, a proof of the conjecture was obtained a few years after the time of these papers, by Iiyori and Yamaki—see [**IY91**]; that is, the Frobenius conjecture is now the Iiyori-Yamaki theorem:

$$\text{If } |E_n(G)| = n, \text{ then the set } E_n(G) \text{ is in fact a subgroup of } G. \tag{5.4.8}$$

The authors proceed using a reduction, from the 1954 Ohio State Ph.D. thesis of R. Zemlin, to the case of G a nonabelian finite simple group. They then invoke the Classification of Finite Simple Groups; and in a sequence of papers, they verify the conjecture for the simple groups in the corresponding list. ♢

[9]This case had also been covered in [**Thé87**, Cor 2.3]).

5.5. Other results on vanishing and non-vanishing

The previous sections were devoted to two theorems which are perhaps the best-known results on the Euler characteristic of subgroups posets. In this section, we will try to give a brief glimpse of some of the various further directions taken by the subsequent literature involving the Euler characteristic.

In fact, our choice of results to present is designed to at least begin to touch on some issues which will be considered in certain of our later chapters—especially in Chapter 8, which discusses Quillen's Conjecture 3.3.8.

A lemma on vanishing mod p. First we continue our theme of vanishing of $\tilde{\chi}$, this time modulo just a single prime p: We present a technical lemma—which is used in the final argument of the Aschbacher-Smith work [**AS93**] on the Quillen Conjecture. (We indicate the actual application of the lemma in the discussion after the statement of Theorem 8.4.4.)

The lemma is essentially a variant of Robinson's influential observation [**Rob88**, Prop 2.3] (which we treat later as Theorem 8.3.1), concerning the fixed points on $\mathcal{S}_p(G)$ of a *q-hyperelementary* subgroup X, for a prime $q \neq p$; this means that X can be written as the semidirect product of a cyclic normal subgroup $\langle x \rangle$ with a q-group Q.

LEMMA 5.5.1 (0.14 in [**AS93**]). *Assume a poset $\mathcal{A}$ satisfies $\tilde{L}(\mathcal{A}) = 0$, and admits action of a q-hyperelementary group X. Then $\tilde{\chi}(\mathcal{A}^X) \equiv 0 \pmod q$.*

PROOF. By hypothesis we may write $X = \langle x \rangle Q$ as above; and also the Lefschetz module is zero, so that the character value on it at x vanishes—and then we apply the Lefschetz formula:

$$0 = Tr(x, 0) = Tr\big(x, \tilde{L}(\mathcal{A})\big) \overset{(5.1.12)}{=} \tilde{\chi}(\mathcal{A}^x) \overset{5.1.1}{=} \sum_{i=-1} (-1)^i \, \#(\mathcal{A}_i^x).$$

Furthermore Q normalizes $\langle x \rangle$, so it permutes the fixed points of x on $\mathcal{A}$, and hence also permutes the fixed chains $\mathcal{A}_i^x$. And non-trivial Q-orbits on $\mathcal{A}_i^x$ have length divisible by q; so arguing just as in (5.3.7) (or using the relative chain complex for the inclusion of $\mathcal{A}^X$ in $\mathcal{A}^x$, as at (5.3.8) in Proof 1 in Section 5.3), we get:

$$\tilde{\chi}(\mathcal{A}^x) \equiv \tilde{\chi}(\mathcal{A}^{\langle x \rangle Q}) = \tilde{\chi}(\mathcal{A}^X) \pmod q,$$

which with the equation above gives the conclusion of the Lemma. □

A formula for $\tilde{\chi}$ in a solvable group. In our earlier discussion before Definition 5.3.15, we had briefly referred to the paper [**KT84**] of Kratzer–Thévenaz, and we will now say a little more:

As a rough summary, that work studies the Möbius function on the lattice $L(G)$ of subgroups of G, emphasizing the context of the Burnside ring of G. Furthermore the authors exploit a well-known combinatorial formula due to Crapo (see [**KT84**, 2.1]), which is based on the set of *complements* to a subgroup H of G—meaning subgroups K with $H \cap K = 1$ and $\langle H, K \rangle = G$.

In a main result [**KT84**, 2.6], they apply this approach to the class of solvable[10] groups G—and are able to express the Möbius function value $\mu(H, G)$ in

[10]Recall this means that the simple quotients in a composition series are of prime order.

terms of numbers of suitable complements in a *chief series* for G: namely subgroups $1 =: G_n < G_{n-1} < \cdots < G_1 < G_0 := G$, where each G_i is normal in G, each quotient G_i/G_{i+1} is abelian, and the series is maximal with these properties.

Below we will just quote the special case $H = 1$—which via (5.3.17) gives the Euler characteristic of the "proper" part of the subgroup lattice: namely to avoid obvious contractibility, we consider the subposet $\hat{L}(G) := L(G)^{<G}_{>1}$ with the unique minimal and maximal elements removed:

(5.5.2) For solvable G, $\tilde{\chi}\big(\hat{L}(G)\big) = (-1)^n \, m_1 m_2 \cdots m_{n-1}$,

where in $\overline{G} := G/G_{i+1}$, m_i is the number of complements to $\overline{G}_i$.

EXAMPLE 5.5.3 (The subgroup lattice of A_4). In the alternating group $G = A_4$, which is solvable of order 12, the proper non-trivial part of the subgroup lattice $L(G)$ consists of 5 connected components: of these, four are points, each given by one of the Sylow 3-subgroups; and the fifth is contractible—a cone, with unique maximal element given by the normal Sylow 2-subgroup V_4, which contains the 3 subgroups of order 2. So by homotopy equivalence, we can replace the cone just by its cone point; and then we calculate directly that $\tilde{\chi}\big(\hat{L}(G)\big) = -1 + 5 = 4$.

On the other hand, the unique chief series, of length $n = 2$, is determined by $G_1 := V_4$—which has 4 complements given by the Sylow 3-subgroups. Thus the formula in (5.5.2) gives $(-1)^2 4$—agreeing with our calculation in the previous paragraph.

We mention that for the subgroup lattice of the alternating and symmetric groups in general, the value of $\tilde{\chi}$ (in the language of the Möbius function) has been studied in the combinatorial literature; see for example Shareshian [**Sha97**]. ◇

Of course the numerical value in the formula is interesting in itself. Furthermore, in the context of vanishing as in our earlier sections in this chapter, we see that the formula in particular gives a criterion for vanishing of $\tilde{\chi}$—namely that at least one term in the chief series is not complemented in the relevant quotient.

At first sight, this condition (characterizing the vanishing of $\tilde{\chi}$) might seem potentially different from the sufficient condition for vanishing given by contractibility: for we recall from Lemma 5.2.3 that in general, contractibility is strictly stronger. However for the special case of $\hat{L}(G)$ for solvable G, the conditions of contractibility and vanishing of $\tilde{\chi}$ can be seen to be equivalent, in Proposition 4.13 of the authors' subsequent paper [**KT85**]. Indeed we mention that the work of [**KT85**] goes beyond the numerical content of the Möbius function, to more explicit topological structure; they show (see Corollary 4.10 there) that $\hat{L}(G)$ has the homotopy type of a suitable bouquet of spheres. (We had mentioned such bouquet structures in discussing the Solomon-Tits argument in the proof of Theorem 3.4.15; and we return to this theme in later Section 9.4.)

Finally we mention that Welker in [**Wel94**] extends the analysis to more general intervals in solvable G, describing values $\mu(H, K)$ for $K < G$. (For more on interval structures, see also Shareshian [**Sha03**], Aschbacher-Shareshian [**AS09**], and papers of Lucchini, including [**BL97**]. And the theme of complementation appears elsewhere in the relevant literature—for example, in [**BW83b**].)

A non-vanishing result toward Quillen's Conjecture. We now turn attention from the lattice $L(G)$ to our more usual p-subgroups poset $\mathcal{S}_p(G)$; and we recall that Quillen's Conjecture 3.3.8 is that $O_p(G) = 1$ should imply that $\mathcal{S}_p(G)$ is not contractible. One approach to the Conjecture is to try to establish the stronger conclusion of non-vanishing of $\tilde{\chi}(\mathcal{S}_p(G))$—since that forces non-contractibility by the contrapositive of Lemma 5.2.2.

This is the approach (though phrased in the language of the Möbius function, via (5.3.18)) taken by Hawkes and Isaacs in [**HI88**, Thm A]—which establishes the indicated result for the special case where G is p-solvable (that is, where the simple quotients in a composition series are either p'-groups or of order p) with an abelian Sylow p-subgroup. We recall that p-solvable means that G has a normal series in which each quotient is either a p-group or a p'-group.

This result is one of the early special cases which we will quote, during our main discussion of Quillen's Conjecture in Chapter 8. But it is also natural to provide, in this chapter, some further discussion of the Hawkes-Isaacs result: partly because of its statement in terms of the Euler characteristic; but also because it involves some further themes which we have touched on already, and will see more of in various later chapters.

Thus we will be discussing:

THEOREM 5.5.4 (Thm A in [**HI88**]). *Assume G is p-solvable, and has abelian Sylow p-subgroups. Then* $O_p(G) > 1 \iff \tilde{\chi}(\mathcal{S}_p(G)) = 0$.

Of course we had already mentioned, after Lemma 5.2.2, that the $\Rightarrow$ direction is a corollary of the contractibility in Lemma 3.3.5. So only the $\Leftarrow$ direction will require proof. We will work toward the contrapositive: that is,

We assume that $O_p(G) = 1$; and we must show that $\tilde{\chi}(\mathcal{S}_p(G)) \neq 0$.

(For an example of this direction, consider $G = A_4$ as in Example 5.5.3, and take $p := 3$: Here G is 3-solvable (indeed even solvable), with $O_3(G) = 1$—and by omitting V_4 from the computation in that Example, $\tilde{\chi}(\mathcal{S}_3(G)) = -1 + 4 = 3 \neq 0$.)

Some easy initial reductions. Since we are assuming $O_p(G) = 1$ (and we may take $G > 1$), the hypothesis of p-solvability guarantees that $O_{p'}(G) > 1$—where we recall this denotes the largest normal subgroup of G of order coprime to p. For consistency with notation for similar configurations in later chapters:

We write $L := O_{p'}(G)$; and A for a Sylow p-subgroup of G. (So $L \cap A = 1$.)

Now let bars denote images in $\bar{G} := G/L$. By conjugacy in Sylow's theorem, $\bar{A}$ contains the normal p-group $O_p(\bar{G})$; and then by the abelian hypothesis, $\bar{A}$ must in fact centralize this subgroup. On the other hand, the standard Hall-Higman lemma [**Hal59**, Thm 18.4.4] states for p-solvable $\bar{G}$ that $O_p(\bar{G})$ contains its own centralizer;[11] and so we must in fact have $O_p(\bar{G}) = \bar{A}$. Thus LA is normal in G; so LA contains *all* Sylow p-subgroups of G, and hence all p-subgroups of G—so that $\mathcal{S}_p(G) = \mathcal{S}_p(LA)$. Furthermore $O_p(LA) = 1$ follows from $O_p(G) = 1$ in view of the normality of LA. Thus it suffices to prove nonzero Euler characteristic for LA, and hence (cf. [**HI88**, 1.2]):

We assume from now on that $G = LA$.

Next we recall from Theorem 4.2.4 that the elementary subposet $\mathcal{A}_p(G)$ is homotopy equivalent to $\mathcal{S}_p(G)$—and hence by Remark 5.1.15, these posets have the same Euler characteristic. Furthermore $\mathcal{A}_p(LA) = \mathcal{S}_p(L\Omega_1(A))$, and $O_p(L\Omega_1(A)) = 1$ just as

[11] Or more generally, one could use here the standard self-centralizing property (8.2.7) of the generalized Fitting subgroup.

before, so that it suffices to prove the assertion for $L\Omega_1(A)$. Hence (cf. [**HI88**, 1.4]):

We assume from now on that $A = \Omega_1(A)$; that is, A is elementary abelian.

At this point, the further analysis of Hawkes and Isaacs emphasizes the combinatorics of the Möbius function—and becomes fairly complicated. So we will indicate instead a fairly elementary group-theoretic argument given later by Alperin.[12]

Alperin's reduction to solvable G. So far, $L = O_{p'}(G)$ is just p-solvable—so it might contain nonabelian simple composition factors of order coprime to p. The following argument of Alperin allows us to avoid this possible complication—by in effect replacing L with a related solvable group L_s; that is, it shows:

we may assume L is solvable.

And this solvable LA is then a configuration which had been treated earlier, e.g. by Quillen.

So we will devote the next few paragraphs to a description of how to construct L_s, so that $\mathcal{S}_p(L_sA)$ has the same Euler characteristic as $\mathcal{S}_p(LA)$.

To this end, we will first develop a formula for $\chi(\mathcal{S}_p(LA))$ which does *not* explicitly depend on the exact isomorphism type of L. We start by decomposing the Euler characteristic according to the end points of chains: For we can partition the chains in the order complex via

$$K(\mathcal{S}_p(G)) = \bigcup_{B \in \mathcal{S}_p(G)} K(\mathcal{S}_p(G))_B ,$$

where the subscripted term on the right denotes chains ending at B. And for $i > 0$, i-chains ending at B correspond—on removal of B—to $(i-1)$-chains in the subposet $\mathcal{S}_p(B)_{<B}$. For $i = 0$, the 0-chain on the right, given by B itself, corresponds to a formal (empty) (-1)-chain in the augmented complex. Thus if we compute the alternating sum on the left for the (non-reduced) Euler characteristic χ, then on distributing that sum over the partition on the right, we obtain reduced Euler characteristics—with reversed sign because of the shift by a single dimension:

$$\chi(\mathcal{S}_p(G)) = - \sum_{B \in \mathcal{S}_p(G)} \tilde{\chi}(\mathcal{S}_p(B)_{<B}).$$

Next, by Sylow's theorem we can partition those various chain end-points B into conjugates of distinct B which lie in our fixed Sylow group A; these conjugates (of which there are $|G : N_G(B)|$) will each contribute the same value of $\tilde{\chi}$. Hence

$$\chi(\mathcal{S}_p(G)) = - \sum_{B \in \mathcal{S}_p(A)} |G : N_G(B)| \, \tilde{\chi}(\mathcal{S}_p(B)_{<B}).$$

And we can now begin to make use of our product structure $G = LA$. Since A is abelian by hypothesis, $A \leq N_G(B) = N_{LA}(B)$; and we can write the latter as the product $N_L(B)A$ using the elementary Dedekind modular law (e.g. [**Asc00**, 1.14]). Furthermore in view of the normality of the p'-group $L = O_{p'}(LA)$, we have

[12]Alperin gives this argument in an unpublished manuscript "Contributions to the Euler characteristic of the Brown complex"—although he there attributes the general technique to a lecture of Isaacs.

that $[N_L(B), B] \leq N_L(B) \cap B \leq L \cap A = 1$, so that $N_L(B) = C_L(B)$. Thus we may re-write the previous formula as:

$$\chi\big(\mathcal{S}_p(LA)\big) = -\sum_{B \in \mathcal{S}_p(A)} |L : C_L(B)|\ \tilde{\chi}\big(\mathcal{S}_p(B)_{<B}\big). \tag{5.5.5}$$

And this is the formula promised earlier: note that it depends just on the values of the centralizer orders $|C_L(B)|$—but not on the exact isomorphism type of L.

Alperin's argument exploits this independence by making use of the *coprime action* of A on L: A convenient statement of standard results—providing essentially a Sylow theory for A-invariant subgroups of L—is 18.7 in [**Asc00**].[13] First, for each prime q dividing the order of L, by (1) of 18.7 there is at least one A-invariant Sylow q-group of L; and we fix one such, denoting it by L_q. Now take any $B \leq A$. As A is abelian it centralizes B, and hence normalizes $C_L(B)$. So by 18.7(1) again, now applied to the action of A on $C_L(B)$, there is an A-invariant Sylow q-group X of $C_L(B)$. And then by 18.7(3), X lies in some A-invariant Sylow q-group Y of L. Further by 18.7(2), $C_L(A)$ is transitive on the A-invariant Sylow groups of L; so on conjugating Y to L_q by an element of $C_L(A) \leq C_L(B)$, we see that we may as well assume that $X \leq L_q$. And now since X is Sylow in $C_L(B)$, we must in fact have $X = C_{L_q}(B)$; that is, we see that $C_{L_q}(B)$ is Sylow in $C_L(B)$.

It follows that we have a set[14] product:

$$C_L(B) = \prod_q C_{L_q}(B). \tag{5.5.6}$$

And now we are ready to define our solvable (even nilpotent) variant L_s of L, as the corresponding *direct* product:

$$L_s := \times_q L_q.$$

We see then that we have a set product $C_{L_s}(B) = \prod_q C_{L_q}(B)$ like that in (5.5.6); so that at the level of group orders, we obtain the equality $|C_{L_s}(B)| = |C_L(B)|$. These orders are the terms of interest in (5.5.5) above—from which we can now conclude that $\chi\big(\mathcal{S}_p(L_sA)\big) = \chi\big(\mathcal{S}_p(LA)\big)$; completing Alperin's reduction to the solvable case.

Obtaining $\tilde{\chi}\big(\mathcal{S}_p(LA)\big) \neq 0$. We now just briefly indicate how to finish the proof—which involves some topics we have touched on earlier, but will which we will be considering in more detail later.

Indeed the most general approach at this point would be to proceed by quoting the later treatment of our group LA, which now is solvable, as a certain standard minimal configuration in the analysis of Quillen for solvable groups. (In particular, we finally use here our earlier reduction to A being *elementary* abelian, say of rank $r \geq 1$, as part of that minimal setup.) For example, at Theorem 9.4.5, we will discuss Quillen's proof that $\mathcal{S}_p(LA)$ has the Cohen-Macaulay property—which in particular guarantees that homology vanishes in all lower dimensions, so that $\tilde{\chi}$ is given (up to sign) by the dimension of $\tilde{H}_{r-1}$; while at Theorem 8.2.9, we will

[13] We emphasize that it suffices as hypothesis there just that one of the two factors in LA (namely A in our case here) should be solvable.

[14] We emphasize that for arbitrary subsets of the primes q, the corresponding set product is *not* necessarily a subgroup of G. We would in fact get such arbitrary products if we already had obtained L solvable—in view of P. Hall's standard extension [**Asc00**, 18.5] of the Sylow theory to subsets of primes in solvable groups.

see his proof that this dimension is nonzero. Hence these later results would then complete the proof of Theorem 5.5.4.

However, we can also give here a more informal sketch, in terms of material at least briefly mentioned earlier: In particular, the Cohen-Macaulay property arises essentially by showing that $\mathcal{S}_p(LA)$ is a bouquet of spheres (in the spirit of our reference in the previous section to [**KT85**, 4.10] for this property for $L(G)$). To expand a little on this remark, we mention that after reductions to a "minimal faithful" situation at the end of the proof of Theorem 8.2.9, the group LA decomposes as a commuting product of r groups L_iA_i, with each A_i of rank 1, and each $O_p(L_iA_i) = 1$. The latter implies that each $\mathcal{S}_p(L_iA_i)$ is a set of (at least 3) isolated vertices—where each pair of such vertices determines a 0-sphere S^0. (We will consider spheres more formally in later Section 8.1.) Now the complex for $\mathcal{S}_p(LA)$ decomposes (cf. Lemma 8.1.2, with a corresponding product formula for $\tilde{\chi}$ in (8.1.1)) as the join of the complexes for the $\mathcal{S}_p(L_iA_i)$—in turn determining spheres of dimension $r-1$ (cf. (8.1.5)). These spheres then give the desired nonzero homology and Euler characteristic (cf. Example 5.1.3). And the above vanishing of lower-dimensional homology follows from "puncturing" each sphere, and observing contractibility of the remaining subcomplex—in the spirit of the Solomon-Tits argument in earlier Section 3.4.

We will see at later Theorem 8.2.12 that the above property of non-zero homology can in fact be obtained directly for p-solvable LA with $O_p(LA) = 1$. But as Hawkes and Isaacs comment, this does not seem to force their conclusion of non-vanishing of $\tilde{\chi}$; and so does not seem to lead to an alternative proof of their result Theorem 5.5.4.

5.6. (Option G): The G-equivariant Euler characteristic

In this section, we give a brief and informal description of the *G-equivariant* Euler characteristic of a G-space such as $\mathcal{S}_p(G)$, as used in Brown's papers [**Bro74**], and [**Bro75**], and in relevant parts of the subsequent literature. For the overall context, we will need to preview some material which we will mainly be discussing in later chapters.

A notion of Euler characteristic $\chi(G)$, for (suitable) groups G. Brown's papers are primarily motivated by an "Euler characteristic" value $\chi(G)$—which can be associated to certain groups G, and is defined in terms of the group cohomology of G. (We will be investigating group cohomology in later Chapter 7; see Definition 7.1.1 for a brief introduction.) So we first need to sketch Brown's definition—which exhibits both similarities and differences, when compared to the Euler characteristic for the chain complex of a finite-dimensional simplicial complex (such as $K(\mathcal{S}_p(G))$) in Remark 5.1.4.

One similarity is the use, for group cohomology, of the natural *co*chain complex for the classifying space BG in Definition 7.1.1.2. However, an important difference is that this cochain complex is infinite-dimensional—and the resulting group cohomology, for our usual case of general finite G, can be non-zero (and non-periodic) in an infinite number of dimensions.[15] Thus in general we could not expect to define

[15] I thank Jacques Thévenaz for helpful clarifications in this area. (Any remaining inaccuracies are mine...)

an Euler characteristic via a finite alternating sum using cochains or cohomology, such as those in Lemma 5.1.14.

But this difficulty can be circumvented for certain classes of (typically) *infinite* groups G; and we now describe this (very roughly, without attempting to indicate (or fully explain) the *exact* hypotheses): The condition of *finite cohomological dimension* (e.g. [**Bro74**, Sec 3]) can be phrased in terms of a suitably-finite projective resolution[16] over $\mathbb{Z}$ for the trivial G-module. (Hence one gets a corresponding finiteness property, when group cohomology is obtained via Ext of the trivial module in Definition 7.1.1.3.) In particular (e.g. [**Bro94**, p 249]) one gets cohomology groups of $\mathbb{Z}$-rank 0 after some fixed dimension—leading to a definition for $\chi(G)$ in terms of a finite alternating sum of cohomology in the spirit of Lemma 5.1.14.

Next, G is said to be of finite *virtual* cohomological dimension (again cf. [**Bro74**, Sec 3]) if G has a subgroup H of finite index, with H of finite cohomological dimension. Here another difference arises: We can extend the definition of χ for H to G, by setting

$$\chi(G) := \frac{\chi(H)}{|G:H|}; \tag{5.6.1}$$

and we see that the resulting value $\chi(G)$ is now in general a rational number—that is, it is *not* necessarily any longer an integer (as would be the case for an ordinary Euler characteristic, given by an alternating sum of dimensions of chain spaces). Indeed, the denominators in $\chi(G)$ are closely connected with any elements in G of finite order; this is a topic of interest for example in the study of various such G which are arithmetic groups.

Finally G is of *finite homological type* if furthermore the finite-index subgroups of finite homological dimension in fact have finitely generated homology. In (most of) the remainder of the section, we implicitly assume that G satisfies these hypotheses.

The G-equivariant Euler characteristic $\chi_G(X)$, for a G-space X. One tool for approaching $\chi(G)$ involves investigating another Euler characteristic, corresponding to the G-equivariant cohomology of a suitable G-space X (such as $\mathcal{S}_p(G)$)—a topic we will discuss at Remark 7.3.1. Indeed our summary below will essentially preview the notion of ampleness of X in Definition 7.3.4, where we can "approximate" the group cohomology of G via the G-equivariant cohomology of X.

Just as in the discussion of $\chi(G)$ above, a complication arises from the infinite-dimensionality of the space $(EG \times X)/G$ involved[17] in the Borel construction for equivariant cohomology in Remark 7.3.1. Here (and again we will indicate the hypotheses only very roughly) for G of finite homological type, and suitably "finite" and contractible X with torsion-free stabilizers G_x for $x \in X$, the pages of the spectral sequence for equivariant cohomology in Remark 7.3.1 are finite; and one gets a suitable finite alternating sum (constant on those pages), defining the G-equivariant Euler characteristic $\chi_G(X)$. Furthermore (see e.g. [**Bro94**, Prop IX.7.3(e′)]) that

[16]We discuss projective modules in the subsequent chapter starting at Definition 6.1.3; for resolutions see e.g. [**Ben98**, Defn 2.4.1].

[17]We mention that this space, and $K(\mathcal{S}_p(G))/G$ a little later, are in fact spaces of G-*orbits*; we provide some further discussion of the role of orbit spaces in later Section 11.1.

definition of $\chi_G(X)$ then agrees with $\chi(G)$; and indeed also with a simpler alternative definition of the equivariant Euler characteristic used in Brown's papers, and indicated below at (5.6.3).

Computing with the G-equivariant Euler characteristic $\chi_G(\mathcal{S}_p(G))$. In Brown's paper [**Bro75**], attention shifts away from a contractible X as above, such that $\chi_G(X)$ exactly computes $\chi(G)$—to a space X with somewhat less restrictive hypotheses, for which $\chi_G(X)$ now computes the rational number $\chi(G)$ only up to an integer.

Indeed Theorem 1 in [**Bro75**] shows that $\chi(G) - \chi_G(X)$ is an integer, under the assumption that X^H is acyclic for each finite nontrivial subgroup H of G.

This is in turn used in the proof of Theorem 2 in [**Bro75**], which shows that

$$\chi(G) - \chi_G(\mathcal{S}_p(G)) \text{ is a } p\text{-local integer.} \tag{5.6.2}$$

Here $\mathcal{S}_p(G)$ now means the poset of *finite* nontrivial p-subgroups of G. We sketch the argument: Since we are assuming G is of finite virtual cohomological dimension, we can find some normal subgroup of N of G which is of finite index, and torsion-free. Let G_p denote the primage of a Sylow p-subgroup of the finite quotient G/N. Because of the multiplicative property of Euler characteristics of groups (cf. (5.6.1)), the values for G are those for G_p divided by $|G : G_p|$. Since this index is coprime to p, we can in effect reduce to the subgroup G_p; that is, it suffices to show that $\chi(G_p) - \chi_{G_p}(\mathcal{S}_p(G))$ is an integer—which we will do via Theorem 1: Because N is torsion-free, any finite nontrivial subgroup H of G_p is in fact a p-group. And then $\mathcal{S}_p(G)^H$ is contractible (the argument for Lemma 3.3.20 works in our present setup), and so is acyclic by 5.2.3. Thus Theorem 1 does indeed complete the proof.

It remains to tie this general result to the assertion for finite groups which we have been calling Brown's Theorem 5.3.1. For this, we now indicate a simpler (but equivalent) formulation of the equivariant Euler characteristic which Brown uses:

$$\chi_G(\mathcal{S}_p(G)) := \sum_{\sigma \in K(\mathcal{S}_p(G))/G} (-1)^{\dim \sigma} \frac{1}{|G_\sigma|}. \tag{5.6.3}$$

Now notice that in the case of finite G, we have $\frac{1}{|G_\sigma|} = \frac{|G:G_\sigma|}{|G|}$, where the numerator on the right is just the number of G-conjugates of σ. So we can re-write the sum over the orbit complex instead as a sum over *all* the simplices of $K(\mathcal{S}_p(G))$; and then the G-equivariant Euler characteristic $\chi_G(\mathcal{S}_p(G))$ becomes just $\frac{\chi(\mathcal{S}_p(G))}{|G|}$, namely the ordinary Euler characteristic divided by $|G|$. On the other hand, the trivial group is just a point, and so has Euler characteristic 1; so $\chi(G)$ is then just $\frac{1}{|G|}$ by (5.6.1). Thus (5.6.2) becomes the statement that $\frac{\tilde{\chi}(\mathcal{S}_p(G))}{|G|}$ is a p-local integer. This is the formulation of Brown's Theorem that we discussed after (5.3.19) during Proof 3 of that result given in Section 5.3. (It appears as Corollary 2 in [**Bro75**].)

CHAPTER 6

The reduced Lefschetz module $\tilde{L}$ and projectivity

Brown's analysis of $\mathcal{S}_p(G)$ in [**Bro75**] led not only to his vanishing result Theorem 5.3.1 for the reduced Euler characteristic, but also to a result giving projectivity for the corresponding reduced Lefschetz module $\tilde{L}(\mathcal{S}_p(G))$. And again, a subsequent version of the proof due to Quillen (who stated the result as [**Qui78**, Cor 4.3]) became widely known; this led in turn to considerable further research on projectivity-related properties of Lefschetz modules. Correspondingly this second "applications" chapter will survey a number of such results on projectivity.

Our first main goal for the chapter will be that Brown-Quillen projectivity result, which we treat below as Theorem 6.2.1—indeed we then go on to a generalized and strengthened form due to Webb [**Web87a**, Thm A′], which we treat as Theorem 6.3.4.

But before doing so, following our usual procedure, we will first present some background material on modular representation theory and projective modules, in the initial Section 6.1. This topic of projectivity essentially continues our earlier theme of vanishing, which ran through the previous Chapter 5: For example, sometimes we will use the elementary fact that a contractible space is acyclic—and hence its reduced Lefschetz module vanishes (as we had indicated in Lemma 5.2.3). But more usually, our eventual goal will instead be the property that the reduced Lefschetz module (though typically not zero) is in fact *projective*; and then the relevant notion of vanishing is instead at the level of the cohomology of the module (as we will observe at Lemma 6.1.5).

Then using this viewpoint, in Sections 6.2 and 6.3, we will discuss the general projectivity results of Brown-Quillen and Webb. Thereafter in Section 6.4, in our Option B, we consider the special case of G a Lie type group: where Quillen observed that for the building geometry, the projective Lefschetz module given by the above general results is in fact the well-known Steinberg (irreducible) module. That result in turn motivated the study of projective modules which are sometimes given by sporadic geometries—the topic of Section 6.5, in our Option S.

In the final Section 6.6, we will indicate some generalizations which deliver the still weaker condition of *relative* projectivity; this extension of the analysis was pioneered by Thévenaz in [**Thé87**].

6.1. Algebraic background: Projectivity and vanishing of cohomology

As usual, the reader familiar with the basics of modular representation theory should be able to skip ahead directly to the applications beginning in the subsequent Section 6.2.

Some basic facts from modular representation theory. So far, our fairly informal applications of the representation theory of our group G have in fact used *ordinary* representations: that is, typically we have considered kG-modules, where k is a field of characteristic 0: this includes chain and homology modules defined initially with coefficients just in $\mathbb{Z}$, so that we can take $k = \mathbb{Q}$; while for the ordinary character theory of more general modules, we usually work over the complex numbers $\mathbb{C}$.

However, we will now begin to devote more attention to the case of *modular* representation theory—where coefficients are taken from a field k of prime characteristic p (often just the field $\mathbb{F}_p$ of order p). When p does not divide the order of G, the theory is not fundamentally different from the ordinary representation theory. But in this book normally p *does* divide the order of G; and then the p-modular representation theory is more complicated.

The study of p-modular representation theory was largely pioneered by Brauer in the 1950s, and has continued to develop substantially since then. As usual, below we will summarize only a small number of facts, which we will want to quote at various points; for more detailed reference, there are many good sources, such as Curtis and Reiner—[**CR06**, Ch XII] or [**CR90**, Ch 2], and Feit [**Fei82**].

The first important difference from characteristic 0 is that (for p dividing the order of G) the group algebra $\mathbb{F}_pG$ is *not* a semisimple G-module: So when we break up $\mathbb{F}_pG$ as far as possible into a direct sum of submodules, the resulting *indecomposable* summands are not typically irreducible.

EXAMPLE 6.1.1 (The group algebra $\mathbb{F}_2S_2$). Consider for example $G := S_2$ and $p = 2$: The group algebra has basis given by the two group elements 1 (identity) and $(1,2)$. And so there is a 1-dimensional (hence irreducible) G-submodule I, spanned by the element sum $1 + (1,2)$—but there is no G-module *complement* to I (that is, no vector-space complement to I is normalized by the G-action). Thus the group algebra $\mathbb{F}_2G$ is in this case indecomposable; there is no decomposition into two or more direct summands. $\diamondsuit$

The indecomposables appearing in $\mathbb{F}_pG$ are called the *principal* (or *projective*) indecomposables. Each is in fact the *projective cover* $P(I)$ of some irreducible I—namely $P(I)$ is indecomposable; and projective in the sense defined below at 6.1.3; and has a unique irreducible quotient, namely I. Furthermore the multiplicity of $P(I)$ as a summand of $\mathbb{F}_pG$ is the dimension $\dim I$.

The case where G is a p-group Q is particularly special: Then there is only one irreducible, given by the (1-dimensional) trivial $\mathbb{F}_pQ$-module. (We denote it by $1_{\mathbb{F}_pQ}$, or more often just by 1 if there is no danger of confusion.) And (cf. the properties in the previous paragraph) the group algebra $\mathbb{F}_pQ$ is indecomposable—it consists just of the single projective cover $P(1)$. Notice that this is indeed the behavior which we observed in Example 6.1.1 above.

In our later discussions, we will make some brief references to the *p-blocks* (or just *blocks*) of $\mathbb{F}_pG$; for details, see e.g. [**CR06**, Sec 85]. Here we will just mention that the irreducibles I are partitioned into blocks—according to the appearance of the $P(I)$ in a decomposition of $\mathbb{F}_pG$ into maximal two-sided ideals. In effect, the partitioning links irreducibles in a graph, where edges are given by pairs which could occur in a *non*-split module extension.

Free and projective modules. As the above glimpse of structure in $\mathbb{F}_pG$ suggests, it is important to study not just the irreducible modules, but also the principal indecomposables that cover them. And we do this in a slightly more general context: namely we first roughly generalize the module behavior of the full group algebra:

DEFINITION 6.1.2 (Free modules). A *free* $\mathbb{F}_pG$-module can be defined as a (finite)[1] direct sum of copies of the group algebra $\mathbb{F}_pG$.

Notice for a subgroup $H \leq G$ that the restriction $\mathbb{F}_pG|_H$ is the direct sum of the G/H-translates of the group algebra $\mathbb{F}_pH$. Hence for a free $\mathbb{F}_pG$-module F, the restriction $F|_H$ is H-free (that is, is free as an $\mathbb{F}_pH$-module). ◊

This in turn allows us to roughly generalize the module behavior of the principal indecomposables:

DEFINITION 6.1.3 (Projective modules). We indicate a number of alternative formulations of the condition of projectivity (see the references for the equivalence of the conditions):

(1) One definition is that a *projective* F_pG-module is a direct summand of a free module.

Hence from our description after Example 6.1.1 of the decomposition of the group algebra, a projective module is just a direct sum of projective covers $P(I)$, for various irreducibles I (appearing with corresponding multiplicities).

Indeed since we saw in particular there that for a p-group Q, the group algebra $\mathbb{F}_pQ$ is indecomposable, it follows that a projective $\mathbb{F}_pQ$-module is in fact always a free module.

Furthermore since we saw in Definition 6.1.2 that freeness is preserved by restriction to a subgroup $H \leq G$, it follows for a projective $\mathbb{F}_pG$-module P that the restriction $P|_H$ is also H-projective[2] (that is, is a projective $\mathbb{F}_pH$-module).

In particular when H is a Sylow p-subgroup Q of G, the restriction $P|_Q$ is projective, and hence free—that is, a sum of copies of the group algebra $\mathbb{F}_pQ$, with dimension a multiple of $|Q| = |G|_p$. (And when P is the reduction mod p of some $\mathbb{Z}G$-module R, the character value on R of nontrivial elements of Q is 0, since they fix no points in the group algebra $\mathbb{Z}Q$; see e.g. [**Fei82**, Cor IV.2.5].)

(2) A standard alternative to the above definition (1) of projectivity of P is given (e.g. [**CR90**, Prop 2.22(iii)] or [**Fei82**, Thm I.4.1(iii)]) by the condition that any module surjection $M \twoheadrightarrow P$ must in fact split—that is, we then get a direct sum $M \cong P \oplus N$ as modules. This can of course be expressed in the language that any short exact sequence $0 \to N \to M \to P \to 0$ should always in fact split.

(3) Here is a partial converse to our final remark in (1) above: Suppose that the subgroup H contains a Sylow p-subgroup of G; then P is $\mathbb{F}_pG$-projective if and only if the restriction $P|_H$ is projective as an $\mathbb{F}_pH$-module.

Of course that remark in (1) gives the $\Rightarrow$ direction. So for the $\Leftarrow$ direction, assume that P is $\mathbb{F}_pH$-projective; and consider any surjection $M \twoheadrightarrow P$. Then the $\mathbb{F}_pH$-projective hypothesis means by (2) that the surjection restricted to H,

[1]Unless otherwise specified, we assume that all modules are *finite*-dimensional.

[2]The reader should contrast this abbreviated terminology with our later usage of "rel-H projective", in Definition 6.6.1 of *relative* projectivity; the meanings are different (even roughly complementary).

namely $M|_H \twoheadrightarrow P|_H$, splits as $\mathbb{F}_pH$-modules. We now summarize the remainder of the argument—previewing the language of module cohomology, which we explore more fully in the subsequent chapter, starting at Definition 7.1.1. Since we have $|G : H|$ coprime to p by our Sylow hypothesis, we can use the standard cohomological Lemma 7.1.2 (or just the more elementary module-theoretic argument [**Fei82**, Cor II.3.10], i.e. Maschke's theorem), to see that this splitting as $\mathbb{F}_pH$-modules conversely implies splitting as $\mathbb{F}_pG$-modules. So here we conclude that the surjection $M \twoheadrightarrow P$ splits as $\mathbb{F}_pG$-modules. Hence as M was arbitrary, we can again use (2) to see that P is indeed an $\mathbb{F}_pG$-projective module. $\diamondsuit$

Here is a useful general instance of (3) above:

EXAMPLE 6.1.4. Assume that K is a p'-subgroup of G; and let P be the permutation module on the cosets G/K. (We might more fully denote this module as $\mathbb{F}_p[G/K]$, or as the induced module $\mathrm{Ind}_K^G(1)$.) Then P is $\mathbb{F}_pG$-projective: For let S denote a Sylow p-subgroup of G; and suppose that $s \in S$ fixes a coset Kg, i.e. $(Kg)s = Kg$. Then the p-element gsg^{-1} lies in the p'-group K, so that we must have $s = 1$. Hence S acts in regular orbits on the cosets G/K; thus the restriction $P|_S$ is a free (hence projective) $\mathbb{F}_pS$-module, and so we see that P is $\mathbb{F}_pG$-projective via (3) of Definition 6.1.3. $\diamondsuit$

In the later sections of this chapter, we will need to consider *virtual* projectivity: namely the extension of the notion of projectivity from ordinary modules to virtual modules; for we will wish to consider the reduced Lefschetz (virtual) module $\tilde{L}(K)$ for a G-complex K.

To approach this extension, we first recall a standard property (e.g. [**CR06**, p 552a]) of ordinary projectivity: that if we tensor a projective module with *any* module, the product is again projective. And then since tensoring gives the multiplication in any of the rings of modules in Definition 5.1.11, we see that the $\mathbb{F}_p$-linear combinations of ordinary projectives determine an ideal, in each of those rings—which we call the ideal of *virtual projective* modules. (Indeed from Definition 6.1.3 we see that the projective covers $P(I)$ of the irreducibles I afford a $\mathbb{Z}$-basis for this ideal.)

Projectivity and vanishing of module cohomology. We close this initial background section by expanding on our theme of vanishing, which we had mentioned more briefly in our summary at the beginning of the chapter.

First we remark that (much as was the case of the reduced Euler characteristic in the previous chapter) our various results in earlier chapters on contractibility of suitable complexes L deliver, via Lemma 5.2.3, some immediate corollaries on the vanishing $\tilde{L}(L) = 0$ of the reduced Lefschetz module. We will be using such corollaries from contractibility[3] in many proofs, in the material from here on.

However, more typically we will be dealing with G-complexes K which are not contractible (though we may study contractible subcomplexes L within K). We may

[3]But for an approach to projectivity not directly phrased in the language of contractibility, see our discussion of [**Rob88**, 2.1] in the final subsection of Section 6.4.

then be able to establish at least that $\tilde{L}(K)$, though non-zero, is still a virtual $\mathbb{F}_pG$-projective. And for this situation, the condition in (2) of Definition 6.1.3 shows essentially that a projective module does not admit any non-split extensions.

That is, the appropriate notion of vanishing for a projective module is at the level of Ext-functors and module cohomology; so once again we anticipate the later language of Definition 7.1.1.3, to state the relevant vanishing property (e.g. [**CR90**, 8.4.v], [**Bro94**, III.6.1.iii′]):[4]

LEMMA 6.1.5. *Assume that P is a projective $\mathbb{F}_pG$-module. Then*

$$Ext^n_{\mathbb{F}_pG}(-, P) = 0 \text{ for } n \geq 1.$$

In particular evaluating this at the trivial module 1, we get the corresponding vanishing of the higher module cohomology of P:

$$Ext^n_{\mathbb{F}_pG}(1, P) \stackrel{7.1.1.3}{=} H^n(G, P) = 0 \text{ for } n \geq 1.$$

We mention that this vanishing of cohomology for $n > 0$ is the condition usually called *acyclicity* for a module; cf. Definition 7.1.1.3. However, sometimes (e.g. Lemma 7.2.2) our arguments will also depend on vanishing at $n = 0$; so for this stronger condition, we will instead use the expression "module cohomology vanishes"—usually adding an explicit emphasis such as "in all dimensions" or "including in dimension $n = 0$". (The literature sometimes seems a little unclear on this point.)

And to briefly pursue this further property: we mention that, since $\text{Ext}^0 = \text{Hom}$ in Definition 7.1.1.3 of module cohomology, we can additionally obtain vanishing for $n = 0$—whenever the projective P does not have as a summand the projective cover $P(1)$ of the trivial module. Thus "usually" in fact our projective modules will have the indicated vanishing of cohomology in all dimensions. Furthermore in later chapters (see e.g. Remark 7.2.4), we will encounter some modules M which are not even projective, but still exhibit this vanishing of cohomology.

(Option S:) In particular in sporadic p-local geometries K, in cases where $\tilde{L}(K)$ turns out not be projective,[5] typically the module still has vanishing of cohomology even for $n = 0$. (And this property is by Lemma 7.2.2 equivalent to a cohomology decomposition—so that projectivity is not necessarily required, in order to obtain a decomposition.)

6.2. The Brown-Quillen result on projectivity of $\tilde{L}(\mathcal{S}_p(G))$

We have now arrived at the initial goal of the chapter, namely the $\mathbb{F}_p$-projectivity of $\tilde{L}(\mathcal{S}_p(G))$. This result is at least implicit in Brown [**Bro75**], and was stated explicitly in Quillen [**Qui78**, Cor 4.3]. We will refer to it as the Brown-Quillen Projectivity Theorem 6.2.1 below.

Motivation and statement. One reason to try to establish projectivity is that it gives a generalization of Brown's Theorem 5.3.1: For we saw in discussing Definition 6.1.3.1 of projectivity that the dimension term $\tilde{\Lambda}(\mathcal{S}_p(G))(1) = \tilde{\chi}(\mathcal{S}_p(G))$ is a multiple of $|G|_p$.

[4] We mention that in this area, we are implicitly using the fact that for a finite group G, projectivity is equivalent to *injectivity*—so that we get for cohomology some properties which are normally stated instead for homology.

[5] Such as in later Example 6.6.5, and its continuation in Example 7.2.13; and indeed in many similar cases in [**BS08**].

(We had in fact mentioned this aspect in the previous chapter, after Surowski's Proof 2 of Brown's Theorem. So for comparison and contrast, we recall that the crucial tool used by Surowski was instead the vanishing (5.3.11), at nontrivial elements of a Sylow p-subgroup S of G, of the character of the underlying $\mathbb{Z}G$-module—whose reduction mod p gives the $\mathbb{F}_pG$-module version of $\tilde{L}(\mathcal{S}_p(G))$ that we focus on in this chapter. Of course Surowski obtains this vanishing property using contractibility results; but we had also noted after 6.1.3.1 that this vanishing could equally well be obtained as a consequence of projectivity—which holds by Theorem 6.2.1 below.)

A still more cogent reason for expecting projectivity is provided, using the viewpoint of module theory, by the free action of the Sylow group S on a basis for the chain spaces for the relative complex $C_*(K, L)$—an argument we saw in (5.3.7) in Proof 1 of Brown's Theorem. This free Sylow action is the formulation of projectivity given in Definition 6.1.3.3, and so it strongly suggests the property for $\tilde{L}(K)$ also. And indeed, such an action is the crucial observation used in Quillen's proof of the projectivity.

The Brown-Quillen result, which we state below, establishes the projectivity of the reduced Lefschetz module $\tilde{L}(\mathcal{S}_p(G))$, as viewed within the Grothendieck group of $\mathbb{F}_pG$-modules in Definition 5.1.11; recall this version of the module is determined just by the Lefschetz character $\tilde{\Lambda}(K)$. The result was later generalized by Webb to projectivity of the Lefschetz module, as viewed within the representation ring of G; we will treat that generalization later, as Theorem 6.3.4.

THEOREM 6.2.1 (Brown-Quillen Projectivity Theorem). *The reduced Lefschetz module $\tilde{L}(\mathcal{S}_p(G))$ is a virtual projective module when viewed in the Grothendieck group of G. That is, the reduced Lefschetz character $\tilde{\Lambda}(\mathcal{S}_p(G))$ agrees with the character of some $\mathbb{Z}$-linear combination of (ordinary) projective modules.*

Details of a particular example. Before describing Quillen's proof of Theorem 6.2.1, we first compute with a small example:

EXAMPLE 6.2.2 (Projectivity of $\tilde{L}(\mathcal{S}_2(A_5))$). We build on our earlier calculation for $G = A_5$ and $p = 2$, which was sketched in Exercise 5.3.3: That $\tilde{\chi}$ computation $-1+(5+15)-15 = 4$ must now be expanded out—from these dimension-terms, to the alternating sum (as in Definition 5.1.9) of the permutation modules for corresponding the chain spaces. Indeed if we let F and T denote representatives of subgroups of order 4 and 2, we can write that sum in terms of the induced modules:

$$\tilde{L}(\mathcal{S}_2(A_5)) = -\mathrm{Ind}_G^G(1) + \left(\mathrm{Ind}_{G_F}^G(1) + \mathrm{Ind}_{G_T}^G(1)\right) - \mathrm{Ind}_{G_{T<F}}^G(1).$$

And in this case, we get a simplification: Since T lies in a unique F of order 4, we have $G_T \leq G_F$, so that $G_{T<F} = G_T \cap G_F = G_T$. Thus the opposite-signed terms above for G_T and $G_{T<F}$ cancel out of the sum, leaving us just with $-1+\mathrm{Ind}_{G_F}^G(1)$. The latter induced module is just the "natural" permutation module for A_5, namely on its 5 Sylow 2-subgroups. One computes directly that this module decomposes as the direct sum of the trivial module and the 4-dimensional (irreducible) orthogonal module V for $A_5 \cong \Omega_4^-(2)$ in earlier Example 2.1.25 (note especially the formulation in the final paragraph there), so that $\tilde{L}(\mathcal{S}_2(A_5)) = V$.

Furthermore it is standard that V is projective: For example, it is the Steinberg module (as in Remark 6.4.1; cf. later Theorem 6.4.2) for A_5 viewed as the Lie type group $\Omega_4^-(2)$ (or $L_2(4)$). Alternatively, a Sylow group F permutes regularly the 4 Sylow groups other than itself; these determine 4 of the 5 singular vectors in Example 2.1.25, which in turn determine a basis for V—and this free Sylow action gives projectivity by Definition 6.1.3.3. Thus we have in fact demonstrated the stronger assertion in later Theorem 6.3.4, namely projectivity for the Lefschetz module itself, in the representation ring (rather than just for its character as in Theorem 6.2.1).

But we can also follow the approach of the Lefschetz character $\tilde{\Lambda}(\mathcal{S}_p(G)) =: \tilde{\Lambda}$. In view of (5.1.12), we need to calculate the values $\tilde{\chi}(\mathcal{S}_p(G)^g)$ for representatives g of the G-conjugacy classes. Since we are considering the character of the $\mathbb{F}_pG$-module, which is a *modular* character, it suffices to consider g a $2'$-element. (If we were considering the module instead in characteristic 0,[6] then for g of order divisible by 2, we would just get character value of 0—by contractibility of $\mathcal{S}_p(G)$ in Lemma 3.3.26, along with the usual 5.2.3.)

Of course for $g = 1$ we have $\mathcal{S}_p(G)^1 = \mathcal{S}_p(G)$; and above we had computed the corresponding dimension term $\tilde{\Lambda}(1)$, as $\tilde{\chi}(\mathcal{S}_p(G)) = 4$. For an element g of order 5 such as (12345), note that g does not normalize any conjugate of F or T; so $\mathcal{S}_p(G)^g$ is empty, giving $\tilde{\chi}(\mathcal{S}_p(G)^g) = -1+0 = -1$. For g of order 3 such as (123), we see g normalizes two conjugates of F (determined by the 4-sets containing $1, 2, 3$), but normalizes no subgroup T of order 2; hence $\tilde{\chi}(\mathcal{S}_p(G)^g) = -1+2 = 1$. In summary, we have the character:

g	1	(123)	(12345)	(13542)
$\tilde{\Lambda}(g)$	4	1	-1	-1

And this agrees with the character of the projective module V above: namely the nontrivial part of the permutation character—obtained by subtracting 1 from the number of fixed points of g in the natural permutation representation on 5 letters.

Alternatively one can find this character denoted by ϕ_4 in the Modular Atlas [**JLPW95**, p 2]. (Of course it is just the reduction mod 2 of the complex irreducible denoted by χ_4 in the Atlas [**CCN$^+$85**, p 2].) And the induced characters defining the Lefschetz character can be easily computed using the group theory language GAP [**GAP00**]. ♢

REMARK 6.2.3 (Blocks of defect 0). A different way of concluding projectivity of V above would be to observe that its $\mathbb{C}$-character χ_4 is an irreducible of degree 4—which is a multiple of (indeed, here equal to) $|G|_2$. For in general, any $\mathbb{C}$-irreducible whose dimension is a multiple of $|G|_p$ is always $\mathbb{F}_p$-projective: namely (in the language of blocks which we briefly mentioned at the end of the first subsection of the previous Section 6.1) it determines a block *of defect* 0. (This is the Brauer-Nesbitt Theorem; see for example [**CR06**, 86.3].) Such a block contains only one complex irrreducible, and the corresponding modular irreducible I—which is projective, hence equal to its own projective cover $P(I)$.

(More generally, any block B has a defect group D, which is a p-group; and we say B is of defect d if $|D| = p^d$.) ♢

[6] As we can: here (and indeed often) our modular character is obtained as the reduction mod p of a $\mathbb{Z}$-character. But more generally, a modular character determines, via "lifting of idempotents", a corresponding *Brauer* character defined in characteristic 0.

Comments on Quillen's proof. Since we will be shortly be presenting a proof of Webb's generalization at 6.3.4, here we will give only a quick sketch of Quillen's proof of Theorem 6.2.1. The argument might be regarded roughly as a way of extending the decomposition for $\tilde{\chi}$ in (5.3.8) to the reduced Lefschetz module $\tilde{L}$: The contractibility of the singular set L this time implies that $\tilde{L}(L) = 0$ using Lemma 5.2.3; so that $\tilde{L}(K)$ is approached via the corresponding alternating sum for the relative complex $C_*(K, L)$. The argument requires some manipulations in homological algebra; for example, passing to a complex where the homology is concentrated in the top dimension—so that one can consider ordinary rather than virtual modules. And these manipulations preserve at least the Lefschetz character, though not necessarily the structure of the Lefschetz module; this is the reason for working in the Grothendieck group rather than the representation ring. Nonetheless, it is the free action of a Sylow p-subgroup S on the relative complex $C_*(K, L)$ that delivers the conclusion of projectivity—essentially via Definition 6.1.3.3.

We recall from Remark 5.1.15 that a homotopy equivalence preserves homology groups, and hence preserves the reduced Lefschetz module; so it follows that the projectivity in the Brown-Quillen Theorem 6.2.1 holds for the reduced Lefschetz module of any member of the standard homotopy type of $\mathcal{S}_p(G)$ in Remark 4.6.1, such as $\mathcal{A}_p(G)$, $\mathcal{B}_p(G)$, etc.; including the building, when G is of Lie type in characteristic p.

But one can also then ask whether projectivity might hold for still other complexes K, which might not be equivalent to $\mathcal{S}_p(G)$. And indeed in the subsequent literature (for example [**RSY90**] and [**AM95a**]), many further interesting complexes K were analyzed via the approach of verifying the hypotheses for Webb's Theorem 6.3.4 below; and we will be discussing various examples of such K, in later sections of this chapter. So we now continue to:

6.3. Webb's projectivity conditions for a more general complex K

We had already mentioned in the previous section that Webb's Theorem 6.3.4 below generalizes the Brown-Quillen Theorem 6.2.1—from projectivity just in the context of the Grothendieck group, to projectivity in the representation ring.

But in a different direction of generalization, Webb in fact considers a more general G-complex K: And as the hypothesis on K, he gives certain possible sufficient conditions—these are essentially variants, abstracted from the special case of $K = K\big(\mathcal{S}_p(G)\big)$, of the condition which is the basis for the proof of projectivity in Theorem 6.2.1. However since these various conditions are phrased in somewhat different terms, we will first pause (prior to discussing Webb's proof), to discuss more explicitly the interrelationship of these conditions, which are relevant to Webb's hypotheses (a) and (b) for Theorem 6.3.4

Contractibility and $\mathbb{F}_p$-acyclicity for fixed points. Recall from our discussion of Theorem 6.2.1 that Quillen's proof used contractibility of the singular set L of $\mathcal{S}_p(G)$; we indicated this result as Proposition 4.4.15—the proof of which in turn depended on contractibility of the fixed-point set $\mathcal{S}_p(G)^Q$ for p-subgroups $Q > 1$, which we gave as Lemma 3.3.20.

REMARK 6.3.1 (Interrelations among some contractibility conditions). Indeed for a more general G-complex K with type-preserving action, let us for comparison juxtapose the corresponding contractibility conditions—now including. as the initial term (I) below, the variant which we had already previewed earlier (e.g. at Example 3.3.23 and Remark 4.4.7):

(I) For each subgroup Q of order p in G, K^Q is contractible.

(II) For each nontrivial p-subgroup Q of G, K^Q is contractible.

(III) For $P \in \mathrm{Syl}_p(G)$, the singular set $K^{P\text{-sing}}$ is contractible.

These conditions have arisen at various earlier points, and they are at least ostensibly different; but our point here will be to emphasize that they are not necessarily *very* different: For example, clearly (II) includes (I)—so that (I) provides a hypothesis which is typically somewhat weaker that (II). And Lemma 4.4.14 with P in the role of "H" shows that (II) in fact also implies (III). (And indeed some of the topological literature focuses on the singular-set hypothesis in (III); for example, this is a feature of the approach of Adem-Milgram in [**AM04**], especially Section V.3 there.)

On the other hand, if we embed the subgroup Q of (I) or (II) in a Sylow group P, we see that K^Q is a subcomplex of $K^{P\text{-sing}}$—and then we might typically expect the contractibility in (III) to lead to contractibility for the subcomplexes in (I) or (II). (For this it is of course necessary to check the actual embedding of that subcomplex.)

(Option G:) There are natural extensions of these observations to the context of Q- and P-contractibility. $\diamondsuit$

Now the contractibility conditions in (I)–(III) above imply, via Lemma 5.2.3 as usual, the corresponding $\mathbb{F}_p$-acyclicity conditions—which appear as (i)–(iii) in Lemma 6.3.3 below. And indeed in various explicit complexes K, in earlier applications as well as many still to come, we will usually be able to establish the stronger contractibility conditions, in order to deduce the $\mathbb{F}_p$-acyclicity conditions. (For example, since we noted that (II) implies (I) and (III), we see that under (II) we get each of (i)–(iii) also.)

However, for the actual proofs of various general results still to come (especially results related to group cohomology in the subsequent Chapter 7), it will turn out to be more natural to state hypotheses in terms of the weaker $\mathbb{F}_p$-acyclicity conditions. In particular,[7] Webb's basic Hypothesis 6.3.4(b) is just the $\mathbb{F}_p$-acyclicity of K^Q in (i) of Lemma 6.3.3 below. And indeed that Lemma indicates why it suffices, in (i), to restrict attention to just those Q of order exactly p. This sufficiency in fact demonstrates one advantage of working in the context of $\mathbb{F}_p$-acyclicity: namely that we can then use the standard (P. A.) Smith Theorem (e.g. [**AM04**, p 163]), which states:[8]

(6.3.2) For a p-group Q on an $\mathbb{F}_p$-acyclic complex K, K^Q is also $\mathbb{F}_p$-acyclic.

This principle allows us to again indicate some interrelations among ostensibly-different conditions:

---

[7]Essentially a variant of Webb's Hypothesis 6.3.4(b) appears in Bouc [**Bou**, Prop 3].

[8]We emphasize that the Smith Theorem requires *finite* dimensionality of K; recall that our K always satisfy this, unless otherwise specified.

LEMMA 6.3.3. *Assume that K is a G-complex with type-preserving action. Consider the following $\mathbb{F}_p$-acyclicity conditions:*

(i) For each subgroup Q of order p in G, K^Q is $\mathbb{F}_p$-acyclic.

(ii) For each nontrivial p-subgroup Q of G, K^Q is $\mathbb{F}_p$-acyclic.

(iii) For $P \in \mathrm{Syl}_p(G)$, the singular set $K^{P\text{-sing}}$ is $\mathbb{F}_p$-acyclic.

Then (i) and (ii) are equivalent, and are implied by (iii).

(Option G:) There are natural extensions of these statements to the context of Q- and P-equivariant $\mathbb{F}_p$-homology.

PROOF. Again (ii) includes (i); so now assume (i). Let $Q > 1$ be a p-subgroup, and consider K^Q. There is some R of order p in Q; and K^Q is a subcomplex of K^R, which is $\mathbb{F}_p$-acyclic since we are assuming (i). Let $R =: Q_1 < Q_2 < \cdots Q_n := Q$ denote a chain of subgroups, each of index p in the next. Then K^{Q_2} is a subcomplex of K^{Q_1}, which we saw is $\mathbb{F}_p$-acyclic. Furthermore Q_2 normalizes Q_1, so that Q_2 normalizes the fixed-point set K^{Q_1}; hence by the P. A. Smith Theorem (6.3.2), $(K^{Q_1})^{Q_2} = K^{Q_2}$ is also $\mathbb{F}_p$-acyclic. And now since Q_3 again normalizes Q_2, we can iterate this argument to see that K^{Q_3} is $\mathbb{F}_p$-acyclic; and eventually, that K^Q is $\mathbb{F}_p$-acyclic. This completes the proof of the converse, namely that (i) implies (ii).

Now assume (iii). Let Q be some nontrivial p-subgroup of G. We embed Q in some $P \in \mathrm{Syl}_p(G)$, so K^Q is a subcomplex of $K^{P\text{-sing}}$, which is in fact $\mathbb{F}_p$-acyclic since we are assuming (iii). Since Q is subnormal in P, we may mimic the above argument, to conclude that K^Q is also $\mathbb{F}_p$-acyclic. This completes the proof that (iii) implies (ii). □

We will see shortly that the $\mathbb{F}_p$-acyclicity in Webb's Hypothesis 6.3.4(b) below in fact implies his Hypothesis 6.3.4(a)—which is stated in yet another context,[9] namely that of the vanishing of the reduced Euler characteristic $\tilde{\chi}(K^H) = 0$, for subgroups H with $O_p(H) > 1$ and $H/O_p(H)$ cyclic. So we mention that, much as for $\mathbb{F}_p$-acyclicity above, we will typically obtain the vanishing in (a) from stronger contractibility results (again via the usual Lemma 5.2.3): For example, we had seen at earlier Lemma 3.3.26 that contractibility of $\mathcal{S}_p(G)^H$ for such H follows from contractibility of $\mathcal{S}_p(G)^Q$ for non-trivial p-subgroups Q (i.e., from a condition of the form (II) in Remark 6.3.1).

In the spirit of the above discussion of conditions which might not in fact be so very different, we conclude with a "warning": namely that the more general complexes K, which will satisfy the hypotheses of Webb's Projectivity Theorem 6.3.4, may not always even be very much more general than the standard poset $\mathcal{S}_p(G)$, which appears in the Brown-Quillen Theorem 6.2.1. This is primarily because (as Webb had observed early on) the usual contractibility of fixed points K^Q under p-subgroups (which as we indicated is the usual way we verify his hypothesis) can sometimes[10] go a long way toward showing that K is even homotopy equivalent to $\mathcal{S}_p(G)$.[11] Thus (as we indicated just before the start of this section) projectivity for such a K could be—and often is—obtained by first showing equivalence of K

[9]I.e. for use in Burnside-ring arguments, namely the appeal to Conlon's induction theorem in the proof below.

[10]Indeed this is often the case in applications to p-local geometries; cf. the discussion of 1.3 and 1.4 in [**SY97**].

[11]E.g. this situation can arise in arguments involving contractibility of the "K^H" in earlier Remark 4.4.7.

with $\mathcal{S}_p(G)$, and then just quoting projectivity for $\mathcal{S}_p(G)$. For example, for the elementary-abelian subposet $K = \mathcal{A}_p(G)$, we get the same Lefschetz module as for $\mathcal{S}_p(G)$, using Quillen's standard equivalence in 4.2.4 (in view of the invariance in 5.1.15).

Nevertheless, we emphasize that the use of Webb's hypotheses for a direct proof of projectivity for K is often very elementary—especially sometimes in cases where equivalence of K with $\mathcal{S}_p(G)$ does not hold (or even is just not obvious).

Statement and proof. So we now turn to Webb's Projectivity Theorem.

In the Brown-Quillen Projectivity Theorem 6.2.1, we had viewed the Lefschetz module $\tilde{L}(\mathcal{S}_p(G))$ in terms of its image just in the Grothendieck group of G-modules. But we can instead consider the Lefschetz invariant $\tilde{\mathcal{L}}(K)$ (for a more general G-complex K) defined via G-set structures in the Burnside ring as in Definition 5.1.11: for we saw in Definition 5.1.9 that $\tilde{L}(K)$ is defined as an alternating sum of permutation representations, and hence $\tilde{L}(K)$ is the image of $\tilde{\mathcal{L}}(K)$ under λ, when we map the Burnside ring into the permutation subring of the representation ring of G.

As usual, for our complex K below we will require the type-preserving property of Remark 1.5.7. We present below a slight re-phrasing of Theorem A$'$ in [**Web87a**]. (But see also our mention of Webb's more general statement in terms of projective *chain complexes* in later Remark 6.4.5.)

THEOREM 6.3.4 (Webb's Projectivity Theorem). *Assume K is a G-complex with type-preserving action, satisfying:*

(a) for any $H \leq G$ with $O_p(H) > 1$ and $H/O_p(H)$ cyclic, we have $\tilde{\chi}(K^H) = 0$.
(In particular (a) holds under the following stronger hypothesis:

(b) for any subgroup Q of order p, the fixed subcomplex K^Q is $\mathbb{F}_p$-acyclic.)
Then $\tilde{L}(K)$ is a virtual projective module in the representation ring of G.

PROOF. We begin by following the initial argument of Webb [**Web87a**, p 148], showing that (b) implies (a): So suppose we are given some subgroup H of G such that $O_p(H) > 1$ with $H/O_p(H)$ cyclic. Since (b) is just (i) of Lemma 6.3.3, the equivalence of (i) with (ii) in that Lemma (via the P. A. Smith Theorem (6.3.2)) shows that the fixed subcomplex $K^{O_p(H)}$ is $\mathbb{F}_p$-acyclic. Hence by 5.2.3, we get vanishing $\tilde{L}(K^{O_p(H)}) = 0$ of the reduced Lefschetz module. Let g denote an element generating the cyclic quotient $H/O_p(H)$, so that $K^H = (K^{O_p(H)})^g$. Then

$$\tilde{\chi}(K^H) = \tilde{\chi}\big((K^{O_p(H)})^g\big) \overset{(5.1.12)}{=} Tr\big(g, \tilde{L}(K^{O_p(H)})\big) = Tr(g, 0) = 0.$$

Thus (a) holds if we assume (b).

Now for the main proof of projectivity, we assume (a): namely that $\tilde{\chi}(K^H) = 0$ for such H. At this point, Webb's argument (as compared to Quillen's homological-algebra argument for Theorem 6.2.1) instead emphasizes the more combinatorial viewpoint of the Burnside ring. In fact, the relevance in that ring of the the hypothesis $\tilde{\chi}(K^H) = 0$ is given a particularly clear and elegant presentation in the related argument of Thévenaz for his generalization [**Thé87**, Thm 2.1] of Webb's result to weaker conditions giving relative projectivity (which we treat later as Theorem 6.6.2). So below, we will largely follow approach the approach of Thévenaz.

First, we recall from Definition 5.1.11 that the Lefschetz invariant $\tilde{\mathcal{L}}(K)$, is given by the expression $\sum_{\sigma \in K/G} (-1)^{\dim \sigma} u_{G/G_\sigma}$: where we are using the basis elements of the form $u_{G/H}$ indicated in our earlier discussion of the Burnside ring, in Proof 3 of Brown's Theorem 5.3.1; and σ varies over representatives of the G-orbits on simplices (we abbreviate this index set as K/G). We also recall that under λ, the image of this invariant in the representation ring is the reduced Lefschetz module $\tilde{L}(K)$ of Definition 5.1.9—in which the G-sets are replaced by permutation modules of the form $\mathrm{Ind}_{G_\sigma}^G(1)$.

The next step is to translate the alternating sum above into terms involving fixed subcomplexes. (As motivation, note that $H \leq G_\sigma$ is equivalent to $\sigma \in K^H$; where we are implicitly using the type-preserving hypothesis, in order to describe the stabilizers of the faces of simplices.) So we now recall, also from that Proof 3 discussion, the mark homomorphisms ϕ_H of the Burnside algebra: Namely ϕ_H takes u_{G/G_σ} to the size $\#\big((G/G_\sigma)^H\big)$ of the fixed-point set of H on the G-orbit of σ in K; and so using (5.4.4) we get:

$$u_{G/G_\sigma} = \textstyle\sum_{H \in L(G)/G} \#\big((G/G_\sigma)^H\big) e_H,$$

using the basis of idempotents e_H of the Burnside algebra. (Here the L, in the index set $L(G)$, refers to the subgroup lattice.) So applying this in the direct sum from the previous paragraph, we obtain (cf. [**Thé87**, 1.2]):

$$\begin{aligned}
\tilde{\mathcal{L}}(K) &= \sum_{\sigma \in K/G} (-1)^{\dim \sigma} \Big(\sum_{H \in L(G)/G} \#\big((G/G_\sigma)^H\big)\, e_H \Big) \\
&= \sum_{H \in L(G)/G} \Big(\sum_{\sigma \in K/G} (-1)^{\dim \sigma} \#\big((G/G_\sigma)^H\big) \Big)\, e_H = \sum_{H \in L(G)/G} \tilde{\chi}(K^H) e_H,
\end{aligned}$$

This expression now gives the specific relevance of Euler characteristics for our present argument via the Burnside ring.

From here on, we consider the image $\tilde{L}(K)$ of this expression under the map λ into the representation ring: Recall from Definition 5.1.11 that this means we identify $u_{G/H}$ and $u_{G/J}$ if the corresponding permutation modules are isomorphic as modules (even though H and J might not define the same G-set, i.e. might not be conjugate in G). Thévenaz now quotes, as 1.6(b) in his paper, Conlon's induction theorem [**Con68**, 6.11]; this essentially examines the kernel of the map λ from the Burnside algebra into the representation ring—and in particular it shows that when $H/O_p(H)$ is not cyclic, we get e_H mapped to 0 in the representation ring. So (see [**Thé87**, 1.7]) this reduces the expression in the previous paragraph—now read for $\tilde{L}(K)$ via images $\lambda(u_{G/H}) = \mathrm{Ind}_H^G(1)$ in the representation ring—to a sum over H with $H/O_p(H)$ cyclic.

Here we finally make use of hypothesis (a), namely that $\tilde{\chi}(K^H) = 0$ for such H which also satisfy $O_p(H) > 1$. This reduces the expression for $\tilde{L}(K)$ to a sum over terms which in addition satisfy $O_p(H) = 1$; that is, to H a cyclic p'-group.

At this point, we can apply the earlier Gluck/Yoshida formula (5.3.19) (or just the more elementary [**Sol67**, Thm 4], cf. our discussion of (5.3.12)), to see that the idempotent e_H is a $\mathbb{Q}$-linear combination of the $u_{G/J}$ for $J \leq H$. So since $\tilde{L}(K)$ is (the reduction mod p of) a $\mathbb{Z}$-linear combination of permutation modules, our expression for $\tilde{L}(K)$ from the previous paragraph is now a $\mathbb{Z}$-linear combination of various $\lambda(u_{G/J})$, for (cyclic) p'-groups J.

But $\lambda(u_{G/J})$ is just the permutation module $\mathrm{Ind}_J^G(1)$; and we saw in Example 6.1.4 that such an induced module from a p'-subgroup J is in fact $\mathbb{F}_p$-projective. So we also obtain (virtual) projectivity for the combination $\tilde{L}(K)$ of such modules—completing the proof. □

The reader may also be interested in the similar treatment of the Brown-Quillen and Webb Theorems at Benson [**Ben91**, 6.7.2]—based on induction theorems indicated in [**Ben98**, Sec 5.6].

Some examples. We first state more formally an observation which we had introduced earlier, expressing perhaps the most "fundamental" class of applications:

COROLLARY 6.3.5. *For K in the standard homotopy type of $\mathcal{S}_p(G)$ in Remark* 4.6.1, *we get $\tilde{L}(K)$ projective by Theorem* 6.3.4.

Of course, we had also begun to explore some particular cases from this general class: In earlier Example 6.2.2, for $\mathcal{S}_2(A_5)$, we had already indicated the stronger conclusion of Theorem 6.3.4, namely projectivity in the representation ring. And for the verification of the $\mathbb{F}_p$-acyclicity in hypothesis (b) of that result: we had also indicated the stronger underlying contractibility results for $\mathcal{S}_p(G)$ (in general, rather than in these explicit examples) in our discussion leading up to Remark 6.3.1—where variant (I) in that Remark is relevant here.

EXAMPLE 6.3.6 (Projectivity of $\tilde{L}(\mathcal{S}_2(S_3 \times S_3))$). We can similarly build on earlier Example 5.3.2, for $K := \mathcal{S}_2(S_3 \times S_3)$: Again as above, our discussion leading up to Remark 6.3.1 delivers the stronger contractibility condition (I) there; which in turn implies hypothesis (b) of 6.3.4. So now we will briefly describe the Lefschetz module and its projectivity.

We had already in 5.3.2 computed the dimension of $\tilde{L}(K)$ as $\tilde{\chi}(K) = -4$. But in contrast to earlier Example 6.2.2 for A_5, this time it would be fairly complicated to analyze chain spaces, since now we have rather larger dimensions in the $\tilde{\chi}$ calculation (namely $-1 + 24 - 27$).

Instead, the calculation will be easier in terms of homology groups, as in the Hopf Trace Formula 5.1.14: Since K is connected, we have $\tilde{H}_0(K) = 0$, as mentioned in Remark 5.1.7. It follows using 5.1.14 and $\tilde{\chi}(K) = -4$ that $\dim H_1(K) = 4$. And we indicate a set of 4 cycles, which gives a basis for that space: In the notation of 5.3.2, a typical such 1-cycle is a quadrangle—take the 4 subgroups of order 2 spanned by $(12, 13)$, $(12, 23)$, (ab), and (ac); and the 4 subgroups of order 4 determined by commuting pairs among them. Indeed we obtain 4 such quadrangle-cycles, when we replace $(12, 23)$ by $(13, 23)$, and/or (ac) by (bc). And it is easy to compute directly that these cycles are linearly independent—and so form a basis for $H_1(K)$, since that space has dimension 4.

Since the Sylow 2-subgroup S generated by (12) and (ab) has free action on this basis, we see that $H_1(K)$ is a virtual projective by Definition 6.1.3.3. (Indeed it a tensor product, where each factor module is the unique 2-dimensional irreducible projective for the corresponding factor group isomorphic to S_3. It is also the Steinberg module, for G regarded as the Lie type group $L_2(2) \times L_2(2)$.)

Here since the module is irreducible, our calculation with homology must give the same result as the calculation with chains that defines the Lefschetz module—that is, we get $\tilde{L}(K)$ equal to $-H_1(K)$ as $\mathbb{F}_pG$-modules (and not just equal at the level of characters), as required by Webb's Projectivity Theorem.

It is in fact easy to compute the character $\tilde{\Lambda}(K) =: \tilde{\Lambda}$; each element of order 3 fixes just one of the 4 cycles, giving character value 1, so that in summary we have:

g	1	(123)	(abc)	$(123)(abc)$
$\tilde{\Lambda}(g)$	4	1	1	1

And we mention that the module could equally well be viewed as the natural orthogonal space in earlier Example 2.1.21 for $S_3 \times S_3$ viewed as $\Omega_4^+(2)$. ◇

We mention that some further examples for the standard poset $\mathcal{S}_p(G)$ are briefly described (in terms of projective covers $P(I)$ of particular irreducible modules I) in Webb's paper [**Web87b**, p 357].

We turn now (and indeed in the subsequent two sections) to examples of Webb's Projectivity Theorem 6.3.4 for complexes K which are not necessarily equal to $\mathcal{S}_p(G)$; indeed in some cases, K is not even equivalent to $\mathcal{S}_p(G)$, and so would not be covered by Corollary 6.3.5.

(Option B:) We mention first that the two examples A_5 and $S_3 \times S_3$ above are in fact groups of Lie type—where we know by Quillen's Theorem 4.4.8 that $\mathcal{S}_2(G)$ is equivalent to the building; so we could easily repeat the analysis in those cases, but instead using the building as our complex K. And indeed, in the subsequent Section 6.4, we will consider a number of further examples of G of Lie type, where we do take K to be the building, and examine projectivity of $\tilde{L}(K)$ in Theorem 6.3.4. (Of course such cases could also be covered "indirectly", via Corollary 6.3.5.)

(Option S:) Furthermore in Section 6.5 thereafter, we will explore projectivity in Theorem 6.3.4 for a number of sporadic geometries K. As a preview, we will begin here the analysis of one such geometry:

EXAMPLE 6.3.7 (Projectivity of $\tilde{L}$ for the C_3-geometry of A_7). We take K to be the geometry for $G = A_7$ and $p = 2$ introduced in Example 2.3.7; as in later Example 3.4.11, it is convenient to work with the external viewpoint on K as a poset with 3 levels, indexed by P, L, π.

At Example 3.4.11 we had verified the contractibility of K^Q for Q of order 2; so via the usual 5.2.3 we have hypothesis (b) for Webb's Theorem 6.3.4, guaranteeing projectivity of the Lefschetz module $\tilde{L}(K)$. So we proceed to explore the projectivity more explicitly.

Since this example is not as small as those above, we will be indicating some of the calculations in fairly abbreviated form. In particular, instead of trying to directly compute the module $\tilde{L}(K)$, we will just compute its character $\tilde{\Lambda}(K) =: \tilde{\Lambda}$. (The calculation below could of course be easily implemented in GAP [**GAP00**].)

First we will obtain the dimension term $\tilde{\Lambda}(1)$, namely $\tilde{\chi}(K)$. For this we use parameter values obtained from 2.3.7:

$$\begin{aligned}\tilde{\chi}(K) &= -1 + (\#P + \#L + \#\pi) - (\#PL + \#P\pi + \#L\pi) - (\#PL\pi) \\ &= -1 + (7 + 35 + 15) - (7 \cdot 15 + 7 \cdot 15 + 35 \cdot 3) - (7 \cdot 15 \cdot 3) \\ &= -1 + 57 - 315 + 315 = 56.\end{aligned}$$

We mention that $\mathcal{S}_2(A_7)$ can be checked to have $\tilde{\chi}$ value given by 176: so it is definitely not homotopy equivalent to K (as we had mentioned at earlier Remark 4.4.12). Hence our present K is not covered by Corollary 6.3.5; and so it really does give a "new" kind of example of Webb's Projectivity Theorem 6.3.4.

Next we compute $\tilde{\chi}(K^g)$ for (conjugacy classes of) nontrivial g of odd order. A 7-cycle g fixes no point or line, and fixes just 1 plane; so $\tilde{\chi}(K^g) = -1 + 1 = 0$. Similarly a 5-cycle g fixes 2 points, but no lines or planes; so $\tilde{\chi}(K^g) = -1 + 2 = 1$. The 3-cycle $g = (123)$ fixes the points $4, 5, 6, 7$, as well as the $\binom{4}{3} = 4$ lines formed from them—a configuration with 3 points per line (and 3 lines per point); and also fixes setwise the line 123, disjoint from this configuration; and fixes no plane. We get $\tilde{\chi}(K^g) = -1 + (1 + 4 + 4) - (4 \cdot 3) = -4$. Finally $g = (123)(456)$ fixes the point 7; and fixes setwise the two lines 123 and 456, as well as the 3 planes containing one[12] of those lines, say 123. So the edges in K^g are from each of the 3 planes to the 2 vertices given by the point 7 and the line 123; and we obtain that $\tilde{\chi}(K^g) = -1 + (1 + 2 + 3) - (3 \cdot 2) = -1$. Thus in summary we have

g	1	(123)	(123)(456)	(12345)	(1234567)	(7654321)
$\tilde{\Lambda}(g)$	56	-4	-1	1	0	0

Next we will sketch the projectivity of $\tilde{\Lambda}$ via the standard *decomposition matrix*, which expresses each complex irreducible as a sum of modular irreducibles (cf. [**CR06**, Sec 84]). We first consider $\tilde{\Lambda}$ with $\mathbb{Z}$-coefficients; as mentioned earlier, the values at g of order divisible by 2 are 0 using 3.3.26. Now consulting the character table for A_7 in the Atlas [**CCN$^+$85**, p 10], we see that $\tilde{\Lambda} = 21 + 35$, where in this expression we have used dimensions as abbreviated names for the characters denoted by χ_8 and χ_9 there. We then reduce these characters mod 2, and decompose them in terms of the modular character table in the Modular Atlas [**JLPW95**, p 13]; obtaining $21 = 1 + 20$ and $35 = 1 + 14 + 20$, where again we use dimensions as names for the modular irreducibles denoted by ϕ_1, ϕ_5. and ϕ_6 there. Furthermore we check that 20 does not appear in the reduction of any other complex irreducible. Consequently the multiplicities of 1 above show that the projective cover $P(20)$ has complex character $21 + 35$—and so we have established the projectivity at least for the character $\tilde{\Lambda}$, as promised.

Furthermore when we reduce $\tilde{\Lambda}$, we see we get composition factors $1, 1, 14, 20, 20$. In fact (see Benson [**Ben84**, p 213]) the module $P(20)$ has a unique composition series, of form $20/1/14/1/20$.

This discussion will suffice, for our purposes of demonstrating the material in this section. But there are some further interesting features of this particular sporadic example K—which we will return to in Example 6.5.3. $\diamondsuit$

In addition to the applications of Webb's Projectivity Theorem 6.3.4 to buildings and sporadic geometries in the following two sections, we also mention (by way of a quick preview) that the conclusion of projectivity was applied by Webb—in the form of the vanishing of corresponding Ext-groups as in Lemma 6.1.5—in establishing his "local method" for decompositions of group cohomology in [**Web87a**];

[12] It fixes the planes above the other line, in the geometry conjugate to K under $S_7 \setminus A_7$—which uses the other orbit under A_7 of 15 out of 30 possible choices for planes (as indicated in the definition of the planes in Example 2.3.7).

we examine this argument later at Theorem 7.2.5, and indeed such decompositions are the main topic of Chapter 7.

Some further directions related to projectivity. We close this section by quickly mentioning a few further topics relevant to projectivity; typically we will not be following them up any further in our later development.

Alternative methods for projectivity results. Bouc in [**Bou91**, Cor, Prop 15] developed another general approach to projectivity, based on "projectors" in the representation ring. In fact the earlier part of that paper works in the Burnside ring—and in a more general context than ours here, in the sense of *not* depending on a fixed prime p.

Furthermore, earlier work of Bouc on "Möbius modules" might be regarded as establishing module-theoretic analogues of the standard connection (5.3.17) of the Möbius function and the Euler characteristic for subgroup posets. And that approach then leads e.g. to [**Bou84b**, Prop 3]—obtaining a virtual $\mathbb{Z}G$-module which is not just projective but even free. This result in turn leads to various numerical generalizations, for a set of primes π, of Brown's Theorem 5.3.1 on the Euler characteristic of $\mathcal{S}_p(G)$. (We had mentioned one such corollary just after Exercise 5.4.3.)

Knörr and Robinson in [**KR89**, Sec 3] defined a "Lefschetz conjugation module" for a G-complex by means of a suitable *local coefficient system* (roughly, using certain spaces defined at each simplex; we discuss such systems in later Definition 10.1.1). In effect, they induce from the p-blocks of the group algebras of simplex stabilizers G_σ. And they correspondingly establish virtual projectivity in [**KR89**, Thm 4.2]. We will mention some work from that paper in later Section 11.2, on the Alperin Conjecture from p-modular representation theory; see especially the discussion of their local system after (11.2.7).

We mention also the approach of Robinson in [**Rob88**]—which begins with a vanishing result 2.1, whose hypothesis is stated not via contractibility, but instead in the language of his chain-pairing method (which we had mentioned at Remark 4.6.2); though Robinson remarks there that it is essentially just a re-phrasing of a contractibility hypothesis. He then obtains a new proof of Webb's projectivity result, as his Corollary 2.2.

Finally Assadi in [**Ass91**] develops a notion of "permutation complexes"; his corresponding projectivity results start with Theorem 3.3 of that work.

The further structure of the projective module. In view of the projectivity of the alternating sum of homology groups in Webb's Theorem 6.3.4, and the behavior of the examples known around that time, it seemed reasonable to ask (see [**Web87b**, Conj 3.2]) whether in fact the *individual* homology groups of such a complex K should then each be $\mathbb{F}_p$-projective. However, a counterexample of Alperin (see [**Ben91**, p 232]) later answered this question in the negative.

Similarly in the examples known at that time, the virtual projective that was the "net" result from the alternating sum always seemed to be ($\pm$) an ordinary module, rather than a sum of both positive and negative parts. I do not know of any counterexample in our present contexts. But the situation can sometimes differ, in the related context of the (non-building) geometry defined by the *non*degenerate subspaces of a unitary space: Here by [**DGM09**], the geometry is usually "spherical" (see later Remark 9.1.8); and so the reduced Lefschetz module is still just ($\pm$) the top homology. However, certain exceptional small cases like $U_3(2)$ instead give a Lefschetz module having components with both positive and negative signs. (I understand from Gramlich that this was computed in GAP by S. Witzel.)

One can also continue analysis like that which we demonstrated e.g. in Example 6.3.7; namely how the projective module $\tilde{L}(K)$ decomposes into projective covers $P(I)$ for various irreducibles I. In at least one standard situation, namely the building as in Theorem 6.4.2 below, the Lefschetz module is projective and irreducible. But in other situations, there seems to be no general result describing the actual distribution of components $P(I)$; for example, the table in Ryba-Smith-Yoshiara [**RSY90**], listing such components for many sporadic geometries, does not seem to suggest any particular pattern.

There is, however, at least a general tool for computing the multiplicity of any $P(I)$ in $\tilde{L}(K)$; this is stated as [**Web87b**, Prop 5.3] for the case of $K = \mathcal{S}_p(G)$. (However, it does not seem to be proved there.) A proof follows from a result of Robinson [**Rob89**] on multiplicities in induced modules—where below I and J are irreducible modules for G and H:

$$\text{(6.3.8)} \qquad \text{mult of } P(I) \text{ in } \mathrm{Ind}_H^G(J) = \text{mult of } P(J) \text{ as a summand of } I|_H.$$

For one just applies this with $I := 1_G$ to the alternating sum of induced modules $\mathrm{Ind}_{G_\sigma}^G(1_{G_\sigma})$ which defines the reduced Lefschetz module $\tilde{L}(K)$—for any K. (The reciprocity formula for the special case $K = \mathcal{S}_p(G)$ is deduced in this way as 5.11 in Grodal [**Gro02**].) Notice that (6.3.8) is a refinement of the usual Frobenius reciprocity formula (as stated for Hom in later (10.2.7)): since requiring $P(I)$ to be a direct summand is a more stringent restriction than just having a module homomorphism from $P(I)$ into $\tilde{L}(K)$.

Of course the effect of this reciprocity formula is to reduce the calculation of the multiplicity of $P(I)$ to corresponding calculations in the stabilizer subgroups G_σ. Most often this has been applied to obtain restrictions on those I for which $P(I)$ can even appear in $\tilde{L}(K)$—typically by giving lower bounds on $\dim I$.

For example, take $I := 1_G$ as indicated above. We recall that the geometries K considered in this book will typically have p dividing the order of G_σ: and then (e.g. using the discussion after (1) of Definition 6.1.3) the projective cover $P(1_{G_\sigma})$ will have dimension divisible by $p > 1$; whereas 1_G (and hence its restriction) has dimension 1—forcing a multiplicity of 0 for this σ. Applying this in (6.3.8) for all σ in the alternating sum, we see that the cover $P(1_G)$ cannot occur in the projective module $\tilde{L}(K)$. This observation is relevant for hypothesis (c) of later Theorem 7.2.5; so we state it here as:

$$\text{(6.3.9)} \qquad \text{If } p \text{ divides each } |G_\sigma|, \text{ then } P(1_G) \text{ has multiplicity 0 in } \tilde{L}(K).$$

On the other hand, in the case of the Lefschetz conjugation modules for blocks in Knörr and Robinson [**KR89**] mentioned above, which have a similar reciprocity formula, it is precisely the multiplicity of $P(1)$ in the projective module from their context that leads to the proof of surprising interrelations. We discuss this work further, in its natural context of the Alperin Conjecture, in our later Section 11.2.

We also mention that Barker in [**Bar96c**] studies the coefficients of the Lefschetz module from the Burnside ring viewpoint via the combinatorial technique of Möbius inversion.

6.4. (Option B): The Steinberg module for a Lie type group

Having established in the previous two sections our basic general results on projectivity (namely Theorems 6.2.1 and 6.3.4), in this section we now focus on the particular case given by G of Lie type, defined over a field of the "same" characteristic p—that is, the prime p is the one for which we are studying our usual p-subgroups poset $\mathcal{S}_p(G)$.

Recall we had given some general background on groups of Lie type in earlier Remark 2.2.2. We make just a few further comments relevant to the material in this section:

REMARK 6.4.1 (The Steinberg module). We assume that G is defined in characteristic p. Among the irreducible $\mathbb{C}G$-modules, Steinberg discovered one (in fact via its character) with a number of remarkable properties—including that it is the unique irreducible whose $\mathbb{F}_p$-reduction is projective, and that it has dimension $|G|_p$; this is called the *Steinberg module* of G. (See e.g. [**Cur71**, Sec 9.2] for more detail.) We may denote[13] this module (in either characteristic 0 or p) by St. From the viewpoint of p-modular representations, St is in turn the only projective indecomposable for G of Lie type which is also an $\mathbb{F}_p$-irreducible. (See e.g. [**Cur71**, Sec 9.2] or [**Cur70**] or [**GLS98**, Sec 2.8] for more detail on the modular case.) $\diamondsuit$

The connection of this special module with the general result on projectivity of $\tilde{L}(\mathcal{S}_p(G))$ in earlier Section 6.2 was observed by Quillen as Theorem 3.1 in [**Qui78**]. (And in particular, this application is perhaps the most significant example of what we called our Building Principle 2.0.3.) Below we give a statement which includes some further details, obtained by quoting certain other sources beyond Quillen:

THEOREM 6.4.2. *Assume G is of Lie type in characteristic p, with building K of rank r. Then*

$$(\pm)\tilde{L}(\mathcal{S}_p(G)) = \tilde{H}_{r-1}(K) \text{ has dimension } |G|_p, \text{ and is the Steinberg module St;}$$

which is projective, and also irreducible over $\mathbb{C}$ or $\mathbb{F}_p$ (giving a block of defect 0—cf. Remark 6.2.3).

PROOF. Using Quillen's homotopy equivalence in Theorem 4.4.8, we obtain that $\tilde{L}(\mathcal{S}_p(G)) = \tilde{L}(K)$.[14] (In particular, we can conclude that $\tilde{L}(K)$ is projective via Corollary 6.3.5.) Next we recall the Solomon-Tits argument in the proof of Theorem 3.4.15: where collapsibility after removal of opposite chambers shows that homology of K vanishes in dimensions $< r-1$; hence $\tilde{L}(K) = \pm\tilde{H}_{r-1}(K)$. Indeed we will see more precisely at the later Solomon-Tits Theorem 9.4.6 that this module has dimension $|G|_p$. (This value in fact gives the number of chambers opposite to a fixed chamber, in the sense indicated after (2.2.21), and demonstrated at (2.2.24).) Furthermore (in details of the proof of that result in [**Sol69**], which we have not reproduced) the Lefschetz character equals the Steinberg character.[15]

[13]There should be no confusion with our use of St also for the open star of a simplex.

[14]In fact Quillen's 3.1 is stated in terms of $\mathcal{A}_p(G)$—which is equivalent to $\mathcal{S}_p(G)$ via 4.2.4.

[15]If I understand the history correctly from Solomon, Curtis had earlier computed the character of the Steinberg module by means of a suitable alternating-sum formula; and Curtis and Feit then realized that that sum agrees with the one in the definition of $\tilde{L}(K)$.

So by irreducibility of the Steinberg module in Remark 6.4.1, we must in fact have equality at the stronger level of modules in the representation ring.

There are several additional ways of seeing the projectivity of $\tilde{L}(K)$: For example, we can just quote projectivity of the Steinberg module from Remark 6.4.1. Alternatively, we could apply Webb's Theorem 6.3.4 directly to the building K (rather than to $\mathcal{S}_p(G)$ as above); for we get hypothesis (b) for K, since we had verified contractibility of the fixed subcomplex K^Q under a nontrivial p-subgroup Q, at Lemma 3.4.16. □

We now examine some particular examples of this general result.

We had already mentioned that the modules in earlier Examples 6.2.2 and 6.3.6 are in fact the Steinberg modules, when those abstract groups $G \cong A_5$ and $S_3 \times S_3$ are regarded as the Lie type groups $\Omega_4^-(2) \cong L_2(4)$ and $\Omega_4^+(2) \cong L_2(2) \times L_2(2)$.

EXAMPLE 6.4.3 (Projectivity of $\tilde{L}$ for the building of $L_3(2)$). Recall that for $G = L_3(2)$, the building K is afforded by the projective plane in Example 2.1.9; and we had provided further information on this geometry in Examples 2.2.5 and 3.4.13. In particular $|G|_2 = 2^3 = 8$ here.

As usual we first compute the Lefschetz character $\tilde{\Lambda}(K) =: \tilde{\Lambda}$. For the dimension term, using the parameters in 2.1.9 we have

$$\tilde{\chi}(K) = -1 + (\#P + \#L) - (\#PL) = -1 + (7+7) - 21 = -8.$$

Next an element x of order 7 does not lie in G_P or G_L in 2.2.5; so that K^x is empty, and hence $\tilde{\Lambda}(x) = \tilde{\chi}(\emptyset) = -1$. Finally an element u of order 3 fixes a single point, and a single line not containing that point (indeed they determine a direct-sum decomposition of 3-space over $\mathbb{F}_2$); so that $\tilde{\Lambda}(y) = \tilde{\chi}(K^y) = -1 + (1+1) = 1$. In summary we get

g	1	y	x	x^{-1}
$\tilde{\Lambda}(g)$	-8	1	-1	-1

And we can recognize $\tilde{\Lambda}(g)$ as the negative of the character of the adjoint module for $L_3(2)$—which is denoted by ϕ_4 in the Modular Atlas [**JLPW95**, p 3] (and is the reduction mod 2 of the $\mathbb{C}$-character denoted by χ_6 in the Atlas [**CCN$^+$85**, p 3].)

At the level of modules, we indicate the construction of $\tilde{H}_1(K)$ in the sprit of the proof above—following notation for chambers as in Example 3.4.13: Fix the initial chamber $PL = 1 \in \langle 1, 2 \rangle$; and order the remaining chambers by increasing distance from PL. Then in the full building graph of (2.1.14), one finds 8 "opposite" chambers at distance 3, for example $3 \in \langle 2, 3 \rangle$. Each of these determines with PL a unique hexagon—which is an apartment, namely a Coxeter complex of the Weyl group $W \cong S_3$. These 8 circuits form a basis for the cycle space $\tilde{H}_1(\Delta)$.

We recall, with regard to hypothesis (b) of Webb's Projectivity Theorem 6.3.4, that we had in fact at Example 3.3.23 given an explicit verification, for this particular example, of the contractibility of the fixed point subcomplex K^z for z of order 2. ◇

EXERCISE 6.4.4 (Further building examples). Mimic Example 6.4.3, but now taking $G = SL_3(4)$, using information from Example 2.2.6. This time the Steinberg module has dimension 64 (it is the tensor product of the two algebraic conjugates of an 8-dimensional adjoint module); it appears as ϕ_6 in [**JLPW95**, p 54] and χ_{10} in [**CCN$^+$85**, p 24].

Also mimic Example 6.4.3 with the (non-simple) group $G = Sp_4(2)$, using information from Examples 2.1.19, 2.2.16, 2.2.25, and 3.4.14—with maximum distance $N = 4$ in the graph, and apartments given by octagons. The viewpoint of $Sp_4(2) \cong S_6$ is indicated in 2.1.24 and 2.2.17; contractibility of fixed points under an involution appears at 3.3.24. Here the Steinberg module has dimension 16; its character can be obtained by summing the two corresponding characters of degree 8 for the simple subgroup $Sp_4(2)' \cong A_6$—which appear as ϕ_4, ϕ_5 in [**JLPW95**, p 4] and as χ_4, χ_5 in [**CCN**$^+$**85**, p 5].

The enthusiastic reader can continue the process, for some or all of the other examples of buildings given in Section 2.2. Indeed, those in Examples 2.1.7, 2.1.18, 2.1.21, and 2.1.25 are essentially already covered by our earlier remarks on A_5 and $S_3 \times S_3$, just before Example 6.4.3. So the remaining "new" computations are:

Example	2.1.15 (+ 2.2.31)	2.1.26 (+ 2.2.33)	2.2.1	2.2.12
G	$L_4(2)$	$Sp_6(2)$	$G_2(2)$	$SL_n(q)$
$\lvert G\rvert_p$	2^6	2^9	2^6	$q^{\binom{n}{2}}$

where of course the bottom row also gives the dimension of the Steinberg module for G. ◇

We also note that, once projectivity of $\tilde{H}_{r-1}(\Delta)$ has been established via Webb's Theorem 6.3.4 in the proof above, there is an alternative argument for identifying it as the Steinberg module: namely one can use the properties of a Lie type group G that every irreducible I other than the Steinberg has dimension less than $|G|_p$—while every simplex stabilizer contains a Borel subgroup of order divisible by $|G|_p$. This shows using the reciprocity formula (6.3.8) that the cover $P(I)$ cannot lie in our projective—so that the Steinberg module itself is the only possible component.

REMARK 6.4.5 (The terminology of "generalized Steinberg module"). The projectivity of the reduced Lefschetz module $\tilde{L}(\mathcal{S}_p(G))$ in Theorems 6.2.1 and 6.3.4 gives an analogue, for any finite G, of the special case in Theorem 6.4.2 for G of Lie type—where the Lefschetz module reproduces the important Steinberg module. This analogy led Webb to refer to $\tilde{L}(\mathcal{S}_p(G))$ as the *generalized* Steinberg module for a general group G; and this usage has gained considerable acceptance in the literature (e.g. Bouc [**Bou84b**] or Benson [**Ben91**, p 224]).[16]

Indeed Webb's later approach, in his Structure Theorem [**Web91**, Thm A], shows that the entire chain complex $C_*(\mathcal{S}_p(G))$ has its homology determined by a direct summand subcomplex of projective modules (whose alternating sum gives the module $\tilde{L}(\mathcal{S}_p(G))$); and correspondingly, this subcomplex is called the *Steinberg complex* for a general group G. There is a particularly intriguing application of this complex in Grodal [**Gro02**, Secs 3,5]. (A recent investigation of the complex appears in Swenson [**Swe09**]; and a relative-cohomology version appears in Symonds [**Sym08**].)

The interested reader is encouraged to follow up these references; since we will only mention the generalized-Steinberg approach occasionally during the remainder of the book. ◇

[16] I personally am a little uneasy with using the name "Steinberg" for this important concept: for the analogy imports only the projective property—while usually not delivering the important irreducibility property—from the usual Lie type Steinberg module.

6.5. (Option S): Analogous projective modules for other simple groups

The examples of particular projective modules $\tilde{L}(K)$ in Webb's paper [**Web87b**, p 357] were in the generalized-Steinberg context of Remark 6.4.5 above; that is, for $K = \mathcal{S}_p(G)$, including some cases where G is a sporadic group. However, Ronan and Smith (and perhaps others) soon verified hypothesis (b) of Webb's Projectivity Theorem 6.3.4 for K given by various of the emerging sporadic geometries. (And some of these are *not* homotopy equivalent to $\mathcal{S}_p(G)$, and so would not be covered by Corollary 6.3.5.)

This sporadic analysis via Webb's Projectivity Theorem 6.3.4 was essentially "completed" by the work of Ryba-Smith-Yoshiara in [**RSY90**], which verified success (or failure) of Hypothesis 6.3.4(b)—among the sporadic geometries which has by then been described in the literature. In the cases of failure, typically it was already clear that the relevant Lefschetz module could not be projective—simply because the dimension term given by the reduced Euler characteristic was not divisible by $|G|_p$. However, the condition was verified in 18 cases; and the corresponding projective module $\tilde{L}(K)$ were analyzed in that paper in terms of projective covers $P(I)$.

We had already previewed such a case, in earlier Example 6.3.7. In this section, we further explore that theme; in particular we consider at least briefly the other geometries introduced in earlier Section 2.3. Of course such projective examples provide an important instance of our Vague Sporadic Principle 2.0.4.

EXAMPLE 6.5.1. Let K be the triples geometry for (non-simple) $G = S_5$, from Example 2.3.5. We had verified at Example 3.3.25 the contractibility of K^Q for Q of order 2; this gives hypothesis (b) of Webb's Theorem 6.3.4, so we get projectivity of $\tilde{L}(K)$. (This could also be obtained from projectivity of $\tilde{L}(\mathcal{S}_2(S_5))$ via Corollary 6.3.5, since we had established homotopy equivalence of K with $\mathcal{S}_2(S_5)$ at Example 4.4.11.)

Now we compute the character $\tilde{\Lambda}(K) =: \tilde{\Lambda}$. First, using the parameters in Remark 2.3.6, the dimension term is given by

$$\tilde{\chi}(K) = -1 + (\#P + \#T) - (\#PT) = -1 + (5 + 10) - 30 = -16.$$

Next an element x of order 5 fixes no point or triple, so that K^x is empty, and hence we have $\tilde{\chi}(K^x) = -1$. And an element $y = (123)$ of order 3 fixes the 2 points $4, 5$; and fixes the triple 123 not containing them—so that $\tilde{\chi}(K^y) = -1 + (2 + 1) = 2$. In summary we have

g	1	(123)	(12345)
$\tilde{\Lambda}(g)$	-16	2	-1

Now we briefly sketch the decomposition-matrix calculation as in Example 6.3.7: The corresponding $\mathbb{C}$-character, using the notation of [**CCN**$^+$**85**, p 2] for the simple subgroup A_5, is the negative of the sum of: the 6-dimensional irreducible $\mathrm{Ind}_{A_5}^{S_5}(\chi_2)$, and the two 5-dimensional irreducibles in $\mathrm{Ind}_{A_5}^{S_5}(\chi_5)$. This reduces mod 2 to the projective cover of the 4-dimensional irreducible $\mathrm{Ind}_{A_5}^{S_5}(\phi_2)$ in the notation of [**JLPW95**, p 2]. This projective cover is mentioned at [**Web87b**, p 357]. (The composition factors are 4 copies of the trivial module, and 3 copies of this irreducible; the module has unique composition series, with structure 4/1/1/4/1/1/4—see e.g. [**Ben84**, p 200].) ♢

Next we indicate one of the easier examples considered in [**RSY90**]; it can be analyzed using permutation-group techniques, much as in Example 6.5.1 above.

EXERCISE 6.5.2 (Projectivity for a characteristic-3 geometry for M_{11}). Consider $G = M_{11}$ with $p = 3$. As points of K, take the 11 letters permuted by M_{11}, and as lines take the $\binom{11}{2} = 55$ pairs of points, with the obvious incidence given by inclusion. (Hence this K is an instance of the conventions in the circle-geometries of Buekenhout [**Bue79**], which we had indicated after 2.3.4.)

From the Atlas [**CCN$^+$85**], G has just one conjugacy class z of elements of order 3, and these have cycle type $3^3 1^2$. It is easy to show that K^z is contractible, and hence $\tilde{L}(K)$ is projective via 6.3.4.

Next compute K^x for x of order coprime to 3; to obtain the reduced Lefschetz character $\tilde{\Lambda}$:

g	1^{11}	$2^4 1^3$	$4^2 1^3$	$5^2 1^1$	$x = 8^1 2^1 1^1$	x^{-1}	$y = 11^1$	y^{-1}
$\tilde{\Lambda}(g)$	-45	3	-1	0	1	1	-1	-1

This corresponds to the $\mathbb{C}$-irreducible denoted by χ_9 in [**CCN$^+$85**, p 18]. As the degree 45 is divisible by $|G|_3 = 9$, the module is of defect 0 as in Remark 6.2.3, and hence projective. Its reduction mod 3 appears as ϕ_8 in [**JLPW95**, p 33]. ◇

We now return to the geometries introduced in Section 2.3.

For K given by the $\tilde{C}_2$-geometry of $U_4(3)$ indicated in Example 2.3.10, the proof of contractibility of the fixed subcomplex K^z for z of order 2 is given at [**RSY90**, p 293]. And we just summarize the further analysis of the projective module $\tilde{L}(K)$: It consists of five characters of defect 0 as in Remark 6.2.3: a pair of complex conjugates of degree $640 = 2^7 \cdot 5$, and 3 copies of a character of degree $896 = 2^7 \cdot 7$. These appear as χ_{17}, χ_{18}, and χ_{20} in [**CCN$^+$85**, p 54], and as ϕ_{10}, ϕ_{11}, and ϕ_{12} in [**JLPW95**, p 125].

Similarly for the 2-local geometry K for M_{24} from Example 2.3.11, the lengthier proof of contractibility of fixed points is given at [**RSY90**, p 289]. This time the projective $\tilde{L}(K)$ is of dimension $21504 = 2^{10} \cdot 3 \cdot 7$, and is the projective cover $P(1792)$ of the $\mathbb{F}_2$-irreducible of dimension 1792 denoted by ϕ_{12} in [**JLPW95**, p 267].

Finally, we now return to the C_3-geometry K for A_7, for which we had established projectivity of $\tilde{L}(K)$ in earlier Example 6.3.7. Recall we saw there that this module is the projective cover $P(20)$ of dimension 56.

We will summarize various further results for this situation; and perhaps these facts raise more questions than they actually answer—one can hope that they will suggest more general patterns for researchers in the future.

EXAMPLE 6.5.3 (Projectivity for the C_3-geometry of A_7, continued). First we consider some features which arise from the containment $A_7 < A_8 \cong L_4(2)$. (In particular we mention that the irreducibles of A_7 over $\mathbb{F}_2$ arise as restrictions of irreducibles for A_8.) Let K_b denote the building for $L_4(2)$. Then as in Exercise 6.4.4, $\tilde{L}(K_b)$ is the Steinberg irreducible module of dimension 64—appearing as ϕ_8 in [**JLPW95**, p 48] and as χ_{13} in [**CCN$^+$85**, p 22]. The restriction of this module to A_7 of course remains projective—but it does not remain irreducible under A_7; instead it is the projective cover $P(14)$ of the 14-dimensional module denoted by ϕ_5 in [**JLPW95**, p 13]. (This cover has a composition series of structure $14/(14 \oplus (1/20/1))/14$; see e.g. [**Ben84**, p 213].)

Notice that K_b and K have the same truncation $K_{L\pi}$ defined by ignoring the points; one checks that $\tilde{L}(K_{L\pi})$ is projective for A_7, and in fact gives the same module $P(20)$ as for the original K. When we truncate the building K_b by ignoring planes, we obtain the same A_7-module $P(20)$; however, when we truncate the C_3-geometry K to K_{PL} by ignoring planes, we again get $\tilde{L}(K_{PL})$ projective—but this time we obtain the same module $P(14)$ as for the full building K_b. (I know of no other example of such an interrelation of natural geometries for a single group.)

Next we summarize the computation of $\tilde{L}(\mathcal{S}_2(A_7))$ of dimension $176 = 2^3 \cdot 3 \cdot 7$, as mentioned in [**Web87b**, p 357]: Notice that $176 = 64 + 56 + 56$; and indeed one checks that this module has structure $P(14) + P(20) + P(20)$. In terms of our various other geometries for A_7 which are *not* equivalent to $\mathcal{S}_2(G)$ (and hence not covered by Corollary 6.3.5), we could choose to express this module decomposition in the form:

$$\tilde{L}(\mathcal{S}_2(A_7)) = \tilde{L}(K_b) \oplus \tilde{L}(K) \oplus \tilde{L}(K^g) \text{ for } g \in S_7 \setminus A_7,$$

where K^g gives a C_3-geometry isomorphic to K, which uses the "other" A_7-orbit of length 15 for its planes, as indicated in the discussion leading up to 2.3.7. (Again I know of no other example of such a representation-theoretic interrelation of geometries.) ◇

We had mentioned at the start of the section that for at least some sporadic geometries, we get failure of the hypothesis (and indeed of the conclusion of projectivity) in Webb's Theorem 6.3.4. The reader might well ask whether any other weaker results might still be available in those situations. And indeed there are:

We consider consider one such weakening in the following section, namely *relative* projectivity of $\tilde{L}(K)$ for a complex K—which has been verified for many sporadic p-local geometries where projectivity fails, in various places in the literature. (However, probably the more important weakening is to replace projectivity by *vanishing of cohomology* of the module $\tilde{L}(K)$; we take up this further theme in the following chapter, at Lemma 7.2.2 and Remark 7.4.11.)

6.6. Weaker conditions on K giving relative projectivity of $\tilde{L}(K)$

Thévenaz in [**Thé87**] observed that the hypotheses of Webb's Projectivity Theorem 6.3.4 can be weakened to cover only *some* nontrivial p-subgroups—and then the conclusion obtained is instead projectivity just *relative to* the remaining p-subgroups. Indeed the approach of Thévenaz via the Burnside ring in that paper delivers the generalization with essentially no further work, as we will see in the proof of Theorem 6.6.2 below.

But first we must introduce the needed terminology (cf. [**CR90**, Sec 19A] or [**Fei82**, Sec I.4]):

DEFINITION 6.6.1 (Projectivity relative to a (p)-subgroup H). For a field k of characteristic p, we say that a kG-module R is projective *relative to*[17] a subgroup H of G, if for any kG-module surjection M on R, splitting of the kH-module surjection of $M|_H$ on $R|_H$ implies splitting of as kG-modules.

[17]We may abbreviate this as "projective rel H" or "rel-H projective".

If we take $H = 1$, then since any extension automatically splits over 1, the condition above reduces to the statement that any surjection *must* split over R; that is, the case of rel-1 projectivity gives just ordinary projectivity, using formulation (2) in Definition 6.1.3. Thus to obtain a proper weakening of ordinary projectivity, we would normally consider rel-H projectivity only for $H > 1$.

On the other hand, if H contains a Sylow p-subgroup S of G, then (as we observed at (3) of Definition 6.1.3) the standard Lemma 7.1.2 shows that H-splitting implies G-splitting: so that *any* kG-module R is automatically rel-H projective for such an H.[18] Thus we usually further restrict our attention to projectivity relative to Q for nontrivial subgroups Q of S (i.e., p-subgroups)—and so (by the previous paragraph) even to *proper* subgroups $Q < S$, in order for relative projectivity to give any real restriction.

Finally if $\mathcal{X}$ is a set of subgroups of G (indeed typically subgroups of the Sylow S), we say R is rel-$\mathcal{X}$ projective if it is a sum of rel-X projectives for various $X \in \mathcal{X}$. $\diamondsuit$

The general result of Thévenaz. We can now state Thévenaz's generalization of Webb's Projectivity Theorem 6.3.4, in which Webb's hypotheses are assumed only for *some* p-subgroups:

THEOREM 6.6.2 (Thm 2.1 in [**Thé87**]). *Assume that K is a G-complex with type-preserving action. Assume further $\mathcal{X}$ is a collection of subgroups of G, closed under subgroups and conjugacy, such that Hypothesis* 6.3.4*(a) holds whenever the H there satisfies $O_p(H) \notin \mathcal{X}$; that is,*

$$\tilde{\chi}(K^H) = 0 \text{ for all } H \text{ with } H/O_p(H) \text{ cyclic and } O_p(H) \notin \mathcal{X}.$$

Then $\tilde{L}(K)$ is a virtual rel-$\mathcal{X}$ projective module when viewed in the representation ring of G.

Furthermore this same conclusion also holds under a similar weakening of Hypothesis 6.3.4*(b)—where we assume that any p-group $Q \notin \mathcal{X}$ contains* some *subgroup P of order p such that K^P is $\mathbb{F}_p$-acyclic.*

(In practice, we obtain this latter hypothesis by first fixing a set B of "bad" conjugacy classes of subgroups of order p, including all those for which the $\mathbb{F}_p$-acyclicity in Hypothesis 6.3.4*(b) does (or even might) fail; and then define $\mathcal{X}$ to consist of those p-subgroups* all *of whose elements of order p come from B.)*

PROOF. (Sketch) Since we followed Thévenaz's argument in the special case of $\mathcal{X}$ given by $\{1\}$, as our proof of Webb's Projectivity Theorem 6.3.4, here we will give only a briefer indication of the fairly straightforward extension of that argument to a more general collection $\mathcal{X}$.

Again we first show how the second hypothesis implies the first—mimicking the logic for "(b) implies (a)" in that earlier proof of 6.3.4. So assume we are given some H with $H/O_p(H)$ cyclic and $O_p(H) \notin \mathcal{X}$. As we are assuming the second hypothesis, there is some P of order p in $O_p(H)$ such that K^P is $\mathbb{F}_p$-acyclic. Then the argument for "(i) implies (ii)" in the proof of Lemma 6.3.3, using the

[18]We re-emphasize our earlier warning about distinguishing between seemingly-similar usages: E.g. for an H containing a Sylow group, we saw in Definition 6.1.3.3 that H-projectivity (that is, projectivity as H-module) is strong enough to guarantee full ordinary projectivity (that is, rel-1 projectivity); while by contrast, *rel-H* projectivity for this H gives no restriction at all.

P. A. Smith theorem (6.3.2), shows that $K^{O_p(H)}$ is $\mathbb{F}_p$-acyclic. And now the remainder of the argument for "(b) implies (a)" in the proof of 6.3.4 goes through—to show that $\tilde{\chi}(K^H) = 0$, establishing the first hypothesis as required.

So for the main proof, we now assume the first hypothesis. We follow the first several paragraphs of that earlier proof: Our hypothesis of vanishing of the Euler characteristic reduces us to considering terms for H with $H/O_p(H)$ cyclic, but now with $O_p(H) \in \mathcal{X}$. Again we apply the Gluck/Yoshida formula (5.3.19), to reduce to a sum of induced modules $\mathrm{Ind}_J^G(1)$ for subgroups $J \leq H$; this time we get only that $J/O_p(J)$ is cyclic (and a p'-group). But note that $O_p(J) \leq O_p(H)$; so by the hypothesis that $\mathcal{X}$ is closed under subconjugacy, we have $O_p(J) \in \mathcal{X}$ also. Now it is standard (essentially an extension of the argument in Example 6.1.4) that the induced module $\mathrm{Ind}_J^G(1)$ is rel-J projective. And then reducing to the Sylow p-subgroup $O_p(J)$ of J, using Lemma 7.1.2 as in our discussion in Definition 6.6.1, we see that it is even rel-$O_p(J)$ projective. So considering terms for all J, which as we saw above have $O_p(J) \in \mathcal{X}$, we conclude that $\tilde{L}(K)$ is rel-$\mathcal{X}$ projective. □

There are various further properties of relative projectivity which we could study at this point. However in the remainder of this section, we will confine ourselves to just one property, namely of the degree term. For there is an analogue of the property of ordinary projectivity of divisibility by $|G|_p$ (cf. Definition 6.1.3.1); namely a standard result due to Green (see e.g. [**CR06**, Thm 19.26]) shows that:

(6.6.3) If R is rel-X-projective, then $\dim R$ is divisible by $\dfrac{|G|_p}{|X|_p}$.

Thus we obtain a generalization of Brown's Theorem 5.3.1:

COROLLARY 6.6.4 (Cor 2.3 in [**Thé87**]). *Assume (one of) the hypotheses of Theorem 6.6.2; and define g, e by $|G|_p = p^g$, with p^e the order of the largest member of $\mathcal{X}$. Then*

$$\tilde{\chi}(K) \equiv 0 \pmod{p^{g-e}}.$$

(Option S:) Some sporadic examples. My understanding is that Thévenaz developed his generalization to relative projectivity essentially for theoretical purposes in modular representation theory. However, the topic is of considerable interest for sporadic geometries such as those we introduced in Section 2.3.

The background is more or less as follows: In Ryba-Smith-Yoshiara [**RSY90**], the authors verified the hypotheses of Webb's Projectivity Theorem 6.3.4 for many sporadic geometries; but in addition, for the remaining such known sporadic geometries K, they also checked the failure of those hypotheses—and indeed of the conclusion of divisibility of $\tilde{\chi}(K)$ by $|G|_p$. However in doing so, they noticed (as was also mentioned later in [**SY97**, p 374]) that $\tilde{\chi}(K)$ still tended to contain "most" of $|G|_p$; and usually this slight difference was soon explained in terms of certain "bad" classes of elements of order p defining a suitable $\mathcal{X}$ for the application of Corollary 6.6.4. (In particular, this allows us to further extend our Vague Sporadic Principle 2.0.4 to non-projective situations: that is, we can move beyond the projective context of earlier Examples such as 6.3.7 and 6.5.1.)

Below we will discuss one such non-projective example at moderate length. We will proceed mainly in overview—but we will at least mention some of the details, for comparison and contrast with the discussion in those earlier projective examples:

EXAMPLE 6.6.5 (Relative projectivity for the 2-local geometry of M_{12}). We take $G = M_{12}$ and $p = 2$; and for K we take the standard 2-local geometry for M_{12}—described for example at [**RS84**, p 63] (cf. also Benson-Smith [**BS08**, p 161].) For reference on the group M_{12} itself, see e.g. [**Wil09**, Sec 5.3]. In summary, the *points* are the 4-subsets of a 12-set acted on by G; and the *lines* are certain 4|4|4-partitions of that set. In fact, below we will instead use a more convenient alternative viewpoint, which is more group-theoretic: for points, one can take the 2-central involutions (the class denoted by $2B$ in the Atlas [**CCN⁺85**]); and for lines, the 4-groups which consist entirely of such involutions. For the parameters as in Section 2.3, we get $\#P = 495 = \#L$ with $\#P/L = 3 = \#L/P$. So we obtain

$$\tilde{\chi}(K) = -1 + (\#P + \#L) - (\#PL) = -1 + (495 + 495) - (495 \cdot 3) = -496 = -2^4 \cdot 31.$$

Since this value is not divisible by $|G|_2 = 2^6$, we see via Definition 6.1.3.1 that $\tilde{L}(K)$ is not projective.

Of course this means that the hypotheses of Webb's Projectivity Theorem 6.3.4 must fail; more specifically we explore the failure of 6.3.4(b): One checks that G has two conjugacy classes of involutions, denoted by $2A$ and $2B$ in the Atlas [**CCN⁺85**]; we will let x and z (respectively) denote representatives. Since the permutation characters of G on points and lines are given in the Atlas, we can simply read off the values at z and x which give the numbers of fixed vertices: namely K^z consists of 31 points and 15 lines; while K^x consists of 15 points and 35 lines.

We turn now to the more detailed analysis. Since most readers will probably not be familiar with M_{12}, and hence with the finer structure of these fixed subcomplexes, below we will just briefly summarize how their treatment proceeds.

First we claim that K^z is contractible (and hence $\mathbb{F}_2$-acyclic by the usual 5.2.3): The points in K^z are just the involutions of type $2B$ in $C_G(z) \cong 2^{1+4}S_3$. Furthermore the extra-special subgroup $E := O_2(C_G(z)) \cong 2^{1+4}$ contains 3 of the lines of the geometry (which lie on z); so these contain $3 \cdot 2 = 6$ of the points beyond z. Furthermore each of these 6 points in turn lies on 2 more lines which extend into $C_G(z) \setminus E$. Thus we have accounted for all $31 = 1 + 6 + (6 \cdot 2) \cdot 2$ points, and for all $15 = 3 + 6 \cdot 2$ lines, of K^z. And then it is easy to see that K^z is a tree rooted at z, hence contractible. (A multi-step collapsibility proof in the spirit of Section 3.4 would apply Lemma 3.4.10, to remove: first the 24 points most distant from z; next the 12 most distant lines; then the 6 closet points; and finally the 3 closest lines—leaving only z.)

Next we check that K^x is *not* contractible (indeed it is not even $\mathbb{F}_2$-acyclic): We observe first that $C_G(x) \cong \langle x \rangle \times S_5$. Now each of the 5 Sylow 2-groups in the A_5 of the indicated S_5 is a 4-group consisting of $2B$-involutions—which gives a line of K^x, with its 3 points; indeed this accounts for all 15 points, and 5 of the lines, of K^x. Furthermore x normalizes, but does not centralize, each of the other 2 lines on any of these points (recall $\#L/P = 3$); so now we have in fact accounted for all $35 = 5 + 15 \cdot 2$ lines of K^x. At this stage, it is easy to check that these give 5 separate connected components, so that K^x is not contractible. (Each component is separately contractible, so $\tilde{H}_0(K^x)$ has dimension 4, and K^x is not even $\mathbb{F}_2$-acyclic.)

In particular we get the non-zero value $\tilde{\chi}(K^x) = 4$ for the reduced Lefschetz character at x; this gives yet another way of seeing (via Definition 6.1.3.1) that the reduced Lefschetz module $\tilde{L}(K)$ is not projective.

However, at this stage we can apply the relatively-projective viewpoint of Corollary 6.6.4: For the hypothesis of Theorem 6.6.2, we define B to consist just of the class of x above—so that $\mathcal{X}$ consists of the 2-subgroups whose involutions are all conjugate to x; and we obtain that $\tilde{L}(K)$ is rel-$\mathcal{X}$ projective. One finds that $\mathcal{X}$ consists just of subgroups which are either of order 2, or elementary abelian of order 4 (e.g. $\langle x, x' \rangle$, for x' an involution from $S_5 \setminus A_5$ in the description of $C_G(x)$ above). Then in Corollary 6.6.4 we have $6, 2$ in the roles of "g, e"—so that $\tilde{\chi}(K)$ should be divisible by $2^{6-2} = 2^4$; thus "explaining" this value which we had in fact verified above.

For the relatively projective module $\tilde{L}(K)$, one can (to a more limited extent) pursue the strucutral analysis of the module as in earlier projective cases: To summarize the computation of the reduced Lefschetz character $\tilde{\Lambda}$, we obtain (cf. Benson-Wilkerson [**BW95**, 3.3]) that the complex character (actually we take its negative, so as to have positive dimension 496) has a constituent of dimension 144 and two copies of a constituent of dimension 176; namely the irreducibles denoted by χ_{14} and χ_{15} in the Atlas [**CCN$^+$85**, p 33]. The modular constituents are two copies each of a pair of complex conjugates of dimension 16, and three copies of a constituent of dimension 144; namely the irreducibles denoted by ϕ_3, ϕ_4, and ϕ_6 in the Modular Atlas [**JLPW95**, p 74].

Furthermore Benson and Wilkerson observe that these characters lie in a non-principal 2-block; indeed that block has defect group $X := \langle x, x' \rangle$. This gives a different proof that $\tilde{L}(K)$ is rel-X projective. We will return to further implications of this observation about blocks in later Example 7.2.13. ♢

EXAMPLE 6.6.6 (Relative projectivity for the 3-local geometry for M_{12}). Use the group $G = M_{12}$ again, but now now take $p = 3$: that is, now take K to be the usual 3-local geometry for M_{12}, described for example at [**RS84**, p 82]. Here points are given by 3-subsets of a 12-set acted on by G, and lines are given by suitable $3|3|3|3$-partitions. (This time we cannot view points as subgroups of order 3; in fact both points and lines correspond to certain distinct conjugacy classes of elementary subgroups of order 3^2.)

We give a much briefer summary than we did for $p = 2$ in Example 6.6.5; but the arguments are in fact similar and even a bit easier. (So the more expert reader may in fact be able to work out the details of the present Example instead as an Exercise).

The point and line stabilizers are the two classes of subgroups $3^2 GL_2(3)$. Hence we obtain $\#P = 220 = \#L$ with $\#P/L = 4 = \#L/P$, and then it follows that $\tilde{\chi}(K) = -1 - 2 \cdot 220 = -441 = -3^2 \cdot 49$.

Again there are two classes z and x of elements of prime order 3 (denoted $3A$ and $3B$ in the Atlas [**CCN$^+$85**, p 33]); and as above K^z is contractible, but K^x is disconnected, and so it is not even $\mathbb{F}_3$-acyclic. Again let B be defined by the class of x, so that $\mathcal{X}$ consists of the 3-subgroups which are "pure x"; this time all members of $\mathcal{X}$ have order 3. So $\tilde{L}(K)$ is rel-$\mathcal{X}$ projective by Theorem 6.6.2; and in Corollary 6.6.4 we have $3, 1$ in the roles of "g, e", so that $\tilde{\chi}(K)$ is divisible by $3^{3-1} = 3^2$—as indeed we had calculated just above. ♢

Finally we mention some larger sporadic examples still more briefly (since the underlying details are somewhat more complicated):

EXAMPLE 6.6.7 (Relative projectivity for some Conway groups). Take $G = Co_1$ with $p = 2$; and K its usual 2-local geometry, described at [**RS80**, p 284]. (For reference on the group Co_1 itself, see e.g. [**Wil09**, Sec 5.4].) Here there are three conjugacy classes of involutions; two have contractible fixed points, so again just one "bad" class x (denoted $2B$ in the Atlas [**CCN**$^+$**85**, p 184]) defines B and hence $\mathcal{X}$. Larger pure-x 2-subgroups can have order up to 2^3 (the largest has structure D_8); so in Corollary 6.6.4 we have $21, 3$ in the roles of "g, e", explaining the value of $2^{21-3} = 2^{18}$ which divides $\tilde{\chi}(K)$. (See e.g. [**RSY90**] for the latter value.)

Next take $G = Co_3$ with $p = 2$, and K its usual 2-local geometry, described at [**Asc86**, p 23]; with classes z, x (denoted $2A, 2B$ in the Atlas [**CCN**$^+$**85**, p 135]. Here only x is bad, and the largest bad subgroup is of type D_8; so shis time in Corollary 6.6.4 we have $10, 3$ in the roles of "g, e", thus explaining the value of $2^{10-7} = 2^7$ which occurs in $\tilde{\chi}(K)$. $\diamondsuit$

We conclude the section by mentioning some more recent further work on relative projectivity and other representation-theoretic properties of the reduced Lefschetz module, for various more general geometries including sporadic examples:

Sawabe in [**Saw05**] considers the topic essentially for the collection of subgroups which are both p-radical and p-centric—we will indicate the latter property at later Definition 7.4.8.[19]

Further sporadic geometries were analyzed by Grizzard in [**Gri09**], in terms of vertices for non-projective summands of $\tilde{L}(K)$.

This latter viewpoint was further pursued in a more general approach by Maginnis and Onofrei, who applied to sporadic geometries their collection of p-subgroups which are "distinguished": For example, they investigate situations where the fixed subcomplex K^x is not contractible for some x; and at [**MO09**, Thm 3.1.abc] they obtain projectivity of $\tilde{L}(K^x)$ as a module for $C_G(x)/O_p(C_G(x))$. (This gives a kind of "local" analogue of results like Theorems 6.2.1 and 6.4.2.) When that centralizer involves a Lie type group, this module may be the usual Steinberg module—or it may involve a suitable extension of the Steinberg module: Indeed in a forthcoming work, they consider an extension H of a Lie type group G by outer automorphisms: and in particular, they describe such extensions to H of the Steinberg module for G. (These "extended" Steinberg modules had arisen in the representation-theoretic literature, e.g. [**Sch92**] [**Fei93**] [**MS95**].)

Finally, we close the chapter with a mention of some work on Lefschetz modules which is not directly about projectivity: namely Segev and Webb in [**SW94**] consider the case of a normal subgroup N of G with order divisible by p—and they then express the Lefschetz module for $\mathcal{A}_p(G)$ in terms of those for $\mathcal{A}_p(N)$ and for the $\mathcal{A}_p(C_N(A))$ where $A \in \mathcal{A}_p(G)$ intersects N trivially.

[19]In this same centric-radical context for sporadics, we mention a recent work of Oda-Sawabe [**OS09**]—which then produces a multiplication giving a *generalized* Burnside ring, in a sense due to Yoshida.

CHAPTER 7

Group cohomology and decompositions

This chapter is primarily devoted to results relevant to group cohomology. Although we have so far been emphasizing a range of applications of subgroup complexes, it is probably fair to say that the most frequent motivation in the literature for the development of the methods was group cohomology.

The theme of group cohomology $H^*(G)$ was explicit in the papers of Brown such as [**Bro75**] and [**Bro74**]; the connections with cohomology are perhaps more implicit in Quillen's paper [**Qui78**].[1] The subsequent work of Webb in [**Web87a**] and [**Web91**] focused attention on the particular aspect of using a G-complex K to obtain a *decomposition* of $H^*(G)$, in terms of the cohomology $H^*(G_\sigma)$ for the stabilizers G_σ for simplices σ of K. And in the years thereafter, researchers in algebraic topology developed a more general *decomposition theory*—where the emphasis is on a corresponding decomposition at the deeper topological level of the classifying space BG. In the present chapter, we will survey some of this work related to decompositions of cohomology.[2]

Following our usual procedure, the chapter begins with an initial background Section 7.1, reviewing certain aspects of group cohomology. The subsequent Section 7.2 develops and discusses the first main goal of the chapter, namely Webb's Decomposition Theorem 7.2.5.

The remainder of the chapter then summarizes some further developments beyond Webb's cohomology decomposition; these in fact involve a number of modern topological methods, which go deeper than the elementary topological framework which we have been maintaining in this book—so we will only very briefly sketch that deeper topological material. First Section 7.3, in the context of our Option G, indicates the alternative approach to $H^*(G)$ via the *G-equivariant* cohomology of the G-complex K. Then Section 7.4 gives a quick overview of the more recent theory of decompositions of the classifying space BG: where the product in the decomposition is in fact a *homotopy colimit* of the the classifying spaces BG_σ of the stabilizers. The final Section 7.5, in our Option S, discusses applications of these methods to at least partially describe the cohomology of sporadic simple groups.

7.1. Topological background: Group cohomology $H^*(G)$ and the classifying space BG

For fuller details on group cohomology, there are many standard sources; we will mention just a few which are particularly suited to the methods in this book, namely Brown [**Bro94**], Adem-Milgram [**AM04**], and Benson [**Ben98**][**Ben91**].

[1]Though of course the elementary abelian members of $\mathcal{A}_p(G)$ (if not necessarily the poset structure) are important for mod p cohomology; e.g. via earlier work of Quillen, such as [**Qui71**].

[2]I thank Jesper Grodal for many valuable suggestions in this chapter.

Furthermore Chapter 2 of Benson-Smith [**BS08**] aims at collecting together (from many sources) the elements of group cohomology which are most relevant to decomposition theory.

DEFINITION 7.1.1 (Group (and module) cohomology). There are a number of equivalent definitions of group cohomology; below we briefly describe those to which we will be referring later. As in earlier Remark 5.1.7, R denote some coefficient ring—in the most basic case, just the integers $\mathbb{Z}$.

(1) In classical group theory: Cohomology was first developed in low dimensions, to describe extensions etc; and was phrased in the language of elementary group theory. The corresponding *bar construction* (see e.g. [**Hup67**, I.16.11]) is based on a map ∂—using the group multiplication to take an n-tuple $(g_1, \cdots, g_n)$ to the linear combination (over $\mathbb{Z}$, or more generally over R) of $(n-1)$-tuples:

$$(g_2, \cdots, g_n) + \sum_{i=1}^{n-1} (-1)^i (g_1, \cdots, g_{i-1}, g_i g_{i+1}, g_{i+2}, \cdots, g_n) + (-1)^n (g_1, \cdots, g_{n-1}) g_n.$$

Then in the later language of elementary homological algebra, as in Remark 5.1.7: we get the image-in-kernel property $\partial^2 = 0$ of (5.1.5), and so ∂ provides the boundary map for a chain complex $C_*(G, R)$ defined via the R-span $C_n := RG^n$ of the tuples; leading in the usual way to cycles, boundaries, and the *group homology* $H_*(G, R)$.

Furthermore we can dualize; for the map ∂ induces a *co*boundary map δ (with $\delta^2 = 0$) on "functions", namely on the R-dual $C^n := \mathrm{Hom}_R(C_n, R)$. This *co*chain complex $C^*(G, R)$ then leads dually to the corresponding definitions of cocycles, coboundaries, and the *group cohomology* $H^*(G, R)$. (There is also a natural definition of *reduced* cohomology, straightforwardly extending from Remark 5.1.7.)

In particular, we recall that a group extension determines a 2-cocycle; and the extension splits if and only if the cocycle is determines the cohomology class of 0.

(2) Next in algebraic topology: Here the usual approach to group cohomology is via the ordinary cohomology of a corresponding topological space—namely the *classifying space* BG of G (e.g. [**BS08**, Sec 2.2]). There are in fact various equivalent constructions for the space BG; and usually one first constructs a contractible space EG admitting a free G-action, with BG obtained as the quotient EG/G. Any EG with these properties yields the same BG, up to homotopy equivalence; see e.g. [**BS08**, 2.2.3]. Indeed in Milnor's version of BG as a CW-complex, the sequence of tuples from the group-theoretic bar construction essentially reappears in EG, in the form of an infinite join of copies of G (cf. [**BS08**, 2.2.5]); and δ reappears as the coboundary map of a cochain complex $C^*(BG; R)$, which is associated to the CW-complex BG, in the spirit of the simplicial-complex case in Remark 5.1.7. It is standard that the corresponding cohomology[3] $H^*(BG; R)$ just reproduces the group cohomology $H^*(G, R)$ as in (1).

(3) Finally in more modern homological algebra: Here cohomology is often defined via free (or just projective) resolutions. Such resolutions are in fact applied

[3]Notice that the notation uses a comma for the special case of group cohomology; but a semicolon for the more general construction of cohomology of a space—the latter is the standard convention that we have been using for homology of a complex since Remark 5.1.7.

for some general (finite-dimensional) RG-module M, giving cohomology *with coefficients in* M;[4] and the ordinary group cohomology then arises from the special case where M is the trivial module R.[5]

In this viewpoint (e.g. [**BS08**, 2.1.2]), the module cohomology of M can be expressed in terms of the Ext-functors as:

$$H^*(G, M) = \mathrm{Ext}^*_{RG}(R, M).$$

The Ext-functors are in fact the *derived functors* of the usual functor Hom. And the computation of Ext is then often expressed in the form $H^*(\mathrm{Hom}_{RG}(F_*, M))$, where F_* denotes a projective resolution of the trivial module R. In fact we typically take F_* to be the *bar resolution*: here again the tuple modules RG^n in the group-theoretic bar construction reappear—since they afford a sequence of *free* (not just projective) RG-modules as the terms of F_*, with the last term RG then surjecting on the trivial module R; and a variant of earlier ∂ connects the terms of the sequence. We direct the reader to the references for fuller details.

We will however record the following fact about the dimension-0 case:

$$H^0(G, M) = \mathrm{Hom}_{RG}(1, M) = M^G \text{ (the } G\text{-invariants in } M).$$

If $H^n(G, M) = 0$ for all $n \geq 0$, then we will by convention use the terminology that the module cohomology of M *vanishes*; and we may for emphasis add the phrase "including in dimension $n = 0$". We do this in order to avoid using the terminology of "acyclic" module;[6] and in particular, to avoid considerations of "reduced" cohomology for a module. For example, any module M with nonzero invariants $M^G \neq 0$ (for example the trivial module $M = 1$) will not, in our convention, have vanishing cohomology. However, we will later be considering suitable modules M where the cohomology does vanish—including in dimension $n = 0$, so that $M^G = 0$.

And within this specific context of module cohomology, we can now alternatively define the group cohomology as follows—where we regard the coefficient ring "R" on the left as the trivial RG-module 1:

$$H^*(G, R) = H^*_{RG}(1, 1) = \mathrm{Ext}^*_{RG}(1, 1);$$

that is, group cohomology is the special case of module cohomology for M, where M is given by the trivial module. ◇

We had already previewed the following standard property (cf. [**Ben98**, Cor 3.6.18]) of module (and in particular, group) cohomology:

LEMMA 7.1.2. *If H is a subgroup of G with p not dividing $|G : H|$, then the restriction map Res^G_H induces for $n > 0$ an injection of the p-part*[7] *$H^n(G, M)_p$ of $\mathbb{Z}G$-module cohomology into $H^n(H, M)_p$.*

Applying this in particular to the 2-cohomology class of an extension of $\mathbb{F}_pG$-modules N by M, the extension splits over G if and only if it splits over H. □

[4] Often we will use the informal language of the *module cohomology* of M.

[5] As in earlier chapters, we sometimes write 1 for the trivial module; e.g. in contexts where group theorists write expressions like 1^G_B for a module induced from the trivial module 1_B.

[6] There seem to be variations in the usage of this term in the topological literature, which are at least potentially confusing.

[7] I.e., the Sylow p-subgroup: it is standard for $n > 0$ that $H^n(G, M)$ is a finite abelian group.

It is fairly complicated to work out examples of group cohomology for explicit groups G; so again we direct the reader interested in such examples to the references (e.g. [**AM04**, Ch II.3]). Instead our discussion will work with the group cohomology of G mainly "indirectly"—notably via decompositions into terms given by cohomology of subgroups of G.

REMARK 7.1.3 (Fusion systems). We will briefly mention here a fairly recent development, which essentially generalizes the foundations of group cohomology.

In rough overview: It has long been known that the p-part of the group cohomology of G is determined by the "p-fusion" pattern—namely the G-conjugations among elements within a Sylow p-subgroup of G. One important approach to abstractly formalizing this observation was begun by Puig—who defined a category on P, where the morphisms are given by both inclusions and G-conjugations.

Indeed one can more generally consider any consistent pattern for the latter type of morphisms—that is, patterns which do not necessarily arise from conjugation within any particular finite group G. Such a structure can be regarded as a kind of purely topological generalization, away from G, of the p-subgroups poset $\mathcal{S}_p(G)$ with its G-action.

In the standard modern development, such a category can be required to satisfy certain further natural restrictions (inspired for example by "extremal" conjugation arising from the properties of Sylow p-subgroups); and the result is called a *fusion system* on P.

Fusion systems are beyond our scope in this book; but we mention that a good introduction to the theory is provided in Broto-Levi-Oliver [**BLO03**]. There are several further refinements of fusion systems in the literature; we mention in particular the notion of a *p-local finite group*—which again is not a group, but a topological structure based on a p-group P. Of course the modern literature establishes for such systems many of the properties of the motivating case of finite groups. ◊

7.2. Webb's decomposition of $H^*(G)$ as an alternating sum over K/G

We saw in Definition 7.1.1.2 that the cochain complex $C^*(BG)$ is infinite-dimensional; so we might initially ask how the *finite*-dimensional G-complexes K which we have been studying could provide information about the group cohomology $H^*(G)$ of G.

Perhaps the most natural answer to this question, at least from a topological viewpoint, is that we can in fact focus on the infinite-dimensional Borel construction of K, given by $(EG \times K)/G$ of K. However, this construction requires some further more advanced topological background; so we will not describe it until the subsequent section, at Remark 7.3.1. In fact, one strong motivation for this equivariant approach was provided by a result of Brown [**Bro94**, X.7.2], which shows that the Borel construction e.g. for $K = \mathcal{S}_p(G)$ leads to the same mod-p cohomology as BG; we will discuss Brown's result in that section, as Theorem 7.3.5. (As a brief preview: This property of K leading to $H^*(BG)$ came to be known as *ampleness*; and a stronger property, in particular implying the decompositions in this section (see (7.3.14)) is called *sharpness*.)

Nonetheless, in the present section, we will still be able to indicate an application of K to $H^*(G)$, which stays within our present framework of more elementary classical topology: Namely, following Webb's analysis in [**Web87a**] of Brown's arguments, we can hope to obtain a direct-sum decomposition of $H^*(G)$, in terms of the cohomology $H^*(G_\sigma)$ of the subgroups stabilizing simplices σ in K.

Webb referred to his formula as a "local method"—it re-focuses attention on the stabilizers G_σ; that is, this reduction to subgroups is in the general spirit of a "global to local" principle. Furthermore we have seen that, for the kinds of complexes K which we have been studying (for example $\mathcal{S}_p(G)$), these stabilizers tend to even be local in the sense of *p-local* subgroups.

The notion of a cohomology decomposition over K/G. As has often been the case in this book, it is convenient for our exposition to subdivide our discussion of Webb's analysis a little more finely than is done in his original papers.

First we will examine just what we mean by a decomposition.

DEFINITION 7.2.1 (Cohomology decomposition over K). We assume that K is a G-complex with type-preserving action (as in Remark 1.5.7). We say that the (mod-p) group cohomology $H^*(G)$ *decomposes over* K (or more precisely, over K/G) if:

$$H^*(G) \cong \bigoplus_{\sigma \in K/G} (-1)^{\dim \sigma} \; H^*(G_\sigma).$$

We emphasize that the isomorphism is as (an alternating sum of) *abelian groups*—it makes no assertion about the *ring* structure of cohomology. ◇

Ideally it should be clear that such decompositions (if they can in fact be established) could be potentially very useful—indeed later we will see that decompositions have in fact been used as an approach to the detailed computation of $H^*(G)$ for various interesting G. And in principle, decompositions should even provide the possibility of an inductive approach, to the cohomology of suitable infinite families of groups G (such as groups of Lie type). However, it should be emphasized that the decomposition itself does not solve the entire problem of cohomology; it just shifts the *location* of the work to the relevant subgroups—where typically some hard work remains to be done.

Next we explicitly separate out an elementary connection between such decompositions and Lefschetz modules (cf. also [**BW95**, 3.1]):

LEMMA 7.2.2. *Assume that K is a G-complex with type-preserving action, and consider cohomology with mod-p coefficients. Then:*

$$H^*(G) \text{ decomposes over } K/G \qquad \text{iff} \qquad H^*\big(G, \tilde{L}(K)\big) = 0 \text{ (in all dimensions).}$$

PROOF. Assume that $H^*\big(G, \tilde{L}(K)\big) = 0$. Using Definitions 7.1.1.3 and 5.1.9, this means that:

$$0 = \mathrm{Ext}^*_{\mathbb{F}_pG}\big(1, \bigoplus_{\sigma \in K/G} (-1)^{\dim \sigma} \mathrm{Ind}^G_{G_\sigma}(1)\big) = \bigoplus_{\sigma \in K/G} (-1)^{\dim \sigma} \mathrm{Ext}^*_{\mathbb{F}_pG}\big(1, \mathrm{Ind}^G_{G_\sigma}(1)\big).$$

Now there is a standard tool (corresponding to Frobenius reciprocity in representation theory; cf. later (10.2.7)) which relates induced and restricted modules; namely the Eckmann-Shapiro Lemma (e.g. [**Ben98**, 2.8.4]), which says for modules A, B for H, G with $H \leq G$ that we have:

$$\mathrm{Ext}^*_G\big(B, \mathrm{Ind}^G_H(A)\big) \cong \mathrm{Ext}^*_H(B|_H, A). \tag{7.2.3}$$

So each Ext-term on the right of the earlier display can be translated using (7.2.3) into $\mathrm{Ext}^*_{\mathbb{F}_p G_\sigma}(1_{G_\sigma}, 1_{G_\sigma})$—which in turn by Definition 7.1.1.3 is just $H^*(G_\sigma)$. And now in the resulting vanishing statement, if we move the term for the empty simplex $\emptyset$, with dimension -1 and stablizer equal to G, over to the left side of the equals-sign—then we have in fact reproduced Definition 7.2.1 of decomposing $H^*(G)$ over K/G. Furthermore notice that these steps are reversible. □

To apply the above lemma towards a decomposition over K/G, note that establishing projectivity of $\tilde{L}(K)$, and then quoting Lemma 6.1.5, gives vanishing of cohomology of $H^n(G, \tilde{L}(K))$ in "most" dimensions—namely for all $n \geq 1$. However, we emphasize that the proof above also crucially requires the vanishing of the module cohomology of $\tilde{L}(K)$ *including in dimension* 0—so that the dimension-0 case must then still be treated. (We will continue the theme of dimension 0, e.g. in our discussion of Theorem 7.2.5 below, and at various later points.)

REMARK 7.2.4 (Decompositions from non-projective Lefschetz modules). On the other hand, our separating-out of Lemma 7.2.2 also shows that such a decomposition does not *require* projectivity of $\tilde{L}(K)$—just the vanishing of its module cohomology (including in dimension 0). We will in fact have projectivity in obtaining the decompositions in Section 7.2 (and hence again in the different viewpoint of Section 7.3); but projectivity need not (and often will not) hold, in the more general decompositions which we will see in Sections 7.4—7.5. ◇

Webb's Cohomology Decomposition. In Webb's original approach (in the paper [**Web87a**]) to the cohomology decomposition which we present as Theorem 7.2.5 below, most of the work was in fact accomplished via what we called Webb's Projectivity Theorem 6.3.4 in the previous chapter.

But since the statement we give below is for group cohomology, whereas Webb's statement of [**Web87a**, Thm A] was in fact given for *Tate* cohomology, we also encounter the dimension-0 consideration mentioned above. So below we have also added the further hypothesis (c), which allows us to explicitly address dimension 0.

THEOREM 7.2.5 (Webb's Decomposition Theorem (Thm A in [**Web87a**])). *Assume that K is a (nonempty) G-complex which has type-preserving action, and also satisfies one of the hypotheses (a) or (b) of Webb's Projectivity Theorem* 6.3.4. *Assume further that:*

(c) For each simplex σ of K, p divides the order of the stabilizer G_σ.
Then the (mod-p) group cohomology $H^*(G)$ decomposes over K/G:

$$H^*(G) \cong \bigoplus_{\sigma \in K/G} (-1)^{\dim \sigma} \, H^*(G_\sigma).$$

PROOF. Since we are assuming the hypotheses of Theorem 6.3.4, from that result we obtain projectivity of $\tilde{L}(K)$. So by Lemma 6.1.5, in the Definition 7.1.1.3 of module cohomology we get:

$$H^n(\tilde{L}(K)) = \mathrm{Ext}^n_{\mathbb{F}_p G}(1, \tilde{L}(K)) = 0 \text{ for } n \geq 1.$$

(In fact this assertion for $n \geq 1$ suffices for Webb's statement of [**Web87a**, Thm A] using Tate cohomology.)

Hence by Lemma 7.2.2, to get the decomposition for ordinary group cohomology, it only remains to establish[8] the analogous vanishing in dimension $n = 0$, namely vanishing of:

$$H^0(\tilde{L}(K)) = \mathrm{Ext}^0_{\mathbb{F}_pG}(1, \tilde{L}(K)) = \mathrm{Hom}_{\mathbb{F}_pG}(1, \tilde{L}(K)).$$

Now since $\tilde{L}(K)$ is projective, the Hom-term vanishes as long as the projective cover $P(1)$ of the trivial module 1 is not a direct summand of $\tilde{L}(K)$. And as K is nonempty, our hypothesis (c) shows, using (6.3.9) (based on the reciprocity formula (6.3.8)), that the multiplicity of $P(1)$ in $\tilde{L}(K)$ is in fact 0 as desired, completing the proof. □

Since we stated the hypothesis above in terms of Webb's Projectivity Theorem 6.3.4, while condition (c) usually holds for the geometries considered in this book, it should now be clear that Webb's Decomposition Theorem also applies to the projective examples considered in the previous chapter, in Sections 6.2—6.5. We will discuss some of these examples in further detail, in a moment.

But first—in view of the role of Hypothesis 7.2.5(c) in the proof above: It is perhaps curious that such conditions do not seem to have been very explicitly mentioned in the literature on decompositions. So below, I indicate some appearances of such considerations, in the literature familiar to me.

Remark 7.2.6 (Hypothesis 7.2.5(c) etc in the literature). Many of the observations collected in this Remark (and in some later continuations of the theme) arose during email conversations with Bob Oliver, while he was reading and reviewing the Benson-Smith book [**BS08**]. Dave and I are grateful to Bob for agreeing to their appearance in the present book.

(1) First for background, we mention that the cohomology decomposition for K in Webb's Theorem 7.2.5 can also be obtained (see later (7.3.14)) from a stronger property for K called *sharpness*—which we will be describing in later Definition 7.3.12. Webb's *later* approach to decompositions in [**Web91**] in effect adopted the viewpoint of sharpness, which indeed became standard in the subsequent literature. And correspondingly, various considerations for decompositions in this section will in effect preview some similar considerations for sharpness, arising in the subsequent Section 7.3.

(2) We have already indicated that Hypothesis 7.2.5(c) was not needed in Webb's proof of [**Web87a**, Thm A], since his statement was for Tate cohomology rather than group cohomology. (Nonetheless, the situation for group cohomology was seemingly at least implicitly well known. In fact, my memory is that Webb and Thévenaz around 1985 also showed me essentially the argument above covering $n = 0$; indeed that argument seems at least suggested at [**Web87b**, p 357].)

There is perhaps one further reason why the hypothesis (c) was not (at least explicitly) prominent in the later literature: namely, as we had mentioned, that (c) automatically holds for p-subgroup complexes, such as those in the standard homotopy type of $\mathcal{S}_p(G)$ in Remark 4.6.1—and those complexes provide the context for much of the literature on decompositions. That is, it is *only* when we want to

[8] I believe that the argument in this paragraph is well known; though it may not appear explicitly in the literature—cf. Remark 7.2.6.2 below.

extend results to more general complexes K, that some condition like (c) needs to be explicitly added to the hypotheses.

Oliver's remarks to us included a further standard way of complementing Webb's Tate cohomology arguments, to treat dimension 0 for decompositions—or more precisely, for the stronger property of sharpness; we indicate this later at Lemma 7.3.17. As a quick preview, that variant instead uses:

(c′) Assume for the orbit complex K/G that $H^n(K/G; \mathbb{F}_p) = 0$ for all $n \geq 1$;

(and indeed (c′) is implied by our Hypothesis 7.2.5(c)). Note that for a decomposition, we may as well reduce to a connected component of our geometry—and so assume that K (and hence K/G) is connected; and then we see that (c′) is in that case just $\mathbb{F}_p$-acyclicity for K/G.

(3) Although Webb did not need hypothesis (c) for his Tate cohomology decomposition, he did in fact make use of the condition of p dividing stabilizer orders elsewhere: namely at [**Web87a**, Prop 8.2], to obtain the condition $\tilde{\chi}(K/G) = 0$. (That concluson was intended as a result in the direction of his conjecture, later established by Symonds as Theorem 3.2.10, of contractibility of $\mathcal{S}_p(G)/G$—which of course implies $\mathbb{F}_p$-acylicity, thus in fact giving (c′) for $\mathcal{S}_p(G)/G$, as noted above.)

(4) Oliver's observations in (2) above in fact arose from his having noticed a mis-statement in [**BS08**]; and we take this opportunity to indicate the needed correction: Namely 5.1.16 in that book was intended to be (and so should be replaced by) Webb's Decomposition Theorem 7.2.5: For 5.1.16 is not correct as stated there—since it omits any hypothesis beyond those of Webb's Projectivity Theorem 6.3.4; and we have seen that something more, such as Hypothesis 7.2.5(c), or (c′), is needed to cover dimension $n = 0$. Hence that further hypothesis is also needed for certain results stated later in [**BS08**] which are related to 5.1.16—namely 5.1.17, 5.6.7, and 5.8.4.

But this omission in fact causes no further problems in that book: For the above results of Webb are only introduced for the background exposition in Chapter 5 there—they are not quoted for the main results, in Chapter 7 of the book, which instead apply later results on decompositions for ordinary group cohomology. (Furthermore, the geometries in Chapter 7 all do satisfy Hypothesis 7.2.5(c); so that the corrected forms of 5.1.16 etc could be quoted—*if* they had been needed.)

The omission presumably arose from too-hasty transcription of Theorem A in [**Web87a**]—which is instead stated for Tate cohomology. Now it is standard (e.g. [**Ben91**, p 210]) that Tate cohomology agrees with ordinary cohomology in dimensions $n \geq 1$—but not, in general, in dimension $n = 0$. Thus the discussion of the above statements in [**BS08**] should also be adjusted to take account of Tate cohomology; and for example, Theorem A of the later paper [**Web91**], in addition to 2.5.1 there, should then be quoted. ◊

Some important applications of Webb's cohomology decomposition. As we had mentioned just after the proof of Webb's Decomposition Theorem 7.2.5, our assumption there of the hypotheses of Webb's Projectivity Theorem 6.3.4 in effect means that we obtain the indicated cohomology decomposition over K/G—for the various examples given earlier in Sections 6.2— 6.5, where $\tilde{L}(K)$ was projective. In particular, when K is a complex of (nontrivial) p-subgroups, then p divides the order of each stabilizer–so that we now get the additional Hypothesis 7.2.5(c). We next make a few further remarks on such examples; we will mainly describe just the decompositions, since the actual detailed structure of group cohomology is beyond the elementary scope of this book.

Cohomology decomposition for a general group G. For instance, we can continue the projectivity theme in our earlier Corollary 6.3.5, for the "fundamental" class of applications:

COROLLARY 7.2.7. *For K in the standard homotopy type of $\mathcal{S}_p(G)$ in Remark 4.6.1, we get the cohomology decomposition over K/G in Theorem 7.2.5*

Webb applied his result, in the case of $\mathcal{S}_p(G)$ in this Corollary, to give cohomology decompositions (at $p = 2$) to a variety of groups G, in [**Web87a**, Sec 7]. He described the resulting group cohomology $H^*(G)$ in terms of the Poincaré polynomial, in which the coefficient of t^n records the dimension of $H^n(G)$. His examples include some smaller alternating and symmetric groups, as well as the sporadic groups M_{11} and J_1.[9] Furthermore the decomposition for the Lie type group $L_3(2)$ is described at the end of Section 3 in his expository paper [**Web87b**], and more fully in Section 2.8 of [**Web91**].

We mention one direction of further research, continuing the application of this Corollary based on $\mathcal{S}_p(G)$, which we will not really follow up in this book: namely the theory of the "generalized Steinberg module", which we had indicated earlier at Remark 6.4.5. The reader wishing to explore this direction may wish to look e.g. at [**Web87b**, 6.8], [**Gro02**, 8.19], and [**Bou91**, 2.7.3].

As a rough approximation, it might be reasonable to say that the literature on cohomology for simple groups had contained rather few fully-computed examples—and those typically for smallish groups; and that the cohomology decomposition in Webb's Theorem 7.2.5 made it more feasible to analyze various somewhat larger groups.

(Option B:) Cohomology decomposition over the building. In a similar way, the natural follow-up to the theme of projectivity for the Steinberg module in earlier Theorem 6.4.2 is:

COROLLARY 7.2.8. *Assume that G is of Lie type in characteristic p, with building K. Then the mod-p cohomology $H^*(G)$ decomposes over K/G as in Theorem 7.2.5.*

In particular we recall from Remark 2.1.4 that G is flag-transitive on K, so that the orbit complex K/G is a single simplex. So the simplices σ_J of the orbit complex correspond to the subsets $\emptyset \neq J \subseteq \Pi$ of a fundamental system as in Remark 2.2.7, with stabilizer given by the parabolic subgroup G_J. Thus the cohomology decomposition takes the particularly elementary "simplex" form:

$$H^*(G) \cong \bigoplus_{\emptyset \neq J \subseteq \Pi} (-1)^{|J|-1} \, H^*(G_J) \ ,$$

in terms of the cohomology of the parabolic subgroups of G.

This general result is an important instance of our Building Principle 2.0.3.

An explicit example (corresponding to our earlier Example 6.4.3) is provided by the decomposition for $G = L_3(2)$ mentioned above, given at [**Web91**, Sec 2.8].

---

[9] For a later treatment, see the corresponding Sections 7.1 and 7.6 in [**BS08**].

(That discussion uses not $\mathcal{S}_2(G)$ but $\mathcal{A}_2(G)$, where the latter poset in fact affords the building K, e.g. via Remark 4.6.1.) In this case, the two maximal parabolics are isomorphic to $2^2L_2(2) \cong S_4$.

Indeed in a number of larger Lie type groups, the decomposition over the building has been used (e.g. by Adem-Milgram) to compute group cohomology.

More generally, this decomposition over the building suggests the possiblity of an inductive approach to describing the cohomology of G in "natural" characteristic. But so far this general problem remains open:[10] Of course a successful inductive proof requires already knowing the overall answer—and no one yet seems to understand the general structure involved.

(Option S:) Some cohomology decompositions over sporadic geometries. Again proceeding as above, the natural follow-up to the theme in earlier Section 6.5 is:

COROLLARY 7.2.9. *For K given by any of the examples in Section 6.5 (or indeed for any of the examples in* [**RSY90**] *where $\tilde{L}(K)$ is shown to be projective), the mod-p cohomology $H^*(G)$ decomposes over K/G as in Theorem 7.2.5.*

Since G is usually flag-transitive on these geometries K, correspondingly we get the elementary "simplex" form of the decomposition as in Corollary 7.2.8—now indexed by subsets $\emptyset \neq J \subseteq I$, where the type set I indexes representatives of the vertices of K.

Of course this result gives important instances of our Vague Sporadic Principle 2.0.4.

We mention, with regard to Hypothesis 7.2.5(c), one feature of the C_3-geometry for $G = A_7$ and $p = 2$ in Example 6.3.7 (this geometry is one of the cases giving a decomposition as in Corollary 7.2.9): Namely the vertex stabilizers $G_p \cong Sp_4(2)'$ and $G_\pi \cong L_3(2)$ are simple—and in particular are not 2-local subgroups; however, they still have order divisible by 2, which is what hypothesis (c) requires.

EXAMPLE 7.2.10 (A cohomology decomposition for M_{11}). As an explicit instance of Corollary 7.2.9, consider $G = M_{11}$ with $p = 2$. Here the 2-local geometry K had not been studied in the early work [**RSY90**]—but it is treated in the standard way in the later [**BS08**, Sec 7.1]. (For reference on the group M_{11} itself, see e.g. [**Wil09**, 5.3.8].) In fact K is afforded by $\mathcal{A}_2(G)$—so we could equally well obtain the decomposition from Corollary 7.2.7; indeed we had mentioned that Webb used M_{11} as an example of the latter viewpoint in [**Web87a**, Sec 7]. The cohomology is described in more detail in [**AM04**, VIII.1.2]); it had originally been computed in a more complicated way by Benson and Carlson in [**BC87**, 7.9]. ◊

We will be continuing this theme of cohomology decompositions for sporadic groups, especially in later Section 7.5.

Consequently it is natural to also recall here, as we had briefly mentioned in our Preface and Introduction, a more general program undertaken by Adem and Milgram (with further collaborators)—aimed at computing the full mod-2 cohomology of the sporadic groups. (Or at least, at beginning with the smallest groups, and then continuing to the larger groups as far as seems feasible—for example, some

[10] By contrast, there are some good general results (notably of Quillen) for $\mathbb{F}_p$-cohomology where say p divides $q - 1$, when q is a power of the characteristic prime for G.

groups of the size roughly of HS have been treated by the present (2011) writing.) This project was in fact one motivation for their book [**AM04**].

And we mention in particular that for some sporadic groups, they do make use of cohomology decompositions of the general type discussed here (that is, as indicated in Definition 7.2.1, corresponding to [**AM04**, V.3.3]). For instance:

EXAMPLE 7.2.11 (Cohomology decompositions for M_{22} and McL). Among the 2-local geometries K with $\tilde{L}(K)$ shown to be projective in [**RSY90**] are those for $G = M_{22}$ and McL. (For group structure, see e.g. [**Wil09**, 5.2.9, 5.5.2].) The corresponding cohomology decompositions as in Corollary 7.2.9 are used in Adem and Milgram [**AM95a**] (though in fact their main methods there work more directly with restrictions to the stabilizers G_σ for $\sigma \in K$) and in [**AM97**]. (See Sections 7.3 and 7.11 in [**BS08**] for a systematic treatment of these groups.) $\diamondsuit$

Transition: Some early sporadic decompositions arising from non-projective $\tilde{L}(K)$. In fact the use of decompositions by Adem-Milgram will lead us next to a preliminary introduction to the topic of decompositions which are more general than those indicated so far in this section—in that they are based on complexes K for which $\tilde{L}(K)$ need *not* in fact be projective. (Instead $\tilde{L}(K)$ will just have vanishing module cohomology, as we had suggested in earlier Remark 7.2.4.) And then following that introduction, the deeper theory developed for these more general decompositions will be the main theme of our later Sections 7.3 and 7.4.

Indeed the desirability of weakening hypotheses related to projectivity seems to have been widely explored in the years after Webb's papers.[11] In particular, these developments typically maintained the viewpoint used in Webb's later paper [**Web91**]—namely, of the G-equivariant cohomology of K; and especially of deducing a decomposition from the property of sharpness (see later (7.3.14)).

However, some of the earliest results on more general decompositions did not require deducing the decomposition from sharpness; instead they were expressed just at the more elementary level of the cohomology decomposition itself. We sample a few such earlier results below.

EXAMPLE 7.2.12 (A cohomology decomposition for M_{12}). The Adem-Milgram program was essentially initiated in the pioneering work of Adem-Maginnis-Milgram in [**AMM91**], which deals with $G = M_{12}$ and $p = 2$. The authors make use of the natural 2-local geometry K for M_{12} indicated in our earlier Example 6.6.5; and in particular, they use the corresponding cohomology decomposition over K/G—but only "eventually", in the following sense:

For recall from Example 6.6.5 that $\tilde{L}(K)$ is only relatively projective,[12] not projective; and hence one cannot just quote Corollary 7.2.9 to obtain the decomposition. (The case for M_{12} in later [**BS08**, Sec 7.2], obtaining that decomposition directly from non-projective $\tilde{L}(K)$, was not yet available.)

[11] For example, this weakening is implicit in the sporadic cases failing projectivity in [**RSY90**]; and it is more explicitly mentioned in [**SY97**, p 377].

[12] In a different direction of generalization, we mention that at [**Thé87**, 2.8], Thévenaz gives an extension of Webb's Theorem 7.2.5—to a decomposition of the *relative* cohomology (corresponding to the relative projectivity in Theorem 6.6.2). Cf. also Symonds [**Sym08**].

So the authors instead begin with $\mathcal{A}_2(G)$, for which the reduced Lefschetz module *is* projective, leading to the decomposition over $\mathcal{A}_2(G)/G$ afforded by Corollary 7.2.7. They then require a fair amount of work in [**AMM91**, Sec 2] to determine the structure of the indexing set, i.e. the orbit complex $\mathcal{A}_2(G)/G$. But once they have done this, they find that certain pairs of cohomology terms $H^*(G_\sigma)$, in adjacent dimensions, are identical—and so cancel out of the alternating sum. And then the remaining terms in the sum are just those of the decomposition over K/G. (Phrased another way: The Lefschetz modules $\tilde{L}(\mathcal{A}_2(G))$ and $\tilde{L}(K)$ are different, but they determine the same "net" decomposition of group cohomology.)

So they can finish by determining the cohomology of the stabilizers G_σ, just for $\sigma \in K/G$ (which still requires some effort). ♢

As the reader might suspect, the method used above is not necessarily easy to generalize—depending as it does on perhaps "accidental" cancellations. In fact there is another approach to the above example, which is able to begin directly with the decomposition over K/G; it exploits a special feature of the particular non-projective module $\tilde{L}(K)$:

EXAMPLE 7.2.13 (An alternative approach to the decomposition for M_{12}). We continue to consider the situation of the previous Example 7.2.12: where $G = M_{12}$, with $p = 2$, and K is the standard 2-local geometry.

We now recall from Example 6.6.5 the determination by Benson-Wilkerson in [**BW95**, 3.3] that $\tilde{L}(K)$ is a module whose indecomposable summands all lie in a non-principal 2-block of G (that is, in a block *not* containing the trivial module 1).

In particular this implies that there can be no non-split extensions (or even nonzero homomorphisms) of 1 with $\tilde{L}(K)$; that is,

$$H^*\big(G, \tilde{L}(K)\big) = \mathrm{Ext}^*_{\mathbb{F}_2 G}\big(1, \tilde{L}(K)\big) = 0 \quad \text{in all dimensions.}$$

Thus by Lemma 7.2.2, we get the decomposition of $H^*(G)$ over K/G. ♢

Again the method does not really generalize: indeed other examples (especially for larger groups) show that $\tilde{L}(K)$ often contains some indecomposable summands which do lie in the principal block (and so which might in principle involve the projective cover $P(1)$).

Another early work on decomposition methods, extending away from projectivity hypotheses (e.g. relaxing contractibility for fixed points just to acyclicity in positive dimensions) appears in Maginnis [**Mag95**]; the methods were applicable at least to certain "medium-sized" sporadic geometries.

7.3. (Option G): Approaching $H^*(G)$ via equivariant cohomology of K

From a topological viewpoint, it is desirable to understand decompositions of group cohomology $H^*(G)$ (such as that in Webb's Decomposition Theorem 7.2.5) in terms of a suitable decomposition at the deeper level of the classifying space BG. In particular, this is the context of the theory of more general decompositions which we will describe in the subsequent Section 7.4.

A first step in the direction of decomposing BG is provided by "approximating" BG (at least at the level of homology) with the Borel construction of K. This is the sense in which, as we had mentioned at the start of the previous Section 7.2, it is perhaps more natural to study cohomology decompositions in the context of the G-equivariant cohomology of K.

So in this section, we will roughly "repeat" the previous section—by developing corresponding results using the viewpoint of G-equivariant cohomology. That is, our hypotheses will still in effect guarantee that $\tilde{L}(K)$ is projective; but the proofs will not explicitly involve the Lefschetz module $\tilde{L}(K)$ itself. (Considerations of projective modules will in fact still be involved in the arguments—but will be shifted to other parts of the proofs.)

Thereafter we will continue the equivariant viewpoint, for the more general decompositions in Section 7.4.

As we had warned in the initial summary at the start of the chapter, we will only be able to give a fairly brief glimpse of the deeper topological material—the reader is urged to consult the indicated references for fuller detail.

Topological background: Equivariant cohomology and ampleness. In fact the material we will need for this section is still fairly "classical"—compared with the further material which we will need to summarize at the start of Section 7.4. The overview that we give in this subsection roughly summarizes the development in Section 2.5 of [**BS08**].

REMARK 7.3.1 (The Borel construction and G-equivariant cohomology). Assume X is a G-space. The *Borel construction* for X is the quotient space[13]

$$(EG \times X)/G,$$

using the free contractible space EG of Definition 7.1.1.2.

The G-*equivariant* cohomology of Xcan be defined (e.g. [**Bro94**, VII.7]) in an elementary was as $H^*_G(X) := H^*\big(G, C^*(X)\big)$. But an alternative approach[14] is given (e.g. [**BS08**, 2.5.14]) via the cohomology of the Borel construction of X:

$$H^*_G(X) \cong H^*\big((EG \times X)/G\big).$$

We mention the standard property (e.g. [**Dwy01**, 6.2][15] or [**Gro02**, 7.2]) that:

(7.3.2) A G-homotopy equivalence preserves Borel constructions (and $H^*_G(-)$).

Here we are are using "preserves" as shorthand, for the more specific statement that if $f : X \to Y$ is a G-homotopy equivalence of G-spaces, then it induces an ordinary homotopy equivalence $(EG \times X)/G \simeq (EG \times Y)/G$ of Borel constructions; and hence an isomorphism $H^*_G(X) \cong H^*_G(Y)$ of equivariant cohomology. ♢

This preservation principle in (7.3.2) is a main motivation for our Option G—and especially for our development of G-equivariant versions of our earlier homotopy equivalence results.

---

[13]We usually write out this quotient; the literature often uses the convention $EG \times_G X$.

[14]More specifically, this is *Borel* G-equivariant cohomology. (We mention the more sophisticated *Bredon* G-equivariant cohomology e.g. in discussing later Lemma 7.3.17.)

[15]Indeed this reference shows that a slightly weaker equivalence suffices for preservation; see also [**BS08**, 2.5.13], and cf. later Remark 7.5.2.

There are a number of important connections of the equivariant cohomology of suitable spaces with ordinary group cohomology: For example, if X is a single point $\star$, then we have $(EG \times \star)/G \cong EG/G$, i.e. BG; so that we obtain:

$$H^*_G(\star) \cong H^*(BG) \cong H^*(G). \tag{7.3.3}$$

Indeed the preservation principle in (7.3.2) above shows that $H^*_G(X) \cong H^*(G)$ still holds even when X is just G-contractible.

However, we are more interested in the case where X might not be contractible—but where just forcibly mapping X to a point $\star$ nonetheless still induces an isomorphism of $H^*_G(X)$ with $H^*(G)$; this is the property called:

DEFINITION 7.3.4 (Ampleness). Assume that K is a G-complex with type-preserving action. We say (using the language of Dwyer's overview indicated in our subsequent Section 7.4) that K is *ample* if the surjection $K \to \star$ induces an isomorphism of mod-p cohomology:

$$H^*_G(K) \cong H^*(G).$$

(In the topological literature, this usage is normally restricted to the situation in which $K = K(\mathcal{C})$ for a *collection* $\mathcal{C}$—namely what we are calling a subgroup complex, i.e. a G-subcomplex of the subgroup lattice $L(G)$.)

From (7.3.2) we see that a G-homotopy equivalence preserves ampleness. $\diamondsuit$

Now ampleness is in fact implied by the stronger property of sharpness (see later (7.3.13))—and sharpness in effect provides the main topic of this section. However, we will first "digress" to briefly explore ampleness: For ampleness can be regarded as a logical (and historical) forerunner of sharpness; and in particular, it provides some very direct motivation for using the context of equivariant cohomology of a complex in the study of the group cohomology $H^*(G)$. (In particular, we will see that many of our usual complexes K are in fact ample.)

Brown's theorem on ampleness. Probably the prototypical result on ampleness for a general p-subgroup complex K is Theorem X.7.2 in Brown [**Bro94**]; we present a version below as Theorem 7.3.5. (Brown deduces the "standard" special case, where K is $\mathcal{S}_p(G)$, as his Corollary X.7.3.)

As usual, we restrict the more general result of Brown to a finite group G. In addition, we focus on mod-p cohomology, rather than the p-part of $\mathbb{Z}$-cohomology—and so we can state our Hypothesis 6.3.4(b+) below in our more usual language of $\mathbb{F}_p$-acyclicity, rather than of ordinary acyclicity as used in Brown's hypothesis.

We also adjust from Brown's statement for Tate cohomology, to ampleness—which involves ordinary mod-p cohomology; so again we must further consider dimension 0. So as in Webb's Decomposition Theorem 7.2.5, to Brown's original hypothesis we again add our earlier Hypothesis 7.2.5(c).

To continue our theme from earlier Remark 7.2.6.2: The role of this hypothesis was perhaps only implicit in the earlier literature on decompositions. And later, I understand that explicit consideration goes back at least to papers of Henn (e.g. [**Hen97**, 0.2]).[16] The necessity of some such further hypothesis was emphasized in Bob Oliver's email to Benson and Smith via the following example:

[16]Note that the "singular set K_s" in Henn, i.e. the union of all simplices with stabilizer order divisible by p, is under (c) just all of K.

Take $p = 2$ and G cyclic of order 2, acting on K given by some triangulation of the "figure 8" graph, with G switching the two loops: Here for the unique subgroup $Q = G$ of order 2, K^Q is the point of intersection, which is contractible; but nonetheless, this complex is not ample. (Notice that the extra hypothesis (c) does not hold here—since some simplices have stabilizer given just by the identity 1, which does not have order divisible by 2.)

THEOREM 7.3.5 (Brown's Ampleness Theorem). *Assume K is a G-complex with type-preserving action, satisfying:*

Hypothesis 6.3.4*(b+): K^Q is $N_G(Q)$-acyclic over $\mathbb{F}_p$, for Q of order p; and*

Hypothesis 7.2.5*(c): Each stabilizer G_σ is of order divisible by p.*

Then K is ample; that is, $K \to \star$ induces a mod-p cohomology isomorphism:

$$H^*_G(K) \cong H^*(G).$$

In particular, this holds for $K = \mathcal{S}_p(G)$, and for the other members of the standard G-homotopy type in Remark 4.6.1.

PROOF. The details of Brown's original proof in [**Bro94**] are beyond the scope of this book. So here we will just "cheat"—by observing that the same hypotheses are used for later Theorem 7.3.16—establishing sharpness, which by (7.3.13) implies ampleness. (The final special case, of the homotopy type of $\mathcal{S}_p(G)$, of course follows using the preservation principle (7.3.2).) □

We get many immediate instances of Theorem 7.3.5: For the "+" in our "Hypothesis 6.3.4(b+)" amounts to just adding equivariance, to the acyclicity in the original Hypothesis 6.3.4(b). Hence making that corresponding (routine) further verification of equivariance in the examples from the previous Section 7.2 (namely examples with projective Lefschetz modules $\tilde{L}(K)$) makes them examples also for Theorem 7.3.5.

We end the subsection by describing (briefly and informally) how one later approach to the proof of Theorem 7.3.5, presented in Adem-Milgram [**AM04**, V.3.1],[17] essentially continues some techniques which we have seen in earlier chapters:

The initial reduction in Adem-Milgram is similar in spirit to that in the final "slight variant" of Proof 1 in our Section 5.3 (i.e. within Quillen's version of the proof of Brown's Euler characteristic result 5.3.1). Their approach begins with the observation that the surjection of K on a point $\star$ determines a short exact sequence of cochain complexes, namely:

$$(7.3.6) \qquad 0 \to C^*(\star) \to C^*(K) \to R \to 0 \; ;$$

where R is the quotient complex $C^*(K)/C^*(\star)$. (This is in roughly the same spirit as the relative-complex sequence in earlier (5.3.9).) The short sequence then leads via $H^*(G, -)$ to a long sequence in G-equivariant cohomology:

$$(7.3.7) \qquad \cdots \to H^{i-1}(G, R) \to H^i_G(\star) \to H^i_G(K) \to H^i(G, R) \to \cdots$$

(So this sequence is in the spirit of the long exact sequence in earlier (5.3.10).) And now showing $H^*(G, R) = 0$ will give $H^*_G(\star) \cong H^*_G(K)$, as desired. In particular, here $H^*(G, R)$ might be regarded as a rough analogue, for the equivariant context, of reduced ordinary cohomology.

[17] Our discussion here similarly extends from the use in Adem-Milgram of just $\mathcal{A}_p(G)$, to more general K; and from their context of Tate cohomology, to ordinary cohomology.

For the proof that $H^*(G, R) = 0$, Adem and Milgram now consider a Sylow group $P \in \mathrm{Syl}_p(G)$; they use the injection in Lemma 7.1.2, to replace $H^*(G, R)$ by its isomorphic image in the P-equivariant cohomology $H^*(P, R)$, so that the proof is now reduced to showing that $H^*(P, R) = 0$. And that deduction is very much in the spirit of our comments at the end of Section 6.2, on Quillen's proof of projectivity of the Lefschetz module in Theorem 6.2.1; indeed since such arguments will recur, in a somewhat simpler form, in our later proof of Theorem 7.3.16, we will just give a brief initial summary of them, as follows: Adem and Milgram examine cohomology from R via indecomposable summands I_σ of permutation modules $\mathrm{Ind}_{G_\sigma}^G(1)$, for σ representing an orbit in K/G; and they then exploit the singular set $L := K^{P\text{-sing}}$. First, for σ outside L, I_σ is projective, giving vanishing in positive dimensions of P-equivariant cohomology; while Hypothesis 7.2.5(c) allows the use of essentially our observation at (6.3.9) (with the single projective I_σ in place of $\tilde{L}(K)$) to give vanishing of dimension-0 cohomology for I_σ. Next, the acyclicity in Hypothesis 6.3.4(b+) is used to show that L is P-equivariantly acyclic, i.e. has the cohomology of a point $\star$. (This is in analogy with contractibility of the singular set in earlier Lemma 4.4.14.) And now for σ in the singular set L, this acyclicity guarantees vanishing of P-equivariant cohomology contributed by any I_σ within R (in the analogy mentioned above with reduced ordinary cohomology, which in effect "avoids" $\star$).

More topological background: Spectral sequences and sharpness. If a G-complex K is in fact ample, so that we know (at least in the abstract) that $H_G^*(K) \cong H^*(G)$, then the reader might still ask how the actual *computation* of $H_G^*(K)$ might offer advantages over direct computation of $H^*(G)$. And in fact, we will see below that we profitably can approach $H_G^*(K)$ via onc of several possible *spectral sequences*.

The actual details of spectral sequences are beyond the scope of this book. So for fuller background, we direct the reader to e.g. [**BS08**, Sec 2.6], as well as the additional references given there. And here, we instead first indicate some of the general context, in a very rough overview: The particular sequences which we will use are defined starting with an E_1*-page*—which is displayed as a doubly-indexed array of suitable terms $E_1^{x,y}$ with $x, y \geq 0$. (In our context of $(EG \times K)/G$, the horizontal and vertical indices x and y can roughly exploit the dimensions of cells in the CW-complex EG, and of simplices in K.) Cohomology from suitable differential maps among these terms (e.g. [**Ben91**, 3.2.3]) leads to a further page called E_2; and then similarly, to pages $E_3, E_4, \cdots$. And in general, we might have to proceed all the way to the "limiting" E_∞-page, to obtain $H_G^n(K)$ from the terms $E_\infty^{x,y}$ with $x + y = n$.

However, in our ideal situation below of sharpness, we will not require such later pages: namely we will get $E_2 = E_3 = \cdots = E_\infty$, with all nonzero terms on the vertical axis (i.e. $x = 0$); so that $E_2^{0,n}$ in fact gives an expression for $H_G^n(K)$.

So within the above general context, we now indicate more specifically the sequences of interest: (The overview that we give in this subsection roughly summarizes the development in Section 2.6 of [**BS08**].)

REMARK 7.3.8 (Spectral sequences for equivariant cohomology). One standard approach to the cohomology of the Borel construction of K is via the *Serre* spectral sequence (e.g. [**BS08**, 2.6.7]). Here the x-indexing in the term $E_1^{x,y}$ is determined (e.g. [**BS08**, 2.6.2]) essentially from dimensions of cells in EG: namely using the corresponding filtration on the Borel construction coming from its natural surjection $(EG\times K)/G \to (EG\times\star)/G \cong BG$ (where this map in turn arises just from the surjection $K \to \star$). But in practice, it is perhaps more usual to define the sequence by instead (e.g. [**Ben91**, p 108]) proceeding directly to the E_2-page—where now the y-indexing is in turn defined using dimensions of simplices in K—namely via cohomology of G with coefficients in the module $H^y(K)$:

$$E_2^{x,y} := H^x\big(BG; H^y(K)\big).$$

In particular (see e.g. [**Ben91**, 2.6.8]) this sequence in the limit E_∞ determines the G-equivariant cohomology $H^*_G(K)$, as noted above.

A different sequence leading in the limit to $H^*_G(K)$ is the *isotropy* spectral sequence: Here in place of $K \to \star$ as above, we instead use $EG \to \star$; leading to $(EG \times K)/G \to (\star \times K)/G \cong K/G$. So the corresponding filtration of the Borel construction now gives an x-indexing determined by dimensions of simplices in K. That is, we can express the y-th row of the E_1-page as $E_1^{*,y} := H^y\big(G, C^*(K)\big)$, and in particular:

$$E_1^{x,y} := H^y\big(G, C^x(K)\big). \tag{7.3.9}$$

Furthermore the differential map on this y-th row leads to an E_2-page whose y-th row has the form $E_2^{*,y} = H^*(\ H^y\big(G, C^*(K)\big)\ )$; that is, we have:

$$E_2^{x,y} = H^x\Big(H^y\big(G, C^*(K)\big)\Big). \tag{7.3.10}$$

The "isotropy" in the name of the sequence refers to isotropy subgroups—these become visible as simplex stabilizers, when we apply, to the E_1-terms in (7.3.9) above, the Eckmann-Shapiro Lemma 7.2.3: this converts cohomology of G, in induced modules of form $\mathrm{Ind}_{G_\sigma}^G(1)$ in $C^*(K)$, into group cohomology of the subgroups G_σ stabilizing the corresponding simplices; that is, we have:

$$E_1^{x,y} \cong \bigoplus_{\sigma\in K_x/G} H^y(G_\sigma). \tag{7.3.11}$$

This is the form of the E_1-page used e.g. in [**Web87a**, 6.1]. ♢

We are mainly interested in the following situation, where the isotropy sequence exhibits a particularly simple behavior:

Definition 7.3.12 (Sharpness). We say (again as in Definition 7.3.4 using the later language of Dwyer's overview) that K is *sharp*,[18] if for the isotropy spectral sequence, with E_2-page as in (7.3.10) above, we get the vanishing behavior:[19]

$$E_2^{x,y} = H^x\Big(H^y\big(G, C^*(K)\big)\Big) = 0 \text{ for all } x > 0.$$

(One can also use the language that the isotropy sequence *collapses to the vertical axis (fiber) on the E_2-page*.) ♢

---

[18]This will correspond to the special case of *normalizer* sharpness, in the viewpoint of decompositions in the following section.

[19]Also see [**Gro02**, Sec 9] for sharpness described from the viewpoint of higher limits.

We first mention the standard property (e.g. [**Dwy01**, 8.2] or [**Gro02**, 9.1]):

(7.3.13) Sharpness implies ampleness.

It also follows from the vanishing under sharpness that $E_2 = E_3 = \cdots = E_\infty$. And in particular, we get $H^n_G(K) = E_2^{0,n}$—since that is the only possible nonzero term of the E_∞-page with indices x, y satisfying $x+y = n$. Indeed we have $H^n_G(K) = H^n(G)$ for this limiting term $E_\infty^{0,n} = E_2^{0,n}$. Furthermore the vanishing, when read along the n-th row of the E_2-page, implies that this term $E_2^{0,n}$ is just the sum (with signs alternating by dimension) of the terms in the n-th row of the E_1-page. So using the view of $E_1^{x,y}$ indicated in (7.3.11) as a sum over x-simplices in K, we see that:

(7.3.14) Sharpness implies a cohomology decomposition over K/G as in 7.2.1.

It also follows using the definition of the isotropy sequence (e.g. [**Gro02**, 7.2]) that:

(7.3.15) Sharpness is preserved by a G-homotopy equivalence.

And cf. also later Remark 7.5.2 regarding preservation under weaker mappings.

Webb's theorem on sharpness. Our treatment in the previous Section 7.2 of Webb's Decomposition Theorem 7.2.5 had emphasized his original approach—via projectivity of $\tilde{L}(K)$—in the paper [**Web87a**]. We now turn to the approach in his subsequent paper [**Web91**]—which involves the viewpoint of G-equivariant cohomology, the theme we are emphasizing in the present section.

In fact, in Section 6 of the earlier paper [**Web87a**], Webb had already begun to indicate aspects of the G-equivariant approach—and indeed in Theorem E there, had established a corresponding sharpness result for the special case of $K = \mathcal{S}_p(G)$ with G of p-rank 2. Then in the later paper, he established the result for more general K as [**Web91**, Thm 2.7.1]. We present a version of this result as Theorem 7.3.16 below. (But see also Webb's more general statement, in terms of projective structure for chain complexes, in Remark 6.4.5.)

Just as with Theorem 7.3.5 above, we adjust from Webb's original statement for Tate cohomology, to ordinary group cohomology and sharpness—which in particular involves further consideration of dimension 0. And again as our focus is mod-p cohomology, we state our hypothesis in terms of acyclicity over $\mathbb{F}_p$, rather than the contractibility in Webb's hypothesis; that is, we continue our hypotheses for Brown's Ampleness Theorem 7.3.5:

THEOREM 7.3.16 (Webb's Sharpness Theorem). *Assume that K is a G-complex with type-preserving action, satisfying:*
Hypothesis 6.3.4*(b+): K^Q is $N_G(Q)$-acyclic over $\mathbb{F}_p$, for Q of order p; and*
Hypothesis 7.2.5*(c): Each stabilizer G_σ is of order divisible by p.*
Then K is sharp; that is, for the isotropy spectral sequence we have:

$$\begin{aligned} E_2^{x,y} &= 0\,, \qquad \textit{for } x \geq 1; \textit{ and} \\ E_2^{0,y} &= H^y(G)\,. \end{aligned}$$

And so from (7.3.14) *we get a cohomology decomposition over K/G:*

$$H^*(G) \cong \bigoplus_{\sigma \in K/G} (-1)^{\dim \sigma}\, H^*(G_\sigma)$$

In particular, this holds for K in the standard G-homotopy type in Remark 4.6.1.

PROOF. (Sketch) Webb in fact deduces his result as a corollary of his Structure Theorem [**Web91**, Thm A], which expresses the cohomology as a split exact sequence of *Mackey functors*; roughly, that approach involves a further generalization of the use of the Burnside ring of projectives in the proof of Webb's Projectivity Theorem 6.3.4. But we will not follow that direction; instead, we will summarize the approach of Adem-Milgram at [**AM04**, Thm V.3.2]—since this continues (with suitable adaptations) some themes we have seen in earlier chapters.[20]

We begin by recalling from (7.3.10) that:

$$E_2^{x,y} = H^x\Big(H^y(G, C^*(K))\Big).$$

We restrict attention to $x \geq 1$; and we must show that $E_2^{x,y} = 0$.

We take $P \in Syl_p(G)$, with singular set $L := K^{P\text{-sing}}$. By the standard injection[21] of Lemma 7.1.2, it will suffice to obtain vanishing (for $x \geq 1$) of:

$$E_2^{x,y}(P) := H^x\Big(H^y(P, C^*(K))\Big).$$

We claim first that we can replace $C^*(K)$ with $C^*(L)$—in the sense that they have the same cohomology as P-complexes (i.e. we get the same P-equivariant cohomology $H_P^*(K) \cong H_P^*(L)$): For suppose we are given σ representing an orbit in $K \setminus L$. Then (cf. our comments at the end of Section 6.2) the free action of P on the orbit of σ implies via Definition 6.1.3.3 that the corresponding induced module $\mathrm{Ind}_{G_\sigma}^G(1)$ is projective—and projective modules make no contribution to cohomology in dimensions $y \geq 1$ (as we saw in Lemma 6.1.5). And for the contribution in dimension $y = 0$, we can consider any indecomposable component I_σ of such a projective module; and now (this time much as in the proof of Theorem 7.2.5) Hypothesis 7.2.5(c)[22] allows us to use essentially our observation at (6.3.9) (with the single projective indecomposable I_σ in place of $\tilde{L}(K)$) to give vanishing of dimension-0 cohomology for I_σ.

We claim next that we can in turn similarly replace L with a point $\star$: namely, we get that L is P-equivariantly acylic over $\mathbb{F}_p$, so L has the same P-equivariant cohomology as a point. This follows using an analogue of earlier Lemma 4.4.14, in which "contractible" is weakened to "F_p-acyclic";[23] in our present context, this analogue says that:

If K^Q is $\mathbb{F}_p$-acylic for all $1 < Q \leq P$, then $L = K^{P\text{-sing}}$ is also $\mathbb{F}_p$-acyclic.

---

[20]But we also mention that Grodal's higher-limits viewpoint leads to a particularly short proof, at [**Gro02**, 8.15].

[21]As Adem and Milgram observe, this injection in fact induces a *split* monomorphism of complexes: since $|G : P|$ is coprime to p, the standard averaging argument of Maschke's Theorem produces a complement to the "image of G". Indeed Webb in [**Web91**] goes farther, producing his split exact sequence of Mackey functors. Cf. also [**Bou91**, 2.8].

[22]We mention that essentially this hypothesis (at least implicitly for all stabilizers σ) appears in the last sentence of the proof of V.3.2 in Adem-Milgram—where they observe that Tate cohomology agrees with ordinary cohomology for groups of order divisible by p. It also appears more explicitly elsewhere in the literature on sharpness, e.g. [**Dwy01**, 11.8].

[23]Furthermore we mention that a proof can be given by tracing back through the proof of 4.4.14: replacing the Quillen closed-set Theorem 4.4.4, and in turn the Quillen Fiber Theorem 4.2.1 in its proof, by similar analogues—in which the hypothesis assumes that the relevant fibers are $\mathbb{F}_p$-acyclic, (rather than contractible); and the conclusion is $\mathbb{F}_p$-homology equivalence (rather than homotopy equivalence). Cf. also our earlier Remark 4.2.3 on fiber-type theorems.

And the hypothesis holds here, by the acyclicity in Hypothesis 6.3.4(b+). Furthermore, the customary extension to P-equivariant acyclicity also follows, using the $N_G(Q)$-equivariance (and hence $N_P(Q)$-equivariance) of the acyclicity there.

At this stage, we have reduced $E_2^{x,y}(P)$ to $H^x(\ H^y(P, C^*(\star))\ )$. Now $C^x(\star) = 0$ for $x \geq 1$; and this vanishing propagates up through the longer expression, showing for $x \geq 1$ that $0 = E_2^{x,y}(P) = E_2^{x,y}$, as required. □

Once again, much as we had commented after Theorem 7.3.5: the examples of ampleness from the previous Section 7.2 (corresponding to projective Lefschetz modules $\tilde{L}(K)$), once we additionally check equivariance of the (known by hypothesis) $\mathbb{F}_p$-acyclicity of fixed points, give examples also of sharpness via Theorem 7.3.16.

Some observations on hypotheses for sharpness. We return to the theme that we had introduced at earlier Remark 7.2.6.2: We recall the background there, namely that the use of Tate cohomology in some of the earlier literature (especially the work of Brown and Webb) had perhaps tended to obscure the role of dimension 0 for ordinary cohomology. This is the reason that we had very explicitly emphasized the use of Hypothesis 7.2.5(c), namely that all stabilizers have order divisible by p.

We now indicate an alternative condition (c′), which—within Webb's projective context—is essentially equivalent to sharpness. This condition arises essentially just from the definition of the isotropy spectral sequence; and so it is presumably well-known in principle, though it does not seem to be emphasized very explicitly in the literature.[24] We are grateful to Bob Oliver for showing us the form given below, as part of his analysis described in 7.2.6.2.

LEMMA 7.3.17. *Assume that K is a (nonempty) G-complex with type-preserving action; and that in the terms for $x \geq 1$ in the E_2-page of the isotropy sequence in (7.3.10), we know for positive dimensions $y \geq 1$ that $E_2^{x,y} = 0$. (For example, this holds under the equivariant form of Webb's Hypothesis 6.3.4(b+) in Theorem 7.3.16.)*

Then the vanishing also for $y = 0$ (and $x \geq 1$; i.e. sharpness) is equivalent to:[25]

(c′) $H^x(K/G; \mathbb{F}_p) = 0$ for all $x \geq 1$.

We recall that in our usual situation where K (and hence K/G) is connected, the condition (c′) just becomes $\mathbb{F}_p$-acyclicity for K/G.

PROOF. (Sketch:) To approach the isotropy spectral sequence, Oliver's argument works in the context of *Bredon cohomology*: we will not try to describe that in any real detail here, but it is fairly standard in the literature—see e.g. Grodal [**Gro02**, Sec 2], or Symonds [**Sym05**]. In particular, this approach emphasizes the role of the orbit complex K/G.

In this viewpoint, the E_2-page of the sequence in (7.3.10) appears in the form (e.g. [**Dwy01**, 8.2] or [**Gro02**, 9.1]):

$$E_2^{x,y} = H_G^x(K; M_y);$$

[24]The idea is at least mentioned e.g. in [**Lib07**, p 517]; cf. also Henn [**Hen97**, 0.2].

[25]We emphasize that K/G here is the orbit *CW-complex*: Thus for example, when K arises as the order complex $K(\mathcal{P})$ for some G-poset $\mathcal{P}$, it is *not* in general sufficient to consider the order complex $K(\mathcal{P}/G)$ of the orbit *poset* $\mathcal{P}/G$.

where H^*_G now denotes Bredon cohomology, and M_y is the local coefficient system (cf. later Definition 10.1.1) on K/G which maps the orbit of a simplex σ (represented by G/G_σ) to the degree-y term of the cohomology of the Borel construction on that orbit. It is standard (cf. [**BS08**, 2.5.19]) that the latter cohomology term can be interpreted as $H^y(G_\sigma)$.

Now by hypothesis, we are assuming for $x \geq 1$ that $E_2^{x,y} = 0$ whenever $y \geq 1$ (for example, from projectivity-related hypotheses such as 6.3.4(b+)). So we want the further vanishing (still for $x \geq 1$), in the remaining case where $y = 0$. And for this dimension, M_0 is just the constant local system, which takes each G/G_σ to $H^0(G_\sigma) = \mathbb{F}_p$, namely the trivial module; that is, the local system is just the ordinary cochain complex $C^*(K/G)$ for the orbit complex K/G. Hence we have:

$$E_2^{x,0} = H^x_G(K; M_0) = H^x(K/G).$$

And vanishing of this expression for $x \geq 1$ is in fact just our hypothesis (c′). □

We mention (especially for later reference in Section 7.5) that the sporadic geometries K in [**BS08**] are all flag-transitive, so that the orbit complex K/G is a single simplex; this is certainly acyclic over $\mathbb{F}_p$, and in particular we thus get Hypothesis 7.3.17(c′).

We conclude with a sketch of an alternative proof of the sufficiency,[26] for sharpness, of our Hypothesis 7.2.5(c) (namely that all stabilizers have order divisible by p); again assuming that we already have Hypothesis 6.3.4(b) of $\mathbb{F}_p$-acyclicity of fixed points. The idea seems to go back at least to Henn [**Hen97**, 0.2], as mentioned earlier; we indicate below an argument in the context of Bredon cohomology, which again was supplied to us by Oliver.

The overall idea is to embed K in a finite-dimensional complex Y, such that simplices σ in $Y \setminus K$ have stabilizers G_σ of order coprime to p. (This can even be done in such a way that G-orbits in $Y \setminus K$ are free.) Namely one can start by embedding $K < Y_0$, adding only free orbits, so that for some n, Y_0 is n-dimensional and $(n-1)$-connected. By [**Oli75**, Lemma 3], $H^n(Y_0)$ is a projective $\mathbb{F}_pP$-module for a Sylow group $P \in Syl_p(G)$, and so is $\mathbb{F}_pG$-projective. Hence it is stably free in the category of infinitely generated $\mathbb{F}_pG$-modules; so one can add free orbits of simplices, to construct $Y > Y_0$ which is $\mathbb{F}_p$-acyclic. Now Hypothesis 7.2.5(c) guarantees that K is the appropriate "singular set"[27] Y_s, in the notation of [**Oli76**] and [**Hen97**, 0.2]. And it follows by [**Oli76**, Prop 2] that $H^*(Y/G, K/G) = 0$. Also, Y/G is acyclic over $\mathbb{F}_p$ in view of [**Oli76**, Thm 1]. Hence we see K/G is also $\mathbb{F}_p$-acyclic. This in particular implies Hypothesis 7.3.17(c′); and so we obtain sharpness from that result.

7.4. Decomposing BG via a homotopy colimit over K/G

In this section, we briefly describe modern *decomposition theory*: in the end, we will still obtain cohomology decompositions, as in the conclusion of Webb's Decomposition Theorem 7.2.5—but these will now in fact be derived (and hence "explained") as a consequence of a corresponding "homotopy decomposition" (one also sees the related term "homotopy approximation"), at the more fundamental

[26] It does not seem clear that it should be a necessary condition, as is the case for (c′) above.
[27] It is in effect "global", including our "local" P-singular sets for all nontrivial p-subgroups P.

level of the classifying space BG. And the terms in that decomposition for BG will be the classifiying spaces BG_σ of the stabilizers G_σ of representatives σ for K/G.

In particular, we will continue the equivariant viewpoint of the previous Section 7.3: involving notably the Borel construction of K, with considerations of ampleness and sharpness; and we will approximate the Borel construction via a suitable homotopy colimit (over terms of form BG_σ). Furthermore, we will also extend away from earlier hypotheses, such as those in Theorem 7.2.5, which would in effect guarantee that $\tilde{L}(K)$ is projective—so that we will obtain new decompositions, which are more general than those we had before.

The theory of decompositions was developed by various authors, primarily during the 1990s. The work in the area was discussed in a unified viewpoint in the work of Dywer—the reader is particularly directed to his expository treatment in [**Dwy01**]. We will in fact mostly refer to the exposition in the background chapters (especially Chapter 5) of Benson-Smith [**BS08**]—which is designed primarily for group theorists, and in particular assumes less topological experience.

Topological background: Homotopy theory and colimit decompositions. The context for decomposition theory is modern homotopy theory—which involves topological material considerably less classical than we sketched in the previous Section 7.3. We begin with a few words about the general setting:

REMARK 7.4.1 (Some aspects of homotopy theory). In much of modern topology, the word "space" usually refers to a *simplicial set*. (We had mentioned this context at earlier Remark 1.2.17; cf. [**BS08**, Ch 3].) This is equivalent to working with topological spaces, by a standard result of Quillen—e.g. [**BS08**, 3.6.10]. In particular, many constructions are developed using the language of categories. (However in our exposition below, the reader can usually still interpret most statements in the more elementary framework of simplicial complexes.)

Since our focus is on $\mathbb{F}_p$-cohomology, we usually replace a space X with its (Bousfield-Kan) *p-completion* $X_p^\wedge$. This is because for our spaces (and indeed for the spaces called $\mathbb{F}_p$*-good* by Bousfield-Kan, cf. [**BS08**, 4.2.7]), the p-completion "isolates" the $\mathbb{F}_p$-cohomology of X: that is, $H^*(X;\mathbb{F}_p) \cong H^*(X_p^\wedge;\mathbb{F}_p)$. In particular for decompositions of BG (which is $\mathbb{F}_p$-good because it has finite fundamental group G), we will in fact restrict attention to the p-completed classifying space $BG_p^\wedge$. (Cf. [**BS08**, 4.3.1].) $\diamondsuit$

The idea of decomposing BG in terms of the classifying spaces BG_σ of the stabilizers is already suggested by the appearance of the stabilizers in Webb's Decomposition Theorem 7.2.5—and especially in the isotropy spectral sequence for the cohomology of the Borel construction of K, notably in the discussion of the decomposition property (7.3.14) of sharpness. Thus we might approximate that Borel construction via some suitable topological gluing of the spaces BG_σ (which allows more flexible structure than just a product). The relevant construction is:

REMARK 7.4.2 (Colimit and homotopy colimit). For fuller details and references, see for example [**BS08**, Sec 4.5–4.8]. Here we give just a very rough overview:

Consider first the special case where we have an inclusion of a space X in spaces Y, Z. Here the usual *colimit* consists of a disjoint union of Y and Z—subject to identifying the images in Y and Z of each point $x \in X$.

By contrast, the (Bousfield-Kan)[28] *homotopy* colimit uses a more relaxed notion of identification. Still in the above special case, it roughly uses a cylinder,[29] to join the images in disjoint Y and Z of each point $x \in X$: this allows for adjusting "up to homotopy". And correspondingly, the *homotopy colimit* construction is preserved by homotopy equivalence in a natural way.

In the general case for the homotopy colimit, we are given some *indexing category* $\mathcal{C}$, and a set of corresponding spaces X_c for $c \in \mathcal{C}$. We will denote the homotopy colimit by:

$$\text{Hocolim}_{c \in \mathcal{C}}\ X_c;$$

where it is implicit in the notation that the assignment X_- is functorial: so that any morphism $c \to d$ in the category $\mathcal{C}$ implies suitable relations between the images of the spaces X_c and X_d. $\diamondsuit$

For us, one category of frequent interest arises from a simplicial complex K, with the morphisms of the category being given just by the face relations $\sigma \subseteq \tau$ among the simplices in K. And then we are interested in a decomposition approximating the Borel construction of K, which is indexed by the quotient category $\mathcal{C}$ corresponding to the orbit complex K/G (cf. earlier (5.1.10)); namely in the form:

$$\text{Hocolim}_{\sigma \in K/G}\ BG_\sigma; \tag{7.4.3}$$

which we then p-complete. (The standard interpretation of BG_σ, as the Borel construction of the orbit of σ, is indicated e.g. in [**BS08**, 2.5.19].)

The homotopy-colimit form in (7.4.3) is of particular interest in that it provides a "decomposed" approach (cf. [**BS08**, 4.7.2]) to the Borel construction $(EG \times K)/G$; so the cohomology of the homotopy colimit is the G-equivariant cohomology $H^*_G(K)$ of K. Thus when K is in fact ample in the sense of Definition 7.3.4, the homotopy colimit provides a space whose cohomology is just the mod-p group cohomlogy $H^*(G)$. Hence we can then regard (the completion of) the homotopy colimit as a decomposition at the level of the p-completed classifying space $BG^\wedge_p$.

In fact in the approach of this book, typically our complex K is (or can be) realized as the order complex $K(\mathcal{P})$, for some poset $\mathcal{P}$ of nontrivial p-subgroups of G. And as indicated at [**BS08**, Defn 5.6.6], the notion of ampleness for the complex K corresponds to the notion of ampleness for the collection $\mathcal{P}$, as developed in the topological literature on decomposition theory.

Now the morphisms in $K(\mathcal{P})$ considered as a category are just the inclusions in $\mathcal{P}$, which are automatically preserved by the G-conjugacy; so the stabilizer of an object P of the category is the full normalizer $N_G(P)$. Suitable intersections then provide stabilizers G_σ of higher-dimensional simplices σ. And this motivates the terminology of the *normalizer* decomposition for the collection $\mathcal{P}$, in the topological literature; which for us takes the form (7.4.3) for $K = K(\mathcal{P})$. (In particular, our term sharpness in Definition 7.3.12 corresponds to sharpness for the normalizer decomposition, that is, normalizer sharpness in the literature.)

---

[28] I thank Pete Bousfield for suggestions in this area.

[29] Indeed this particular special case, predating the general homotopy colimit, is called the *double mapping cylinder* or *homotopy pushout*.

Next, we will be seeing some additional indexing categories other than K/G, for which the corresponding stabilizers are instead provided by certain typically-proper subgroups of the normalizers—leading to the terminology of "centralizer" and "subgroup" decompositions.

Decompositions based on elementary abelian p-subgroups. We will now summarize the *centralizer* decomposition, introduced by Jackowski and McClure in [**JM89**] [**JM92**]; we use the viewpoint of Dwyer, as in [**BS08**, Sec 5.4].

We let $\mathcal{A}$ be a G-subposet of $\mathcal{A}_p(G)$; that is, a G-poset of nontrivial elementary abelian p-subgroups.[30] Rather than considering the category given by the order complex $K(\mathcal{A})$ (where the morphisms are just inclusions), we instead use the corresponding *conjugation* category $\mathcal{C}_{\mathcal{A}}$ on $\mathcal{A}$. Very roughly, this means we take as morphisms not just the inclusions, but also the G-conjugations among these subgroups. More precisely, as in [**BS08**, 5.3.1], an object of $\mathcal{C}_{\mathcal{A}}$ is an injection $i : E \to G$ of an abstract elementary p-group E, with the natural notion of morphisms. Since such an injection distinguishes the individual elements of E, the stabilizer of the object i is the centralizer in G of the image $i(E)$—which we somewhat abusively denote by by $C_G(E)$. (This is in contrast to the stabilizer of $i(E)$ in the normalizer decomposition for the collection $\mathcal{A}$, which would be $N_G(E)$.) Hence in the *centralizer decomposition*, i.e. the corresponding homotopy colimit over $\mathcal{C}_{\mathcal{A}}/G$, the term for an orbit with that stabilizer is the classifying space $BC_G(E)$.

The analysis of [**JM89**] goes on to investigate conditions on the poset $\mathcal{A}$ under which this centralizer decomposition should be ample and sharp (hence, *centralizer-sharp*). For example, in [**JM92**, Thm 1.3] they initially obtained ampleness and centralizer sharpness for $\mathcal{C}_{\mathcal{A}}$ given by the full collection $\mathcal{A} = \mathcal{A}_p(G)$. This can be compared with ampleness and normalizer sharpness for $K(\mathcal{A}_p(G))$, which we saw in Theorems 7.3.5 and 7.3.16. (The latter corresponds to a decomposition of BG of the form (7.4.3).)

Furthermore, their methods in [**JM92**, Thm 7.7] allowed for the possibility that the results should hold for a suitable subcollection $\mathcal{A} \subseteq \mathcal{A}_p(G)$, where $\mathcal{A}$ might be strictly smaller than $\mathcal{A}_p(G)$.

For simplicity of exposition, below we will restrict such a general $\mathcal{A}$ to a more specialized form $\mathcal{E}_p(G)$ studied by Benson, which is group-theoretically very natural. Let C denote some conjugacy class of *p-central* elements of order p in G: that is, any $x \in C$ lies in the center of some Sylow p-subgroup of G.

For the moment, we assume a fairly common special case—namely that C is closed under commuting products:

(7.4.4) Hypothesis: For commuting $x, y \in C$ with $xy \neq 1$, we have $xy \in C$.

Then the set of all nontrivial elementary abelian p-subroups E, such that each nontrivial element of E lies in the class C, defines (for this special case) the desired subcollection $\mathcal{E}_p(G)$.

And now we describe $\mathcal{E}_p(G)$ in the general case: If (7.4.4) fails, then we enlarge C by adding the classes of any xy not in the original C. Then we again

---

[30] We could consider more general p-subgroups; but it is convenient to have an object E lie inside the stabilizer $C_G(E)$ below—which motivates the assumption that each E should be abelian.

compute the commuting products, this time using the enlarged C, to see if (7.4.4) now holds. We can enlarge C again if necessary; eventually, this process of closing up under commuting products must terminate—and the final result defines the desired subcollection $\mathcal{E}_p(G)$. Of course, it can (and often does) happen that this process leads to the full collection $\mathcal{A}_p(G)$. But we will also shortly be seeing some groups G where $\mathcal{E}_p(G)$ is proper in (and not homotopy equivalent to) $\mathcal{A}_p(G)$.

Below we state the result [**JM92**, Thm 7.7], as specialized from a more general $\mathcal{A}$ down to $\mathcal{E}_p(G)$. (Of course as we mentioned earlier, ampleness and centralizer sharpness for $\mathcal{A}_p(G)$ are given by [**JM92**, Thm 1.3].) And we have added the analogous normalizer sharpness as in (7.4.3), which was established in Grodal-Smith [**GS06**, 1.2]—since that is the main form we will use in later applications.

THEOREM 7.4.5. *The centralizer decomposition for $\mathcal{E}_p(G)$, corresponding to the classifying-space decomposition given by (the p-completion of) the homotopy colimit*

$$\mathrm{Hocolim}_{E \in \mathcal{C}_{\mathcal{E}_p(G)}/G}\ BC_G(E),$$

is ample and sharp. And the collection $\mathcal{E}_p(G)$ is also normalizer-sharp; that is, we may also use the form:

$$\mathrm{Hocolim}_{E \in K\left(\mathcal{E}_p(G)\right)/G}\ BN_G(E).$$

(Option S:) We illustrate the result for some sporadic geometries; these examples preview a main theme of the subsequent Section 7.5.

EXAMPLE 7.4.6 (A classifying-space decomposition for M_{12}). We take $G = M_{12}$ with $p = 2$; thus we revisit the cohomology decompositions indicated in earlier Examples 7.2.12 and 7.2.13, now "explaining" them via a decomposition at the level of the classifying space (cf. [**BS08**, Sec 7.2]). Hypothesis (7.4.4) can be verified for the unique 2-central class, denoted in the Atlas [**CCN**$^+$**85**, p 33] by $2B$; indeed using the character table of G, this can be checked by means of the standard formula for the product of the class-sum of C with itself—for example, using the ClassMultiplicationCoefficient function in GAP [**GAP00**]. Thus by Theorem 7.4.5, we obtain sharpness of the normalizer decomposition corresponding to $\mathcal{E}_2(G)$—giving our cohomology decomposition for M_{12} from the earlier Examples. Furthermore $\mathcal{E}_2(G)$ is homotopy equivalent to the 2-local geometry K for M_{12}; and we saw in Example 6.6.5 that $\tilde{L}(K)$ is not projective, so that $\mathcal{E}_2(G)$ is not homotopy equivalent to $\mathcal{A}_2(G)$. Thus from $\mathcal{E}_2(G)$ we get a decomposition which is new, i.e. not available from Theorem 7.2.5. ♢

EXAMPLE 7.4.7 (A classifying-space decomposition for Co_3). We take $G = Co_3$ with $p = 2$; cf. [**BS08**, Sec 7.13]. (For the group Co_3 itself, see e.g. [**Wil09**, 5.5.3].) Benson in [**Ben94**] verifies Hypothesis (7.4.4) for the unique 2-central class, denoted by $2A$ in the Atlas [**CCN**$^+$**85**, p 135]. Thus by Theorem 7.4.5, he obtains sharpness of the normalizer decomposition corresponding to $\mathcal{E}_2(G)$, and the resulting cohomology decomposition. Again $\mathcal{E}_2(G)$ is the 2-local geometry K for G, with $\tilde{L}(K)$ not projective, so that $\mathcal{E}_2(G)$ is not homotopy equivalent to $\mathcal{A}_2(G)$; hence our decomposition is again in fact a new one, i.e. not available from Theorem 7.2.5. ♢

These Examples essentially led to the Benson-Smith book [**BS08**]: For Benson had observed that the sporadic projective modules $\tilde{L}(K)$ in [**RSY90**] lead to decompositions of cohomology; while Theorem 7.4.5 provides decompositions for at least some of the remaining sporadics, for which $\tilde{L}(K)$ is not projective. So the authors in the mid-1990s began a long-term project to treat more—and eventually all—of the sporadic groups. (Indeed the treatment of the largest groups sometimes had to await further improvements in the technology of decomposition theory.)

Decompositions from radical and centric p-subgroups. Next we summarize the *subgroup* decomposition, introduced in work of Jackowski-McClure-Oliver [**JMO92**]; again we use the viewpoint of Dwyer, as in [**BS08**, Sec 5.5].

We let $\mathcal{B}$ be a G-subposet of $\mathcal{S}_p(G)$; that is, a G-poset of nontrivial p-subgroups. Again rather than considering the category given by the order complex $K(\mathcal{B})$ (where the morphisms are just the inclusions), this time we focus instead on the corresponding *orbit* category $\mathcal{O}_{\mathcal{B}}$ defined on $\mathcal{B}$. Very roughly, we take as morphisms not just the inclusions, but also the G-set maps among these subgroups. More precisely, as in [**BS08**, 5.3.1], by an object $H \in \mathcal{B}$ we mean the G-set G/H; with the natural morphisms among G-sets. Of course the stabilizer of H (viewed as a coset), under multiplication by elements of G, is just H itself. (This is in contrast to the stabilizer of H in the normalizer decomposition for the collection $\mathcal{B}$, which would be $N_G(H)$, since there the G-action is conjugation.) Hence in the *subgroup decomposition*, i.e. the corresponding homotopy colimit over $\mathcal{O}_{\mathcal{B}}/G$, the term for the G-set G/H with stabilizer H is the classifying space BH.

The analysis of [**JMO92**] (continued e.g. by Dwyer and Grodal as indicated below) investigated conditions on the poset $\mathcal{B}$ under which this subgroup decomposition should be ample and sharp (hence, *subgroup-sharp*). For example, [**JMO92**] initially obtained ampleness and subgroup-sharpness for $\mathcal{O}_{\mathcal{B}}$, where $\mathcal{B}$ is given by the collection $\mathcal{B}_p(G)$ of p-radical subgroups along with 1 (which they called the *p-stubborn* subgroups); thereafter work of Dwyer [**Dwy98**, 10.4] allowed reduction to just $\mathcal{B}_p(G)$ itself. This can be compared with ampleness and normalizer sharpness for $K(\mathcal{B}_p(G))$, which we saw in Theorems 7.3.5 and 7.3.16. (The latter corresponds to a decomposition of BG of the form (7.4.3).)

Some subsequent development of the literature on decompositions focused attention on some alternative collections $\mathcal{B}$; especially certain natural subcollections which might be strictly smaller that $\mathcal{B}_p(G)$:

DEFINITION 7.4.8 (p-centric subgroups). Dwyer in [**Dwy97**][**Dwy98**] emphasized the property:

A p-subgroup P is *p-centric* if $Z(P)$ is Sylow in $C_G(P)$.

We write $\mathcal{C}e_p(G)$ for the poset of p-centric subgroups of G. (Since for us p usually divides the order of G, the identity subgroup 1 is then not p-centric.) We write $\mathcal{B}_p^{cen}(G)$ for the subposet of the p-radical subgroups $\mathcal{B}_p(G)$ given by the members which are also p-centric.

Grodal in [**Gro02**] emphasized the subcollection $\mathcal{D}_p(G)$ of $\mathcal{C}e_p(G)$ given by P satisfying $O_p\big(N_G(P)/PC_G(P)\big) = 1$. These subgroups are necessarily radical, and determine the *principal* p-radical subgroups, a subcollection of $\mathcal{B}_p^{cen}(G)$. (The subcollection should be "minimal" for decompositions, in a sense mentioned after Conjecture 8.4.1.) ♢

Dwyer obtained subgroup-sharpness for $\mathcal{C}e_p(G)$. Furthermore he showed (in analogy with the work of Bouc in Theorem 4.3.4) that removal of G-orbits of members which are *not* radical gives a G-homotopy equivalence. Hence by (7.3.15), the same subgroup-sharpness result holds for any resulting subcollection, down to $\mathcal{B}_p^{cen}(G)$. It will be important for our applications that this latter subposet may be proper in, and not homotopy equivalent to, the full poset $\mathcal{B}_p(G)$.

And among the results of Grodal in [**Gro02**] are various analogues, for normalizer sharpness as in (7.4.3), of the above subgroup-sharpness results. For example, normalizer sharpness for subposets from $\mathcal{C}e_p(G)$ down to $\mathcal{B}_p^{cen}(G)$ (and even $\mathcal{D}_p(G)$) follows using [**Gro02**, 9.1] (confirming the suspicion indicated in [**SY97**, p 377]).

We summarize these results in the following form:[31]

THEOREM 7.4.9. *For $\mathcal{B}$ given by any of $\mathcal{B}_p(G)$, $\mathcal{C}e_p(G)$, $\mathcal{B}_p^{cen}(G)$, and $\mathcal{D}_p(G)$, the subgroup decomposition, which corresponds to the classifying-space decomposition given by (the p-completion of)*

$$\mathrm{Hocolim}_{H \in \mathcal{O}_{\mathcal{B}}/G}\ BH,$$

is ample and sharp. And the collection $\mathcal{B}$ is also normalizer-sharp; that is, we may also use the form:

$$\mathrm{Hocolim}_{H \in K(\mathcal{B})/G}\ BN_G(H).$$

(Option S:) As before, we illustrate the result for a sporadic geometry.

EXAMPLE 7.4.10 (A classifying-space decomposition for M_{12}, revisited). As in Example 7.4.6, we take $G = M_{12}$ with $p = 2$; and we now get an approach which is even easier than in that Example: For in [**BS08**, Sec 7.2], it is observed that $\mathcal{B}_2^{cen}(G)$ gives the same poset as the standard 2-local geometry K for G. Thus by Theorem 7.4.9, we obtain sharpness of the normalizer decomposition corresponding to K—again giving our earlier cohomology decomposition for M_{12}. As before, $\tilde{L}(K)$ is not projective, so that the approach via $\mathcal{B}_2^{cen}(G)$ gives a new decomposition, not available from Theorem 7.2.5. ◊

The approach via $\mathcal{B}_2^{cen}(G)$ will in fact apply to quite a few of the sporadic groups in Section 7.5.

Summary of available normalizer-sharp decompositions. In a warning after the proof of Lemma 6.3.3, we had mentioned that the conditions for K in that Lemma are often more or less equivalent to K being in the standard G-homotopy type determined by $\mathcal{S}_p(G)$ (as in Remark 4.6.1). Hence the same restriction applies to the ampleness and sharpness results in earlier Sections 7.2 and 7.3, since they in effect require such conditions, related to projectivity of $\tilde{L}(K)$.

Of course the results of the present section, which do not require projectivity of $\tilde{L}(K)$, apply to some more general G-homotopy types, beyond the standard type in Remark 4.6.1.

So below, we collect together the various G-homotopy types for which we have in particular indicated normalizer sharpness. (As we will indicate in a moment, we are essentially excerpting from [**GS06**, 1.1, 1.2].)

[31] In fact they can all be obtained from [**Gro02**, 1.5].

REMARK 7.4.11 (The normalizer-sharp super-type). We let $\mathcal{N}$ (standing for the *normalizer-sharp "super-type"*) denote the set of complexes K, such that K is G-homotopy equivalent to (at least) one of:

$\mathcal{S}_p(G)$ (recall Remark 4.6.1 indicates some G-equivalent complexes),

$\mathcal{E}_p(G)$,

$\mathcal{B}_p^{cen}(G)$,

$\mathcal{D}_p(G)$.

And then in view of Theorems 7.3.5, 7.3.16, 7.4.5, and 7.4.9, we know—for any complex K in this set, that is, which is G-homotopy equivalent to one of the above four posets—that K is ample and normalizer-sharp. ◊

Correspondingly the systematic approach to sporadic groups in [**BS08**], described in Section 7.5 below, is to show that the usual 2-local geometry K for each sporadic group G is in fact G-equivalent to one of the particular posets indicated in Remark 7.4.11.

Indeed for a general finite group G, we mention that the nine well-known collections considered by Grodal and Smith in [**GS06**] are all in fact in $\mathcal{N}$—that is, the result 1.1 in that paper shows that each is G-equivalent to one of the four particular collections indicated in Remark 7.4.11; and their normalizer sharpness is observed in 1.2 of that paper. Of course, normalizer sharpness does not guarantee subgroup or centralizer sharpness for those collections; indeed the strengths of the particular equivalences, which do hold among the normalizer, centralizer, and subgroup decompositions for those collections, are specified in 1.1—and 1.2 indicates whether the corresponding sharpness holds for general G.

We also mention that Maginnis and Onofrei in [**MO08**] demonstrate suitable equivalences of their further collection of "distinguished" p-subgroups with members of the above standard collections (hence they also obtain ample and sharp decompositions).

7.5. (Option S): Applications to cohomology of sporadic groups

In Section 7.2, we had mentioned some cohomology decompositions over certain sporadic geometries—in Corollary 7.2.9 and the subsequent discussion. However, to methodically cover *all* sporadic groups (at $p = 2$), the treatment in the Benson-Smith book [**BS08**] used the equivariant viewpoint from Section 7.3—as we had suggested via some further examples in Section 7.4.

We next sketch the overall work in [**BS08**]; this is essentially the culmination of the track that we have called Option S in the present book.

The Benson-Smith results. We had already previewed ampleness and sharpness results for some of the sporadic groups G: namely in earlier Example 7.2.10 for $G = M_{11}$ (and indeed for that group as well as J_1, after Corollary 7.2.7); in Example 7.4.6 for M_{12} (with variations at Examples 7.4.10, 7.2.12, and 7.2.13); in Example 7.4.7 for Co_3; and in Example 7.2.11 for M_{22} and McL.

In fact each individual sporadic group is treated in a corresponding section in Chapter 7 of [**BS08**]. We express the common features of all those individual results in the form:

THEOREM 7.5.1. *Take $p := 2$. For each of the 26 sporadic simple groups G, there is a suitable 2-local geometry K (namely a G-complex K with type-preserving action, corresponding to the normalizers of suitable 2-subgroups) with the following properties:*[32]

(1) G is flag-transitive on K. Hence the orbit CW-complex K/G is a single simplex. So we may index the orbits in K/G in the form σ_J: where $\emptyset \neq J \subseteq I$, for a set I indexing representatives of the vertices. We let G_J denote the stabilizer of σ_J.

(2) K is G-equivalent[33] *to one of the collections in the standard normalizer-sharp super-type of Remark 7.4.11; hence K is ample and normalizer-sharp. So we get decompositions in the elementary "simplex" form, namely:*

(3) We have a decomposition of the 2-completed classifying space $BG_2^\wedge$, given by (the 2-completion of) the homotopy colimit

$$\mathrm{Hocolim}_{\emptyset \neq J \subseteq I}\, BG_J; \text{ and}$$

(4) We have a decomposition of the mod-2 group cohomology

$$H^*(G) \cong \bigoplus_{\emptyset \neq J \subseteq I} (-1)^{|J|-1}\, H^*(G_J)$$

This uniform behavior is a striking instance of the Sporadic Principle 2.0.4.

We mention that one motivation for the Grodal-Smith work [**GS06**] was the application mentioned in Theorem 7.4.5—in effect allowing the uniform choice, in (3) above, of the normalizer decomposition (rather than centralizer or subgroup decompositions). This decomposition is often more convenient computationally—since the normalizers typically have smaller index in G than the subgroups themselves or their centralizers; so that the resulting induced modules are smaller.

We next indicate more specifically the member(s) of the super-type of Remark 7.4.11, which are used in (2) above to treat the various sporadic groups G; in particular, note that some groups can be treated via more than one approach—typically when for that G, differently-defined posets turn out to be G-equivalent. First the 11 groups G where $\tilde{L}(K)$ is projective lie in the standard G-homotopy type determined by $\mathcal{S}_2(G)$, in Remark 4.6.1; these groups are:

$$M_{11}, M_{22}, M_{23}, M_{24}, J_1, J_3, J_4, McL, Co_2, Th, Ly.$$

The projectivity of $\tilde{L}(K)$ was established in Ryba-Smith-Yoshiara [**RSY90**] (except for Ly, where the relevant K had not yet been described). There are 6 further groups where Hypothesis 7.4.4 holds[34] (i.e., can be verified as in Example 7.4.6), so that K is in fact G-equivalent to $\mathcal{E}_2(G)$; these groups are:

$$M_{12}, J_2, HS, Suz, Co_3, Ru.$$

[32]Further K is chosen with dimension minimal for such a cohomology decomposition.

[33]See later Remark 7.5.2 for more detailed discussion.

[34]We mention that Matt Bardoe checked the condition for the 2-local geometry in Example 2.3.10, for the non-sporadic group $G = U_4(3)$. So this group is normalizer-sharp for the $\mathcal{E}_2(G)$ decomposition; as well as for the usual $\mathcal{S}_2(G)$ decomposition, in view of [**RSY90**].

(Recall that normalizer sharpness then follows using [**GS06**, 1.2].) Finally there are 13 groups where K is G-equivalent to $\mathcal{B}_2^{cen}(G)$,[35] namely:

$$M_{12}, J_2, Suz, Co_1, F_{22}, F_{23}, F'_{24}, HN, B, M, He, Ru, O'N.$$

(Recall that normalizer sharpness then follows using using [**Gro02**, 9.1].)

We mention that the equivalence proofs are largely carried out via the methods (notably 1.3 and 1.4) in Smith-Yoshiara [**SY97**].

REMARK 7.5.2 (Equivalences for normalizer sharpness). We take this opportunity to fill a gap in [**BS08**]—continuing with some observations made by Bob Oliver, which we had already mentioned in earlier Remark 7.2.6:

(1) We note first that for the homotopy equivalences in the various sections of Chapter 7 of [**BS08**], the extension to G-homotopy equivalences could in fact have been written out in detail—using the usual methods of [**TW91**]. And then those equivalences will preserve normalizer sharpness, by (7.3.15).

However, the authors intended to circumvent those rather lengthy details for extending to G-equivariance: Instead, the statement on p 157 of [**BS08**] claims that the preservation of normalizer sharpness can be deduced from a weaker type of equivalence, as in [**BS08**, 2.5.13]; namely via a G-map which induces an ordinary homotopy equivalence.[36] The claim appears at [**BS08**, 5.8.11]; but as Oliver pointed out, the claim does not actually follow from the references indicated as the proof there. So proceeding via this weaker equivalence, rather than supplying details for the G-homotopy equivalences, results in a gap.

(2) The gap was filled when the claim was in fact subsequently established: Oliver provided a proof, and we now sketch his argument. (Full details are of course beyond the scope of this book.)

Thus we assume that $f : K \to L$ is a G-map which induces an ordinary homotopy equivalence of (nonempty) finite-dimensional CW-complexes. And we wish to show that f induces an isomorphism $H_G^x(K; M_y) \cong H_G^x(L; M_y)$ of Bredon cohomology, where M_y is the local coefficient system for the isotropy sequence of the normalizer decomposition, as indicated in the proof of Lemma 7.3.17. (And then since vanishing of this cohomology for $x \geq 1$ determines normalizer sharpness, K is sharp if and only if L is).

On replacing L by the the standard mapping cylinder for f, we may as well assume that f is an inclusion $K \subseteq L$. This map induces (for each row y above) a long exact sequence in Bredon cohomology, in the spirit of (5.3.10); so that isomorphism of the inner terms for K and L will follow from vanishing of outer terms for the quotient CW-complex L/K and a point.

Observe that the homotopy equivalence induced by f in the hypothesis guarantees that K and L have the same ordinary cohomology; thus L/K has the cohomology of a point, and in particular, L/K is $\mathbb{F}_p$-acyclic. We can use this fact to follow the argument in the two cases for y in the context of Bredon cohomology in Lemma 7.3.17.

[35]For sporadic groups, it seems that the (suitably minimal) subcollection $\mathcal{D}_p(G)$ is not in practice any smaller than $\mathcal{B}_p^{cen}(G)$.

[36]This weaker map preserves the Borel construction, as we had noted in Remark 7.3.1; but it need not preserve the corresponding isotropy spectral sequence, so that preservation of sharpness is not automatic—and certainly can fail for centralizer or subgroup sharpness. (Also recall earlier Warning 3.5.5.)

First for $y \geq 1$: From the hypothesis of finite-dimensionality for K and L, we have finite-dimensionality for L/K. So for each nontrivial p-subgroup P of G, we may apply the P. A. Smith Theorem (6.3.2), to see that $(L/K)^P$ is also $\mathbb{F}_p$-acyclic. This gives an acyclicity hypothesis essentially of the form 6.3.4(b+); and indeed following the proof of Theorem 7.3.16, as interpreted via Bredon cohomology in Lemma 7.3.17, we obtain vanishing in positive dimensions $y \geq 1$ of the Bredon cohomology $H^*_G(-;M_y)$ for L/K (and for a point).

It remains to consider dimension $y = 0$. Here from $\mathbb{F}_p$-acyclicity of L/K, it follows from the Conner Conjecture (cf. [**Oli76**]), proved by Conner in the case of finite G which we have here, that the orbit complex $(L/K)/G$ is also $\mathbb{F}_p$-acyclic. And now, again following the proof of Lemma 7.3.17, this gives vanishing of $H^x_G(-;M_0)$ in positive dimensions $x \geq 1$, for L/K and a point.

This completes the proof of the desired isomorphism $H^x_G(K;M_y) \cong H^x_G(L;M_y)$.

(3) We also sketch a different deduction of the isomorphism in (2) above; the argument was provided by Jesper Grodal, and is presumably well-known to experts. The more general context is that of Mackey functors, as in [**Web00**].

For the map $f : K \to L$ above, it suffices to assume that f induces an isomorphism of $\mathbb{F}_p$-homology. And the conclusion is isomorphism $H^*_G(K;F) \cong H^*_G(L;F)$ of Bredon cohomology—where F is a Mackey functor with values in $\mathbb{Z}_{(p)}$-modules, and F is projective relative to p-subgroups. (An example of such an F is given by $H^n(-;M)$, $n \geq 0$, where M is a $\mathbb{Z}_{(p)}G$-module; so that the result covers our local coefficient systems M_y for normalizer sharpness.)

The argument proceeds via the standard property (cf. [**Web00**]) for Mackey functors F which are projective relative to p-groups, that $H^*_G(K;F)$ is a retract of $H^*_S(K;F)$ for a Sylow p-subgroup S of G. Hence it is enough to establish the isomorphism for G replaced by a finite p-group S—that is, for $H^*_S(-;F)$.

Now as for all subgroups P of S, since K and L are assumed finite-dimensional, we may apply P. A. Smith theory (underlying (6.3.2)) to conclude that each map of fixed points $f^P : K^P \to L^P$ is also an isomorphism of $\mathbb{F}_p$-homology, and hence of homology with $\mathbb{Z}_{(p)}$-coefficients.

And then by the standard invariance property of Bredon cohomology, this in turn implies that $H^*_S(K;F) \to H^*_S(L;F)$ is an isomorphism in all degrees. $\diamondsuit$

We had mentioned (after Corollary 7.2.9) that taking K as the C_3-geometry in Example 6.3.7 gives a cohomology decomposition over K/G for $G = A_7$. And we had seen in Remark 4.4.12 that K is not equivalent to $\mathcal{A}_2(G)$. Now for this G, it turns out that the other possibilities $\mathcal{B}^{cen}_2$ and $\mathcal{E}_2(G)$ in the normalizer-sharp super-type of Remark 7.4.11 are just equivalent to $\mathcal{A}_2(G)$. So the example of K shows that we can sometimes get ampleness and normalizer sharpness, even away from that super-type.

Some further directions. We close the section with an indication of some other work related to cohomology of sporadic groups.

In addition to results on decomposition of cohomology (indicated earlier in this section), the discussion of the individual sporadic groups in [**BS08**, Ch 7] also attempts to give an indication of the current status (as of 2008) of results on the explicit computation of the mod-2 cohomology of the groups. (In particular there is some discussion of connections of sporadic cohomology with Dickson invariants, in the sections on J_1, Co_3, Th, and Ly; for Co_3 there is a relationship observed by Benson with the exotic Dwyer-Wilkerson space $DI(4)$. For a connection in the case of M_{12} and groups of type G_2, see also [**AMM91**] and [**BW95**, Sec 5].)

The literature also contains some results on the mod-p cohomology of sporadic groups for odd primes p. We mention for example Milgram-Tezuka [**MT95**], on M_{12} for $p = 3$,[37] and the more general techniques in Tezuka-Yagita [**TY96**].

[37] Computations of stability under restriction can sometimes be simplified, by using geometric properties to desecribe product structure of double cosets. An approach was developed in Smith-Umland [**SU96**], and applied to streamline the proof of [**AMM91**, 3.1] for M_{12} and $p = 2$. A similar application to M_{12} with $p = 3$ was given by Umland (unpublished).

Part 4

Some More Advanced Topics

The final chapters provide a briefer view of some more specialized areas of research in the area of subgroup complexes.

Correspondingly, as has already been the case for some more advanced material, the results introduced will more frequently be just discussed, rather than proved in detail.

CHAPTER 8

Spheres in homology and Quillen's Conjecture

In this chapter, we return our main focus to posets of p-subgroups such as $\mathcal{S}_p(G)$, and $\mathcal{A}_p(G)$—which are of course equivalent via Quillen's Theorem 4.2.4; and we study situations where we can establish that these posets have nonzero homology. Such results are of particular interest when $O_p(G) = 1$, because nonvanishing in particular then gives (via the usual Lemma 5.2.3) the non-contractibility of $\mathcal{S}_p(G)$ required for Quillen's Conjecture 3.3.8.

From a topological viewpoint, perhaps the easiest way to establish nonzero homology is to construct spheres in the top dimension. Correspondingly, our usual initial topological background Section 8.1 discusses spheres—notably via the standard elementary construction in terms of a join of 0-spheres.

Then in the rest of the chapter, we consider applications of this join-construction (and others)—to establish non-zero homology for $\mathcal{A}_p(G)$, for suitable classes of groups G. In particular we adopt the viewpoint of Aschbacher-Smith [**AS93**], namely that most groups G should have the "Quillen dimension" property—i.e. that they should have $\tilde{H}_d(\mathcal{A}_p(G)) \neq 0$ for the top dimension d of the poset; and Section 8.2 is devoted to a number of such nonvanishing results—for example, this holds for solvable groups, and also for most simple groups. Thereafter Section 8.3 considers results with the somewhat different non-vanishing condition $\tilde{L}(\mathcal{A}_p(G)) \neq 0$ for the reduced Lefschetz module (which of course also implies non-contractibility)—via a technique introduced by Robinson, which we refer to as "Robinson subgroups".

Finally (in particular using these two previous sections as background) Section 8.4 discusses the main result of [**AS93**]—which establishes Quillen's Conjecture for "most" finite groups G. (The Conjecture remains open in full generality—but the structure of a minimal counterexample becomes strongly restricted.)

8.1. Topological background: Homology via top-dimensional spheres

In order to exhibit nonzero homology for a complex K, in this chapter we will focus primarily on what is usually regarded as the most elementary construction, namely spheres in the highest dimension. That is, we will typically make the following two restrictions:

First, we consider just the *top* dimension d of K: For then (recall Remark 5.1.7) the incoming boundary map is zero, so that $\tilde{H}_d(K) = \tilde{Z}_d(K)$. Thus to exhibit nonzero homology, it will suffice to just locate nonzero *cycles*.

And then second, to find such cycles, ordinarily we will just look for the simplest-possible geometric configurations with zero image under the boundary map—namely *spheres*[1] in that highest dimension d.

[1]But see also later Remark 8.2.18 for cycles given by tori and Klein bottles.

Hence in this initial section, we will recall some standard topological background on spheres. In fact we will begin with some additional features of the join of complexes, since that will be used in our usual construction of spheres.

Some properties of the join. Recall we indicated the join at Definition 1.6.9; indicating there that it provides a more natural product structure for complexes than the usual Cartesian product; we now expand a little on that earlier remark.

For example, one useful feature of the join is the following skew-multiplicative formula for the reduced Euler characteristic: (Cf. [**AS93**, 0.18.iii], which in turn quotes [**Mil56**, 2.1].)

(8.1.1) For complexes K and L, we have $\tilde{\chi}(K * L) = - \tilde{\chi}(K)\, \tilde{\chi}(L)$.

We sketch the reason for the minus sign: One standard characterization of the join is as the suspension of the smash product. (See e.g. [**Bjö95**, 9.5], and also [**Ben91**, 1.1.2] for the product; and recall that the *suspension* of K is the double cone $v*K*w$ on K defined using new vertices v and w.) The suspension in effect provides a shift by one dimension upward, which in turn reverses signs in the Euler characteristic.

A second convenient feature is that for the elementary poset $\mathcal{A}_p(G)$, the join arises naturally for the direct product of groups:

LEMMA 8.1.2 (Prop 2.6 in [**Qui78**]). *For the direct product of groups H and J,*

$$\mathcal{A}_p(H \times J) \text{ is homotopy equivalent to } \mathcal{A}_p(H) * \mathcal{A}_p(J).$$

In particular, $\tilde{\chi}\big(\mathcal{A}_p(H \times J)\big) = - \tilde{\chi}\big(\mathcal{A}_p(H)\big)\, \tilde{\chi}\big(\mathcal{A}_p(J)\big)$.

PROOF. (Outline:) From the relation with the Cartesian product in [**Qui78**, 1.9], the join is equivalent to the subset $\mathcal{S}$ of $\mathcal{A} := \mathcal{A}_p(H \times J)$ given by products $A_H \times A_J$ (with at least one factor nontrivial). We define a map r from $\mathcal{A}$ to $\mathcal{S}$ by taking each A to the product $\mathrm{proj}_H(A) \times \mathrm{proj}_J(A)$ of its projections. Since A is contained in this product, letting i denote the inclusion of $\mathcal{S}$ in $\mathcal{A}$, we get the poset-map dominance property $\mathrm{Id}_{\mathcal{A}} \le ir$; this shows via Lemma 3.1.7 that ir is homotopic to $\mathrm{Id}_{\mathcal{A}}$—while for the reverse composition we have equality $ri = \mathrm{Id}_{\mathcal{S}}$, establishing the equivalence. Of course the final statement then follows from the multiplicativity in (8.1.1). □

There is a suitable extension of this result to the *central product* of H and J: namely the set product HJ when the factors commute; notice it then follows that:

$$H \cap J \le Z(HJ).$$

One standard notation for the central product is $H*J$; and this use of $*$ for groups is in fact consistent with the usage of $*$ for the join of the complexes $\mathcal{A}_p(H)$ and $\mathcal{A}_p(J)$ as in 8.1.2 above, in the following situation (which is standard in [**AS93**]):

COROLLARY 8.1.3 (0.18 in [**AS93**]). *The conclusion of Lemma* 8.1.2 *holds for the central product $H * J$, if $H \cap J$ is a p'-group.*

PROOF. (Outline:) Set $\overline{HJ} := HJ/(H \cap J)$. Then $\overline{HJ}$ is in fact the direct product $\bar{H} \times \bar{J}$, to which we can apply Lemma 8.1.2. Furthermore $H \cap J$ is a p'-group by hypothesis, and we saw above that $H \cap J \le Z(HJ)$ in the central product $H * J$; so for any p-subgroup A of HJ, the preimage of $\bar{A}$ is the direct product $A \times (H \cap J)$, with unique Sylow p-subgroup A. It follows from this unique preimage (cf. [**AS93**, 0.11]) that $\mathcal{A}_p(\overline{HJ})$ is isomorphic to $\mathcal{A}_p(HJ)$; and similarly for the factors in the join. □

A standard construction of the sphere via joins. The usual (geometric) viewpoint on the n-sphere S^n is as the boundary ∂B^{n+1} of the $(n+1)$-ball B^{n+1}.

Thus the 0-sphere S^0 is the boundary of an interval, and so consists just of two isolated vertices; in particular:

(8.1.4) $$\tilde{\chi}(S_0) = -1 + 2 = 1.$$

Notice further that the suspension $v * K * w$ of a complex K is in fact given by the join $K * S^0$ with the 0-sphere.

Thus the suspension of S^0 itself is given by the join $S^0 * S^0$—this is essentially a square, which in turn is a triangulation of the 1-sphere S^1. And $S^0 * S^0 * S^0$ is an octahedron, which is a triangulation of the 2-sphere S^2; and so on. This method of successive suspensions gives the standard join-construction of spheres (cf. [**Mun84**, Exer 33.3]):

(8.1.5) We can obtain S^n as the $(n+1)$-fold join: $\underbrace{S^0 * S^0 * \cdots * S^0}_{n+1}$.

In particular from (8.1.4) and the multiplicativity in (8.1.1) we get a simple formula for the reduced Euler characteristic:

(8.1.6) $$\tilde{\chi}(S^n) = (-1)^n.$$

Indeed since the sphere S^n has dimension n, and is a cycle, while no proper subcomplex is a cycle, we have (e.g. [**Mun84**, 31.8]):

(8.1.7)
$Z_n(S^n) = H_n(S^n)$ has dimension 1 (and lower reduced homology groups vanish).

8.2. Quillen dimension: Non-vanishing top homology for $\mathcal{A}_p(G)$

The join-construction of spheres has been applied in many calculations of homology in the literature. Here is an elementary instance:

EXAMPLE 8.2.1 (Quillen's example of $\mathcal{A}_p(S_{2p})$). We quickly summarize Example 2.7 in [**Qui78**]: We take $p \geq 3$, with G the symmetric group S_{2p}. First note that $\mathcal{S}_p(G) = \mathcal{A}_p(G)$, and the complex has dimension 1. Furthermore $\mathcal{A}_p(G)$ is in turn is a disjoint union of subcomplexes $\mathcal{A}_p(H)$, where $H \cong A_p \times A_p$. Then by Lemma 8.1.2, each such subocomplex is homotopy equivalent to $\mathcal{A}_p(A_p) * \mathcal{A}_p(A_p)$; where $\mathcal{A}_p(A_p)$ is a set of $(p-2)!$ isolated vertices. This value is just 1 when $p = 3$, in which case the join-subcomplex is just a single edge $v * w$, from the single vertex v and w in each factor. But for $p \geq 5$, each $\mathcal{A}_p(A_p)$ contains at least $3! = 6 > 2$ vertices, so in particular contains (multiple copies of) S^0. Hence the join-subcomplex contains copies of the sphere $S^0 * S^0 = S^1$ as in (8.1.5). (And correspondingly we get nonzero homology in the top dimension, namely 1, of $\mathcal{A}_p(G)$.) $\diamondsuit$

In a moment, we will concentrate further on the top-dimensional aspect of this example; but before doing so, we mention some more general homological calculations which use this method (among others).

For instance, Bouc in [**Bou92**] also considers symmetric groups G, and some posets related to $\mathcal{A}_2(G)$; he obtains various homological results such as his Propositions 5, 7, and 8 there. Further results for the symmetric group are developed in Alperin [**Alp89**, pp 52–57].

For solvable groups, homology was described by Quillen in results in Section 12 of [**Qui78**]. This homology is in turn deduced from results earlier in the paper, on topological "sphericality"; these led to a corresponding stream of subsequent research (including e.g. [**KT85**]), which we will mainly consider in our later Section 9.4.

The notion of Quillen dimension. The specific feature of Example 8.2.1 on which we now focus is that (for $p \geq 5$) we got nonzero homology in the *highest* dimension of the complex—namely dimension 1 in that case. This is the feature which we will call "Quillen dimension" below.

The relevance of this property to Quillen's Conjecture 3.3.8 is immediate: namely in the contrapositive, we assume $O_p(G) = 1$—and so it will suffice if we can construct spheres (and hence nonzero reduced homology) in that top dimension[2] of $\mathcal{A}_p(G)$.

Now this is not in fact possible for *all* G with $O_p(G) = 1$. However, the viewpoint of Aschbacher-Smith in [**AS93**, p 474] is that it should be possible fairly generically—that is, for "most" groups G; and that mindset then provides the framework for their general approach to the Conjecture. Hence we will also adopt that viewpoint in the remainder of this Chapter. (And in particular, we will be using that view in describing some earlier examples and results, which were developed before the terminology of Quillen dimension was introduced.)

So we now recall the definition given in Aschbacher-Smith [**AS93**, p 474]:

DEFINITION 8.2.2 (Quillen Dimension ($\mathcal{QD}_p$)). Assume that $O_p(G) = 1$, and that G has p-rank $m_p(G) =: r \geq 1$. We say that G has *Quillen dimension* at p if $\tilde{H}_{r-1}(\mathcal{A}_p(G)) \neq 0$.

We abbreviate this condition as $\mathcal{QD}_p$; or just $\mathcal{QD}$ when the prime p is clear. $\diamondsuit$

In the remainder of the section, we consider examples, and also discuss some results for general classes of groups G, which exhibit (or occasionally, fail to exhibit) the $\mathcal{QD}$ property.

For instance, in Example 8.2.1, we got $\mathcal{QD}_p$ for $p \geq 5$. However, we do not get $\mathcal{QD}_3$: since for $p = 3$, the contractibility of the join-subcomplexes of form $v * w$ exhibits nonzero (reduced) homology in dimension 0 rather than 1. (After contracting these $v * w$ to points, one computes that the resulting (equivalent) complex consists of 10 isolated vertices; so certainly there are copies of S^0 here. And since the latter complex has dimension 0, this homology suffices for the conclusion of Quillen's Conjecture in this case; but that conclusion is not established via $\mathcal{QD}_3$.)

We also consider some variations on the idea in Example 8.2.1:

EXAMPLE 8.2.3 ($\mathcal{QD}_2$ for $S_3 \times S_3$, and some generalizations). Take $p := 2$. Consider first the case $G := S_3 \times S_3$. (In earlier Example 2.1.21, we had drawn a picture of a geometric version of the poset $\mathcal{A}_2(G)$; and so one could independently verify there the spheres we are about to construct.) Then $\mathcal{A}_2(G)$ is by Lemma 8.1.2 equivalent to $\mathcal{A}_2(S_3) * \mathcal{A}_2(S_3)$. Now each factor consists of three isolated vertices, and so contains copies of S^0; so the join contains copies of $S^0 * S^0 = S^1$ as in (8.1.5). That is, $\mathcal{QD}_2$ holds.

[2]We recall from Definition 3.3.9 that this dimension is given by $m_p(G) - 1$.

We can generalize this Example in various ways: For G of 2-rank 2, it suffices to find a subgroup $H \cong S_3 \times S_3$. We can also consider some more general 2-rank r, now using r factors—which might more generally be dihedral of twice odd order. (Such a product arises in Thompson's fundamental Dihedral Lemma [**Tho68**, 5.34], in a minimal faithful action of an elementary 2-group of rank r.)

Indeed we can consider analogues with p odd, where each factor has an element of order p acting nontrivially on a normal p'-subgroup. (Again such configurations arise in the context of a minimal faithful "coprime" action; reductions can lead more generally to the *critical* subgroups defined by Thompson, e.g. [**Asc00**, p 108].) $\diamondsuit$

One viewpoint on the $\mathcal{QD}$ condition is that $\mathcal{A}_p(G)$ provides the largest dimension where, for some arbitrary finite group G (with $O_p(G) = 1$), we might normally expect to find nonzero homology—since Quillen's equivalence in Theorem 4.2.4 of $\mathcal{S}_p(G)$ with $\mathcal{A}_p(G)$ shows that homology definitely vanishes in any dimensions for $\mathcal{S}_p(G)$ which exceed that of $\mathcal{A}_p(G)$.

(Option B:) For G of Lie type in characteristic p, similarly Quillen's further equivalence with the building in Theorem 4.4.8 provides a notable non-example of $\mathcal{QD}$: for usually[3] the building has lower dimension than that of $\mathcal{A}_p(G)$. Correspondingly this situation gives the main conclusion (1) in the list in later Theorem 8.2.15 of the comparatively few simple groups which fail to satisfy $\mathcal{QD}_p$. Of course because of the equivalence with $\mathcal{A}_p(G)$, the nonzero homology[4] for the building in Theorem 6.4.2 provides a route, different from $\mathcal{QD}_p$, toward establishing Quillen's Conjecture for this class of groups.

REMARK 8.2.4 (Quillen's ordering: "solvable then simple"). In the treatment in [**Qui78**], Quillen's Conjecture and various other possible properties for a general finite group G are often first investigated in the special case where G is solvable; as for example in Corollary 12.2 there.

So after that, it remains to consider groups G having at least one composition factor which is (nonabelian) simple. And correspondingly, Quillen then typically continues by at least beginning to consider the basic subcase where G is just a single simple group—as for example in [**Qui78**, Thm 12.4].

(And then to proceed from the simple case to a general group G, in modern group theory one usually considers how (quasi)simple groups can arise as components of the generalized Fitting subgroup $F^*(G)$—a topic we develop further starting at Remark 8.2.6 below.)

Correspondingly we will tend to follow that order of treatment in this chapter—and also for the properties considered in the subsequent Chapter 9. $\diamondsuit$

$\mathcal{QD}_p$ for solvable and p-solvable groups. One of the major results in Quillen's paper [**Qui78**] was his Corollary 12.2—which in our present language establishes $\mathcal{QD}_p$ for solvable groups. (This was later extended to p-solvable groups.)

[3] For example, if the field of definition of G is $\mathbb{F}_{p^a}$, then the dimension of $\mathcal{A}_p(G)$ increases with a, unboundedly—whereas the dimension of the building is fixed, independently of a.

[4] We mention that the spheres giving this homology in Theorem 6.4.2 in fact come from apartments; and these apartment-spheres are *not* usually obtained from the join-construction which we are emphasizing in the present Section.

In this subsection, we discuss some of that development. First we state Quillen's result for solvable groups:

THEOREM 8.2.5 (Cor 12.2 in [**Qui78**]). *Assume G is solvable, with $O_p(G) = 1$, and with p-rank $m_p(G) =: r > 0$. Then $\tilde{H}_{r-1}(\mathcal{A}_p(G)) \neq 0$. (That is, $\mathcal{QD}_p$ holds.)*

*More generally (Thm 12.1 in [**Qui78**]), if s is the rank of any elementary subgroup which is maximal under inclusion in $\mathcal{A}_p(G)$, then $\tilde{H}_{s-1}(\mathcal{A}_p(G)) \neq 0$.*

Notice that this result of $\mathcal{QD}_p$ in particular establishes Quillen's Conjecture 3.3.8 for solvable G. (And the second statement generalizes the first: since an elementary group of the maximal p-rank given by $r = m_p(G)$ is automatically maximal under inclusion in $\mathcal{A}_p(G)$.)

In fact Quillen's homological argument for a general solvable G, which we will sketch later, is based on the homology of a minimal subconfiguration (to which we will be able to apply the join-construction of spheres) given by LA: where we set $L := O_{p'}(G)$, and A is elementary of rank s (and maximal under inclusion in $\mathcal{A}_p(G)$).

So here we will review only Quillen's initial discussion leading to LA. Indeed, we can carry out the first part of the argument under the weaker assumption that G is just p-solvable: Since by hypothesis $O_p(G) = 1$, using the definition of p-solvability (namely, composition factors are p-groups or p'-groups) we have $L = O_{p'}(G) > 1$. Indeed these hypotheses also give $F^*(G) \leq L$—where we are using some standard language of group theory which we quickly review:

REMARK 8.2.6 (The generalized Fitting subgroup $F^*(X)$). We define the *generalized Fitting subgroup* $F^*(X)$ of a finite group X (cf. [**Asc00**, Sec 31]) as the (commuting) product $E(X)F(X)$: where $F(X)$ (the classical *Fitting* subgroup) is the product of all subnormal nilpotent subgroups of X; and $E(X)$ is the product of all subnormal quasisimple subgroups of X. (Minimal such subgroups are called *components*, and they commute pairwise.) Probably the most crucial feature of $F^*(X)$ is the self-centralizing property (e.g. [**Asc00**, 31.13]):

$$C_G(F^*(X)) = Z(F^*(X)) = Z(F(X)). \tag{8.2.7}$$

This shows roughly that most of X (that is, the quotient modulo any elements in $Z(F(X))$) is represented *faithfully* as automorphisms of $F^*(X)$. ◇

Returning now to our p-solvable group G with $O_p(G) = 1$: we see using (8.2.7) and the condition $F^*(G) \leq L$, along with the fact that our A is a p-group, that:

$$C_A(L) \leq C_A(F^*(G)) \leq A \cap Z(F(G)) \leq O_p(G);$$

so we conclude that $C_A(L) = 1$.

From this point, we will temporarily leave G aside, and work just with LA—where we have obtained $C_A(L) = 1$. Indeed since $O_{p'}(LA) = L$, we can argue much as above that $C_A(L) = O_p(LA)$, and hence:

(8.2.8) For L a normal p'-subgroup, $O_p(LA) = 1$ is equivalent to $C_A(L) = 1$.

So just as earlier we had $O_p(G) = 1$, we now also have $O_p(LA) = 1$.

(However before continuing, we do mention that if our original G is in fact solvable, then of course so also is the L here; whereas for more a general p-solvable G,

L might involve non-abelian simple p'-groups; we will encounter this situation later, e.g. in the proof of Theorem 8.2.12.)

$\mathcal{QD}$ for Quillen's minimal solvable case LA. In fact this particular configuration for LA arises in several different ways in the material for this book:

For instance, such an LA appears as a minimal situation in the proof of the Hawkes-Isaacs Theorem 5.5.4, which establishes Quillen's Conjecture for this LA—not by obtaining $\mathcal{QD}_p$, but instead by exhibiting nonzero reduced Euler characteristic $\tilde{\chi}(\mathcal{A}_p(LA)) \neq 0$.

Alternatively, Quillen's treatment of LA at [**Qui78**, 11.2] in fact establishes (using a fairly detailed inductive argument) that the poset $\mathcal{A}_p(LA)$ has property of Cohen-Macaulayness, which is stronger than $\mathcal{QD}_p$; we indicate that result in our corresponding later chapter as Theorem 9.4.5.

However, for the more limited goal of obtaining the result $\mathcal{QD}$ (that is, concerning just the homology of $\mathcal{A}_p(LA)$),[5] Alperin observed that a much simpler induction suffices, using some standard group-theoretic reductions—with the proof culminating in the join-construction of spheres in the top dimension. Thus we will establish:

THEOREM 8.2.9 (essentially 11.2(b) in Quillen [**Qui78**]). *Assume that $G = LA$ with $O_p(G) = 1$, where L is a solvable normal p'-subgroup, and A is an elementary p-subgroup of rank $s > 0$. Then $\tilde{H}_{s-1}(\mathcal{A}_p(G)) \neq 0$. (That is, $\mathcal{QD}_p$ holds for LA.)*

PROOF. (Alperin) The result is stated by Alperin as [**Alp90**, Thm 1], in the extension to the p-solvable case given in our later Theorem 8.2.12 (where L is not required to be solvable). But as far as I know, the proof appears only in his unpublished lecture notes [**Alp89**]. In fact pp 49–51 there indicate the argument for solvable L, essentially as given below.

The bulk of the argument involves of a series of reductions; in effect, replacing LA by a sub-configuration in a product form essentially as in earlier Example 8.2.3, from which we will obtain the needed spheres via the join-construction.

For the reductions, recall that $O_p(LA) = 1$ is equivalent to $C_A(L) = 1$ by (8.2.8). So assume that H is some subgroup of L, normalized by A and also satisfying $C_A(H) = 1$. Since $s-1$ is the top dimension, where cycles in fact give homology, we have $\tilde{H}_{s-1}(\mathcal{A}_p(HA)) \subseteq \tilde{H}_{s-1}(\mathcal{A}_p(LA))$; so we see that it suffices to establish the nonzero result for HA. And in that case, we may as well just replace our original L by H. In particular, the inductive content of the argument now amounts to assuming that L is *minimal* subject to the conditions $A \leq N_G(L)$ and $C_A(L) = 1$ (that is, subject to admitting a faithful A-action).[6] So we investigate consequences of this minimality:

(1) *We may assume $L = F(L)$ (so L is nilpotent):*

For this we refine our earlier argument using the self-centralizing property (8.2.7): Since now L is solvable, the generalized Fitting subgroup $F^*(LA) = F^*(L)$ is just

[5]This more limited result on $\mathcal{QD}$ is also reproduced as [**AS93**, 0.10].

[6]So at this point, the more experienced reader could just skip down to (6) at the end of the proof, by quoting essentially the standard reduction of Thompson to a *critical* subgroup, e.g. [**Asc00**, 23.6]. However, we have given a fairly detailed deduction here—primarily for the reader interested in seeing just how the group theory eventually leads to a product form, which gives the join-construction of spheres.

the Fitting subgroup $F(L)$.[7] As before we get $C_A\big(F(L)\big) = 1$; so we have $F(L) = L$ by minimality, establishing (1).

By nipotence in (1), L is a direct product of q-groups, for primes $q \neq p$. And then the quotient:

$$\bar{L} := L/\Phi(L)$$

is elementary abelian (e.g. Remark 3.3.11)—so we now examine its A-module structure, using standard facts about coprime action (e.g. [**Gor80**, Ch 5]). For example, by [**Gor80**, 5.1.4] we have $C_A(\bar{L}) \leq C_A(L)$; so as $C_A(L) = 1$, we get:

$$C_A(\bar{L}) = 1.$$

Also, $\bar{L}$ is completely reducible under A (e.g. [**Gor80**, 3.3.1]); so we may write $\bar{L}$ as a direct sum of irreducible modules $\bar{L}_i$, for i in some index set i. And then setting $C_i := C_A(\bar{L}_i)$, we have $\cap_{i\in I} C_i \leq C_A(\bar{L}) = 1$. We will need the following further reduction:

(2) *We may assume* $[\bar{L}, A] = \bar{L}$ *and* $C_{\bar{L}}(A) = 1$*:*

For by [**Gor80**, 5.2.3], we have a factorization:

$$\bar{L} = C_{\bar{L}}(A) \times [\bar{L}, A].$$

Now $C_A([L, A]) \leq C_A([\bar{L}, A])$, which by the factorization lies in $C_A(\bar{L})$; and we saw $C_A(\bar{L}) = 1$ above, so we get $C_A([L, A]) = 1$. Hence by minimality we have $[L, A] = L$. So $\bar{L} = \overline{[L, A]} = [\bar{L}, A]$; and now the factorization gives $C_{\bar{L}}(A) = 1$, completing (2).

From $C_{\bar{L}}(A) = 1$ in (2), the irreducible submodules $\bar{L}_i$ are now all nontrivial. It is then standard essentially via Schur's lemma (e.g. [**Gor80**, 3.2.3]) that:
(8.2.10)

An abelian p-group, faithful and irreducible in characteristic $\neq p$, is cyclic.

Hence each quotient A/C_i is cyclic; and since A is elementary, in fact each A/C_i is of order p—that is, C_i is a "hyperplane" of A. Since A has rank s, and we saw $\cap_{i\in I} C_i = 1$, we must have $|I| \geq s$. And indeed if we choose an s-sequence from I, by requiring that the intersection of all the terms chosen so far strictly decreases each time we choose a new term C_i, then we obtain:

$$\cap_{i=1}^{s} C_i = 1.$$

This leads to the reduction:

(3) *We may assume* $\bar{L} = \bar{L}_1 \oplus \cdots \oplus L_s$*; that is,* $I = \{1, \cdots, s\}$*:*

From $\cap_{i=1}^{s} C_i = 1$ it follows that A is faithful on the sum on these $\bar{L}_i$; so that sum must by minimality give all of $\bar{L}$, as in (3).

We now set $A_i := \cap_{j\neq i} C_j$. Then each A_i is of order p, with C_i the direct product of the A_j for $j \neq i$. Also $A = A_1 \times \cdots \times A_s$, where A_i acts faithfully on $\bar{L}_i$, and centralizes $\bar{L}_j$ for $j \neq i$.

At this point, we have obtained a factorization of the quotient $\bar{L}A$ as the direct product of the $\bar{L}_i A_i$; we wish to pull this back to a suitable central-product factorization of LA in terms of the preimages $L_i A_i$. For this, we will need one further reduction:

(4) *We may assume each* A_i *centralizes* $\Phi(L)$ *and also centralizes* L_j *for* $j \neq i$*:*

If A_i does not centralize $\Phi(L)$, then the condition $\cap_{j=1}^{s} C_j = 1$ again shows (much as in (3) above) that A acts faithfully on $\Phi(L)(\Pi_{j\neq i} L_j)$—contrary to the minimality of L; hence all such composition factors are trivial. So by [**Gor80**, 5.3.2], A_i must

[7] And the self-centralizing property (8.2.7) is the classical Fitting lemma (e.g. [**Asc00**, 31.10]).

centralize $\Phi(L)$. And we saw A_i centralizes $\bar{L}_j$ for $j \neq i$, so by the same result it centralizes L_j, as required for (4).

The remainder of the proof will no longer appeal to minimality of L.

(5) *We have* $\Phi(L) \leq Z(LA)$:
Since A centralizes $\Phi(L)$ by (4), we just need to establish that $\Phi(L) \leq Z(L)$. And (4) gives $[[L, \Phi(L)], A] \leq [\Phi(L), A] = 1$ and $[[\Phi(L), A], L] \leq [1, L] = 1$. So by the Three-Subgroups Lemma (e.g. [**Gor80**, 2.2.3(ii)]) we get $[[A, L], \Phi(L)] = 1$; so from $[A, L] = L$ in (2) we get $[L, \Phi(L)] = 1$, as required.

(6) *LA is the central product of the L_iA_i, with $Z(LA)$ a p'-group:*
It remains only to show for $i \neq j$ that L_i and L_j commute; so suppose $[L_i, L_j] \neq 1$. Since $[L_i, L_j]$ commutes with L_iL_j by (5), a standard calculation with commutators (e.g. [**Gor80**, 2.2.2]) shows, since A_i is faithful on L_i but trivial on L_j, that A_i must involve a faithful module on $[L_i, L_j]$—and this contradicts (4). So we conclude that the L_i in fact commute.

The configuration in (6) is essentially the product structure at the end of earlier Example 8.2.3—extended to a central product as in Corollary 8.1.3. Thus the faithfulness of A_i on L_i gives us copies of S^0 in $\mathcal{A}_p(L_iA_i)$; and then the join of these structures in $\mathcal{A}_p(LA)$ gives us copies of the sphere S^{r-1}, affording the nonzero homology required to complete the proof. □

$\mathcal{QD}$ for solvable groups. We now indicate how to use the above minimal case of LA (with A elementary of rank r) to complete the proof of Theorem 8.2.5 for solvable G.

In fact the first assertion of the Theorem, namely of $\mathcal{QD}_p$ for G, is immediate: For we can take A of rank $s := r = m_p(G)$; and since $r - 1$ is the top dimension of $\mathcal{A}_p(LA)$ and $\mathcal{A}_p(G)$, we have $\tilde{H}_{r-1}(\mathcal{A}_p(LA)) \subseteq \tilde{H}_{r-1}(\mathcal{A}_p(G))$, where we have just seen in Theorem 8.2.9 that the former is nonzero.

So we turn to the more general second assertion, where A of rank s is assumed maximal under inclusion in $\mathcal{A}_p(G)$ (but not necessarily of maximal p-rank r):

PROOF. Recall that after the statement of Theorem 8.2.5, we had summarized the initial steps of the proof: setting $L := O_{p'}(G)$, and showing $C_A(L) = 1$ (or equivalently, $O_p(LA) = 1$ by (8.2.8)). Thus from the minimal-case result in Theorem 8.2.9, we have $\tilde{H}_{s-1}(LA) \neq 0$.

And as above, it would again suffice to show $\tilde{H}_{s-1}(\mathcal{A}_p(LA)) \subseteq \tilde{H}_{s-1}(\mathcal{A}_p(G))$; which is not necessarily obvious now, since $(s-1)$ may be lower than the top dimension of $\mathcal{A}_p(G)$, so that the incoming boundary map might be nonzero. Quillen in fact establishes the inclusion via homological algebra, using an appropriate Mayer-Vietoris sequence.

However, we will instead sketch the more elementary viewpoint indicated in the two paragraphs after the statement of [**AS93**, 0.10]: We had constructed explicit $(s-1)$-cycles for LA in the proof of Theorem 8.2.9; and it will now suffice to show that they remain nonzero in $\tilde{H}_{s-1}(\mathcal{A}_p(G))$—that is, that they are not in the image of the boundary map for $\mathcal{A}_p(G)$. Now since A is Sylow in LA, we may assume that such a cycle z involves A—that is, that z contains as a summand an inclusion-chain c of length $(s-1)$, in which A appears; and since A has rank s, which is the full p-rank of LA, the chain c must be "full" (Definition 0.9 [**AS93**])—that is, of form $A_1 < A_2 < \cdots < A_s = A$, with each A_i of rank i. Now if z is in the image of the boundary map, c must be a subchain of some length-s inclusion chain d from $\mathcal{A}_p(G)$; and since each A_i has rank i, this means that:

(8.2.11) d is of form $A_1 < \cdots A_s < B$ (where $A_s = A$); so B lies in $\mathcal{A}_p(G)_{>A}$.

However, this contradicts our hypothesis that A is maximal under inclusion in $\mathcal{A}_p(G)$. So we conclude that z is not in the image of the boundary map for $\mathcal{A}_p(G)$; and this completes the proof that $\tilde{H}_{s-1}(\mathcal{A}_p(G)) \neq 0$. □

The argument we have just presented, for A of rank s which is just maximal under inclusion in $\mathcal{A}_p(G)$, is not really required for the work of the present section, which only involves the property $\mathcal{QD}_p$ for the top dimension of $\mathcal{A}_p(G)$.

However, a suitable generalization of the argument in (8.2.11) will be relevant for the more general work of Aschbacher-Smith [**AS93**] on the general Quillen Conjecture, to be described in later Section 8.4.

And we also mention a later application of the nonzero homology in various possible dimensions $(s-1)$ in Theorem 8.2.5: The situation of different ranks s for A maximal under inclusion in $\mathcal{A}_p(G)$ is used in later Example 9.3.2, to exhibit a solvable group G with various exceptional properties (for instance, $\mathcal{A}_p(G)$ is not Cohen-Macaulay).

Extension to $\mathcal{QD}_p$ for p-solvable groups. Quillen had observed, as [**Qui78**, Problem 12.3], that the general p-solvable case of $\mathcal{QD}_p$ (namely Theorem 8.2.5) would in fact follow from his method of reduction, if one could establish $\mathcal{QD}_p$ for LA as in Theorem 8.2.9, but with L not assumed solvable. This extension seems to have been fairly well known, around 1990, to various researchers in the area (some of whom we will indicate below); I do not know if anyone claims specific priority for the extension.

We mention that the particular proof which we will now indicate in fact quotes the Classification of Finite Simple Groups (CFSG)—as is also the case in the more general work on Quillen's Conjecture in Aschbacher-Smith [**AS93**]. (However, the use of the CFSG just below is comparatively mild, in the sense of not involving particularly deep detail on the various simple groups.)

For the proof, it will be convenient to state the extension also for the minimal case LA in Theorem 8.2.9:

THEOREM 8.2.12. *We have the following extensions of earlier results:*
(1) $\mathcal{QD}_p$ holds for LA in Hypothesis 8.2.9 *even when L is not solvable.*
(2) $\mathcal{QD}_p$ holds for G in Hypothesis 8.2.5 *even when G is just p-solvable.*

PROOF. We first verify that (2) follows from (1); so assume that G is p-solvable. We had indicated after the statement of Theorem 8.2.5 that the first part of the argument there also applies, just assuming p-solvability; and it gives $C_A(L) = 1$ for $L := O_{p'}(G) > 1$, which is the hypothesis for (1). (Though now in the absence of solvability, we may have nonabelian simple groups involved in L—as we will see later on in the proof.)

So assume further that (1) holds, namely $\tilde{H}_{s-1}(\mathcal{A}_p(LA)) \neq 0$, where A is of rank s. We now observe that our proof just above, showing that 8.2.9 implies 8.2.5, still goes through: That is, it gives $\tilde{H}_{s-1}(\mathcal{A}_p(G)) \neq 0$, using just the maximality of A under inclusion in $\mathcal{A}_p(G)$ (applied as there, or more generally as in the Mayer-Vietoris sequence in Quillen's proof); and does *not* use the solvability of G.

Thus (1) indeed implies (2). (For another approach to this implication, see later Remark 8.2.14.)

Thus it remains to establish (1). Here also we can adapt some earlier arguments,[8] now from the proof for solvable L in Theorem 8.2.9: Namely p-solvability rather than solvability suffices in the initial argument for this result also; showing that we may assume that L is minimal subject to admitting faithful action by A. And similarly we may also then apply the argument in the first reduction there, using the self-centralizing property (8.2.7) to show that we may assume $L = F^*(L)$. Of course now in the absence of solvability, we have more generally that $F^*(L) = F(L)E(L)$, where $E(L)$ is a central product of quasisimple groups of order coprime to p.

It will now suffice to construct inside L a solvable subgroup L_0 having faithful A-action: for then $L_0 = L$ by minimality, while the result holds for L_0 by Theorem 8.2.9.

And we will define L_0 as the product of $F(L)$ with certain groups T_i defined as follows: For the product E_i of the members of some A-orbit on the components in $E(L)$, say with representative given by quasisimple L_i, we will find a Sylow subgroup S_i of L_i (for some prime q_i depending on L_i) with the property that $\mathrm{Aut}_A(L_i) = N_A(L_i)/C_A(L_i)$ acts faithfully on S_i; that is:

$$(8.2.13) \qquad N_A(L_i) \leq N_A(S_i) \text{ with } C_{N_A(L_i)}(S_i) \leq C_A(L_i).$$

We then define T_i as the product of the A-orbit of S_i. To see that A is faithful on the resulting L_0, assume that $a \in C_A(L_0)$: Then as a centralizes T_i, in particular it centralizes each A-conjugate of S_i; and so it must normalize each A-conjugate of L_i in the orbit E_i. But now $a \in C_{N_A(L_i)}(S_i) \leq C_A(L_i)$ by construction in (8.2.13); so in fact a must centralize all the A-conjugates of L_i (using the fact that A is abelian), whose product gives E_i. Thus a centralizes $F(L)$ and all E_i, so that A centralizes $F^*(L) = L$; and then $a \in C_A(L) = 1$—so that L_0 indeed suffices to complete the proof.

And now to obtain our L_0, it now remains just to locate the desired Sylow group S_i of L_i. If we happen to have $Aut_A(L_i) = 1$, i.e. $N_A(L_i) = C_A(L_i)$, then we can choose S_i to be *any* Sylow group of L_i (for any prime), and (8.2.13) will automatically hold.

So we turn to the more interesting case where $N_A(L_i) > C_A(L_i)$. Here we exploit the fact that the order of L_i is coprime to p. From the "Sylow theory" for this situation that we indicated after (5.5.5), there is a Sylow q-subgroup invariant under $N_A(L_i)$ for each prime q dividing the order of L_i. Furthermore as $N_A(L_i)$ acts nontrivially on L_i, it must act nontrivially on at least one such Sylow group; so we fix, as our choice of S_i, one such Sylow group with nontrivial action by $N_A(L_i)$. And in particular, we already have the first statement $N_A(L_i) \leq N_A(S_i)$ in (8.2.13). We claim now—and will prove in a moment—that $\mathrm{Aut}_A(L_i)$ is cyclic, and hence of order p since A is elementary. Since a generator of this group of prime order is nontrivial on S_i, so also is every nontrivial element. That is, only the trivial element of $\mathrm{Aut}_A(L_i)$ centralizes S_i—giving the second statement $C_{N_A(L_i)}(S_i) \leq C_A(L_i)$ of (8.2.13), as required.

[8] In fact we are deviating here from [**Alp89**], which for the nonsolvable case has on p 51 just a brief reference to the Sylow intersections complex. (In particular, it is not clear if the argument intended there would require quoting the CFSG.)

So now it only remains to show that $\mathrm{Aut}_A(L_i)$ is cyclic. We may as well assume, by standard results such as [**Gor80**, 5.3.5]), that L_i is not just quasisimple, but even simple. And this is where we will invoke the CFSG: that result (recall our Remark 2.0.1; cf. [**GLS94**, p 6]) states that a finite simple group is either an alternating group, a group of Lie type, or a sporadic group. Outer automorphisms of alternating and sporadic groups are indicated e.g. in Chapter 5 of [**GLS98**] and in Section 2.4 of [**Wil09**]. These have order 2 or 3, and so are ruled out for our A, since these orders are not coprime to the order of L_i of those indicated types. Outer automorphisms of Lie type groups are indicated for example in [**GLS98**, 2.5.1] and in Chapters 3 and 4 of [**Wil09**]. Each is a product of diagonal, field, and graph automorphisms. Diagonal automorphisms arise from diagonal matrices with determinant not 1, and so are again ruled out for A, as not being of order coprime to that of L_i; and graph automorphisms are similarly eliminated, since they have order 2, or 3 for (possibly twisted) type D_4 (whereas only Suzuki groups, of twisted type ${}^2B_2$, have order coprime to 3). This leaves only field automorphisms, which arise from Galois automorphisms of the field of definition; and this Galois group is of course cyclic (e.g. [**DF99**, Ex 7, Sec 14.1]), as required. This completes the proof. □

REMARK 8.2.14 (A different deduction for p-solvable $\mathcal{QD}_p$). The general p-solvable case in Theorem 8.2.12.2 is stated as 0.5 in [**AS93**]; and a variant proof can be obtained from 1.6 in that paper, by following the proof down to nonzero homology for $\mathcal{A}_p(G)$ in the next-to-last line. (The final line on being a counterexample to 1.6 is not relevant for the present purpose.) In effect, this argument represents a different deduction of "(1) implies (2)" from that we gave above—where the reference to 0.10, a few lines before the end, contains the minimal case LA of (1). The deduction in 1.6 proceeds via 0.27 there, using analysis of $C_G(LA)$—and so is more typical of the generic arguments in that paper (which we will be discussing in Section 8.4).

In fact, we take this opportunity to correct an oversight in [**AS93**]:[9] Namely we had inadvertently given the statement of 0.10 there only for the case of solvable L (as in Theorem 8.2.9); instead, it *should* have been stated for L any p'-group—along with a proof of the extension from the solvable case, such as that given just above for Theorem 8.2.12.1. (Indeed we had informally circulated this additional argument to many early readers of [**AS93**]—though perhaps the deduction was already reasonably well known to experts at that time.) ♢

$\mathcal{QD}$ for most simple groups. The introduction of the notion of Quillen dimension in [**AS93**] was largely motivated by the observation that, in the results in the literature at that time on the Quillen Conjecture, the conclusion usually seemed to actually be the nonzero top homology as in $\mathcal{QD}_p$. (Essentially the only exception being the Lie-type groups in characteristic p mentioned after Example 8.2.3.)

For example (as we had indicated in Remark 8.2.4), after treating the solvable case in Theorem 8.2.5, Quillen began consideration of simple groups G with his [**Qui78**, Thm 12.4]—which we discuss later as Theorem 9.4.7. That result in fact shows Cohen-Macaulayness of $\mathcal{A}_p(G)$, for the ("nearly simple") Lie-type

[9] We thank Alberto Delgado for bringing this to our attention.

group $G = GL_n(q)$, where q is a suitable prime power of characteristic not p; and $\mathcal{QD}_p$ in particular is implicit in the conclusion. (The weaker assertion just of nonzero top homology when $O_p(G) = 1$ in fact follows from Theorem 8.2.15 below.)

And indeed, the basic approach of Aschbacher-Smith in [**AS93**] to the Quillen Conjecture for general G is founded on separating off possible simple components in $F^*(G)$ into a generic inductive case, in which $\mathcal{QD}_p$ holds—along with an exceptional case, where components without $\mathcal{QD}_p$ must then be handled by non-inductive methods (such as Robinson subgroups, described in the subsequent Section 8.3).

Consequently that work involves an intermediate technical result, showing that most simple groups do indeed satisfy $\mathcal{QD}$—by specifying explicitly which simple groups fail the property; this analysis of course requires the Classification of the Finite Simple Groups.

The analysis is in fact made for G which is *almost simple*: that is, $F^*(G)$ is simple, so that by the self-centralizing property (8.2.7) we have $G \leq \mathrm{Aut}(F^*(G))$. Indeed the relevant minimal situation (in parallel with Quillen's minimal solvable LA in Theorem 8.2.9) has a semidirect product $G = LB$, where $L := F^*(G)$ is simple, and B is an elementary abelian p-subgroup of *outer* automorphisms of L. (And since $\mathcal{QD}_p$ was established for the case of L a p'-group in Theorem 8.2.12, we may as well assume that p divides the order of L. In that case, B will no longer be maximal in $\mathcal{A}_p(LB)$—it will lie in larger subgroups having the form $A = B \times C$, where $C \in \mathcal{A}_p(C_L(B))$.)

Here is a rough statement of the relevant result—we have simplified by omitting some technical details on some of the groups appearing as exceptions:

THEOREM 8.2.15 (3.1 in [**AS93**]). *Assume that p is odd, L is a (nonabelian) simple group, and B is an elementary abelian p-subgroup of Out(L). Then either LB satisfies $\mathcal{QD}_p$, or possibly L is one of:*

(1) a group of Lie type in the same characteristic p;
(2) $U_n(q)$ with $q \equiv -1 \pmod p$ and $n \geq q(q-1)$;
(3) ${}^3D_4(q)$ or $E_8(q)$ for certain q, with A inducing field automorphisms;
(4) the Monster M for $p = 7$, or J_4 for $p = 11$;
(5)–(7) several further cases for $p = 5$;
(8)–(12) several further cases for $p = 3$.

PROOF. (Sketch:) Here is a very rough overview of the method of proof: We embed B in an elementary p-subgroup A of maximal rank in LB, so that $LA = LB$. Then using the subgroup structure of the known group L (that is, by examining the simple groups from the CFSG), we try to find a subgroup L_0A, for which it will suffice (as in arguments above) to establish $\mathcal{QD}$. And in order to treat such a candidate L_0A, we can typically proceed via a product construction, much like those we have already seen in this section: That is, we try to break L_0A into a product $L_1A_1 \times L_2A_2$; and then product homology for the join in Lemma 8.1.2 reduces us to estabishing $\mathcal{QD}$ for the factors. Indeed often we can iterate this process until all factors have p-rank 1, so that we end up with the standard join-construction of spheres. In other cases, we may be able to stop earlier—if we have already inductively established $\mathcal{QD}$ for the factors. □

We mention that the hypothesis of odd p in the statement of Theorem 8.2.15 is made only because the Aschbacher-Smith main result (see later Theorem 8.4.4) is proved only for primes $p > 5$. Certainly a result similar to 8.2.15 could be established for $p = 2$—though the list of exceptional groups would be longer.

And concerning the word "possibly" used in the conclusion of Theorem 8.2.15: Certainly $\mathcal{QD}$ does fail for most of the Lie-type groups in conclusion (1); but in conclusions (2)–(12), we did not typically check for failure of $\mathcal{QD}$. Thus further improvements might be possible here.

This is particularly relevant with regard to the unitary groups in conclusion (2): For in the Aschbacher-Smith proof, the exceptional groups in the conclusions (1) and (3)–(12) do not really cause a problem. However the unitary groups in (2) do manage to evade the analysis, and so they have to be excluded in the main result Theorem 8.4.4.

Thus the analysis would in fact go through, if it could be demonstrated that the unitary groups in conclusion (2) typically do *not* really afford an exception to $\mathcal{QD}$; that is, we believe (perhaps somewhat wishfully) that:

CONJECTURE 8.2.16 (Conjecture 4.1 in [**AS93**]). *Assume that p is odd; and when $p = 3$, exclude the cases $q = 2, 8$. Then $\mathcal{QD}_p$ should hold for LA, where L is a unitary group $U_n(q)$ and A is a nontrivial elementary p-group.*

As partial evidence, we mention that the result holds at least "asymptotically" (if we fix n):

LEMMA 8.2.17 (4.4(b) in [**AS93**]). *Conjecture* 8.2.16 *holds when $q(q-1) > n$.*

(Indeed this result is what allows the restriction $n \geq q(q-1)$ in the unitary conclusion (2) in Theorem 8.2.15.)

This section has focused on nonzero homology in the top dimension of $\mathcal{A}_p(G)$, primarily via construction of spheres. We close the section with a brief digression on other possibilities.

REMARK 8.2.18 (Some examples of cycles other than spheres). Of course the topological literature also includes various boundary-free structures other than spheres; and these sometimes arise in the context of subgroup complexes (though they do not seem to have been much exploited).

For example, in earlier Remark 2.3.9, we indicated that some finite geometries of interest arise as finite quotients of infinite affine buildings. Now the apartment in that affine building is not a sphere, but instead an affine (i.e. "flat" Euclidean) space in the top dimension; and one then checks that its image in the finite quotient can determine a torus—providing a cycle in the top dimension of the finite geometry. For example, in Kantor's $\tilde{C}_2$-geometry for $U_4(3)$ of Example 2.3.10, we can get an ordinary 2-torus.

Moreover, in at least one 2-dimensional geometry, Klein bottles are known to arise: Namely consider the Neumaier C_3-geometry for the group A_7 in earlier Example 2.3.7: This is a 2-dimensional complex; and [**SY95**, Prop 9] shows that the smallest 2-cycles are in fact Klein bottles (defined using 36 maximal faces). ◇

8.3. Robinson subgroups: Non-vanishing Lefschetz module for $\mathcal{A}_p(G)$

In this section, we describe a different method for establishing—for many G with $O_p(G) = 1$—the non-contractibility required in Quillen's Conjecture. Whereas the condition $\mathcal{QD}$ of the previous section obtains non-contractibility by exhibiting nonzero homology specifically in the top dimension of $\mathcal{A}_p(G)$, the approach of Robinson in [**Rob88**] instead just exhibits it in *some* dimension—via a nonzero reduced Lefschetz module $\tilde{L}(\mathcal{S}_p(G)) \neq 0$ (that is, homology somewhere does not cancel out in the alternating sum).

From the viewpoint of Remark 8.2.4, since Quillen's Conjecture for solvable groups is handled by Theorem 8.2.5, it makes sense to shift the main focus to simple groups G. And we will see in Theorem 8.3.3 below that simple groups failing $\mathcal{QD}$ as in Theorem 8.2.15 (again with the vexing exception of unitary groups) can be given an alternative treatment via Robinson's method. (This is the crucial secondary method used in the Aschbacher-Smith work [**AS93**], as we will see in the subsequent Section 8.4.)

So below we will briefly sketch some of the influence of Robinson's method in the literature on Quillen's Conjecture.

The pre-history of the method begins with Thévenaz, who around 1986 observed a special case of the congruence in Lemma 5.5.1 (i.e. $\tilde{\chi}(\mathcal{S}_p(G)^X) \equiv 0 \pmod q$ in suitable circumstances), namely where "X" is given just by a q-group Q for some prime q. And he applied that result in the contrapositive: For notice that if Q normalizes no member of $\mathcal{S}_p(G)$, then we have $\mathcal{S}_p(G)^Q = \emptyset$, hence $\tilde{\chi}(\mathcal{S}_p(G)^Q) = -1$. Since $q \geq 2$, we cannot have this value of -1 congruent to 0 mod q, as in the conclusion of the Lemma; hence the hypothesis of the Lemma must fail, giving $\tilde{L}(\mathcal{S}_p(G)) \neq 0$. Thévenaz thought of using this approach to verify Quillen's Conjecture for simple groups G, by finding some Sylow q-group Q with $\mathcal{S}_p(G)^Q = \emptyset$; and Lyons checked this was so in the various simple groups G in the CFSG—but details were never published. (A similar argument appeared later, as Theorem 2 in Aschbacher-Kleidman [**AK90**], which we indicate as Theorem 8.3.2 below.)

We turn to the more general method as developed by Robinson in [**Rob88**]. Recall that the congruence in Lemma 5.5.1 is stated for a subgroup X which is q-hyperelementary—that is, a semidirect product $X = \langle x \rangle Q$ of a normal cyclic subgroup $\langle x \rangle$ with a q-group Q. Arguing contrapositively as above on the congruence gives the form below, which is useful for Quillen's Conjecture: (Notice that $\mathcal{S}_p(G)^X = \emptyset$ in particular gives the hypothesis $O_p(G) = 1$ of the Conjecture.)

THEOREM 8.3.1 (2.3 in [**Rob88**]). *Assume that G contains a q-hyperelementary p'-subgroup X such that $\mathcal{S}_p(G)^X = \emptyset$. Then $\tilde{L}(\mathcal{S}_p(G)) \neq 0$.*

Robinson used this method to show in [**Rob88**, 2.6] that many simple groups G—and consequently many further classes of larger groups G with $O_p(G) = 1$—must contain such a subgroup X (and hence must satisfy Quillen's Conjecture). Of course this work implicitly uses the CFSG to identify and examine the individual simple groups. And as we had mentioned just above, Aschbacher and Kleidman later carried the technique still further:

THEOREM 8.3.2 (Theorems 2, 3 in [**AK90**]). *Assume that G is almost simple (that is, $F^*(G)$ is simple). Then either G has a q-hyperelementary p'-subgroup X with $\mathcal{S}_p(G)^X = \emptyset$ (so $\tilde{L}(\mathcal{S}_p(G)) \neq 0$ by Theorem 8.3.1), or else $F^*(G) = L_3(4)$ and $p = 2$.*

Furthermore any almost-simple G satisfies Quillen's Conjecture (i.e. $\mathcal{S}_p(G)$ is not contractible).

We had also mentioned earlier that Robinson's method turned out to be crucial in the Aschbacher-Smith work [**AS93**] on the Quillen Conjecture: namely it provided an alternate way of handling (most of) those simple components L of $F^*(G)$ which might happen to fail the condition $\mathcal{QD}$ (that is, that appear among the exceptions listed in Theorem 8.2.15). Note in fact that the above contrapositive argument on the congruence in Lemma 5.5.1 does not require $\mathcal{S}_p(G)^X$ to be empty—it suffices if we just have $\tilde{\chi}(\mathcal{S}_p(G)^X) = \pm 1$ (for example, if $\mathcal{S}_p(G)^X$ is a sphere—recall (8.1.6)). We call such subgroups *Robinson* subgroups; they arise in:

THEOREM 8.3.3 (Theorem 5.3 in [**AS93**]). *Assume the hypotheses of Theorem* 8.2.15, *that is: We assume that p is odd (indeed further that $p \geq 5$), L is a (nonabelian) simple group, and B is an elementary abelian p-subgroup of $Out(L)$. Assume furthermore that LB fails $\mathcal{QD}_p$, and so L is one of the groups listed in the conclusions of* 8.2.15*—except that we exclude the unitary groups in conclusion (2) there.*

Then L contains a 2-hyperelementary p'-subgroup of the form $X = \langle x \rangle \times Q$ with $O_2(X) > 1$, which satisfies either:
(i) $\mathcal{S}_p(L)^X = \emptyset$ (so that $\tilde{\chi}(\mathcal{S}_p(L)^X) = -1$), or
(ii) L is some linear group $L_n(p^a)$ with $\tilde{\chi}(\mathcal{S}_p(Aut(L))^X) = +1$.

We will say more about how this result allows us to use the properties of both $\mathcal{QD}$ and Robinson subgroups, in the subsequent Section 8.4, where we outline the proof of the Main Theorem in [**AS93**].

8.4. The Aschbacher-Smith result on Quillen's Conjecture

In this final section of the chapter, we sketch the work of Aschbacher-Smith in [**AS93**], which established the most general result known so far on Quillen's Conjecture.

For background, we quickly recall that Quillen showed (in the result we called Lemma 3.3.5) that if $O_p(G) > 1$, then $\mathcal{S}_p(G)$ is contractible; and he conjectured that the converse should hold—which we had correspondingly recorded as Conjecture 3.3.8.

Of course Quillen's result in Theorem 4.2.4 shows that $\mathcal{S}_p(G)$ is homotopy equivalent to the elementary subposet $\mathcal{A}_p(G)$; and indeed most of the literature on the Conjecture works with $\mathcal{A}_p(G)$. Furthermore the literature usually approaches the Conjecture via the contrapositive—so for convenience in this section, we now re-state the Conjecture in that form:

CONJECTURE 8.4.1 (Conjecture 2.9 in [**Qui78**]). *If $O_p(G) = 1$, then $\mathcal{A}_p(G)$ is not contractible.*

Since non-contractibility is not always easy to demonstrate directly, most results on the Conjecture in fact establish one of the stronger conditions in Lemma 5.2.3, such as $\tilde{H}_*(\mathcal{A}_p(G)) \neq 0$, or $\tilde{\chi}(\mathcal{A}_p(G)) \neq 0$.

The Conjecture has considerable direct esthetic appeal—and this is perhaps the main motivation for group theorists to study it. However we also mention, with regard to the area of decomposition theory in algebraic topology (indicated in the previous Chapter 7), that a strengthened form of the Conjecture has implications for Grodal's principal p-radical collection in Definition 7.4.8 being minimal for giving decompositions; see e.g. Remarks 3.9 and 3.12 in [**Gro02**].

(Option G:) We had mentioned in Example 3.5.7 that the proof of Lemma 3.3.5 extends to showing when $O_p(G) > 1$ that $\mathcal{S}_p(G)$ is in fact G-contractible. And conversely, assuming the stronger condition of G-contractibility would in particular imply (cf. [**Ben91**, p 229]) a fixed point on $\mathcal{S}_p(G)$, so that $O_p(G) > 1$. That is, a weaker analogue of the Conjecture holds, namely $O_p(G) = 1$ implies that $\mathcal{S}_p(G)$ is not G-contractible. (But this doesn't really help with the original Conjecture.)

The treatment in [**AS93**] begins with an overview (in Section 0 there) of results in the literature on the Conjecture up to about 1991—some of those results were then used later in the paper. Correspondingly we next provide a similar (but briefer) summary.

Cases originally established by Quillen. After stating the Conjecture in [**Qui78**], Quillen went on to establish important evidence for it. Indeed much of the general theory developed later in the paper can in fact be applied to the Conjecture, in particular giving:

THEOREM 8.4.2. *Conjecture* 8.4.1 *holds for G in any of the following cases:*
(i) [**Qui78**, Prop 2.10] *G has p-rank $m_p(G) \leq 2$;*
(ii) [**Qui78**, Cor 12.2] *G is solvable;*
(iii) [**Qui78**, Thm 3.1] *G is of Lie type in (the same) characteristic p;*
(iv) [**Qui78**, Thm 12.4] *$G = GL_n(q)$ for $q \equiv 1 \pmod p$.*

PROOF. (Sketch:) For the solvable case (ii), we gave a fairly full discussion earlier as Theorem 8.2.5. With solvable groups treated, it is then natural to examine simple groups; and of course most simple groups (recall Remark 2.0.1) are of Lie type over a finite field $\mathbb{F}_q$, of some characteristic r—as in cases (iii) and (iv).

In case (iii), where the characteristic r is equal to p, we indicated the conclusion of nonzero homology in Theorem 6.4.2. (Of course this involves Quillen's homotopy equivalence in Theorem 4.4.8 of $\mathcal{S}_p(G)$ with the building.)

And for characteristic r different from p, case (iv) treats the standard Lie-type family of (nearly-simple) linear groups $G = GL_n(q)$, for $q \equiv 1 \pmod p$—so that $\mathbb{F}_q$ contains a primitive p-th root of unity. In fact [**Qui78**, Thm 12.4] establishes the stronger result of Cohen-Macaulayness for $\mathcal{A}_p(G)$, and we discuss that result in our corresponding later chapter as Theorem 9.4.7. (For the Quillen Conjecture, it suffices just to verify the condition $\mathcal{QD}_p$, i.e. that the top-dimensional homology is nonzero; and this is one of the easier special cases of Theorem 8.2.15, which was established later in [**AS93**].)

For (i), we can in fact argue directly for the converse of Lemma 3.3.5; namely we assume that $\mathcal{A}_p(G)$ is contractible, and we will show that $O_p(G) > 1$—by producing a vertex of $\mathcal{A}_p(G)$ fixed under the G-conjugation action. We recall from Definition 3.3.9 that our hypothesis that $m_p(G) \leq 2$ implies $\dim \mathcal{A}_p(G) \leq 1$. If $\dim \mathcal{A}_p(G) = 0$, then by our assumption of contractibility, $\mathcal{A}_p(G)$ is a single point—necessarily fixed by G, and we are done. Thus we may as well assume $\dim \mathcal{A}_p(G) = 1$, so that $\mathcal{A}_p(G)$ is a (bipartite) graph; and then it is standard from contractibility (e.g. [**Mun84**, 7.1]) that $\mathcal{A}_p(G)$ is a tree. At this point, for [**Qui78**, Prop 2.10] Quillen simply refers to a well-known result of Serre that a finite group acting on a tree (with the condition we are calling type-preserving) must fix a point—and this completes the proof.

However, for our finite G, it is easy to provide below an elementary argument for such a fixed point: This will be in the spirit of successive collapses in the Solomon-Tits argument in Theorem 3.4.15. Since our complex is a tree, it has at least one "end"—that is, a vertex v lying on just one edge E; and then E is free over v in the sense of Definition 3.4.3. So the removal of v and E is by Lemma 3.4.6 a homotopy equivalence. (Option G: Indeed note that by the type-preserving property (recall Remark 1.5.7), the other vertex of E is not in the G-orbit of v; hence the simultaneous removal of the entire G-orbit of the pair v, E gives a G-homotopy equivalence.) If this collapsing reduces us to a single vertex, it is G-fixed and we are done; otherwise we can repeat the process of removing a G-orbit of ends. By finiteness, the process must in fact terminate—that is, eventually we must reduce to a single vertex, fixed by G. □

Cases in the subsequent literature. In the years following the appearance of Quillen's paper [**Qui78**], further cases of the Conjecture were established by a number of researchers. In fact we have already indicated the following results in earlier sections; we collect them here as:

THEOREM 8.4.3. *Quillen's Conjecture* 8.4.1 *holds for* G *in any of the following cases:*
(i) (Thm A in [**HI88**]*)* G *is* p*-solvable with abelian Sylow* p*-groups;*
(ii) (various authors) G *is* p*-solvable;*
(iii) (Thm 3 in [**AK90**]*)* G *is almost-simple (that is,* $F^*(G)$ *is simple).*

To quickly review the relevant earlier discussion: For (i), we had indicated as Theorem 5.5.4 the result of Hawkes-Isaacs in [**HI88**] giving nonzero Euler characteristic for $\mathcal{A}_p(G)$, which suffices for Conjecture 8.4.1 for LA.

To obtain (ii), which we had indicated as Theorem 8.2.12.ii, we saw that Alperin and Aschbacher-Smith (and possibly others) began by establishing the more specific condition $\mathcal{QD}_p$ for LA as in (i) (namely our Theorem 8.2.12.i), using properties of the known simple groups in the CFSG. (Recall Remark 8.2.14 for some details concerning the derivation from the result for LA.)

And we had indicated (iii) as Theorem 8.3.2.

The Aschbacher-Smith result for more general G. Aschbacher and Kleidman were originally hopeful that their adapation of Robinson's method, as developed for [**AK90**, Thm 3] (i.e. Theorem 8.3.2 above), would further extend to a proof of Quillen's conjecture for general G. For since they had covered the case where $F^*(G)$ consists of a single component L, it can be assumed that $F^*(G)$ has at least two components—and ideally then they might be able to apply induction in appropriate subgroups. However, the difficulty is that the main hypothesis $O_p(G) = 1$ is not necessarily inherited by arbitrary subgroups.

Furthermore, it seems in retrospect that the nonzero Lefschetz module in the conclusion of the approach via Robinson subgroups, which gives non-zero homology only in some unknown dimension, does not provide enough leverage for an inductive proof. However, using the more specific notion of $\mathcal{QD}_p$ (namely nonzero homology in the *top* dimension of $\mathcal{A}_p(G)$), Aschbacher and Smith were able in early 1992 to develop a fairly general inductive approach. Below we state a corollary (with a less technical hypothesis) of the Main Theorem on p 474 in [**AS93**]; notice that it is general in the sense that it emphasizes which groups might *not* satisfy the

Conjecture—in contrast to the above results, which establish particular classes of groups which do satisfy the Conjecture.

THEOREM 8.4.4 (Theorem, p 474 in [**AS93**]). *Assume that $p > 5$, and $F^*(G)$ has no unitary component $U_n(q)$ with $q \equiv -1 \pmod p$.*

Then $O_p(G) = 1$ implies $\tilde{H}_*(\mathcal{A}_p(G)) \neq 0$ (and so Quillen's Conjecture 8.4.1 *holds for G).*

We mention that the hypothesis $p > 5$ should not really be necessary—but for smaller primes, the case analysis required in the method would have to be much more detailed. Furthermore, if Conjecture 8.2.16 above (that unitary groups should satisfy $\mathcal{QD}$) in fact holds, then there would be no need to exclude unitary groups in the hypothesis.

The techniques used in the proof are mainly rather elementary. However, as we have mentioned, the argument at several points uses the Classification of Finite Simple Groups; and this of course can sometimes lead to inspection of the detailed structure and properties of the individual simple groups.

In the remainder of this section, we will provide a brief overview of the work in [**AS93**].

After introducing some preliminaries in Section 0 (some of which we have summarized earlier in this chapter), the main logic sequence for the proof of the Main Theorem in [**AS93**] is carried out in Section 1 there—which in turn refers to some more technical material developed in later sections, such as $\mathcal{QD}_p$ and Robinson subgroups.

In fact within Section 0, an outline of the proof[10] is given in pp 476–479 of [**AS93**], particularly in the discussion starting at the bottom of p 477. Our discussion below will be simplified still further over that outline.

As a brief initial summary: In case $F^*(G)$ contains a component L with $\mathcal{QD}$, we are able to make an inductive argument to produce non-zero homology for $\mathcal{A}_p(G)$, and we are done. In the remaining case, all components L of $F^*(G)$ are in the list of exceptions in 8.2.15. And for those exceptional components (again recalling that we have excluded unitary groups), we can construct Robinson subgroups as in Theorem 8.3.3; and then as usual in Robinson's method, we obtain a nonzero Lefschetz module via Theorem 8.3.1, completing the proof in the remaining case.

Preliminaries. We already summarized some of these earlier in the chapter.

In particular, we emphasize again that 0.27 in [**AS93**] is roughly an analogue of Quillen's reduction, in the solvable case in Theorem 8.2.5, to the minimal situation of LA in Theorem 8.2.9: but now L is a component in $F^*(G)$ satisfying $\mathcal{QD}_p$, and we actually apply induction to the subgroup $C_G(LA)$—as we will describe in a little more detail shortly.

We had also mentioned the initial reduction in [**AS93**, Prop 1.6]—which in particular shows that we may assume $O_{p'}(G) = 1$. (Hence $F^*(G) = E(G)$; and components L in $E(G)$ are in fact simple, rather than just quasisimple—and in addition they have order divisible by p.) The proof requires a non-trivial inductive argument; however, we know by Theorem 8.2.12[11] that $O_{p'}(G)A$ has $\mathcal{QD}$. And

[10]However, we mention a confusing misprint at the top of p 479: which incorrectly refers to the outline as applying to the proof of 0.10—whereas of course it applies instead to the proof of the entire main result.

[11]Recall also the discussion in Remark 8.2.14.

that inductive argument is essentially a simpler case of the main argument for components with $\mathcal{QD}$, which we will now describe.

Treating components with $\mathcal{QD}$. For such an LA and its centralizer, the main reduction (namely 1.7 in [**AS93**]) considers the case in which $F^*(G)$ contains a (simple) component L, such that $\text{Aut}(L)$ satisfies $\mathcal{QD}_p$.

We wish to construct an elementary subgroup which has a maximality property, suitably analogous to the maximality under inclusion of the "A" which was exploited in (8.2.11). To this end, we first choose an elementary A in $N_G(L)$ which is maximal subject to being faithful on L; i.e. $C_A(L) = 1$. Such an A exists, since p-elements inside simple L are faithful on L. And since any A normalizes at least one Sylow p-subgroup of L, and hence centralizes some elements of order p there, we may assume by maximality that the corresponding inner automorphisms are in fact induced in $A \cap L$—so that $A \cap L > 1$.

This procedure is designed to provide an appropriate factorization of any B from $\mathcal{A}_p(G)$ which contains A, as follows: Since B in particular centralizes $A \cap L > 1$, it must then normalize the component L of $F^*(G)$, so that $B \leq N_G(L)$. Since A is faithful on L, we have a direct product $A \times C_B(L)$ inside B, and then we may write $B = B' \times C_B(L)$ for some complement $B' \geq A$. However, by the maximal-faithful choice of A, we must have $B' = A$. Thus $B = A \times C_B(L) = A \times C_B(LA)$. So setting $H := LA$ and $K := C_G(H)$, we have $H \cap K = Z(H) = 1$ since L is simple and A is faithful. And so we have established that $B \in \mathcal{A}_p(H \times K)_{\geq A}$; this is our analogue of earlier (8.2.11).

Now the choice of A may be made with the further restriction that $A \cap L$ has maximal order. This is used to show that we may assume A is involved in some cycle α for $\mathcal{A}_p(H)$ which exhibits $\mathcal{QD}_p$.

And we also wish to obtain some cycle β of $\mathcal{A}_p(K)$ which has a nonzero image in homology; for this, we want to apply induction to K, so in particular we need $O_p(K) = 1$. Notice (recall Remark 8.2.6) that $C_G(L)$ contains at least all the other components of $F^*(G)$; but $K = C_G(LA)$ contains only the subgroups of those components centralized by A—these give the crucial configuration to examine. A corresponding technical result [**AS93**, Thm 2.4], also requiring the CFSG and inspection of outer automorphisms of the simple groups, shows that we can if necessary replace our original A by another suitable complement to L in $H = LA$, so as to guarantee that $O_p(K) = 1$ (and hence also to guarantee the existence of the desired β).

The remaining argument is then an elementary exercise in product homology for $H \times K$ applied to the join of the complexes $\mathcal{A}_p(H)$ and $\mathcal{A}_p(K)$; it involves a somewhat more specific analysis than that given in earlier Lemma 8.1.2: Namely the Eilenberg-Zilber description of product homology shows that we may form the "shuffle product" $\alpha \times \beta$ (see [**AS93**, Defn 0.21]) to get a cycle of $\mathcal{A}_p(H \times K)$. Now we use the maximality property of our A discussed above (in analogy with the earlier argument at (8.2.11)): The cycle α involving A is in the top dimension of $\mathcal{A}_p(LA)$, so in particular involves a "full" inclusion-chain leading up to A (in the sense of [**AS93**, Defn 0.9]). Thus if $\alpha \times \beta$ is a boundary, then any longer inclusion-chain involved must in fact involve an elementary subgroup $B > A$—where by the above argument B factors over $H \times K$ as $A \times C_B(LA)$, where $C_B(LA) \in \mathcal{A}_p(K)$. A corresponding factorization in homology in [**AS93**, 0.27] shows that β would then have to be a boundary for $\mathcal{A}_p(K)$, contradicting its choice as representing a nonzero element of homology. Consequently $\alpha \times \beta$ is nonzero in $\tilde{H}_*(\mathcal{A}_p(G))$, completing the proof in this case.

Treating non-$\mathcal{QD}$ components. In the remaining case (treated in the final part of Section 1 of [**AS93**]), we may assume that all components L_i of $F^*(G) = E(G)$ fail $\mathcal{QD}$. Since our main hypothesis excludes unitary groups, by Theorem 8.3.3 inside each L_i we can find a "Robinson subgroup" X_i, so that $\tilde{\chi}(\mathcal{S}_p(L_i)^{X_i}) = \pm 1$.

From their product we form a 2-hyperelementary group X. Some further argument shows that $\mathcal{S}_p(G)^X$ is just $\mathcal{S}_p(E(G))^X$, and the latter decomposes as the join of the $\mathcal{S}_p(L_i)^{X_i}$. Hence by the multiplicativity in Lemma 8.1.2 (where we may replace $\mathcal{A}_p$ by $\mathcal{S}_p$ using Theorem 4.2.4), we have $\tilde{\chi}(\mathcal{S}_p(G)^X) = \Pi_i \, \tilde{\chi}(\mathcal{S}_p(L_i)^{X_i}) = \pm 1$. On the other hand, if we had $\tilde{L}(\mathcal{S}_p(G)) = 0$, then Lemma 5.5.1 would (much as in the proof of Theorem 8.3.1) give $\tilde{\chi}(\mathcal{S}_p(G)) \equiv 0 \pmod 2$. This contradiction forces $\tilde{L}(\mathcal{S}_p(G)) \neq 0$, and completes our outline of the proof in [**AS93**].

Some further investigation related to the possibility of allowing unitary groups in the hypothesis of Theorem 8.4.4 appears in Welker [**Wel95a**].

We close this section on Quillen's Conjecture by mentioning that Weidner and Welker in [**WW93**] consider a "dual" situation, namely the poset $\mathcal{S}^p(G)$ of subgroups of G which have nontrivial p-power *index*. (Of course there is no dual to the Sylow theory for these subgroups—in particular such subgroups can be regarded as more "rarely occurring" than p-subgroups.) Weidner and Welker are able to establish in particular the result naturally dual to the Quillen Conjecture—namely that $\mathcal{S}^p(G)$ is contractible if and only if $O^p(G) < G$. (Here $O^p(G)$ is the smallest *normal* subgroup of p-power index.) In [**WW97**] there is a suitable extension from a single prime p to a set π of primes.

A different variation from p-subgroups is to consider the poset of nilpotent subgroups; see e.g. Lucido [**Luc95**].

CHAPTER 9

Connectivity, simple connectivity, and sphericality

In the previous Chapter 8, we considered spheres just as possible substructures of a complex, sufficient to guarantee the existence of *some* homology. This chapter is largely devoted to more specific notions of "sphericality"—in which the spheres essentially determine the fuller structure of the complex (and even *all* its homology).

One fundamental role of spheres within a complex K arises in the definition of the homotopy groups $\pi_n(K)$—whose triviality in dimensions $n = 0$, 1, and ≥ 2 defines the notions of (path) connectivity, simple connectivity, and "higher" connectivity. Correspondingly, our usual initial topological background Section 9.1 quickly reviews the definition and some basic properties of homotopy groups; these then lead us to n-connectivity, and notions related to sphericality.

Thereafter, Section 9.2 briefly discusses the case $n = 0$, where the usual expectation for K given by $\mathcal{A}_p(G)$ is path-connectivity. So the main focus is then in fact on the exceptional situation of disconnectedness; where corresponding group-theoretic conditions have been of considerable independent significance, especially in the Classification of Finite Simple Groups.

Similarly Section 9.3, devoted to the case $n = 1$, focuses on the less usual situation where $\mathcal{A}_p(G)$ is not simply connected—namely where that complex has a proper topological covering. Again the corresponding group-theoretic conditions have been of independent importance in the CFSG; and furthermore, methods for establishing simple connectivity have been useful in giving geometrically-flavored uniqueness proofs for simple groups.

The final Section 9.4 considers n-connectivity mainly for $n \geq 2$. Indeed for a complex K of dimension d, the literature usually focuses on d-sphericality, that is, $(d-1)$-connectivity: this reflects a standard property of the structure given by a bouquet of d-spheres—which we had briefly touched on at various earlier points. We especially consider the local version, when the links of simplices satisfy sphericality in the appropriate dimension; this is the *Cohen-Macaulay* property, which was motivated by the study of Cohen-Macaulay rings, in commutative algebra and algebraic geometry. In fact Quillen (starting in Section 8 of [**Qui78**]) provided considerable material on Cohen-Macaulayness for K given by $\mathcal{A}_p(G)$; and a substantial subsequent research literature has developed, which our treatment can only briefly overview and occasionally sample. In particular the study of Cohen-Macaulayness led to various further related conditions, especially in the combinatorial literature—notably *shellability*, which we will indicate at the end of the chapter.

9.1. Topological background: Homotopy groups, n-connectivity, and sphericality

Our review of homotopy groups and higher connectivity below will be quick and very informal; the reader should if needed follow up the indicated references.

Definitions and basic properties. We recall:

REMARK 9.1.1 (Homotopy groups). For a complex K, the *n-th homotopy group* $\pi_n(K)$ is defined[1] as the group $[S^n, K]$ of homotopy classes of maps of the sphere S^n into K. ◊

Probably the most basic question about the group $\pi_n(K)$ is when a map $f : S^n \to K$ should determine the identity; that is, when is f homotopic to the trivial map taking S^n to a point? Since the sphere S^n is the boundary of the ball B^{n+1}, this amounts to asking when the image $f(S^{n+1})$ can be continuously "filled out" to a ball in K (and hence then contracted to a point). This is the standard statement (e.g. [**Spa81**, 1.6.7]):

LEMMA 9.1.2. *A map $f : S^n \to K$ represents the trivial element in $\pi_n(K)$ if and only if it can be continuously extended over B^{n+1}.*

Correspondingly are particularly interested in situations where $\pi_n(K)$ is just the trivial group—in which case the extension in Lemma 9.1.2 is automatic.

REMARK 9.1.3 (n-connectivity). For example in case $n = 0$: Since S^0 is a pair of points, the extension to B^1 (i.e. a line segment) gives a path between the points; so that $\pi_0(K) = 1$ says K is *path-connected*.

And when $n = 1$, since S^1 is a triangulation of a circle, the extension to B^2 fills out a disk. Here $\pi_1(K)$ is the usual *fundamental group* (e.g. [**Spa81**, Sec 1.8]) parametrizing the topological *coverings* of K; and triviality of $\pi_1(K)$ makes loops contractible to a point, so that K is *simply* connected.

More generally, we say K is *n-connected* if $\pi_i(K) = 1$ for $0 \leq i \leq n$. ◊

We will return to notions related to higher connectivity (n-connectivity for $n \geq 2$) in a moment.

But first, we will recall some other basic properties of homotopy groups: In overview, we will recall some earlier facts about homology groups—and then indicate their corresponding extensions to homotopy groups.

We saw in Remark 5.1.15 that a homotopy equivalence preserves homology groups; similarly, it is standard (e.g. [**Spa81**, 7.6.20], or cf. Lemma 9.1.2) that

(9.1.4) a homotopy equivalence also preserves homotopy groups.

We mention that the converse is not true in general: that is, a *weak* (homotopy) equivalence—namely a map preserving homotopy groups—does not necessarily induce a homotopy equivalence. However, for CW-complexes (and hence for simplicial complexes), the two kinds of maps are in fact the same, via a standard theorem of Whitehead—see e.g. [**Spa81**, 7.6.24].

We saw at Lemma 5.2.2 that contractibility of K is sufficient for vanishing of its homology groups; similarly contractibility suffices (e.g. [**Spa81**, 1.8.5]) for triviality of homotopy groups:

LEMMA 9.1.5. *If K is contractible, then $\pi_n(K) = 1$ for all n. (Hence K is n-connected.)*

[1] In fact, relative to some fixed basepoint—which we omit here, but cf. [**Ben91**, 1.2.2] or [**Spa81**, p 43].

Furthermore the condition of trivial homotopy groups in turn essentially implies vanishing of homology groups; for the standard Hurewicz theorems, see e.g. [**Spa81**, Secs 7.4–7.5] or [**Ben91**, 1.3.6]. For example, when we regard the homotopy groups additively:[2]

LEMMA 9.1.6. *If* $\pi_i(K) \otimes \mathbb{F}_p = 0$ *for all* $0 < i < n$, *then also* $H_i(K; \mathbb{F}_p) = 0$ *for all* $0 < i < n$.

Thus we can use the triviality of homotopy groups as a sufficient condition for vanishing of homology—in the spirit e.g. of the hierarchy of conditions related to contractibility in Lemma 5.2.3.

Higher connectivity and sphericality. We now consider more specific conditions, related to the situation where our complex K might have structure determined by sphere subcomplexes—for example, where K of dimension d might have the homotopy type of a bouquet of d-spheres.

For a single d-sphere S^d, the homotopy groups are roughly as elementary as the homology groups in earlier (8.1.7): intuitively, puncturing the sphere at a point produces a contractible subcomplex—so that homotopy groups in dimension below d are trivial (e.g. Lemma 9.1.5). In particular

(9.1.7) $$S^d \text{ is } (d-1)\text{-connected.}$$

This latter property then holds also for a bouquet of such spheres; and this motivates the definition of sphericality (cf. [**Qui78**, Sec 8]):

REMARK 9.1.8 (d-spherical complexes). We say that a simplicial complex K of dimension d is *d-spherical* (or just *spherical*) if it is $(d-1)$-connected. ♢

We make a few elementary observations; several in fact refer to the Cohen-Macaulay property—we had informally mentioned this earlier (and we indicate it more formally at Remark 9.1.9 just below). Notice that a 0-dimensional complex is vacuously Cohen-Macaulay, as links of vertices are empty; while for a 1-dimensional complex, the only requirement for Cohen-Macaulayness is path-connectivity. And by convention, we regard the empty complex as (-1)-spherical. By Lemma 1.6.13, a contractible complex K of dimension d is in fact d-spherical in the above sense. (These remarks are of no real depth—but they turn out to be useful in later inductive situations.)

In fact d-sphericality is of more substantial interest when the structure of a complex K of dimension d is actually determined by spheres of dimension d (so in particular, homology and homotopy groups in dimension d are nontrivial).

For example, the Solomon-Tits collapsibility argument in the proof of Theorem 3.4.15 essentially shows that a building of dimension d is in fact d-spherical.

Indeed since in buildings, the links of simplices are also buildings of smaller dimension (as we saw in Remark 2.2.29), we see that the building even has an inductive "local" sphericality property. Thus buildings provide an instance of the Cohen-Macaulay (CM) property—which was developed in connection with the standard notion of Cohen-Macaulay rings in commutative algebra and algebraic geometry

[2]The groups are in fact abelian for $n \geq 2$; while for $n = 1$, $H_1(K)$ is the maximal abelian quotient of $\pi_1(K)$. (Cf. [**Ben91**, 1.2.3] and [**Spa81**, Exer 4.H.3].)

(cf. [**Ben91**, 5.4.9]). Indeed as our posets are always finite, (homology) Cohen-Macaulayness holds when the *Stanley-Reisner ring* which is associated to the poset (e.g. as in [**Bjö95**, 11.5]) is Cohen-Macaulay as a ring.

REMARK 9.1.9 (Cohen-Macaulay complexes). The more standard definition in the combinatorial literature (e.g. [**Sta86**, p 123]) is in terms just of vanishing of homology groups: We say that a complex K of dimension d has the *(homology) Cohen-Macaulay* property if $\tilde{H}_j(K) = 0$ for all $j < d$; and if for each simplex $\sigma \in K$ of dimension i, we have $\tilde{H}_j\big(\mathrm{Lk}_K(\sigma)\big) = 0$ for $j < d-i-1$. (Coefficients for homology are assumed to be over $\mathbb{Z}$, unless we indicate a specific coefficient ring R.)

However Quillen in [**Qui78**, Sec 8]; uses the following stronger (cf. 9.1.6) formulation in terms of triviality of homotopy groups (though he also warns that the stronger property is not necessarily invariant even under homeomorphisms): We say that a simplical complex K of dimension d has the *homotopy* Cohen-Macaulay property if it is d-spherical, and for each simplex σ of dimension i, the link $\mathrm{Lk}_K(\sigma)$ is $(d-i-1)$-spherical. ♢

We will be discussing some of the literature on these and other higher connectivity conditions in Section 9.4. But we also mention here beforehand that (much as for the hierarchy in Lemma 5.2.3 of conditions related to contractibility) there is a natural hierarchy of implications among various conditions related to sphericality. We now mention some of these, as they are reasonably prominent in the relevant combinatorial literature (though our later treatment will refer only occasionally to them).

REMARK 9.1.10 (A hierarchy of conditions related to sphericality). We first mention informally some implications among conditions related just to d-sphericality for a complex of dimension d; these are indicated after [**Bjö95**, 11.2]:

vertex-decomposable $\Longrightarrow$ shellable $\Longrightarrow$ constructible $\Longrightarrow$ d-spherical.

See [**Bjö95**] for the conditions of vertex-decomposability and constructibility—which we will not follow up here. However, we will give some discussion of shellability later, starting at Remark 9.4.8.

Now it turns out that the three above conditions on the left imply not only sphericality, but the inductive local sphericality of the Cohen-Macaulay condition; e.g. after (11.5) in [**Bjö95**] we see (for any field k):

constructible $\Longrightarrow$ homotopy CM $\Longrightarrow$ CM / $\mathbb{Z}$ $\Longrightarrow$ CM / k $\Longrightarrow$ CM / $\mathbb{Q}$.

(Indeed the reader can find a discussion of various further conditions at that point in [**Bjö95**].) ♢

9.2. 0-connectivity: Disconnectedness of $\mathcal{S}_p(G)$ and strong p-embedding

We recall that triviality of $\pi_0(K)$ gives path-connectedness. (Since our space K is a simplicial complex, this amounts to just ordinary topological connectedness.) We briefly examine connectivity in this section.

In rough overview: When our K is a p-subgroup complex such as $\mathcal{A}_p(G)$, aside from certain "small" cases, we can usually expect connectedness. Correspondingly, here we will mainly investigate just the exceptional situation of disconnectedness.

This turns out to be related to some important purely group-theoretic conditions—which have been independently studied, and indeed are even fairly well understood.

Example: p-subgroup complexes of dimension 0. We consider first the special case where our complex K of p-subgroups has dimension 0. For example, if a Sylow p-subgroup P of G is of order exactly p, then $\mathcal{S}_p(G)$ will be 0-dimensional.

And since a complex of dimension 0 is a set of isolated vertices, here disconnectedness is automatic—except when the complex happens to consist of just one point: which corresponds to a normal p-subgroup. So in the remainder of this discussion of disconnectedness, we will assume that $O_p(G) = 1$ (in view of the usual Quillen contractibility result 3.3.14).

For the elementary subposet $\mathcal{A}_p(G)$, we see (recalling Definition 3.3.9):

$$\dim\big(\mathcal{A}_p(G)\big) = 0 \iff G \text{ has } p\text{-rank } m_p(G) = 1. \tag{9.2.1}$$

Here the Sylow group P has a unique subgroup of order p; and this strong group-theoretic restriction guarantees (e.g. [**Gor80**, 4.10(ii)]) that P is cyclic—except when $p = 2$, in which case P could also be a generalized quaternion group.

And in classical finite group theory, it has long been a topic of independent interest to determine groups G with a Sylow group P of p-rank 1. Indeed the case $p = 2$ provided an important first step in the Classification of Finite Simple groups (e.g. [**ALSS11**, 1.4.1, 1.4.2]); while for p odd, the determination emerged only from the full CFSG. In fact p-rank 1 was part of the treatment of the more general strongly p-embedded situation, which we will discuss next.

Disconnectedness and strong embedding. Away from the rank-1 special case in (9.2.1), we have p-rank $m_p(G) \geq 2$—so that $\mathcal{A}_p(G)$ and $\mathcal{S}_p(G)$ have dimension at least 1.

And here we can "usually" expect connectedness. As partial motivation for this expectation, we indicate connectivity for a standard possible subconfiguration of top dimension: Since we are assuming $O_p(G) = 1$, we might expect to encounter the minimal-faithful situation LA occurring at the end of the reductions, in (6) of the proof of Theorem 8.2.9: There for A of rank s, the final product of terms L_iA_i gives a join of subsets $\mathcal{A}_p(L_iA_i)$; resulting in a complete multipartite graph with s parts—which is connected since we may take $s = m_p(G) \geq 2$.

Indeed when we approach disconnectedness more formally, we find it can arise only in a rather restricted way—which generalizes a certain aspect of the dimension 0 situation above:

First consider the situation where the Sylow group P of G has prime order p. Then distinct conjugates of P must intersect trivially. More generally, any subgroup X of G whose distinct conjugates intersect trivially is called a *trivial intersection* set (or TI-set). And in fact TI-subgroups have many special properties, which have long been exploited in finite group theory (e.g. for Frobenius groups—cf. [**Asc00**, 35.22]).

A natural further generalization of the TI-condition is as follows: Assume that for P a Sylow p-subgroup of G, the intersection of P with its distinct conjugates has order which is not divisible by p. Then the normalizer $N_G(P)$ (proper in G as we are assuming $O_p(G) = 1$) is called (e.g. [**ALSS11**, A.1.2]) a *strongly p-embedded* *subgroup*. This notion provides the standard group-theoretic characterization of disconnectedness for $\mathcal{S}_p(G)$ (e.g. [**Qui78**, 5.2], or [**Asc00**, 46.6]):

(9.2.2) $\mathcal{S}_p(G)$ is disconnected $\Longleftrightarrow$ G has a strongly p-embedded subgroup.

(In particular this includes the case $m_p(G) = 1$ with $O_p(G) = 1$.)

The situation of strong embedding has considerable independent signficance in group theory. For example, in a conjugation family in Alperin's sense (e.g. [**Asc00**, Sec 38]) of affording control of p-fusion, the normalizers of the p-groups in the family can be taken to have disconnected quotients. (Quillen notes this in [**Qui78**, Sec 6], referring to an observation of Puig; and cf. also the earlier result of Goldschmidt in [**Gol70**].)

And strong embedding also provided a minimal situation whose treatment was fundamental in the Classification of the Finite Simple Groups. From the viewpoint of earlier Remark 8.2.4, we mention that strong embedding can occur in both solvable and simple groups: In the solvable case, we must in fact have p-rank $m_p(G) = 1$. In the simple case, the main examples (beyond the case $m_p(G) = 1$) are provided by G of Lie type in characteristic p, with Lie rank 1: these are the groups $L_2(q)$, $U_3(q)$, Suzuki groups ${}^2B_2(2^{odd})$ (also denoted $Sz(2^{odd})$) and Ree groups ${}^2G_2(3^{odd})$. And thessee groups of rank 1 then further arise as quotients in minimal parabolic subgroups of larger-rank Lie-type groups (recall Remark 2.2.7). They also arise similarly in analogous "minimal parabolics" of sporadic simple groups (see e.g. [**RS84**]). And McBride introduced (cf. [**ALSS11**, 3.1.3]) an influential related notion of "abstract" minimal parabolics in an arbitrary finite group G.

The treatment of the strongly embedded situation began with the case $p = 2$: the determination of the possibilities for groups with strongly 2-embedded subgroups by Bender and Suzuki (e.g. [**GLS94**, 20.1]) was an early, and particularly fundamental, step in the overall CFSG. For odd p, the corresponding groups were first determined inductively in known groups (see [**GL83**, I.24-9(4)]); and then the final determination followed with the completion of the CFSG. (Indeed the case division in the Classification weakened the relevant minimal situation to a suitable notion of *almost* strongly p-embedded; and then the final step in the CFSG logic was Aschbacher's elimination (cf. [**ALSS11**, 8.0.2]) of the Uniqueness Case—showing that a 2-local subgroup in a sufficiently generic (and in particular connected) situation could not be almost strongly p-embedded for an odd prime p.)

Thus in view of these results, the restricted situation in (9.2.2) where $\mathcal{S}_p(G)$ is disconnected is reasonably well understood from the group-theoretic viewpoint.

(We mention that connected components for the proper part of the subgroup lattice $L(G)$ are considered in Lucido [**Luc03**]; and for some other posets in Ramras [**Ram05**].)

9.3. 1-connectivity: Simple connectivity (and its failure) for $\mathcal{A}_p(G)$

In this section, we indicate some of the literature related to simple connectivity for suitable complexes K of p-subgroups; we recall that simple connectivity corresponds to $\pi_1(K) = 1$. (Indeed we are mainly interested in 1-connectivity, which further entails $\pi_0(K) = 1$, so that K is connected.)

In fact we will first restrict attention to the case $K = \mathcal{A}_p(G)$. We make some initial remarks, which are much in the spirit of those at the start of the previous Section 9.2: Since the disconnected case was described in (9.2.2), we may as well assume that $\mathcal{A}_p(G)$ is connected.

And as contractible spaces are automatically simply connected (e.g. Lemma 9.1.5),

we may as well assume that $O_p(G) = 1$,

And as before, aside from small cases, the usual expectation will be 1-connectivity for $\mathcal{A}_p(G)$. So in this section, we will again largely focus on the exceptional situation, where $\mathcal{A}_p(G)$ is a proper quotient of some universal cover. (But we will also indicate some relations of the more standard situation of simple connectivity with group-theoretic considerations, such as uniqueness proofs.)

The case where $\mathcal{A}_p(G)$ has dimension 0, namely $m_p(G) = 1$ as in earlier (9.2.1), is now ruled out—using our hypotheses above of 0-connectedness and $O_p(G) = 1$. But we can also see independently that the dimension-0 situation is really irrelevant: For dimension 0 implies that links of vertices are empty—so that there is no notion of local isomorphism for a covering; or alternatively, there are no paths, and in particular no loops to be tested for contracting.

Thus the first situation of any interest is where $m_p(G) = 2$, so that $\mathcal{A}_p(G)$ is of dimension 1—and so the complex is just a (connected) graph. If there are no loops, again $\pi_1\big(\mathcal{A}_p(G)\big) = 1$ vacuously; but then the graph is just a tree, and in particular contractible—contrary to our assumption of $O_p(G) = 1$ (since Quillen's Conjecture holds when $m_p(G) = 2$ by Theorem 8.4.2.i). On the other hand, if there are loops—then dimension 1 still implies that there are no disks to fill out the "interior" of the loop. Consequently such loops have covers given by affine lines; and the fundamental group $\pi_1(K)$ is infinite, and in particular nontrivial. So we see that failure of simple connectivity includes the connected cases where $m_p(G) = 2$ with $O_p(G) = 1$.

Hence the first situation of really substantial interest for simple connectivity is $\mathcal{A}_p(G)$ of dimension ≥ 2, so that $m_p(G) \geq 3$. Here the expectation will be that K is simply connected; and as in the previous section, we provide partial motivation via possible subconfigurations of top dimension which are simply connected: Again since $O_p(G) = 1$, we might expect to encounter the minimal-faithful situation LA in (6) at the end of the proof of Theorem 8.2.9: which gives a join of s subsets $\mathcal{A}_p(L_iA_i)$ of dimension 0. We may take $s = m_p(A) \geq 3$; and it is standard that this join-subconfiguration is simply connected—indeed it is a bouquet of $(s-1)$-spheres. From a slightly different viewpoint, note that each set $\mathcal{A}_p(L_iA_i)$, consisting of isolated vertices, is trivially a bouquet of 0-spheres—and so is vacuously 0-spherical. Hence—e.g. by the standard observation in [**Qui78**, Ex 8.1], roughly generalizing the multiplicativity in earlier (8.1.1) from homology to the homotopy groups of a join—the subconfiguration is $(s-1)$-spherical; that is, it is $(s-2)$-connected with $(s-2) \geq 1$.

Exceptions to simple connectivity arising from simple groups. In fact the situation for 1-connectivity is not quite as neatly characterized as 0-connectivity is in (9.2.2). But in that direction, we will now summarize the approach of Aschbacher in [**Asc93**]: which for $m_p(G) \geq 3$ shows—modulo a Conjecture for which there is considerable evidence—that exceptions to simple connectivity are confined to cases arising essentially from some *simple* groups. (And we emphasize that a number of such exceptions do arise; we will be indicating some of these a little later on, e.g. at Example 9.3.11.)

We mention that Aschbacher does not in fact work directly with $\mathcal{A}_p(G)$, but instead with the commuting complex $C_p(G)$ on subgroups of order exactly p, which we had indicated in earlier Remark 4.5.3.3. We observed there the equivalence of this complex with $\mathcal{A}_p(G)$; and so for convenience, we will instead use $\mathcal{A}_p(G)$ in stating Aschbacher's results below.

We note first that certain conditions turn out to be *necessary* for simple connectivity: Namely under the hypotheses indicated above, Aschbacher shows that in addition to $m_p(G) \geq 3$ as above, a further "local" connectivity condition is also required:

THEOREM 9.3.1 (Thm 2 in [**Asc93**]). *Assume that $O_p(G) = 1$ and $\mathcal{A}_p(G)$ is connected. If $\mathcal{A}_p(G)$ is also simply connected, then:*
(a) $m_p(G) \geq 3$, and
(b) for all A of order p, $\mathcal{A}_p(G)_{>A}$ is connected.

Aschbacher's further work then in effect shows conversely that (a) and (b) are "almost" sufficient for simple connectivity; this is [**Asc93**, Thm 1], which we indicate as Theorem 9.3.4 below.

The solvable case. We continue to follow the viewpoint of Remark 8.2.4, in which solvable groups are considered first. In fact Aschbacher's treatment does not separate out the solvable subcase; but here we will briefly summarize how to extract, from his argument for general G, a somewhat more elementary proof of the sufficiency of (a) and (b) for simple connectivity, under the assumption that G is solvable.

First observe that no solvable groups appear among the exceptions in Theorem 9.3.4 below. And correspondingly in the proof of [**Asc93**, Thm 1], any solvable candidates for exceptions are in fact eliminated by the initial reductions in the proof, to the case where $O_{p'}(G) = 1$ and $O_p(G) = 1$. We had of course already indicated, at the start of the section, the reduction (via Lemma 9.1.5) to $O_p(G) = 1$.

But the reduction to $O_{p'}(G) = 1$ is deeper; it requires an earlier result [**Asc93**, 6.5] relating $\mathcal{A}_p(G)$ with $\mathcal{A}_p\big(G/O_{p'}(G)\big)$. So we now summarize how to check, in the proof of Aschbacher's 6.5, that Conjecture 9.3.3 is not in fact needed for the case of solvable G: In the minimal situation arising there, G is of the usual form LA, where $L := O_{p'}(G)$. Aschbacher uses essentially the inductive[3] approach of Quillen, which we will indicate in the proof of later Theorem 9.4.5. The only part of the argument requiring comment for our present purpose is the base step (which corresponds to having the minimal normal subgroup "K" in the proof of 9.4.5 equal to L), where A is irreducible[4] on L; and here, solvability of G implies that L is an elementary q-group for $q \neq p$. In particular, L cannot contain a product of simple p'-groups—which is the only configuration which requires the Conjecture. So this completes the analysis of Aschbacher's argument required to cover the solvable subcase.

---

[3]In particular in the inductive step, Aschbacher uses his preliminary lemma [**Asc93**, 1.4] as the appropriate analogue of Quillen's [**Qui78**, 9.7]. (These are fiber-type results, in the spirit of earlier Remark 4.2.3.)

[4]We mention that Aschbacher's proof implicitly uses this irreducibility to invoke Schur's Lemma, e.g. (8.2.10)—applied to the Sylow A of $LA = G$, which is noncyclic by the hypothesis that $\mathcal{A}_p(G)_{>B}$ is connected for B of order p (i.e. condition (b) of Theorem 9.3.1).

The following example illustrates the necessity of condition 9.3.1(b) for simple connectivity (and it will also be useful in the subsequent Section 9.4 for demonstrating a case of failure of Cohen-Macaulayness):

EXAMPLE 9.3.2 (An exceptional solvable configuration). We take $p = 3$; and we begin by constructing a suitable 3-group P (which later will be Sylow in our larger solvable group G). We use the wreath product $\mathbb{Z}_3$ wr $\mathbb{Z}_3$: so that $P = AQ$ with A elementary of rank 3, and Q of order 3 permuting a set of generators of A. Then $Z(P) = C_A(Q)$ is of order 3; and $B := Z(P)Q$ is of rank 2 and self-centralizing in P, and in particular is maximal under inclusion in $\mathcal{A}_3(P)$.

In fact Aschbacher's [**Asc93**, 6.3] indicates (when (4a) there fails) the relevance of such a self-centralizing B of rank 2 to the failure of condition 9.3.1(b). (But we also emphasize that such maximal B of rank 2 are of considerable independent significance in group theory: as the obstruction to the connectedness of the commuting graph on elementary subgroups of rank 2, and hence to the construction of signalizer functors—see e.g. [**ALSS11**, B.4.8].)

Of course $\mathcal{A}_3(P)$ is contractible by Lemma 3.3.14, and so it is simply connected by Lemma 9.1.5; hence to demonstrate failure of simple connectivity, we need a group G with $O_3(G) = 1$. So we can take $G := LP$, where L is a $3'$-group admitting faithful action by P (recall (8.2.8)); for specificity, we will take L to be an elementary 7-group of rank 3 (though we might take say an elementary 2-group of rank 6).

We can now apply Quillen's Theorem 8.2.5: since both A and B (of ranks 2 and 3) are maximal under inclusion in $\mathcal{A}_3(G)$, we see that $\tilde{H}_1(\mathcal{A}_3(G)) \neq 0$ and also $\tilde{H}_2(\mathcal{A}_3(G)) \neq 0$.

We have condition 9.3.1(a) since $m_3(P) = 3$. But as $H_1(\mathcal{A}_3(LP)) \neq 0$, where H_1 is the abelian quotient of π_1 (e.g. [**Spa81**, Exer 4.H.3]), we must in fact have $\pi_1(\mathcal{A}_3(LP)) \neq 1$, so that $\mathcal{A}_3(LP)$ is not simply connected. And correspondingly condition 9.3.1(b) fails: namely $\mathcal{A}_3(LP)_{>Z(P)}$ is disconnected—for it is the disjoint union of conjugates of B, along with connected components containing A and its rank-2 subgroups over $Z(P)$. ♢

The minimal nonsolvable configuration. We now turn to the troublesome configuration in Aschbacher's inductive argument for his 6.5, which requires the Conjecture for general G—or even for p-solvable G. Here $L = F^*(G)$ is the product of an A-orbit determined by a component K, which has order *coprime* to p; this is fairly similar to the minimal nonsolvable situation we saw in the proof of p-solvable $\mathcal{QD}_p$ in Theorem 8.2.12.1. Correspondingly Aschbacher formulates his Conjecture:

CONJECTURE 9.3.3 (see [**Asc93**]). *Assume that $G = F^*(G)A$, where A is an elementary p-subgroup of rank ≥ 3, and $F^*(G)$ is the direct product the A-conjugates of a simple p'-group K.*

Then $\mathcal{A}_p(G)$ should be simply connected.

The obvious approach to the Conjecture of course involves studying the individual possibilities for the simple group K in the CFSG. And indeed Aschbacher in [**Asc93**, Thm 3] treated all K except for Lie type groups of rank 1 and some sporadic groups; while later Segev in [**Seg94**] dealt with many of the Lie rank 1 cases. However,

I believe that the remaining cases have resisted treatment, so that the Conjecture remains open in general.

The general case. Inductive arguments using the Conjecture (and also involving the general uniqueness theory developed by Aschbacher and Segev in e.g. [**AS92b**]) lead to the following result—which essentially reduces the failure of simple connectivity to certain configurations involving simple groups. (Below we simplify the statement, by omitting some details in Aschbacher's conclusions.)

THEOREM 9.3.4 (Thm 1 in [**Asc93**]). *Assume that Conjecture* 9.3.3 *holds. If $\mathcal{A}_p(G)_{>A}$ is connected for all A of order p, then either $\mathcal{A}_p(G)$ is simply connected, or possibly one of the following holds for $\bar{G} := G/O_{p'}(G)$:*

(1) $F^*(\bar{G})$ is the product of a pair of simple groups, each of which contains a strongly p-embedded subgroup.

(2) $F^*(\bar{G})$ is a single simple group.

We expand on several aspects of this statement:

First observe that while the hypothesis above includes condition (b) of earlier Theorem 9.3.1, it does not mention condition (a) there, namely $m_p(G) \geq 3$; so as stated, some cases with $m_p(G) = 2$ and $O_p(G) = 1$ also appear in conclusions (1) and (2).

But we could just as well add (a) to the hypothesis: since Theorem 9.3.1 shows that $m_p(G) \geq 3$ is necessary for simple connectivity, away from the restricted situation of $\mathcal{A}_p(G)$ disconnected; where we saw in (9.2.2) we saw that disconnectedness in turn requires a strongly p-embedded subgroup—which can be described using the CFSG.

Of course that same description of the strongly embedded situation can also be used to determine the possibilities for the product of two components in conclusion (1) above.

So attention is focused on which of the simple possiblities in conclusion (2) give actual exceptions. Here for $m_p(G) \geq 3$, many simple groups G are known to have $\mathcal{A}_p(G)$ simply connected—while comparatively few are known to fail simple connectivity. But as far as I know, there has been no methodical treatment of all simple groups. So in the remainder of the section, we will indicate some examples, both of simple connectivity and of its failure.

Indeed we first mention in particular that Sections 7 and 8 of [**Asc93**] treat simple connectivity for various simple groups—for example determining when failure occurs because of disconnectedness of some $\mathcal{A}_p(G)_{>A}$.

(Option B:) Simple connectivity for Lie type groups via buildings. We consider the case of G simple of Lie type in characteristic p. From Quillen's equivalence in Theorem 4.4.8, $\mathcal{A}_p(G)$ is homotopy equivalent to the Tits building K. And indeed it is the Lie rank of G (and hence the geometric rank of K), rather than the p-rank $m_p(G)$,[5] that is actually relevant for homotopy groups. Thus for simple connectivity, we will focus on K of rank at least 3, and Tits's corresponding theorem.

However we first digress briefly on smaller Lie rank: We had mentioned in the previous Section 9.2 that groups of Lie rank 1 exhibit strongly p-embedded

[5] which of course is a multiple of a, where F_{p^a} is the field of definition of G—and so can be arbitrary large for a fixed Lie rank.

subgroups, so that $\mathcal{A}_p(G)$ is in fact disconnected. Similarly for groups G of Lie rank 2, we have K, and hence $\mathcal{A}_p(G)$, not simply connected. (We mention that it is essentially this failure of simple connectivity which requires special methods for rank 2 in finite group theory (and especially the CFSG): for example, the theory of *amalgams*; and the Fong-Seitz and Tits-Weiss approaches to characterizing G, which augment the building structure K with some additional hypotheses—see e.g. [**AS04a**, Ch F].)

Simple connectivity for finite buildings of rank ≥ 3. The fundamental work of Tits in [**Tit74**] gave a definition of buildings which is independent of particular simple groups; a main result was then to characterize finite buildings of rank $r \geq 3$ (and hence dimension $(r-1)$) as those naturally arising from Lie type groups.

In particular, he showed that a finite building K is necessarily *spherical*—namely its apartment is a triangulation of an $(r-1)$-sphere (recall our discussion around (2.2.21)). And of course when $(r-1) \geq 2$, that apartment is then simply connected via (9.1.7).

This property of the apartment in fact extends to the global geometry of K: Namely Tits's characterization in [**Tit74**, 11.4] (which assumes irreducibility, that is, having a connected Dynkin diagram) in particular implies:

(9.3.5) A finite irreducible building K of rank $r \geq 3$ is simply connected.

The simple connectivity can also be explicitly deduced from the property that K has the homotopy type of a bouquet of $(r-1)$-spheres, obtained in the Solomon-Tits argument in Theorem 3.4.15.

The relevance of simple connectivity of certain inifinite buildings. We now expand somewhat on a theme which we had briefly mentioned at earlier Remark 2.3.9: Namely it turns out that for certain geometries K for some simple groups G, K is not simply connected—and the universal cover of K is in fact given by a suitable infinite building—corresponding to an *affine* Dynkin diagram (where the apartment is not a sphere, but instead a triangulation of an infinite Euclidean space).

Here we can give only a brief sketch of the underlying general theory due to Tits; the reader interested in fuller details may wish to consult the indicated papers of Tits, or [**Ron89b**, Ch 4].

The general background for our discussion comes from Tits's later "local approach" to buildings in [**Tit81**]. That approach does not really involve e.g. apartments;[6] instead, the overall geometric context is that of diagram geometries: roughly as we indicated in Definition 2.3.4—but assuming more specifically as in Definition 2.2.34 that rank-2 links are always generalized polygons; Tits called this class the *geometries of type M*. (But nowadays most other authors use the term *Tits geometries*.) Now much as for the finite case above, away from small rank we have simple connectivity for buildings—and a main result [**Tit81**, Thm 1] shows that this property essentially characterizes the subclass of buildings within the larger class of geometries. For simplicity, we will give here only a very informal statement:

THEOREM 9.3.6 (Tits). *Buildings of rank* ≥ 3 *are (roughly) the simply connected geometries of type M.*

---

[6] Tits explains in the introduction of [**Tit81**] that he is actually returning to a view of buildings which predated the definition and axioms in his earlier work [**Tit74**].

And of course it follows from this that:

(9.3.7)

The universal cover of a geometry of type M of rank ≥ 3 is (usually) a building.

Here the parenthetical qualifications "roughly" and "usually" indicate that there might be at least some exceptions: Tits's arguments in fact required the further assumption that any subgeometries of types C_3 or H_3 should have buildings as their universal covers. And indeed there is at least one actual exception corresponding to this hypothesis; as we will see at Theorem 9.3.12, it turns out that the C_3-geometry for A_7 of Neumaier in Example 2.3.7 is simply connected (hence equal to its own universal cover)—without being a building.

Next there is some background which is more specific to the affine subcase of buildings: First, Bruhat and Tits gave a construction of certain affine buildings, which "come from" algebraic groups; see e.g. [**Ron89b**, Ch 9–10] or [**AB08**, Ch 11] for details. Very roughly: Given a spherical diagram D, and a local field F,[7] let $\tilde{G}$ denote the corresponding algebraic group over F—which has a spherical building of type D. Then the Bruhat-Tits construction produces for $\tilde{G}$ a building whose type is the affine extension $\tilde{D}$, and whose field of definition is the residue field $\bar{F}$ determined by F. Later, Tits in [**Tit86**, Cor, p 175] gave a corresponding classification, which essentially shows conversely that most affine buildings must in fact be of that particular type:

THEOREM 9.3.8. *Affine buildings of rank ≥ 4 (usually) come from algebraic groups.*

We make just a few comments on the proof: Since the proper subdiagrams of an affine diagram are spherical, Tits can make inductive use of his classification of spherical buildings (underlying Theorem 9.3.6). And given K of affine type $\tilde{D}$, he constructs a corresponding spherical building K_∞ "at infinity" of the spherical type D—and shows that this is the usual building for a suitable algebraic group $\tilde{G}$ as above.

We will continue this theme in a moment, when we describe some finite geometries covered by affine buildings in Kantor's Theorem 9.3.10.

(Option S:) Simple connectivity for some sporadic groups and geometries. We now turn to simple connectivity for other cases of simple groups.

Of course the buildings above are associated with $\mathcal{A}_p(G)$ when G is of Lie type in the "same" characteristic p. For G of Lie type in characteristic $q \neq p$, simple connectivity of $\mathcal{A}_p(G)$ has been discussed for some cases in papers of K. M. Das (e.g. [**Das95**] [**Das98**] [**Das00**]). For alternating groups, and more generally for symmetric groups, simple connectivity of $\mathcal{A}_p(G)$ is discussed in papers of R. Ksontini (e.g. [**Kso04**]). (The complexes for symmetric groups are further studied in Shareshian-Wachs [**SW09**].)

Here we will focus mainly on sporadic groups; but as in earlier Section 2.3, we will more generally consider sporadic geometries, including unexpected non-building geometries for some Lie-type groups.

[7] defined via a discrete valuation—e.g. if F is the p-adics $\mathbb{Q}_p$, then the residue field $\bar{F}$ is $\mathbb{F}_p$.

Some finite sporadic geometries covered by infinite affine buildings. In fact it will be convenient to begin with certain sporadic geometries in characteristic 2, many of them coming unexpectedly from groups of Lie type in odd characteristic. A further unexpected feature is that these particular finite geometries have affine diagrams, which usually correspond instead to infinite groups. And indeed it turns out that these geometries K admit proper covers, corresponding to buildings for infinite groups; so the nontriviality of $\pi_1(K)$ exhibits failure of simple connectivity—where in at least some cases, K is equivalent to $\mathcal{A}_2(G)$.

A number of these affine geometries were found within a relatively brief period during the early 1980s; we summarize the discoveries in a table. The first column indicates the affine type of a geometry K, with flag-transitive action by the group G in the second column. As in our discussion before Remark 2.3.9, the tilde in the type indicates the affine extension of the corresponding type without the tilde:

(9.3.9)

Type	Group	Reference
$\tilde{A}_2$	A_7	Ronan [**Ron84**], Köhler-Meixner-Wester [**KMW84**]
$\tilde{C}_2$	$\Omega_6^-(3)$	Kantor [**Kan81**], Ronan-Smith [**RS80**]
$\tilde{C}_2$	$U_6(2)$	Kantor [**Kan81**]
$\tilde{C}_2$	Suz	Ronan-Smith [**RS80**]
$\tilde{G}_2$	Ly	Kantor [**Kan81**]
$\tilde{G}_2$	$G_2(3)$	Cooperstein, see [**AS83**]
$\tilde{A}_3$	$\Omega_6^-(3)$	Smith [**AS83**]
$\tilde{B}_3$	$\Omega_7(3)$	Aschbacher [**AS83**]
$\tilde{D}_4$	$\Omega_8^+(3)$	Aschbacher [**AS83**]

Soon thereafter Kantor, who had discovered some of the examples, gave an elegant explanation of most of the examples—in terms of Tits's results indicated just above: For note that by (9.3.7), the universal cover of each K in Table (9.3.9) should be a building—indeed of the indicated affine type; and by Theorem 9.3.8, the building comes from an algebraic group. In fact Kantor showed that the relevant algebraic group is defined over the 2-adics; below we state the special case for characteristic 2 of a more general classification theorem:

THEOREM 9.3.10 (Kantor [**Kan85**]). *For the Lie types $T := C_2, G_2, A_3, B_3, D_4$, the group $\tilde{G}$ of spherical type T defined over the 2-adics $\mathbb{Q}_2$ has a discrete subgroup $\tilde{G}_{\mathbb{Z}}$, which acts flag-transitively on the affine building of type $\tilde{T}$ for $\tilde{G}$.*

In particular for each odd p, there is a surjection of $\tilde{G}_{\mathbb{Z}}$ on the finite group $G(p)$ of type T defined over $\mathbb{F}_p$; so that $G(p)$ has a finite geometry of type $\tilde{T}$.

Thus Kantor's result, in the special case $p = 3$, gives the universal cover of the geometries K in Table (9.3.9) for which G is a Lie type group over $\mathbb{F}_3$. (And in addition it gives similar examples for other odd p.)

Similar methods were used to treat the $\tilde{C}_2$ geometry for $G = U_6(2)$; and were also applied in the case for $G = A_7$ in the cited papers [**Ron84**] and [**KMW84**].

So the main cases from Table (9.3.9) where the universal cover is not understood are those for the sporadic groups Suz and Ly (which have characteristic 2 and 5, respectively). Here as we saw, the cover must be some building; but little seems to be known about the nature of those buildings. Presumably they would arise from some kind of "free construction" of buildings, as in the work of Ronan-Tits [**RT87**].

And of course in all these cases, the proper universal cover of K shows that K is not simply connected. So to return to our main theme of simple connectivity for $\mathcal{A}_p(G)$, we observe:

EXAMPLE 9.3.11 (Failure of simple connectivity for $\mathcal{A}_2(U_4(3))$). Let K denote the $\tilde{C}_2$ geometry for $G = U_4(3)$, as in Example 2.3.10. Thus K appears in Table (9.3.9) for $\Omega_6^-(3) \cong U_4(3)$. Furthermore as we had mentioned in Remark 4.4.12, K is homotopy equivalent to $\mathcal{A}_2(G)$. Thus we conclude that $\mathcal{A}_2(U_4(3))$ in not simply connected—whereas we might have expected simple connectivity here, since $m_2(U_4(3)) = 4$ (indeed we had seen rank-4 subgroups given by left and right quads, in our discussion leading up to Example 2.3.10.) ♢

However, equivalence with $\mathcal{A}_2(G)$ does not hold for the other geometries K indicated in Table (9.3.9); I am not sure if simple connectivity of $\mathcal{A}_2(G)$ has been checked for the corresponding groups G. (It fails for example for $G = G_2(3)$—since [**SY97**] shows that $\mathcal{A}_2(G)$ is equivalent to a rank-2 truncation of the K in Table (9.3.9), where that rank-2 geometry as usual has a proper universal cover; whereas we might have expected simple connectivity since $m_2(G_2(3)) = 3$.)

We mention that some 3-adic analogues of Kantor's 2-adic work can be found in [**KMW90**]. The general area is described further in [**Kan86**] and [**KLT87**].

Simple connectivity for various sporadic geometries. We turn now to some sporadic geometries where simple connectivity has been established. Of course this is the situation we more usually expect when $m_p(G) \geq 3$.

However we digress briefly to consider a case with $m_2(G) = 2$, so that certainly $\mathcal{A}_2(G)$ is not simply connected, namely $G = A_7$: Here, for our frequent example of a sporadic geometry given by the Neumaier C_3-geometry K for A_7 of Example 2.3.7, we had already indicated at Example 6.3.7 that K is *not* equivalent to $\mathcal{A}_2(G)$. So the issue of simple connectivity for K is not answered by its failure for $\mathcal{A}_2(G)$; and indeed Ronan showed:

THEOREM 9.3.12 (Ex 2 in Sec 4.4 of [**Ron89b**]). *The C_3-geometry for A_7 is simply connected.*

Ronan's proof in fact shows directly that any circuit can be homotopically shrunk to a point.

The literature contains a fair number of simple connectivity proofs for various sporadic geometries; below I will just indicate a sampling of some results known to me.

We had mentioned after Definition 2.3.4 the geometries studied by Buekenhout in [**Bue79**], involving "circle" subgeometries. Ronan in [**Ron81**, Table 1] studied many of these geometries, and we summarize his results: He obtained simple connectivity for the Buekenhout geometries[8] of:

$$M_{11}, M_{12}, M_{22}, M_{23}, M_{24}, Suz, Co_1, F_{22}, F_{23}, HS, McL.$$

By contrast, he found non-trivial fundamental groups (in fact finite, and of orders 2, 2, 3) for:

$$Co_2, Co_3, F'_{24}.$$

[8]where the relevant geometry for Suz is the case with parameters 1, 9, 3 in Buekenhout.

Infinite universal covers occur[9] for:

$$J_2, Suz, Ru, He.$$

Usually these Buekenhout geometries are not equivalent to $\mathcal{A}_p(G)$ for any p; so these results need not give information about simple connectivity of $\mathcal{A}_p(G)$.

Ronan also in [**Ron81**, Table 1] establishes simple connectivity in one further case for $G = M_{24}$, for a geometry not taken from Buekenhout's examples; instead it is the 2-local geometry for M_{24}, viewed in the minimal-parabolic version given in [**RS84**]—recall we indicated this geometry K at Example 2.3.11. Now K is equivalent to $\mathcal{A}_2(G)$, as we had observed in Remark 4.4.12; so we see that $\mathcal{A}_2(G)$ is indeed simply connected, as we might have expected from $m_2(G) = 6$.

In fact this geometry for M_{24} is the first in the sequence of "tilde geometries" (in effect built up using the $\Box$/truncation indicated in our discussion leading up to Example 2.3.11). These geometries, along with the "Petersen-geometries" (in effect built up using the triples geometry for S_5 in Example 2.3.5), were studied methodically by Ivanov and Shpectorov; their results are summarized in [**IS94a**, Table 1]. In particular, that list includes simple connectivity of the tilde geometries for:

$$M_{24}, He, Co_1, M.$$

As for Petersen geometries, they obtain simple connectivity for the cases:

$$M_{23}, J_4;$$

while the geometries for:

$$M_{22}, Co_2, B$$

are in fact finitely covered by group extensions having the structures $3 \cdot M_{22}$ and $3^{23} \cdot Co_2$ and $3^{4371} \cdot B$. (These tilde and Petersen geometries are usually not equivalent to $\mathcal{A}_2(G)$.) We remark that the simple connectivity proof by Ivanov for the Monster M was the final step needed to establish the Conway-Norton conjecture, on a presentation of the "BiMonster" (i.e. the wreath product M wr 2) using the diagram they denote by Y_{555}; see [**Iva91**] for details.

We mention that coverings of complexes are also considered in the work of Alperin-Glauberman in [**AG94**].

Uniqueness proofs via simple connectivity. We conclude the section by noting that simple connectivity proofs of the above type can be useful in giving uniqueness proofs, especially for sporadic groups.

Indeed historically, many of the sporadic groups were first shown to be uniquely determined (by suitable natural properties) using a corresponding uniqueness proof for a graph on which the group acts. This method has been axiomatized and systematized by Aschbacher and Segev—a useful summary is given in [**Asc94**, Sec 17]; indeed we have already mentioned this approach in our discussion leading up to Theorem 9.3.4.

In essence, their proofs proceed via consideration of simple connectivity of the clique complex of the graphs; the method was originally developed in [**AS92a**]. In the later chapters of the Aschbacher book [**Asc94**], the approach is used to give new uniqueness proofs in the standardized approach for the groups:

---

[9]where this time the relevant geometry for Suz is the case with parameters $1, 4, 4$ in Buekenhout.

$$L_5(2), M_{24}, He, U_4(3), J_2, Suz, Co_1.$$

The method has also been influential in the subsequent literature.

We also emphasize that uniqueness proofs (and hence the underlying simple connectivity) have been important historically for their use in "recognition" theorems—for example, as the last step in a classification problem: to identify a target group, once sufficient internal structure has been established. Perhaps the most well-known case is given by the Curtis-Tits presentation for groups of Lie type, with the variant due to Phan for unitary groups; see for example the discussion in [**ALSS11**, Sec A.5]. Indeed recently there has been renewed interest in Phan-type presentations—see for example the survey in Gramlich [**Gra04**].

REMARK 9.3.13 (Symonds' argument for contractibility of $\mathcal{S}_p(G)/G$). This section is also a fairly natural place to give a sketch of the proof of earlier Theorem 3.2.10: namely the argument of Symonds in [**Sym98**], establishing Webb's conjecture on the contractibility of the orbit CW-complex $\mathcal{S}_p(G)/G$. For in fact Symonds does not directly prove contractibility of $\mathcal{S}_p(G)/G$—instead, he establishes simple connectivity; along with vanishing of homology groups, so that by the Whitehead theorem (e.g. [**Spa81**, 7.6.24]), these conditions suffice for contractibility of a CW-complex.

Below we will just indicate that argument for simple connectivity: since it demonstrates the basic reduction, to elementary properties of p-groups (and then requires only a little further adjustment, to establish the vanishing of homology).

Furthermore Symonds also does not work directly with $\mathcal{S}_p(G)$—but instead with Robinson's complex $N_p(G)$ (indicated in Remark 4.6.2) of "normal" chains of p-subgroups—which is G-homotopy equivalent to $\mathcal{S}_p(G)$.

To establish simple connectivity, we assume the existence of a nontrivial closed path $\bar{c}$ in $N_p(G)/G$ of length ≥ 2, and work to a contradiction. We may as well let our base point be the orbit $\overline{P}$ of a Sylow p-subgroup P of G. We consider some "lift" c of $\bar{c}$ to $N_p(G)$: this is a sequence $P, Q_1, Q_2, \cdots, Q_n, P'$ of p-groups, where P' is a G-conjugate of P, and each successive pair defines an edge in $N_p(G)$—meaning that one member of the pair is normal in the other. We have $n \geq 1$ since $\bar{c}$ is of length ≥ 2. By nontriviality of $\bar{c}$ we may as well assume that $P \neq Q_1$; and hence $|P| > |Q_1|$.

We argue by induction as follows: We restrict attention to lifts c in which some Q_j attains the minimal possible order (which we saw just above is less than $|P|$); and among those lifts, to those for which that value of $|Q_j|$ occurs the minimal number of times. We will now proceed to construct a homotopic lift c', from which we have removed such a Q_j—contrary to the minimality of the multiplicity of $|Q_j|$.

First by minimality of $|Q_j|$, we must have Q_j normal in Q_{j-1} and Q_{j+1}; and indeed by minimality of multiplicity, we must have Q_j *properly* contained in each. Now we can embed Q_{j-1} in some Sylow p-subgroup R of $N_G(Q_j)$. Similarly there is some $g \in N_G(Q_j)$ such that $g^{-1}Q_{j+1}g \leq R$. In fact, if we let d denote the path where conjugate by g all the terms in c from Q_{j+1} on, then the orbit $\bar{d}$ is still our original $\bar{c}$. Since conjugation preserves order, we may as well replace our minimal lift c by d; that is, we can assume $Q_{j+1} \leq R$. It now follows from the proper-normalizer property (3.3.17) of p-groups that we can form a sequence $Q_{j-1}, S_1, \cdots S_s, R$ with each group normal in the next; and we get a similar sequence $Q_{j+1}, T_1, \cdots T_t, R$.

Note that each adjacent pair in these sequences, taken along with Q_j, forms a 2-simplex in $N_p(G)$. Thus we can homotopically (that is, alterning only by a path representing the trivial element in $\pi_1(N_p(G))$) replace the edge Q_{j-1}, Q_j in our lift c with the edges Q_{j-1}, S_1 and S_1, Q_j in a new lift; and since $S_1 \geq Q_{j-1} > Q_j$, our new vertices have not introduced any new occurrences of the minimal value $|Q_j|$. We can continue this process with the next pair S_1, S_2, to replace the edge S_1, Q_j; and so on, all the way up to R—and then, similarly down the chain of groups T_i. The final step replaces the edge T_1, Q_j with the path from T_1 to Q_{j+1} to Q_j. Since our lift already contains the edge from Q_j back to Q_{j+1}, on removing this redundancy, we have produced a lift c' not containing Q_j—and we have added only vertices of order greater than $|Q_j|$. This contradicts the minimal-occurrence property of $|Q_j|$ in c, and so establishes the simple connectivity of $N_p(G)/G$. ♢

9.4. n-connectivity: Spherical and Cohen-Macaulay complexes

In this final section, we consider higher connectivity, namely n-connectivity for $n \geq 2$: in particular, we study conditions related to the structure of a bouquet of spheres. For a complex K of dimension d, we mainly focus on d-sphericality in the sense of Remark 9.1.8—that is, $(d-1)$-connectivity. We also consider the inductive local version of this condition given by homotopy[10] Cohen-Macaulayness in Remark 9.1.9.

As an initial summary: For different kinds of groups G, there is considerable variation in when these conditions hold for p-subgroup complexes such as $\mathcal{A}_p(G)$; so our treatment here will just sketch some results in the area for various G.

Much of the material has in fact been developed in the combinatorial literature; the reader interested in further detail may wish to consult references such as [**BGS82**].

We mention that for Tits buildings, the Cohen-Macaulay property (which we had essentially indicated at Theorem 3.4.15) was already well known to researchers by the 1970s, and provided considerable motivation for the study of the property in more general circumstances. However, we will postpone further discussion of the building case to later in the chapter—in order to maintain the "usual order" indicated in Remark 8.2.4, in which solvable groups are considered first.

But prior to that, we will record some preliminaries.

We recall from Lemma 8.1.2 that a direct product of groups leads to a join of the $\mathcal{A}_p$-posets of the factors. Using the corresponding multiplicativity property of homotopy groups over the join in his 8.1, Quillen as [**Qui78**, 10.3] obtains a similar result for Cohen-Macaulayness over the join:

(9.4.1) If $\mathcal{A}_p(H)$ and $\mathcal{A}_p(K)$ are Cohen-Macaulay, then so is $\mathcal{A}_p(H \times K)$.

[10] In particular in this section, we will usually use "Cohen-Macaulay" to mean the homotopy Cohen-Macaulay property; and so we will more specifically write "homology Cohen-Macaulay", when we mean the original definition of the property via vanishing of homology groups.

The multiplicativity in Quillen's 8.1 is also relevant to his observation in [**Qui78**, 8.5, 8.6]: For links in a poset $\mathcal{P}$ are joins of intervals, as we had indicated in Lemma 1.6.13; and correspondingly Quillen shows:

LEMMA 9.4.2. *Cohen-Macaulayness for a poset $\mathcal{P}$ is equivalent to appropriate-dimensional connectivity for $\mathcal{P}$ itself, as well as for open intervals of the forms $P_{>y}$, $\mathcal{P}_{<y}$, and $\mathcal{P}_{>x}^{<y}$, for all $x < y \in \mathcal{P}$.*

For our standard case where $\mathcal{P} = \mathcal{A}_p(G)$, Quillen notes in [**Qui78**, 10.1] that the sufficient condition in Lemma 9.4.2 simplifies still further: For assume that we are given $X, Y \in \mathcal{A}_p(G)$ with $X < Y$. Observe first that $\mathcal{A}_p(G)_{<Y} = \mathcal{A}_p(Y)_{<Y}$—and this is just the projective space of Y, regarded as a vector space over $\mathbb{F}_p$. Hence it is empty when $\dim Y = 1$ (and so is (-1)-spherical, i.e. (-2)-connected); while in the case $\dim Y \geq 2$, it is a building of some dimension $d \geq 0$, and so it is $(d-1)$-connected essentially via the Solomon-Tits argument in the proof of earlier Theorem 3.4.15. (Indeed it is even Cohen-Macaulay, as noted in later Theorem 9.4.6.) Observe next that $\mathcal{A}_p(G)_{>X}^{<Y} \cong \mathcal{A}_p(Y/X)_{<Y}$—and this in turn is the projective space of Y/X; hence the same two cases arise, according to $\dim Y/X = 1$ or ≥ 2, and are similarly handled. So we see that we are reduced to just upper intervals:

(9.4.3) Appropriate connectivity for $\mathcal{A}_p(G)$ and all $\mathcal{A}_p(G)_{>Y}$ suffices for CM.

Before investigating solvable groups, we first consider Cohen-Macaulayness just in the case of certain kinds of p-groups. For example, we can apply the above sufficiency to get a positive result in the most basic case, given by an elementary p-group A:

LEMMA 9.4.4. *For A an elementary p-group, $\mathcal{A}_p(A)$ is Cohen-Macaulay (of dimension $m(A) - 1$).*

PROOF. We first recall our earlier observation after Remark 9.1.8, that empty and contractible structures are vacuously spherical of the relevant dimensions.

By (9.4.3) we need to consider intervals of the form $\mathcal{A}_p(A)_{>Y}$. We may regard this as $\mathcal{A}_p(A/Y)$—which is empty when $Y = A$; and which when $Y < A$ is a cone on its unique maximal element A, and hence contractible. And this latter argument also deals with $\mathcal{A}_p(A)$ itself. □

We mention that Quillen in [**Qui78**, 10.4] further extends Cohen-Macaulayness to the case of $\mathcal{A}_p(E)$, for a (nonabelian) extraspecial p-group E.

However, we also observe that for the 3-group P of order 3^4 in earlier Example 9.3.2, $\mathcal{A}_p(P)$ is not Cohen-Macaulay: For we saw there that $m_3(P) = 3$, so that $\mathcal{A}_3(P)$ has dimension 2. And we note that the 1-dimensional link of $Z(P)$, namely $\mathcal{A}_3(P)_{>Z(P)}$, is not 0-connected. (Similarly, as B there (of rank 2) is maximal under inclusion in $\mathcal{A}_3(P)$, its link is just $\mathcal{A}_3(P)_{<B}$—which is just the (disconnected) union of its $p+1$ subgroups of order p.)

Thus we cannot expect Cohen-Macaulayness for all p-groups—much less for all solvable groups.

Sphericality and Cohen-Macaulayness in solvable groups. In summary, we will see for solvable G (with $O_p(G) = 1$) that $\mathcal{A}_p(G)$ at least conjecturally has the structure of a bouquet of spheres—but these spheres need not all have the same dimension. And certainly when two or more different dimensions do occur, we will not have Cohen-Macaulayness.

Cohen-Macaulayness for Quillen's minimal LA. We had seen that Quillen's analysis of $\mathcal{A}_p(G)$ for solvable G begins with a minimal case of form LA, where we set $L := O_{p'}(G)$, and A is elementary of rank s: Indeed in earlier Theorem 8.2.9, we had presented part (ii) of Quillen's result 11.2 in [**Qui78**]—giving the homological conclusion $H_{s-1}(\mathcal{A}_p(LA)) \neq 0$, which does not require an inductive proof. Now we present part (i) of 11.2, which gives Cohen-Macaulayness for $\mathcal{A}_p(LA)$; in particular we will describe Quillen's inductive argument to establish that property:

THEOREM 9.4.5 (11.2(i) in [**Qui78**]). *Assume $G = LA$, with L a solvable normal p'-subgroup, and A an elementary p-subgroup.*
Then $\mathcal{A}_p(G)$ is Cohen-Macaulay (of dimension $m(A) - 1$).

PROOF. (Sketch:) We may assume that $A > 1$, since otherwise $\mathcal{A}_p(G) = \mathcal{A}_p(L)$ is empty, and hence vacuously Cohen-Macaulay.

And in view of Lemma 9.4.4, we may also assume that $L > 1$. Hence we may let K denote a minimal nontrivial normal subgroup of G which is contained in L. From solvability of $L = O_{p'}(G)$, it is standard (e.g. [**Asc00**, 8.4(2)]) that K is an elementary abelian q-group for some prime $q \neq p$.

(Inductive step:) Consider first the case that $K < L$. We set $\bar{G} := G/K$; since $\bar{G}$ inherits the hypotheses on G, we may assume by induction on the order of G that $\mathcal{A}_p(\bar{G})$ is Cohen-Macaulay. Quillen considers the map $f : \mathcal{A}_p(G) \to \mathcal{A}_p(\bar{G})$, which is clearly a poset map; and he applies his technical result [**Qui78**, 9.7], to "pull back" the Cohen-Macaulay property from $\bar{G}$ to G. The result 9.7 is analogous to his Fiber Theorem 4.2.1, in the spirit of our earlier Remark 4.2.3; so here we will only indicate how the hypotheses of 9.7 are satisfied: Since the kernel K of the group surjection is a p'-group, f preserves the orders of members of $\mathcal{A}_p(G)$, and hence preserves strict containments $X < Y$ in $\mathcal{A}_p(G)$; that is, f is *strictly* increasing in Quillen's sense. Furthermore any $\bar{Y} \in \mathcal{A}_p(\bar{G})$ is the isomorphic image of (at least one) $Y \in \mathcal{A}_p(G)$; and we see that $f^{-1}(\mathcal{A}_p(\bar{G})_{\leq \bar{Y}}) = \mathcal{A}_p(KY)$. Since KY inherits the hypotheses of Theorem 9.4.5, again by induction on the order of G we get that $\mathcal{A}_p(KY)$ is Cohen-Macaulay (of dimension $m(B)-1$). And this completes the verification of the hypotheses on f needed to guarantee via Quillen's 9.7 that $\mathcal{A}_p(G)$ is Cohen-Macaulay.

(Base step:) So we turn to the remaining case $K = L$; in particular, A is irreducible on the elementary q-group L. If A is trivial on L, we have $G = L \times A$—so that $\mathcal{A}_p(G) = \mathcal{A}_p(A)$, which is Cohen-Macaulay by Lemma 9.4.4. So we may assume that A is nontrivial on L; and then it is standard (e.g. using (8.2.10)) for $A_0 := C_A(L)$ that A/A_0 is cyclic, hence of order p as A is elementary. Thus we may write $A = A_1 \times A_0$ for some complement A_1 of order p, and indeed we have $G = LA_1 \times A_0$. Now $\mathcal{A}_p(A_0)$ is Cohen-Macaulay by 9.4.4; and $\mathcal{A}_p(LA_1)$ has dimension 0, and hence is vacuously Cohen-Macaulay. It follows by (9.4.1) that $\mathcal{A}_p(G)$ is Cohen-Macaulay. □

Quillen asks in [**Qui78**, Problem 12.3] whether the Cohen-Macaulay conclusion above might in fact extend to the case where L is a nonsolvable p'-group. I am not aware of any further results in that direction.

However, recall that a nonsolvable L is allowed in the Hawkes-Isaacs Theorem 5.5.4—which establishes just the property of nonzero Euler characteristic for $\mathcal{A}_p(LA)$. And in fact that weaker property follows, when L is solvable, from Theorem 9.4.5 above: For once we know that $\mathcal{A}_p(LA)$ (and hence $\mathcal{S}_p(LA)$) is Cohen-Macaulay, so that appropriate homotopy groups are trivial, then in particular we get the weaker homology Cohen-Macaulay property (e.g. via Lemma 9.1.6)—namely vanishing also of homology in all dimensions below $d := m(A) - 1$. And from this it then follows that $\tilde{\chi}(\mathcal{S}_p(LA)) = (\pm) \dim \tilde{H}_d(\mathcal{S}_p(LA))$. Thus when $O_p(LA) = 1$, Quillen's Theorem 8.2.9 shows that this homology is nonzero.

A weaker sphericality property for solvable groups? Quillen in [**Qui78**] does not explicitly consider Cohen-Macaulayness for solvable groups more general than LA above; though at least implicitly his results demonstrate the difficulty of further extensions: For instance, we saw in the group LP of Example 9.3.2, where P is non-abelian of order 3^4, that Theorem 8.2.5 guarantees that $\mathcal{A}_3(LP)$ of dimension 2 has nonzero homology in dimension 1 as well as 2. This shows that $\mathcal{A}_3(LP)$ is not even homology Cohen-Macaulay (of dimension 2). (However further conditions beyond solvability, to guarantee Cohen-Macaulayness, are considered in Matucci [**Mat09**].)

In particular, this behavior is not consistent with the homotopy type of a bouquet of 2-spheres. However, it *is* consistent with the structure of a bouquet of spheres of dimensions 1 and 2. And correspondingly Thévenaz asked whether, for a solvable group G with $O_p(G) = 1$, such a bouquet of spheres of varying dimensions should be the appropriate sphericality structure.

Subsequently Pulkus and Welker in [**PW00**] gave an explicit decomposition of $\mathcal{A}_p(G)$—which reduced the problem to the study of certain upper intervals in $\mathcal{A}_p(G/O_{p'}(G))$. However, some difficulties in extending this work (to obtain a bouquet structure) are indicated by Bornand [**Bor09**].

Bouquet structures also arise for p-groups, e.g. in the poset of elementary subgroups of rank at least 2, in Bouc-Thévenaz [**BT08**]. And for groups which are not necessarily solvable: Shareshian showed in [**Sha04**] that the bouquet structure does not hold for $\mathcal{A}_2(S_{13})$.

The subgroup lattice $L(G)$ for a solvable group G. We digress briefly from our focus on $\mathcal{A}_p(G)$, to describe some parallel results on sphericality in the literature[11] on the subgroup lattice $L(G)$ for a solvable group G. Of course to avoid obvious contractibility, we instead consider the reduced version $\hat{L}(G) := L(G)^{<G}_{>1}$—from which the unique minimal element 1 and maximal element G have been removed.

We recall that at several points in earlier chapters, it had been relevant to mention some of the technical material in the Kratzer–Thévenaz papers [**KT84**][**KT85**]. And this is now the natural place to describe the main result 4.10 in [**KT85**]—which shows that $\hat{L}(G)$ for a solvable group G has the homotopy type of a bouquet of spheres. The precise description given there in fact underlies their expression for the Euler characteristic, which we had given earlier at (5.5.2): namely the spheres have dimension $(s - 2)$, where s is the number of quotients in a chief series; and the number of such spheres is the product of the number of complements in the successive factors.

This result was also obtained via some different approaches: In [**Thé85**], Thévenaz supplied an easier proof, and strengthened the sphericality property

[11] I thank Jacques Thévenaz for discussion of references in this area.

to the G-homotopy type. Björner and Walker in [**BW83b**] more generally obtain the homotopy type of a bouquet of suitable suspensions, in the situation of a lattice in which the complements of some point form an anti-chain; and in the solvable subcase, this gives [**KT85**, 4.10]. Subsequently Welker in [**Wel95b**] extended [**BW83b**] to the G-homotopy type, in particular giving another proof of [**Thé85**].

(Option B:) Cohen-Macaulayness for buildings. Having discussed solvable groups, following the viewpoint of Remark 8.2.4, we now turn to simple groups in our investigation of sphericality.

We should mention at the outset that there will of course be simple groups G for which $\mathcal{A}_p(G)$ is *not* Cohen-Macaulay. For example, the groups failing simple connectivity in Theorem 9.3.4 will also fail Cohen-Macaulayness when $\mathcal{A}_p(G)$ has dimension ≥ 2 (that is, when $m_p(G) \geq 3$). And indeed there seem to be various further situations of failure. However, we will focus for a while on cases where Cohen-Macaulayness can in fact be established.

For simple groups, as usual the model case is given by a group G of Lie type in characteristic p; and in view of Theorem 4.4.8, we replace $\mathcal{A}_p(G)$ by the building K.

Here the bulk of the work of establishing Cohen-Macaulayness, namely the basic higher connectivity proof, has in fact already been indicated in the earlier Solomon-Tits argument, in the proof of Theorem 3.4.15. So below, we just make the relevant adjustments to that earlier statement; obtaining the celebrated Solomon-Tits Theorem. (In fact we follow essentially the idea of Tits indicated at [**Sol69**, p 217], rather than the more formal argument given by Solomon in that paper. See also Benson [**Ben91**, 6.8.5], who refers in turn to an approach of Quillen.)

THEOREM 9.4.6 (Solomon-Tits Theorem—Thms 1,2 in [**Sol69**]). *Assume G is of Lie type in characteristic p, with BN-rank r.*

Then the building K has the homotopy type of a bouquet of $(r-1)$-spheres. Furthermore K is Cohen-Macaulay of dimension $(r-1)$.

Indeed if $\mathbb{F}_q$ is the field of definition, and N is the maximal distance in K, then the number of such spheres is q^N. So $\tilde{H}_n(K) = 0$ for $n < r-1$, but $\tilde{H}_{r-1}(K) \cong \mathbb{Z}^{q^N}$.

PROOF. We assume the setup in the proof of Theorem 3.4.15, and interpret that result in our present context:

Namely we fix a chamber σ of K. And then (cf. our discussion around (2.2.21)) a single apartment A on σ is a triangulation of an $(r-1)$-sphere; and in particular contains a unique chamber τ at distance N from σ. Furthermore it is standard that a full unipotent group U of order q^N, in a Borel subgroup stabilizing σ, is regular on those apartments A on σ (and hence on those opposite chambers τ); so that the number of such spheres is q^N.

Thus restricting to $K^{\leq N-1}$ in the proof of 3.4.15 amounts to "puncturing" those spheres at q^N points opposite to σ; and then the proof there of collapsibility of $K^{\leq N-1}$ implies contractibility, which in turn gives $(r-2)$-connectedness of K.

And since the link in K of a simplex of rank s is in fact a building of dimension $(r-1-s)$ (as we had seen in Remark 2.2.29), we can apply this connectedness result to the link. This gives the local connectedness showing that K is Cohen-Macaulay of dimension $(r-1)$.

The contractibility argument also establishes the desired vanishing of homology in dimensions below $(r-1)$. It follows that $\tilde{L}(K)$ is just $(\pm)\tilde{H}_{r-1}(K)$; and then the projectivity of $\tilde{L}(K)$ (recall Theorem 6.4.2) guarantees that this dimension is a multiple of $|U| = q^N$. Hence the q^N apartments on σ are linearly independent cycles in this homology group, which then has dimension exactly q^N. An alternative calculation of this value of q^N for the dimension appears in [**Car89**, Secs 8.6,14.3].

In particular, the q^N cycles given by the apartments on σ give an explicit basis for $\tilde{H}_{r-1}(K)$. And similarly K has the homotopy type of a bouquet of q^N spheres of dimension $(r-1)$. □

Before proceeding to other simple-group posets, we mention some further work related to buildings and sphericality properties. Abels and Abramenko in [**AA93**] establish the homotopy type of a bouquet of spheres for certain subcomplexes of buildings (which occur as links in certain filtrations of the affine building for SL_n). Vogtmann in [**Vog82**] establishes such a spherical homotopy type for the poset of orthonormal frames in an orthogonal space. And Lehrer and Rylands in [**LR93**] establish Cohen-Macaulayness in certain cases, and conjecture it more generally, for the *split* building of a reductive group of Lie type (defined in effect via Levi complements, rather than by full parabolics as in Definition 2.2.9).

Cohen-Macaulayness for some matrix groups in coprime characteristic. Since the building is equivalent to $\mathcal{A}_p(G)$ when G is of Lie type characteristic p, we can next consider $\mathcal{A}_p(G)$ when G is defined in characteristic different from p. Quillen makes a start in this direction by investigating the almost-quasisimple general linear group; we will give the proof, since it illustrates methods both from this chapter and from earlier chapters.

THEOREM 9.4.7 (Thm 12.4 in [**Qui78**]). *Assume that $G = GL_n(q)$, and we have $q \equiv 1 \pmod p$. Then $\mathcal{A}_p(G)$ is Cohen-Macaulay (of dimension $(n-1)$).*

PROOF. By the hypothesis $q \equiv 1 \pmod p$, we get an elementary p-subgroup D of diagonal matrices (i.e. on a split torus) of the maximal rank n; thus $m_p(G) = n$, so that $\mathcal{A}_p(G)$ has dimension $(n-1)$ (as in Definition 3.3.9). In particular D contains a subgroup Z of order p, consisting of scalar matrices.

We observe first since $Z \leq Z(G)$ that we get the relations $X \leq XZ \geq Z$ for $X \in \mathcal{A}_p(G)$; this gives conical contractibility for $\mathcal{A}_p(G)$ by (3.3.2) (and in particular $(n-2)$-connectivity).

Therefore by (9.4.3), it remains to investigate intervals of form $\mathcal{A}_p(G)_{>A}$, for some $A \in \mathcal{A}_p(G)$. Note that $\mathcal{A}_p(G)_{>A} = \mathcal{A}_p\big(C_G(A)\big)_{>A}$—so that we might be able to apply induction in $C_G(A)$. Correspondingly we consider the action of A on the natural module for $G = GL_n(q)$, namely on a vector space V of dimension n over $\mathbb{F}_q$. The congruence on q also guarantees that irreducible A-submodules on V have dimension 1—so these are subspaces on which A acts via a suitable character λ. Collecting together the subspaces for each λ, we get a vector-space decomposition $V = \oplus_\lambda V_\lambda$; and a corresponding direct-product group decomposition $C_G(A) = \prod_\lambda GL(V_\lambda)$.

Suppose first that two or more distinct λ occur. Then by induction on n, we may assume that each $\mathcal{A}_p\big(GL(V_\lambda)\big)$ is Cohen-Macaulay—and then $\mathcal{A}_p\big(C_G(A)\big)$ is Cohen-Macaulay by (9.4.1). It follows using Lemma 9.4.2 that the interval given by $\mathcal{A}_p\big(C_G(A)\big)_{>A} = \mathcal{A}_p(G)_{>A}$ is also connected in the appropriate dimension; which completes the proof in this case.

In the remaining case, we assume that only one λ occurs. Then A consists of scalar matrices, so that A is the central subgroup Z of order p; hence we are considering the particular interval $\mathcal{A}_p(G)_{>Z}$. Here Quillen applies a variant of his closed-set Theorem 4.4.4:

We use the interval $\mathcal{A}_p(G)_{>Z}$ as the first poset, in the stabilizing-pairs setup of Example 4.4.5: in particular, all the members of the interval are nontrivial groups. For the second poset there, we take the building K of G; and in particular, we have seen (in the discussion of types of simplices late in Section 2.2) that the G-action is type-preserving. So we take the standard closed set $\mathcal{S}$ of stabilizing pairs as in that Example.

Then for a simplex $\sigma \in K$, we get $\mathcal{S}_\sigma = \mathcal{A}_p(G_\sigma)_{>Z}$; now our goal will be the appropriate connectivity condition (rather than the usual goal of contractibility in Example 4.4.5). We consider the Levi decomposition in (2.2.8) of the parabolic G_σ as $U_\sigma L_\sigma$: The unipotent radical U_σ is a nilpotent p'-group (and so in particular is solvable). The Levi complement L_σ is a product (possibly with just one factor) of general linear groups, of dimensions strictly smaller than n; and so, again using induction on n and (9.4.1), much as in the argument in the case of at least two λ above, we obtain that $\mathcal{A}_p(L_\sigma)$ is Cohen-Macaulay. And now Quillen's argument for the inductive step in the proof of Theorem 9.4.5 gives Cohen-Macaulayness of $\mathcal{A}_p(G_\sigma)$. So again using Lemma 9.4.2, we get the appropriate connectedness also for the interval $\mathcal{A}_p(G_\sigma)_{>Z}$.

Next for a subgroup $Y \in \mathcal{A}_p(G)_{>Z}$, we have $\mathcal{S}_Y = K^Y$. By the contrapositive of an observation we made above, $Y > Z$ implies that there is more than one character λ' for the action of Y on V. Now the poset $\mathcal{P}_\lambda$ of *all* subspaces of the component $V_{\lambda'}$ has unique minimal and maximal elements, and so is contractible, and hence Cohen-Macaulay; and then so is their join by (9.4.1). Further K^Y is the open interval in this join above 0 and below V—and this interval also has the appropriate connectedness by Lemma 9.4.2.

So now for the closed set $\mathcal{S}$, Quillen makes a variant of the argument for Theorem 4.4.4: Here in place of the contractibility which is used as hypothesis in his Fiber Theorem 4.2.1 which underlies the proof of 4.4.4, he instead uses as hypothesis the appropriate connectedness for $\mathcal{S}_\sigma$ and $\mathcal{S}_Y$—and now the corresponding underlying fiber-type theorem (in the spirit of earlier Remark 4.2.3) is [**Qui78**, Prop 7.6]. And this time the conclusion is that the property of connectedness below the top dimension is transferred from K back to $\mathcal{A}_p(G)_{>Z}$. This completes the proof in the remaining case, corresponding to just one character λ for the action of A. □

We mention that Das in [**Das98**] establishes a similar Cohen-Macaulayness result for the symplectic group $G := Sp_{2n}(q)$ with $q \equiv 1 \pmod p$.

(Option S:) Cohen-Macaulayness for some sporadic situations. The occurrence of Cohen-Macaulayness for sporadic geometries seems variable; and in particular there seems to be no systematic literature on the question. So here we will indicate just a few sample results.

First consider the (non-sporadic) alternating group $G := A_7$. Here $m_2(G) = 2$, so that $\mathcal{A}_2(G)$ has dimension 1; thus Cohen-Macaulayness amounts just to the observation that $\mathcal{A}_2(G)$ is connected. The question is more substantial for the sporadic geometry K of type C_3 for G in Example 2.3.7, which has dimension 2 (and is not equivalent to $\mathcal{A}_2(G)$, as we saw in Example 6.3.7). Here the bulk of the work[12] amounts to observing that K itself is connected, and simply connected by Theorem 9.3.12. (As for the requirements on links of simplices: they are buildings, and hence Cohen-Macaulay by Theorem 9.4.6.) We also mention that this geometry K, along with the usual model case of buildings, is used as an example of an approach to homology Cohen-Macaulayness in [**Smi90**], which is based on the linear algebra of induced modules;[13] that viewpoint turns out to be partly parallel to the poset-ring approach of Garsia in [**Gar80**].

A similar analysis applies for the sporadic group $G := M_{24}$ and its 2-local geometry K in Example 2.3.11—which we saw is equivalent to $\mathcal{A}_2(G)$ in Remark 4.4.12: We had indicated the simple connectivity in our discussion after Theorem 9.3.12; and this time, links are truncations of buildings—which still inherit Cohen-Macaulayness from the building in Theorem 9.4.6.

On the other hand, there are certainly sporadic geometries for which Cohen-Macaulayness fails: One example is provided by $G := U_4(3)$ with K the $\tilde{C}_2$-geometry of Example 2.3.10. Recall we saw in Example 9.3.11 that K has dimension 2 (and indeed is equivalent to $\mathcal{A}_2(G)$), but is not simply connected.

We had mentioned at the beginning of the chapter that the Cohen-Macaulay condition for $\mathcal{A}_p(G)$ corresponds to Cohen-Macaulayness of an associated poset ring. We now digress briefly on Cohen-Macaulayness for the ring structure of group cohomology $H^*(G)$—which has some important relations with $\mathcal{A}_p(G)$. Indeed the Quillen-Venkov Theorem (e.g. [**Ben91**, 5.2.1]) shows that nilpotency in mod-p cohomology is determined by restriction to members of $\mathcal{A}_p(G)$. A standard consequence is that a group G, with subgroups of different ranks which are maximal under inclusion in $\mathcal{A}_p(G)$, does not have the equi-dimensionality property (cf. [**Ben91**, 5.18.1]) of a Cohen-Macaulay cohomology ring. Thus our earlier Examples 9.3.2 and 9.3.11—which were designed to exhibit failure of Cohen-Macaulayness for the complex $\mathcal{A}_p(G)$—also exhibit failure of Cohen-Macaulayness for the cohomology ring.

For simple groups, and especially sporadic groups, Cohen-Macaulayness for the cohomology ring is discussed for example in [**AMM91**] and [**AM95b**].

A notion of Cohen-Macaulayness in the context of chain complexes arises in Mathas [**Mat94**].

Shellability. We conclude the section with a brief discussion of a sphericality property which has been prominent in the combinatorial literature, namely shellability; as we had indicated in Remark 9.1.10, this condition in fact implies (homotopy) Cohen-Macaulayness. For further background on shellability, the reader may wish to consult e.g. [**Bjö80**] or [**Bjö95**, Sec 11].

Here is the definition given at [**Bjö95**, 11.2]—which involves a refinement of the notion of elementary collapse in earlier Definition 3.4.3:

REMARK 9.4.8 (Shellability). Assume that K is a complex of dimension d (with all maximal faces of dimension d).

If a k-simplex σ is contained in exactly one d-simplex τ, then the removal of faces between σ and τ (inclusive) is called a *(k, d)-collapse*.

[12]This general viewpoint is indicated by Garst in his U. Wisconsin Ph.D. thesis [**Gar79**].

[13]The decomposition there of $\mathrm{Ind}_B^G(1)$ was extended to $\mathrm{Ind}_U^G(1)$ in Fisher [**Fis93**].

If $k < d$, such a collapse is a homotopy equivalence (much as in Lemma 3.4.6). But it is also of interest to allow $k = d$, in a more general procedure which is not in general an equivalence:

We say K is *shellable* if it can be reduced to a single simplex, via a sequence of (k, d)-collapses for $0 \leq k \leq d$. ♢

For some history of shellability in the literature (including the alternative definition from the viewpoint of anti-collapsing), see the discussion after [**Bjö95**, 11.2].

(For the "nonpure" case, see e.g. Björner-Wachs [**BW96**] [**BW97**].)

We will now follow the usual order of Remark 8.2.4:

Shellability for some solvable groups. The treatment of shellability for $\mathcal{A}_p(G)$ of suitable solvable groups G seems unfortunately to have remained unpublished so far; we will indicate the sources referred to in the literature:

For Quillen's usual solvable minimal case LA as in Theorem 9.4.5, the conclusion there of Cohen-Macaulayness can in fact be strengthened to shellability. This was proved for the subcase where L is abelian by Kutin and Özaydin in a preprint [**KÖzaydin**]. The corresponding result for a general solvable p'-group L was given by Pulkus in in his 1994 Erlangen Diplomarbeit [**Pul**].

(Option B:) Shellability for buildings. As usual the model case for a simple group is provided by G of Lie type in characteristic p, with geometry given by the building K.

Here the Cohen-Macaulayness for K in Theorem 9.4.6 was strengthened to shellability by Björner in [**Bjö84**, Thm 4.8] (which also covers the case of type-selected subcomplexes). The proof essentially refines the Solomon-Tits argument for Theorem 3.4.15: From the Solomon-Tits viewpoint, the initial removal of all chambers at maximal distance N from a fixed chamber σ (corresponding to puncturing the spheres given by the apartments on σ) requires (d, d)-collapses, and of course is not a homotopy equivalence. However the subsequent removals in the Solomon-Tits argument—of chambers, along with parts of their boundaries—are (k, d)-collapses for $k \leq (d - 1)$, and collapse the remaining subcomplex homtopically down to a single simplex.

(Option S:) Shellability for sporadic geometries? We had mentioned earlier that some sporadic geometries are not Cohen-Macaulay—and hence are certainly not shellable. And little seems to be known about positive results on shellability for sporadic geometries; indeed, it seems conceivable that this stronger property might not extend much beyond the above model case for buildings.

For example, we had indicated earlier that the C_3-geometry K for A_7 of Example 2.3.7 is Cohen-Macaulay. There is partial evidence suggesting that K should not be shellable: For example, one can check using criteria in the literature that K does not have suitably stronger properties of *lexicographic* shellability (see e.g. [**BW83a**]). Furthermore some (incomplete) experimental computation which I did in the early 1980s failed to locate a shell-ordering. (Though it might be possible that with today's computer speeds, the backtrack algorithm of Danaraj-Klee [**DK78**] could terminate, to settle this question one way or the other.)

Finally we emphasize that there is a very subtantial further literature on shellability; see for example various papers of Björner, Wachs, Welker, and Shareshian.

In particular, for the subgroup lattice $L(G)$, Björner in [**Bjö80**] shows that Cohen-Macaulayness is equivalent to shellability, which is in turn equivalent to supersolvability for G itself. This is extended by Shareshian in [**Sha01**], who shows that "sequential" Cohen-Macaulayness of $L(G)$ is equivalent to nonpure shellability, which is in turn equivalent to solvability for G itself. And see also the recent work of Woodroofe [**Woo08**].

CHAPTER 10

Local-coefficient homology and representation theory

In this chapter, we consider some applications of homology for a local coefficient system $\mathcal{F}$ on a G-complex K: such a system in effect assigns, to each simplex $\sigma \in K$, a module $\mathcal{F}_\sigma$ for the stabilizer G_σ; this gives a chain complex (generalizing the usual chain complex for K, where all the $\mathcal{F}_\sigma$ would take the constant value $\mathbb{F}_p$)—and the homology of this chain complex can lead to interesting modules for G.

Our usual initial topological background Section 10.1 briefly reviews the definition and basic properties of local systems. And then we focus on the particular system given by the *fixed-point presheaf* $\mathcal{F}_V$ for a G-module V; this is the basic tool in the viewpoint of Ronan-Smith [**RS85**], which we adopt in the remainder of the chapter.

Section 10.2 is then devoted to our standard model case—namely where K is the building, for G of Lie type in characteristic p. We begin by reviewing some "classical" work of Lusztig [**Lus74**], and of Curtis-Lehrer [**CL81**]. Their results in particular provided inspiration for the approach of [**RS85**]—and we indicate some applications of local systems to the modular representation theory of G (for instance, geometric presentations and embeddings).

The approach of presheaves for buildings can also be developed more general geometries, such as p-local geometries; and Section 10.3 surveys some similar applications to modular representation theory for sporadic groups and geometries.

10.1. Topological background: Coefficient systems and their homology

As usual, our review will be fairly brief and informal; so we first indicate some general references for fuller details:

For local systems in more general algebraic topology, see for example [**Spa81**, Exer 1.F, 6.F] or [**Bro94**, Sec VII.4]. For the more specific approach followed in this chapter, see [**CL81**] or [**RS85**]; an expository treatment of the latter also appears in Benson [**Ben91**, Ch 7].

Some basics of local systems. In its general form, a *local coefficient system* $\mathcal{F}$ on a complex K is just a functor from K to some category $\mathcal{C}$. For $\sigma \in K$, we usually write just $\mathcal{F}_\sigma$ for the object of $\mathcal{C}$ which is the image $\mathcal{F}(\sigma)$. The functoriality implies the existence of suitable *connecting maps*[1] $c_{\sigma,\tau}$ when $\sigma \subseteq \tau$: that is, $c_{\sigma,\tau} : \mathcal{F}_\sigma \to \mathcal{F}_\tau$ should be a morphism of the category $\mathcal{C}$—and in particular, should satisfy the corresponding transitivity under compositions.

[1]sometimes also called *restriction* maps

We will not follow up local systems in this generality; though we do pause to recall that at various earlier points, they have been at least in the background of our discussion:

(1) Local systems are prominent in the technical underpinnings of Quillen's paper, starting in [**Qui78**, Sec 7]. Systems in which the target category $\mathcal{C}$ consists of abelian groups are used for example in his proof of the Fiber Theorem 4.2.1, and indeed in other fiber-type theorems in the spirit of Remark 4.2.3. (The combinatorial literature contains some similar applications; we mention for example work of Yuzvinsky such as [**Yuz87**]—where the terms of the system even have ring structure.)

(2) During our sketch of the proof of Lemma 7.3.17, we had mentioned that one standard approach to the isotropy spectral sequence for the G-equivariant cohomology of K is via Bredon cohomology—which involves a coefficient system M defined on the orbit complex K/G. (We also mention here that Grodal, in his work on higher limits (see [**Gro02**, p 406]), in fact uses local systems defined on the original complex K—which underlie the systems M on K/G in the more usual Bredon approach. Indeed though Grodal's systems are designed for equivariant cohomology, they are at least partly in the spirit of the *non*-equivariant presheaves studied below.)

Instead, in the remainder of this chapter we will concentrate on a more specific type of local system, in which the terms $\mathcal{F}_\sigma$ have a G_σ-module structure; this situation is of particular interest for the modular representation theory of the group G acting on K. (Indeed we had also already mentioned such a system of modules, toward the end of Section 6.3, in our discussion of the work of Knörr and Robinson [**KR89**, Sec 3]: there the module $\mathcal{F}_\sigma$ is in fact a sum of block subalgebras of the group algebra $\mathbb{F}_p G_\sigma$. However, we postpone discussion of their system $\mathcal{F}$ to later, after (11.2.7)—since it is related to Alperin's Conjecture, which is usually discussed in the context of the orbit complex K/G.)

Since we normally assume a type-preserving action of G on K, when $\sigma \subseteq \tau$ we get the reverse inclusion $G_\tau \leq G_\sigma$ of stabilizers; and so to reflect the latter order, it is natural for our purposes here to use a *contravariant* functor: i.e. $c_{\sigma,\tau} : \mathcal{F}_\tau \to \mathcal{F}_\sigma$.

We indicate below our more restricted notion of local system; we express it fairly informally here—for details, see e.g. [**RS85**, p 321] or [**Ben91**, 7.1.1].

DEFINITION 10.1.1 (Local coefficient systems and their homology). For a G-complex K, a *local coefficient system* (we may add the prefix "G-equivariant", or the suffix "of $\mathbb{F}_p$-modules") is a contravariant functor from K to the category of $\mathbb{F}_p$-vector spaces, satisfying certain further restrictions:

(a) For $\sigma \in K$, the space $\mathcal{F}_\sigma$ has a G_σ-module structure.

(b) The connecting map $c_{\sigma,\tau} : \mathcal{F}_\tau \to \mathcal{F}_\sigma$ is an $\mathbb{F}_p G_\tau$-module map.

(c) There is a G-action permuting the spaces G_σ, commuting appropriately with the maps in (b).

Following [**RS85**], we will often use the topological term *presheaf* (cf. [**Spa81**, Sec 6.7]) for such a system consisting of modules.

Collecting together the spaces for σ of a fixed dimension defines the chain complex $C_*(\mathcal{F})$ for the system—in which the boundary map is defined by the alternating sum of the appropriate connecting maps. So we get cycles and boundaries, leading to the homology $H_*(\mathcal{F})$ of the system. ◇

The following standard initial example demonstrates that this construction gives a generalization of the ordinary homology of the complex K:

EXAMPLE 10.1.2 (The constant system). We let $\mathcal{K}$ denote *constant* system: in which each $\mathcal{K}_\sigma$ is just a copy of the field $\mathbb{F}_p$ (with trivial action by G_σ); and each connecting map $c_{\sigma,\tau}$ is the identity. It should be clear that the chain complex $C_*(\mathcal{K})$ just reproduces the usual chain complex $C_*(K)$ of the complex K.

Similarly given some more general G-module M, we can define the corresponding constant system $\mathcal{K}_M$, in which all terms are M. $\diamondsuit$

Of course the real reason for introducing local systems is to allow for the more interesting possibility of *non*-constant local coefficients. And indeed we will turn to such examples in a moment.

But before doing so, we indicate some motivation for studying the homology $H_*(\mathcal{F})$ of a system. It is standard (e.g. [**RS85**, p 325] or [**Ben91**, 7.2.1]) that the bottom homology module H_0 can be described in the topological language of a *direct limit*; the corresponding universality can also be expressed in terms of a presentation for the module:

LEMMA 10.1.3. *For a local system $\mathcal{F}$ on the G-complex K, the bottom homology $H_0(\mathcal{F})$ is the module presented by: generators corresponding to the spaces $\mathcal{F}_v$ for the vertices $v \in K$; subject to relations determined, when v, w are on an edge E, by identifying the images in $\mathcal{F}_v$ and $\mathcal{F}_w$ of the edge space $\mathcal{F}_E$.*

Notice that if K is of dimension 0, then there are no edges—and hence no relations in the Lemma; in that case, $H_0(\mathcal{F})$ consists just of induced modules $\mathrm{Ind}_{G_v}^G(\mathcal{F}_v)$.

Now consider K of dimension at least 1, with a constant sheaf $\mathcal{K}_M$: Then the paths identify corresponding copies of M; hence $H_0(\mathcal{K}_M)$ is a direct sum of copies of M, one for each connected component of K. In particular if K is connected, then $H_0(\mathcal{K}_M)$ just reproduces M.

(Option B:) The fixed-point presheaf $\mathcal{F}_V$ for a Lie module V. With the convenience of historical hindsight, it is possible to see that various ostensibly-different applications of local systems on buildings in fact arise essentially from one construction. Below we will give a definition of that system, in the form eventually axiomatized by Ronan-Smith in [**RS85**, p 322].

The system from fixed points of unipotent radicals. For the rest of the section, we assume that G is of Lie type in characteristic p, with K denoting its building. For a simplex $\sigma \in K$, we recall from (2.2.8) the Levi decomposition $G_\sigma = U_\sigma L_\sigma$ of the parabolic subgroup which stabilizes σ—where U_σ is the unipotent radical, and L_σ is a Levi complement.

We let V denote some kG-module where k is a field of the natural characteristic p: for technical purposes, k should be a splitting field for the representation theory of G—but usually it will be sufficient to think of k as being just $\mathbb{F}_p$. At various points, we will quote facts about such V from the theory of modular representations for G. (In earlier Remark 6.4.1, we had indicated some standard background references for the theory, such as [**Cur71**], [**Cur70**], and [**GLS98**].) Ronan and Smith in [**RS85**, p 322] define:

DEFINITION 10.1.4 (The fixed-point presheaf $\mathcal{F}_V$). Given a kG-module V, we define the *fixed-point presheaf* $\mathcal{F}_V$ using the fixed points of the various unipotent radicals:

$$\mathcal{F}_\sigma := V^{U_\sigma}$$

Notice, since U_σ by definition acts trivially, that we may in fact regard $\mathcal{F}_\sigma$ as a module for the quotient $G_\sigma/U_\sigma \cong L_\sigma$ (that is, for a Levi complement).

Sometimes it is natural to replace the usual chain complex $C_*(\mathcal{F}_V)$ with a suitably *augmented* complex $\tilde{C}_*(\mathcal{F}_V)$: Recall this means that we also use a formal empty simplex $\sigma = \emptyset$ in dimension -1; this is stabilized by G itself—which we regard as a parabolic with Levi decomposition as in (2.2.8) given by $U_\emptyset = 1$ and $L_\emptyset = G$. Consequently it is natural to augment the chain complex with a term $C_{-1}(\mathcal{F}_V) := V^1 = V$. Correspondingly we obtain the (suitably-)reduced Lefschetz module $\tilde{L}(\mathcal{F}_V)$. ◇

We begin with a small and particularly straightforward instance—which we can then use as a running example, as the chapter proceeds.

EXAMPLE 10.1.5 (The natural-module presheaf for $SL_3(2)$). Take $G := SL_3(2)$. Recall that the building K for G is the projective plane $\mathbb{P}^2(2)$ described in Example 2.1.9; and the parabolics of G are described in Example 2.2.5.

We take V to be the natural module for G, of dimension 3 over $\mathbb{F}_2$; we will see that the fixed-point presheaf $\mathcal{F}_V$ is essentially a version of the projective space $\mathbb{P}(V)$—i.e. of the usual view of the building K:

For from the matrix forms of parabolics in 2.2.5, note that U_P fixes the 1-subspace corresponding to the projective point P, and U_L fixes the 2-space for the projective line L; while the radical U_{PL} for the edge $P \in L$ fixes the same 1-space as U_P. So for the corresponding sheaf terms, we see that the connecting map $c_{P,PL}$ is an isomorphism of the 1-spaces $\mathcal{F}_{PL}$ and $\mathcal{F}_P$; while $c_{L,PL}$ is an injection of the 1-space $\mathcal{F}_{PL}$ into the 2-space $\mathcal{F}_L$. ◇

EXERCISE 10.1.6. Mimic Example 10.1.5 for the next larger case $GL_4(2)$. Recall that the building K is described via the projective space of the natural module V in Example 2.1.15; and the parabolics of G are sketched in Exercise 2.2.15.

Indeed, argue similarly for any $GL_n(q)$. ◇

Notice in the Example (and Exercise) above that for each σ, the term $\mathcal{F}_\sigma$ is irreducible under L_σ. This feature in fact holds for any irreducible V; for earlier Smith in [**Smi82**] had shown:

(10.1.7) If V is irreducible for G, then each V^{U_σ} is irreducible for L_σ.

Indeed the consequent irreducibility[2] property of $\mathcal{F}_V$ (cf. [**RS85**, 2.1]) was one important motivation for axiomatizing Definition 10.1.4.

---

[2]Webb considered a different notion of irreducibility—in which one type of term is irreducible, but terms for other types are 0.

Some initial features of $H_*(\mathcal{F}_V)$. The subsequent Section 10.2 will be largely devoted to the homology of $\mathcal{F}_V$ for Lie type groups. But in this section we will preview a few basic properties.

First, we will sketch how (10.1.7) leads to a sharpened version of the universal property in earlier Lemma 10.1.3:

Using the standard description of the irreducibles V in terms of highest weights (cf. later Remark 10.2.5) we can begin by strengthening (10.1.7) to [**RS85**, 2.2]:

(10.1.8) Irreducibles V correspond $1:1$ with irreducible presheaves $\mathcal{F}_V$.

Now we fix some irreducible V; and abbreviate $H_0(\mathcal{F}_V)$ by H. From the universal property in Lemma 10.1.3, the spaces in $\mathcal{F}_V$ appear in H—so that we get a natural inclusion $\mathcal{F}_V \subseteq \mathcal{F}_H$ of presheaves. Indeed we even get $\mathcal{F}_V \subseteq \mathcal{F}_{\bar{H}}$, for any quotient $\bar{H} := H/M$ modulo a proper submodule M. In view of the $1:1$ correspondence in (10.1.8), this eliminates any irreducible quotients of H other than V—giving [**RS85**, 2.3]:

(10.1.9) For irreducible V, $H_0(\mathcal{F}_V)$ has V as its unique irreducible quotient.

We illustrate this in our running example:

EXAMPLE 10.1.10 (Homology of the natural presheaf for $SL_3(2)$). We continue the setup of Example 10.1.5 above, with V the natural module for $G := SL_3(2)$.

Here it is easy to check that $H := H_0(\mathcal{F}_V)$ is in fact just V itself: Indeed since $\mathcal{F}_V$ is essentially a version of the projective space $\mathbb{P}(V)$, we are really just recovering the linear space V from its projective space. Hence the argument could be regarded as part of the classical coordinatization of projective space; indeed as the easier part, since we do not have to construct a division ring—as we are already working over the relevant field $\mathbb{F}_2$.

We now sketch a fuller argument, using the viewpoint of the presentation in Lemma 10.1.3: Identifying the images of $\mathcal{F}_{PL}$ in H leads to an inclusion of the 1-space $\mathcal{F}_P$ in the 2-space $\mathcal{F}_L$. In particular, each $\mathcal{F}_L$ is spanned by the spaces $\mathcal{F}_P$ for a pair of points P in L; so H can in fact be spanned by just the 7 point 1-spaces. And we do not even need all 7 of those: For in the notation of Example 2.1.9, the three points 1, 2, 12 of the line $L := \langle 1, 2 \rangle$ correspond to three vectors in H which sum to zero in the 2-subspace $\mathcal{F}_L$; so that this space is spanned just by the 1-spaces for 1 and 2. Indeed, it follows that H is spanned by the three 1-spaces corresponding to the points 1, 2, and 3: since the spaces for the other points $12, 13, 23, 123$ are spanned via line relations as above. Thus H has dimension 3; and so in view of (10.1.9), H can then only be V itself. (Or, it is easy to check directly, e.g. via the action of elements $g \in G$, that H is the original V—rather than the other (dual) 3-dimensional module V^*.)

We indicate one further calculation, as a preview of higher homology groups (and a notion of "duality") in the subsequent Section 10.2: Namely using the geometric parameters from Example 2.1.9, along with the dimensions of the spaces $\mathcal{F}_\sigma$, we can compute the Euler characteristic of the chain complex $C_*(\mathcal{F}_V)$:

$$\chi(\mathcal{F}_V) = \dim C_0 - \dim C_1 = (7 \cdot 1 + 7 \cdot 2) - (21 \cdot 1) = 0.$$

By the Hopf Trace Formula 5.1.14, we get the same value of 0 for the alternating sum in homology; so we have $\dim H_0 = 3 = \dim H_1$. However, H_1 of dimension 3 here is in fact the dual (i.e. contragredient) V^* of V: For we can check that the action of elements of order 7 in G on H_1 is different from that on V. (This can be implemented via a Lefschetz-module computation as in (5.1.12); but the larger field $\mathbb{F}_8$ is needed to record the character value on elements of order 7.) Alternatively, we can obtain H_1 as the *reduced* Lefschetz module $\tilde{L}(\mathcal{F}_V)$—corresponding to the augmented form of the chain complex indicated in Definition 10.1.4 (from which V then cancels with H_0 in the alternating sum $\tilde{L}$).

We will see, starting in later Remark 10.2.5, that top homology suggests a notion of a certain "dual" to V—which in general differs from the usual vector-space dual V^* which happens to arise in this small example. ♢

EXERCISE 10.1.11. Mimic Example 10.1.10 above, in the setting of earlier Exercise 10.1.6: That is, obtain $H_0(\mathcal{F}_V) = V$ for $GL_4(2)$—or indeed for any $GL_n(q)$. (We will examine a case where V is a *proper* quotient of $H_0(\mathcal{F}_V)$, in later Example 10.2.11.) ♢

Next, we continue our preview by demonstrating the "ideal" structure for homology of $\mathcal{F}_V$—which we will also see in a number of special cases for G and V thereafter. (The situation for general G and V can be somewhat more complicated—as will emerge at later Theorem 10.2.10.)

REMARK 10.1.12 (Homology of the trivial pressheaf $\mathcal{F}_1$). If we take V to be the trivial module 1, then the fixed-point presheaf $\mathcal{F}_1$ just reproduces the constant system $\mathcal{K}$ of Remark 10.1.2; this gives the chain complex $C_*(K)$ of the building K—whose homology of course we have already computed (in particular encountering in top dimension the Steinberg module St). We now re-state the content of several earlier results, in a format designed for later comparison with $\mathcal{F}_V$ for other V. Namely for K a building of rank r (where we assume $r \geq 2$ to guarantee connectedness), we have:

$$H_i(\mathcal{F}_1) = \begin{cases} 1 & \text{for } i = 0 \\ 0 & \text{for } i = 1, \ldots, r-2 \text{ (when } r \geq 3) \\ \mathrm{St} & \text{for } i = r-1 \end{cases}$$

Here $H_0(\mathcal{F}_1) = 1$ just reflects the connectedness of K; usually in view of connectedness, we have just been ignoring dimension 0, via the notation $\tilde{H}_0(\mathcal{F}_1) = 0$ of reduced homology. The vanishing for intermediate dimensions comes from the (homology) Cohen-Macaulay property, in the Solomon-Tits Theorem 9.4.6 (i.e. essentially from Theorem 3.4.15). And of course this vanishing means that the top dimensional homology $H_{r-1}(\mathcal{F}_1)$ is in effect computed via the reduced Lefschetz module $\tilde{L}(K)$, corresponding to the augmented complex $\tilde{C}_*(K)$; the identification of this Lefschetz module with the Steinberg module St was indicated in Theorem 6.4.2. ♢

10.2. (Option B): Presheaves on buildings

We continue the setup above, in which G is of Lie type in characteristic p, and K is its building. This section surveys some work involving local systems on K, typically systems of the fixed-point form $\mathcal{F}_V$ for a suitable irreducible V.

Lusztig's discrete-series modules for $GL_n(q)$. To the best of my knowledge, local-system techniques were first introduced into the representation theory of Lie-type groups by Lusztig—in his work on the groups $GL_n(q)$ in [**Lus74**].

Certain aspects of that work provided obvious inspiration for the viewpoint of [**RS85**], which we are following in this chapter. But we should emphasize from the outset that Lusztig's goals were by no means confined to our restricted viewpoint:

For example, though he works initially with representations in the characteristic p of G, his eventual object is the description of certain representations in characteristic 0. Indeed these give an analogue for finite G of the "discrete series" for a (continuous) Lie group in characteristic 0, corresponding to the discrete part of the spectrum of suitable eigenvalues. These representations might be considered less obvious, when contrasted with the representations in the "principal series", namely those appearing in the module $\mathrm{Ind}_B^G(1)$ induced from a Borel subgroup B.

Also, though that work involves only the elementary geometry of simplicial complexes, the reader may well be aware of Lusztig's celebrated later work on characteristic 0 representations which exploits deep algebraic geometry, starting e.g. in Deligne-Lusztig [**DL76**]. (After this later work, the discrete/principal contrast has been more customarily replaced by the distinction between cuspidal and unipotent representations.)

Lusztig in [**Lus74**] considers the particular Lie type family $G := GL_n(q)$, where q is a power of p (though he later extended the analysis to other classical groups in [**Lus75**]).

The discrete series in natural characteristic. We first sketch the beginning part of Lusztig's work, which deals with representations in characteristic p.

He considers a certain natural local system on the building K—which, in the later language of our Definition 10.1.4, is just the fixed-point presheaf $\mathcal{F}_V$ corresponding to the natural module V for G. Of course the building K is just the projective space $\mathbb{P}(V)$; and since K for $SL_n(q)$ has rank $(n-1)$, its topological dimension is $(n-2)$. Hence we will also assume that $n \geq 3$, so that K is connected.

Lusztig's fundamental result below shows that the homology of $\mathcal{F}_V$ exhibits a structure very similar to the Cohen-Macaulay situation in earlier Remark 10.1.12: For here, we also get vanishing of homology in the intermediate dimensions; that is, aside from the top dimension—and also aside from the bottom dimension, which of course is usually just ignored for a connected simplicial complex, via the formalism of reduced homology. Lusztig's notation below of $D(V)$ for top homology presumably stands for "discrete series"; it will also be appropriate for the notion of a certain "dual" to V, which we indicate starting at later Remark 10.2.5.

THEOREM 10.2.1 (Thm 1.12 (also p 24) in [**Lus74**]). *For the finite linear group $GL(V)$, with $n := \dim V \geq 3$:*

$$H_i(\mathcal{F}_V) = \begin{cases} V & \text{for } i = 0 \\ 0 & \text{for } i = 1, \ldots, n-3 \text{ (when } n \geq 4) \\ D(V) & \text{for } i = n-2 \end{cases}$$

where $D(V)$ is a certain module of dimension $(q^{n-1}-1)(q^{n-2}-1)\cdots(q-1)$.

We briefly summarize a few aspects of the proof: We had in fact already sketched the essentially-classical argument for $H_0(\mathcal{F}_V) = V$ in Exercise 10.1.11 (via Example 10.1.10). Lusztig's discussion of the vanishing of homology in intermediate dimensions in fact makes use of an appropriate version of the Nerve Theorem 4.5.2. And this vanishing means that the top-dimensional homology module $D(V)$ can be described (at least up to character) as essentially the Lefschetz module $\tilde{L}(\mathcal{F}_V)$ from the suitably-augmented chain complex $\tilde{C}_*(\mathcal{F}_V)$ indicated in Definition 10.1.4.

Obtaining irreducibles in characteristic 0. We will now be digressing, for a few pages, from our usual focus on characteristic p—to consider representations in characteristic 0. We begin this digression by describing the remarkable continuation of Lusztig's characteristic-p work above, in which he goes on to produce irreducible modules in characteristic 0:

We saw above that Lusztig's work started with the natural module V for G in characteristic p; note now that V is not a priori visible from the viewpoint of the *complex* representation theory of G. However, Lusztig now shows that the module $D(V)$, obtained from the local information in $\mathcal{F}_V$, in fact does admit a suitable "lifting" to a characteristic-0 module $\tilde{D}(V)$—and the latter is even *irreducible*. Indeed, he extends this suitably to vertex stabilizers: Roughly, given a subspace V_i of dimension i in V, Theorem 10.2.1 can be applied[3] to the corresponding factor in a Levi complement $L_{V_i} \cong GL(V_i) \times GL(V/V_i)$; so for the full stabilizer G_{V_i}, we get a lifting $\tilde{D}(V_i)$ to characteristic 0—which we can in turn induce to a module for G. (Thus the original case of $D(V)$ corresponds to $i = n$: namely V_i given by V, with G itself as the stabilizer.) He obtains:

THEOREM 10.2.2 (Cor, Sec 5.2 in [**Lus74**]). *Let V_i be an i-subspace $(1 \leq i \leq n)$ in V. Then the characteristic-*0 *module* $\mathrm{Ind}_{G_{V_i}}^G(\tilde{D}(V_i))$ *is irreducible (except for the case where $i = 1$ and $q = 2$— when a single further trivial constituent arises).*

We will only examine this result within our running example:

EXAMPLE 10.2.3 (The discrete series for $GL_3(2)$). We continue the setup of earlier Example 10.1.5, for $G := GL_3(2)$.

For contrast, we begin by indicating the principal series: The module $\mathrm{Ind}_B^G(1)$ induced from a Borel subgroup B has dimension 21—and determines the irreducible characters of degrees 1, 6 (occurring twice), and 8; these are denoted by χ_1, χ_4, and χ_6 in the Atlas [**CCN**$^+$**85**, p 3].

Since we have $q = 2$, for $i = 1$ we get the exceptional case in Theorem 10.2.2: Here $\tilde{D}(V_1)$ is the trivial module, so that inducing it gives the permutation module on the 7 projective points. That module decomposes as an irreducible of dimension 6, along with the extra 1-dimensional trivial constituent indicated in the Theorem. (And of course because of the triviality of $\tilde{D}(V_1)$ for G_{V_1} containing B, these are principal-series modules, which we already saw in the previous paragraph.)

We turn to the (non-exceptional) discrete series modules in Theorem 10.2.2: First consider the "original" case for V itself, namely $i = n = 3$. Here $D(V)$ is of dimension $(2^2 - 1)(2 - 1) = 3$. And indeed we had independently calculated in Example 10.1.10 that $D(V)$ is the 3-dimensional module V^* dual to V. Now Theorem 10.2.2 shows that this module lifts to $\tilde{D}(V)$ of dimension 3 in characteristic 0—corresponding to the degree-3 irreducible character denoted by χ_3 in the Atlas [**CCN**$^+$**85**, p 3]. (And similarly starting with V^* instead of V would

[3] In fact because of the hypothesis $n \geq 3$ in Theorem 10.2.1, the cases $i = 1, 2$ require special treatment. Chapter 3 of [**Lus74**] gives a suitable definition of $\tilde{D}(V_1)$ of dimension 1, and $\tilde{D}(V_2)$ of dimension $(q - 1)$. (These are nontrivial modules—except for $\tilde{D}(V_1)$ when $q = 2$.)

lead to the other degree-3 irreducible, denoted by χ_2.) Finally consider the remaining case $i = 2$: Here $\tilde{D}(V_2)$ is the nontrivial character of $GL(V_2) \cong S_3$ of dimension $(2-1) = 1$; and the corresponding induced module has the degree-7 irreducible character denoted by χ_5 in the Atlas [**CCN**$^+$**85**, p 3]. (In this particular Example, we have in fact obtained *all* irreducible characters of G.) ◊

EXERCISE 10.2.4. Mimic Example 10.2.3, in the case of $GL_4(2)$ in Example 10.1.6. The principal-series irreducibles have dimensions 1, 14, 20, ..., 64. The discrete-series irreducibles for $i = 2, 3, 4$ have dimensions 35, 45, 21. ◊

Curtis-Lehrer duality for unipotent $\mathbb{C}$-modules. Next we continue our digression on characteristic-0 representations, by discussing a result of Curtis-Lehrer in [**CL81**]—which obtains a Cohen-Macaulay-like structure in homology, similar to that in Remark 10.1.12; but now for $\mathcal{F}_X$ determined analogously by certain *complex* representations X. And in this context of Curtis and Lehrer, G can now be *any* Lie type group.

In fact the conclusion of the Curtis-Lehrer Theorem 10.2.8 also shows that the relation of X with $D(X)$ is *symmetric*—giving a certain notion of "duality". So we briefly review some background for this concept:

As I understand the history:[4] Remark 10.1.12 relates the trivial module 1 and the Steinberg module St, by comparing bottom and top dimensions in homology. And this suggests an analogue, for our building K (which is not a manifold), of the classical *Poincaré* duality[5] for manifolds—relating low with high dimensions (cf. [**Spa81**, 6.17]). And with this particular homological viewpoint, Curtis and Lehrer were able to explain some earlier observations about relations of irreducible modules appearing in the induced module $\mathrm{Ind}_B^G(1)$ in characteristic 0.

Indeed underlying this characteristic-0 relationship is a natural pairing among the irreducibles for G in natural characteristic—so we switch back, for a moment, to characteristic p; to observe:

REMARK 10.2.5 (Duality with respect to the Steinberg weight ρ). Assume that G is of Lie type with rank r, and defined over $\mathbb{F}_q$ in characteristic p. Then it is standard (e.g. [**GLS98**, 2.8.2]) that an irreducible V in characteristic p corresponds to a *highest weight* λ, which is in the *(q-)restricted range*: that is, λ is a linear combination $\sum_{i=1}^{r} c_i\lambda_i$ of the *fundamental weights* λ_i, with each integer coefficient c_i between 0 and $(q-1)$. We may write $V = V(\lambda)$ when we wish to specify λ.

In particular, the trivial module 1 is $V(0)$; and the Steinberg module St is $V(\rho)$, where ρ has maximal coefficients $c_i = q-1$. This pair $V(0)$ and $V(\rho)$ gives an example of what we will be calling *ρ-duality*, relating $V(\lambda)$ and $V(\rho-\lambda)$. Since we have $\rho-(\rho-\lambda) = \lambda$, this relation is in fact symmetric; this is one of the motivations for the terminology of duality. A different characterization of the relation is that the tensor product of $V(\lambda)$ with its ρ-dual $V(\rho-\lambda)$ contains a component given by the Steinberg module $\mathrm{St} = V(\rho)$.

---

[4]See also the beginning of [**CL81**, Sec 5].

[5]as opposed to just vector-space duality for modules

This duality was in fact already relevant to $\mathcal{F}_V$ earlier: for it is implicit in Lusztig's Theorem 10.2.1 that $D(V(\lambda))$ there has a unique irreducible submodule given by the ρ-dual $V(\rho-\lambda)$. Indeed this feature holds for general G and V: for Ronan showed in [**Ron89a**, Thm 2] that

(10.2.6) $H_{r-1}(\mathcal{F}_{V(\lambda)})$ is the dual module (i.e. contragredient) to $H_0(\mathcal{F}_{V(\rho-\lambda)})$;

this is relevant to later Theorem 10.2.10.

Finally, to return to our present theme of representation theory in characteristic 0: We can extend the above relation to a pair of characteristic-0 modules X, Y—when their reductions mod p involve a pair of irreducibles which are in ρ-duality. ◇

So we can now turn to the Curtis Lehrer result. Here G can be of any Lie type; with rank $r \geq 2$, so that K is connected.

We consider any irreducible summand X of the characteristic-0 module $\mathrm{Ind}_B^G(1)$. Even though we are no longer in characteristic p, we can still define the fixed-point-presheaf $\mathcal{F}_X$ as in Definition 10.1.4. Here $\mathcal{F}_\sigma = X^{U_\sigma}$ might be 0 for some more general module X; but in view of Frobenius reciprocity (e.g. [**Ben98**, 3.3.1]):

$$\mathrm{Hom}(A, \mathrm{Ind}_H^G(B)) \cong \mathrm{Hom}(\mathrm{Res}_H^G(A), B), \tag{10.2.7}$$

we see since X here is a quotient of $\mathrm{Ind}_B^G(1)$ that the restriction of X to G_σ does indeed have some constituent trivial on the Borel subgroup $B = UH \geq U_\sigma$. We also recall the (suitably-)augmented chain complex $\tilde{C}(\mathcal{F}_X)$ indicated in Definition 10.1.4. The result is:

THEOREM 10.2.8 (Cor 5.7 in [**CL81**]). *Assume that G has Lie rank $r \geq 2$, and X is an irreducible summand of $\mathrm{Ind}_B^G(1)$ in characteristic 0. Then*

$$H_i(\mathcal{F}_X) = \begin{cases} X & \text{for } i = 0 \\ 0 & \text{for } i = 1, \ldots, r-2 \text{ (when } r \geq 3) \\ D(X) & \text{for } i = r-1 \end{cases}$$

Here $D(X) = \tilde{L}(\mathcal{F}_X)$ is also (up to sign) an irreducible in $\mathrm{Ind}_B^G(1)$; and $D(X)$ is ρ-dual to X.

We sketch certain aspects of the proof: It is standard, essentially as in (10.1.9) for characteristic p, that X is a quotient of $H_0(\mathcal{F}_X)$; and since here we are working in characteristic 0, it follows using complete reducibility that this quotient is just X itself. The remaining assertions are then obtained by translating the problem to a simpler setting: Namely Curtis and Lehrer consider the endomorphism algebra E of $H_*(\mathcal{F}_X)$. This algebra is based on double cosets of B in G; and from the standard decomposition $G = BWB$ using the Weyl group W of B (e.g. [**Car89**, 8.2.3]), E is isomorphic to the endomorphism ring E' of the homology of a corresponding presheaf $\mathcal{F}_Y$, for a suitable irreducible Y of W. In this smaller setting, it is then easier to establish for $\mathcal{F}_Y$ the homological structure (including vanishing) as in the Theorem; in particular, top homology is now given by $D(Y)$, which turns out to be the (irreducible) tensor product of Y with the 1-dimensional sign representation[6] of W. And now E' is 2-dimensional, since $H_*(\mathcal{F}_Y)$ contains just the two distinct irreducibles Y and $D(Y)$. Translating the problem back via E to G, we get vanishing of intermediate homology; and just one further irreducible, namely $D(X)$, in top

[6]This use of the sign representation is important in the Curtis-Lehrer viewpoint on duality.

dimension. The symmetry of the relation between X and $D(X)$ follows from the role of the sign representation above; it also follows essentially via the technique of Möbius inversion (cf. [**Sta86**, Sec 3.7]).

We also summarize the content of Theorem 10.2.8 in our running example:

EXAMPLE 10.2.9. We continue the setup of Example 10.1.5; and we recall from Example 10.2.3 that $\mathrm{Ind}_B^G(1)$ can be described in terms of irreducible dimensions as $1+6+6+8$. We have already suggested the calculation of $D(1) = 8 = \mathrm{St}$, in Remark 10.1.12.

We sketch the converse calculation that $D(8) = 1$. The 8-dimensional irreducible W can be realized via 3×3 matrices of trace 0. And there the terms of $\mathcal{F}_W$ are essentially visible using the description of the maximal unipotent radicals U_P and U_L in Example 2.2.5 (along with the observation that $U_{PL} = U_P U_L$); we get terms of dimensions $2, 2, 1$ for P, L, PL. Thus for the Euler characteristic:

$$\tilde{\chi}(\mathcal{F}_8) = -8 + (7 \cdot 2 + 7 \cdot 2) - (21 \cdot 1) = -1.$$

As G is perfect, its only 1-dimensional module is trivial. (Or in more detail, we compute for all $g \in G$ that the Lefschetz character is trivial.)

Finally we sketch the calculation that $D(6) = 6$; i.e. that the 6-dimensional irreducible W is self-ρ-dual. We can realize W as the nontrivial irreducible (i.e. vectors with coefficient sum 0) in the natural 7-dimensional permutation module on projective points. And then the dimension of a fixed space of U_σ is given by the number of orbits (e.g. [**Gor80**, 4.3.5]) on the points. We can compute those orbits, again using the above matrix forms, but now in their action on the 7 nonzero vectors of V; we get presheaf terms of dimension $3, 3, 2$ for P, L, PL. So the Euler characteristic is:

$$\tilde{\chi}(\mathcal{F}_6) = -6 + (7 \cdot 3 + 7 \cdot 3) - (21 \cdot 2) = -6.$$

Or in more detail, the Lefschetz character just reproduces the original 6-dimensional module W. ◊

Finally we mention that duality is also studied in work of Alvis, e.g. [**Alv79**].

Applications of $H_*(\mathcal{F}_V)$ to modular representations. We now return from our digression on characteristic-0 representations—to resume our more usual focus on representations for G in characteristic p.

In particular, we mention that a main motivation for the work in Ronan-Smith [**RS85**] was to move beyond the particular case in Lusztig's Theorem 10.2.1, in order to explore—for any G of Lie type, and any (irreducible) module V—how the homology of $\mathcal{F}_V$ can contribute to the natural-characteristic representation theory.

The structure of $H_*(\mathcal{F}_V)$. For general G and V, there are only limited analogues of the Cohen-Macaulay-like structure of homology in Theorem 10.2.1; for comparison, we give a fairly informal statement of the general result, in the form:

THEOREM 10.2.10 (see 2.3 in [**RS85**] and Thm 4 in [**Ron89a**]). *Assume that G is of Lie type in characteristic p, with building K of rank $r \geq 2$; and that $V = V(\lambda)$ is irreducible for G over a splitting field in characteristic p. Then:*

$$H_i(\mathcal{F}_V) \text{ has: } \begin{cases} \text{for } i = 0\text{:} & \text{has unique irreducible quotient } V; \\ \text{for } 1 \leq i \leq r-3\text{:} & \text{may be nonzero;} \\ \text{for } i = r-2\text{:} & \text{has unique irreducible submodule } V(\rho - \lambda). \end{cases}$$

We had in fact previewed the quotient property for H_0 earlier at (10.1.9). The submodule property for top homology then follows from the quotient property for H_0 using Ronan's duality result (10.2.6).

Before proceeding to more theoretical applications, we indicate a few examples of the structure of homology in Theorem 10.2.10.

First, we present some cases where V is a *proper* quotient of $H_0(\mathcal{F}_V)$:

EXAMPLE 10.2.11 (Natural-presheaf homology of $Sp_4(2)$). Take $G := Sp_4(2)$, with V the natural 4-dimensional module. The calculation will be very similar to the rank-2 case of $GL_3(2)$ in Example 10.1.10. (But the proper quotient will demonstrate a subtlety that arises in extending Lusztig's Theorem 10.2.1 from linear groups to other classical groups.)

To begin, we can similarly use the matrix forms of parabolics given in Example 2.2.16—to determine that as before, we get presheaf terms $\mathcal{F}_\sigma$ of dimensions $1, 2, 1$ for $\sigma = P, L, PL$. However, here the argument will give $H := H_0(\mathcal{F}_V)$ of dimension 5—exceeding the dimension 4 of V.

The beginning of the spanning argument is just as before: The line 2-spaces are in the span in H of the 1-spaces for the points; and we will not need all 15 of the point spaces—since we get relations for example among the 1-spaces corresponding to the 3 points of the line $L = \langle 1, 2\rangle$. However, in contrast to Example 10.1.10, this time we can only use the *totally isotropic* lines in V; we do not get relations for the ordinary projective lines corresponding to 2-spaces such as $\langle 1, 4\rangle$, which are not isotropic in the polar space of Example 2.1.19. Thus if we hoped to get $H = V$, we could try to span H with the 1-spaces for the four points $1, 2, 3, 4$. As consequences we would then get the spaces for the further points $12, 13, 24, 34, 1234$—but not for the remaining 6 points. On the other hand, as soon as we add one of these remaining six 1-spaces, say that for the point 14, as a fifth member of our candidate for a spanning set—we then do get the rest of the 15 generating spaces as consequences. Thus H is in fact of dimension 5 rather than 4. And in particular, H has a unique maximal (trivial) submodule of dimension 1; the quotient is then the original V of dimension 4. (In fact H is just the natural module for $\Omega_5(2) \cong Sp_4(2)$, which we described in earlier Remark 2.1.24.)

Finally, we do still get an exact analogue of the final calculation in Example 10.1.10; that is, we again get vanishing of the Euler characteristic:

$$\chi(\mathcal{F}_V) = \dim C_0 - \dim C_1 = (15 \cdot 1 + 15 \cdot 2) - (45 \cdot 1) = 0.$$

So as before, the Hopf Trace Formula 5.1.14 gives the same value of 0 for the alternating sum in homology; and then by the 5-dimensionality of H, we see that $J := H_1(\mathcal{F}_V)$ is also 5-dimensional (but now with a unique trivial quotient rather than submodule).

In fact the highest weight λ for V is the fundamental weight λ_1, with character denoted by ϕ_2 in the Modular Atlas [**JLPW95**, p 4]. And we find (e.g. computing the Lefschetz character) that the unique irreducible submodule of J is of dimension 4. Since $q = 2$, we have $\rho = \lambda_1 + \lambda_2$, in the notation of Remark 10.2.5; then J corresponds to highest weight $\rho - \lambda = \lambda_2$, and its character is denoted by ϕ_3 in [**JLPW95**, p 4]. ◇

EXERCISE 10.2.12 (Homology of the natural presheaf for $Sp_6(2)$). Mimic Example 10.2.11 above, but using the next-higher-rank symplectic group $G := Sp_6(2)$.

So now V is the 6-dimensional natural module (again for weight λ_1, with character denoted by ϕ_2 in [**JLPW95**, p 110]). We had described the building K via the polar space of V in Example 2.1.26; and the parabolics of G were sketched in Exercise 2.2.18. In particular, we obtain presheaf terms of dimensions $1, 2, 3$ for vertices of types P, L, π.

Show then (much as in Example 10.2.11) that $H_0(\mathcal{F}_V)$ is the 7-dimensional orthogonal module for $\Omega_7(2) \cong Sp_6(2)$; with trivial submodule, and (unique irreducible) quotient V. (The general case of $\Omega_{2n+1}(2) \cong Sp_{2n}(2)$ is contained in [**RS85**, 4.3]; the calculation is classical and well known.) ◊

Next we examine an instance of duality in the top dimension:

EXAMPLE 10.2.13 (Homology of the natural presheaf for $SL_4(2)$). We recall the setup of earlier Exercise 10.1.6, for $G = GL_4(2) = SL_4(2)$, and V the 4-dimensional natural module. Of course in this case, we can apply Lusztig's Theorem 10.2.1: We get $H_0(\mathcal{F}_V) = V$ and $H_1(\mathcal{F}_V) = 0$; and we had commented in Exercise 10.2.4 that the top homology module is Lusztig's $D(V)$ of dimension 21.

Here the highest weight for V is the fundamental weight λ_1, and the character of V is denoted by ϕ_2 in the Modular Atlas [**JLPW95**, p 48]. Computation of the Lefschetz character of $D(V)$ gives the sum of the 20-dimensional character (denoted by ϕ_7) with the trivial character. Thus the unique irreducible submodule of $D(V)$ in Theorem 10.2.10 is the 20-dimensional irreducible for weight $\rho - \lambda_1 = \lambda_2 + \lambda_3$; and the quotient modulo this irreducible is trivial.

Or, using the duality viewpoint more directly: It is shown in [**RS89**, 1.1] that $H_0\big(\mathcal{F}_{V(\lambda_2+\lambda_3)}\big)$ is of dimension 21—so its unique maximal submodule must be trivial. Hence by (10.2.6), the dual module $H_2\big(\mathcal{F}_{V(\lambda_1)}\big)$ has trivial quotient modulo its unique irreducible submodule of dimension 20. ◊

Finally we sketch a case of non-vanishing of intermediate homology:

EXAMPLE 10.2.14 (Homology of the orthogonal presheaf for $SL_4(2)$). We continue with the group $G := SL_4(2)$ as in Example 10.2.13 above; but now take V to be the 6-dimensional orthogonal module for $\Omega_6^+(2) \cong SL_4(2)$—corresponding to highest weight λ_2, and with character denoted ϕ_4 in the Modular Atlas [**JLPW95**, p 48]. We obtain presheaf terms of dimensions $3, 1, 3$ for vertices of types P, L, π; and then dimensions $1, 2, 1, 1$ for simplices of types $PL, P\pi, L\pi, PL\pi$.

We get $H_0(\mathcal{F}_V) = V$, using e.g. later Remark 10.2.15.2. Also $\rho - \lambda_2 = \lambda_1 + \lambda_3$; and for $W := V(\lambda_1 + \lambda_3)$, we get using e.g. Remark 10.2.15.4 that $H_0(\mathcal{F}_W)$ is the adjoint module of dimension 15. Hence by duality in (10.2.6), $H_2(\mathcal{F}_V)$ is also of dimension 15. Now we compute the Euler characteristic of $C_*(\mathcal{F}_V)$:

$$\chi = (15 \cdot 3 + 35 \cdot 1 + 15 \cdot 3) - \big(105 \cdot (1 + 2 + 1)\big) + (315 \cdot 1) = 20 \ .$$

Since we get the same result for the alternating sum in homology by the Hopf Trace Formula 5.1.14, we conclude that $\dim H_1 = 6 + 15 - 20 = 1$. ◊

We mention that, so far at least, there seems to be no deeper understanding of the significance of such nonzero homology in intermediate dimensions.

In the remainder of the section, we will survey some of the literature on applications of the structure of the homology of $\mathcal{F}_V$ in Theorem 10.2.10. Most of these applications are related to the property that $H_0(\mathcal{F}_V)$ has unique irreducible quotient V.

Geometric presentations of modules. In those cases where we can show $H_0(\mathcal{F}_V)$ is just V itself, then the universal property in Lemma 10.1.3 can be interpreted as saying that $\mathcal{F}_V$ gives us a *geometric* presentation (i.e., based on the geometry of K) for the module V. The term *local recognition* of V also appears in the literature.

We have already demonstrated several examples of geometric presentations; so below we will just summarize, in tabular form, a number of such results which have appeared in the literature. (Experience seems to suggest that we can normally hope to get that $H_0(\mathcal{F}_V)$ is just V itself only when V (or its highest weight λ) is in some suitable sense fairly "small"—although the Steinberg module below is an exception to this observation.)

REMARK 10.2.15 (Some results on geometric presentations). This overview is fairly informal; while each result is essentially of form $H_0(\mathcal{F}_V) = V$, the reader should consult the individual references for details of hypotheses and statements.

	Lie type G	Module V	Reference
(1)	(classical)	natural	Lusztig [**Lus74**] [**Lus75**]
(2)	(any)	λ minimal	Ronan-Smith [**RS85**, Thm 4.1]
(3)	(any)	Steinberg	Ronan-Smith [**RS85**, Cor 2.4]
(4)	(most)	adjoint	Smith-Völklein [**SV89**], Völklein [**Völ89b**]
(5)	${}^3D_4(2)$	adjoint	Frohardt-Smith [**FS92**]
(6)	G_2	natural 7	Segev-Smith [**SS86**]
(7)	F_4	natural 26	Cohen-Smith [**CS90**]
(8)	C_n, D_n	λ_2	Cohen [**Coh94**]

(The special *minimal* condition in (2) (sometimes called *minuscule*) requires all weights of the characteristic-0 (Weyl) module for λ to lie in a single orbit under the Weyl group.)

For example, we have seen (in Example 10.2.11 and Exercise 10.2.12) that for the natural module V of $Sp_{2n}(2)$, the isomorphism with $\Omega_{2n+1}(2)$ contributes a further trivial submodule to $H_0(\mathcal{F}_V)$—which does not arise for the same symplectic group in odd characteristic. ◊

Embeddings and universal presheaves. The presentations above involve relations from the geometry of K; and this is also relevant to the classical notion of an *embedding*: Roughly, given objects in a geometry called "points" and "lines", is there some vector space V in which they can be realized as projective points and lines in $\mathbb{P}(V)$? (That is, as linear 1- and 2-subspaces of V?)

For our present case of a building K, the *existence* of a suitable V is usually clear:[7] for example for classical groups G, in Section 2.1 we saw an obvious embedding of the geometry within the projective space of the natural module V for G.

Indeed we saw in those cases (e.g. Example 10.1.5) that the fixed-point sheaf $\mathcal{F}_V$ for this V has the property that $\mathcal{F}_P$ and $\mathcal{F}_L$ are linear subspaces of dimensions 1 and 2; and it follows that K embeds in the (sometimes larger) module $H_0(\mathcal{F}_V)$.

[7]However, this is not necessarily clear for more general geometries—recall we saw in earlier Remark 2.3.8 that the C_3 geometry for A_7 is not embeddable. We will return to this point in the subsequent Section 10.3.

This observation suggests the search for a most general embedding space: that is, a *universal* embedding—presented by point and line subspaces, related by their inclusions. Notice this is closely related to the presentation property in Lemma 10.1.3—except that now only vertices of types "point" and "line" are used. (Hence $H_0(\mathcal{F}_V)$ for an embedding in V is in turn a quotient of the universal embedding.)

Indeed the process lends itself to treatment via the properties of presheaf homology:

REMARK 10.2.16 (Universal presheaves). Suppose our K has rank at least 3, with a further vertex type of "planes". Then there is a natural notion of a universal term $\mathcal{F}_\pi$—given by $H_0(\mathcal{F})$ for the group G_π, where $\mathcal{F}$ is the "partial" presheaf from terms $\mathcal{F}_P$ and $\mathcal{F}_L$ above. Continuing in this way, we can build up a full presheaf $\mathcal{U}$, having suitable universal properties with regard to embedding for K. But this notion of building up is not restricted just to the context of embeddings; we could just specify $\mathcal{F}$ at *minimal* parabolics, and build up $\mathcal{U}$ in this same way to all parabolics. These notions for embeddings are described in [**Ron87**], and for more general universal presheaves in [**RS86**]. ♢

The context of presheaves is exploited in a number of papers, within the substantial geometric literature on embeddings. For example, in Remark 10.2.15 above on geometric presentations, the results on natural modules in cases (1), (6), and (7) determine embeddings—usually universal.

So below, we will simply describe a universal embedding in one further building case:

EXAMPLE 10.2.17 (Embedding the dual polar space of $Sp_6(2)$). We recall the setup of earlier Example 10.2.12 for $G := Sp_6(2)$; where the usual view of the building is as the polar space of a 6-dimensional symplectic space, as in Example 2.1.26. However, we now change the viewpoint on K, to get the *dual* polar space: namely we re-name the previous vertices of plane type π as the new "points" P. And now the old vertices of point type will be called "quads" Q—this is because the link of a new point has the structure of a generalized quadrangle for $L_Q \cong Sp_4(2)$, as in Example 2.1.19.

Can these new points (and old lines) be embedded in a suitable V? Yes: For certainly it is classically known that this dual polar space arises in the 8-dimensional irreducible module V—which has highest weight given by the fundamental weight λ_3 (and has character denoted by ϕ_3 in the Modular Atlas [**JLPW95**, p 110]). Indeed V arises as the classical *spin* module, for the orthogonal group $\Omega_7(2) \cong Sp_6(2)$.

Now for $\mathcal{F}_V$, we find sheaf terms of dimensions $4, 2, 1$ for vertices of the new types Q, L, P. In fact we even get $H_0(\mathcal{F}_V) = V$ using Remark 10.2.15.2 (applied to λ_3 for $\Omega_7(2)$).

However, this V is not the universal embedding; and correspondingly $\mathcal{F}_V$ is not the universal presheaf $\mathcal{U}$ for embedding purposes: For the universal term in $\mathcal{U}$ at Q is not the 4-dimensional irreducible term from $\mathcal{F}_V$; but instead the 5-dimensional indecomposable module for $L_Q \cong Sp_4(2)$ regarded as $\Omega_5(2)$—as we see from Example 10.2.11, in view of its presentation by by point and line spaces. Furthermore the trivial submodule of this term at Q at least suggests the possibility of generating, in the universal embedding, a further natural presheaf as in Exercise 10.2.12.

And this turns out to be the case: In fact it was shown by various authors (including Ronan-Smith, Wells, and Brouwer) that $H_0(\mathcal{U})$ has dimension 15—with unique maximal submodule of dimension 7 from the natural presheaf as in Exercise 10.2.12; and quotient V of dimension 8.

The nature of the universal embedding for the dual space of the general symplectic group $Sp_{2n}(2)$ is more subtle, and remained open for some time; eventually the conjecture of Brouwer, in which the embedding has dimension $\frac{1}{3}(2^n+1)(2^{n-1}+1)$, was established by Li in [**Li01**]. ♢

We conclude the section with a brief mention of some other applications of the property that $H := H_0(\mathcal{F}_V)$ has V as its unique irreducible quotient; let M denote the corresponding unique maximal submodule of H.

First if $M \neq 0$, the uniqueness implies a nonsplit extension of V over suitable sections of M. Correspondingly, calculating with the presheaf formalism has been useful in deciding existence or non-existence of various such extensions: see for example Smith [**Smi85**] and Völklein [**Völ89a**]. (The latter in fact involves K-theory, namely $K_2(F)$—where F is the field of definition for our Lie type group G; though this is just zero for finite fields, which is our usual situation in this book.)

Finally we mention that the presheaf context suggests (see [**RS85**, Sec 3]) an essentially inductive computational algorithm for constructing the irreducible module V, based on the input of the "local" information in $\mathcal{F}_V$: For this information, plus essentially the permutation action of G on its parabolics, leads to the construction of $H = H_0(\mathcal{F}_V)$. But as M above is a submodule of H, we get a corresponding presheaf inclusion $\mathcal{F}_M \subseteq \mathcal{F}_H$. And since $\mathcal{F}_V$ is also a subpresheaf of $\mathcal{F}_H$, we even get a splitting $\mathcal{F}_H = \mathcal{F}_M \oplus \mathcal{F}_V$. Then we can similarly construct M inside H, using the local information from $\mathcal{F}_M$; so that eventually we obtain V as H/M.

10.3. (Option S): Presheaves on sporadic geometries

In contrast to the case of Lie type groups in the previous Section 10.2, there is no theory of modular representations which is specially tailored to sporadic groups. Nonetheless, the modular representations for these groups are important; and in this section, we survey some of the literature which uses the presheaf formalism, to obtain at least limited analogues of the results indicated above for Lie type groups and buildings.

Analogues for fixed-point presheaves. So we now consider some sporadic geometry K (with its group G), in some characteristic p—in the spirit of the diagram geometries considered in Section 2.3.

We saw there (e.g. Remark 2.3.3.2c) that in many examples of such geometries, we will have simplex stabilizers G_σ which have a product structure $U_\sigma L_\sigma$—similar to the Levi decomposition (2.2.8) of the parabolics in a group of Lie type. In this case, given a modular representation V for our group G, we can form the straightforward analogue $\mathcal{F}_V$ of the fixed-point presheaf in Definition 10.1.4.

EXAMPLE 10.3.1 (The code presheaf for M_{24}). We recall from Example 2.3.11 the 2-local geometry K for $G := M_{24}$—where the stabilizers G_σ do have the desired structure $U_\sigma L_\sigma$.

Now the "external" viewpoint there considered K as a subposet of the projective space $\mathbb{P}(V)$, for V given by the 11-dimensional irreducible quotient of the

Golay code—with the vertices of types O, T, S obtained as certain subspaces of V of dimensions $1, 2, 4$. And in fact, on computing fixed points under the subgroups U_v for vertices v, we just recover these same subspaces—now viewed as the presheaf terms $\mathcal{F}_O, \mathcal{F}_T, \mathcal{F}_S$ of the fixed-point presheaf $\mathcal{F}_V$. (The remaining presheaf terms are then essentially appropriate copies of these: For example, for O incident with T, since $U_{OT} = U_O U_T$ we check that $V^{U_{OT}} = (V^{U_T})^{U_O} = V^{U_O}$ of dimension 1, so that $\mathcal{F}_{OT}$ reproduces $\mathcal{F}_O$; etc.) $\diamondsuit$

However, we also saw in certain geometries that a subgroup U_σ might sometimes be trivial. In the above convention, this would force $\mathcal{F}_\sigma = V^1 = V$. Now since we usually think of a local system as containing "local" information (rather than the "global" case of V itself), we might correspondingly wish to re-define $\mathcal{F}_\sigma$ as some suitable G_σ-submodule which is proper in V—if in fact such a choice is available.

EXAMPLE 10.3.2 (Some presheaves on the C_3-geometry of A_7). We recall from Example 2.3.7 the Neumaier geometry K for $G := A_7$, with Dynkin diagram of type C_3. Here we saw that the point and plane stabilizers have $U_P = 1 = U_\pi$.

First take V to be the 6-dimensional irreducible contained in the natural permutation module for A_7 on the 7 points; we may realize V as the even subsets of a 7-set. This module admits an orthogonal structure, from the inclusion $A_7 < A_8 \cong \Omega_6^+(2)$. The general structure is considered in Remark 2.1.27; but in this case, it is more simply described as in Remark 2.1.23—in particular, the 4-sets give the singular vectors of V. And indeed, we can compute that our usual definition of $\mathcal{F}_L$, namely V^{U_L}, is just the (singular) 1-space given by a 4-set—namely the complement of the 3-set defining L. Now at this point, for π over L it is natural (since we want connecting maps in our presheaf to reflect natural inclusions of subspaces of V) to let $\mathcal{F}_\pi$ be defined by the *span* of the G_π-conjugates of $\mathcal{F}_L$; then $\mathcal{F}_\pi$ is a (totally singular) 3-subspace, affording the natural module for $G_\pi \cong SL_3(2)$. And we can apply the same spanning process to define $\mathcal{F}_P$: Under action of $G_P \cong A_6$, we get all the 4-sets in a 6-set—namely the complement of the point defining P; then $\mathcal{F}_P$ is the orthogonal 5-space for A_6 of Remark 2.1.24 (which contains a nonsingular radical of dimension 1). Once these choices are made, we get further presheaf terms of dimensions $1, 2, 1, 1$ for simplices of types $PL, P\pi, L\pi, PL\pi$.

We have less choice when we consider the two 4-dimensional irreducible modules W, arising from $A_7 < A_8 \cong SL_4(2)$. In either case, the usual definition via W^{U_L} produces a 2-subspace $\mathcal{F}_L$. We now take W to be the particular 4-dimensional module in which our plane stabilizer G_π centralizes a 1-space (as opposed to normalizing a 3-subspace). Then for π over L, in fact the 1-space fixed by G_π lies in $\mathcal{F}_L$—so we define that 1-space to be $\mathcal{F}_\pi$. (And then $\mathcal{F}_L$ is just what we would get from $\mathcal{F}_\pi$ under the above spanning process.) However, we have now run out of flexibility: For the span of the G_P-conjugates of $\mathcal{F}_L$ (or $\mathcal{F}_\pi$) is all of V—so that we are forced to choose $\mathcal{F}_P = V$ (for each P). That is, no proper G_P-submodule is available. $\diamondsuit$

We now briefly summarize the work in **[RS89]**—which demonstrates for some individual sporadic geometries that one may "in practice" obtain a reasonable approximation of the theoretical results for Lie type groups and buildings in the previous Section 10.2:

In fact that work examines the sporadic geometries for A_7 and M_{24} in the Examples above; and also the triple cover of the $Sp_4(2)$ quadrangle, which is involved in "tilde geometries". (The classical building case of $L_4(2)$ is also explored there, for comparison.)

In those cases, for each irreducible V, the corresponding presheaf $\mathcal{F}_V$ is indicated. One observes for example that $\mathcal{F}_V$ is then usually reasonably close to the ideal of irreducibility in (10.1.7).

And in the reverse direction, first the possible irreducible presheaves $\mathcal{F}$ are exhibited—abstractly, i.e. not necessarily assuming the form $\mathcal{F}_V$ from some known V. Then each corresponding homology module $H := H_0(\mathcal{F})$ is described, and in some cases even fully determined. (Cf. the results on geometric presentations in the previous section.) Here it is possible that $H = 0$; but we find that each nonzero H usually has a unique irreducible quotient—in rough analogy with (10.1.9). (And one can even explore limited analogues of ρ-duality; though there is now no underlying irreducible with the suitable "Steinberg module" properties.)

Embeddings of sporadic geometries. In fact much of the geometric literature involving modular representation theory has been focused on problems related to embeddings, in the sense of Remark 10.2.16.

Embeddability. We had already mentioned that for sporadic geometries, the *existence* of an embedding is not necessarily guaranteed.

For comparison, we first recall the case of the building K for $G := Sp_6(2)$, with Lie type C_3. We had seen in Example 2.1.26 that the usual view of K, with vertices of types called P, L, π, is embeddable—as the polar space of the natural symplectic module; and in Example 10.2.17, that the dual polar-space geometry, where the vertex types are called Q, L, P, is also embeddable—in the spin module.

Next consider K the sporadic C_3-geometry for $G := A_7$ from Example 2.3.7. Here we had seen in Remark 2.3.8 that the usual view of the geometry, with vertices called P, L, π, is not embeddable; while by the discussion of $\mathcal{F}_W$ in Example 10.3.2, the dual view of the geometry, with vertices called Q, L, P, is embeddable e.g. in the 4-dimensional module W.

We mention finally that a similar situation occurs for the much-larger 2-local geometry of the Monster M: That geometry has the structure of a *parapolar* space (namely a certain weakening of the notion of a polar space), in which the "points" are 2-central involutions, with stabilizer $2^{1+24}Co_1$; and Ivanov and Shpectorov in [**IS94b**] showed that this parapolar space is *not* embeddable.[8] On the other hand, the corresponding *dual* parapolar structure, where a "point" is stabilized by $2^{10+16}\Omega_{10}^{+}(2)$, is shown in [**Smi94a**] to be embeddable in the mod-2 reduction of the 196884-dimensional Griess module.

Universal embeddings. Many of the sporadic geometries are in fact known to be embeddable; so typically the more relevant problem is the determination of the universal embedding, in the sense of Remark 10.2.16. There is a considerable literature on such problems.

For instance, to continue Example 10.3.1 for the 2-local geometry of M_{24}, we saw that the embedding in V of dimension 11 was reflected in the corresponding irreducible presheaf $\mathcal{F}_V$. However, the universal presheaf $\mathcal{U}$ for an embedding instead has a 5-dimensional sextet term $\mathcal{U}_S$, in view of Example 10.2.11; and it is

[8] I.e. in their language, does not have a "natural" representation.

shown in [**RS89**, Sec 4] that the corresponding universal embedding $H_0(\mathcal{U})$ is of dimension 23.

As in the previous section, we will now just summarize, in tabular form, a number of relevant works which have appeared in the literature: that is, at least some of the results involve universal embeddings:

Group G	Reference
$3S_6, A_7, M_{24}$	Ronan-Smith [**RS89**]
M_{12}	Smith [**Smi88a**]
$U_4(3), U_6(2)$	Yoshiara [**Yos90**]
J_2	Frohardt-Smith [**FS92**]
Co_1	Smith [**Smi94b**]
M_{22}, Co_2, B, M	Ivanov-Shpectorov [**IS94b**]
M_{24}, Co_1, M	Ivanov-Shpectorov [**IS94a**]
$U_4(3), Suz, Co_1$	Bardoe [**Bar95**] [**Bar96b**] [**Bar96a**] [**Bar99**]

We mention that the latter three papers of Bardoe deal with geometries determined by 2-central involutions: and typically these can be embedded in a suitable analogue of a Lie-type adjoint module; roughly, the embedding is in the exterior square of a "natural" module—which in turn embeds the more usual geometry for that group.

CHAPTER 11

Orbit complexes and Alperin's Conjecture

In this final chapter, we briefly indicate some topics which did not fit in naturally earlier.

The initial Section 11.1 reviews orbit posets and complexes—which have been involved at least in the background of some of our material. Then Section 11.2 gives a glimpse of Alperin's Conjecture in modular representation theory—which in most versions involves a sum over an orbit poset like $\mathcal{B}_p(G)/G$.

11.1. The role(s) of the orbit complex

Recall we had briefly introduced orbit posets and complexes starting at Definition 3.2.6. (Indeed we mentioned there some contractibility results, such as Symonds' Theorem 3.2.10.) Here we will try to give a retrospective overview of the appearance of orbit structures throughout our earlier chapters.

Expressions indexed by the orbit complex. We have perhaps encountered the orbit complex K/G most frequently in the comparatively "passive" role just of the natural *index* set for various important constructions.

For example, K/G appears in the standard expression (5.1.10) for the reduced Lefschetz module of K—and hence also for the reduced Euler characteristic; these of course are crucial tools in a great deal of our treatment. In those expressions, the reason for using the orbit complex is that it is natural to collect together the simplices σ into G-orbits: each of which can be expressed as the induced module $\mathrm{Ind}_{G_\sigma}^{G}(1)$ from the stabilizer G_σ. This is an elementary application of the principle underlying Frobenius reciprocity (10.2.7).

Essentially as a consequence, the orbit complex also indexes expressions such as Webb's formula in Theorem 7.2.5, and indeed the underlying homotopy colimit in Remark 7.4.2. And then the reason is the appropriate corresponding extension of the Frobenius reciprocity principle, such as the Eckmann-Shapiro Lemma (7.2.3), and the interpretation of the Borel construction of G/G_σ as BG_σ in the proof of Lemma 7.3.17; cf. [**BS08**, 2.5.20].

In these expressions, the actual structure of the orbit complex K/G plays no theoretical role. But of course for *practical* computation of the expressions, the complexity of K/G is an issue: And here it is extremely relevant that many of the complexes K which we study in fact admit flag-transitive action by G—so that the orbit complex K/G is a single simplex. And in that case, the expressions reduce to computation over the subsets J of an indexing set I. (Cf. [**BS08**, 5.6.11].)

The significance of acyclicity of K/G. In the above flag-transitive situation, the orbit complex K/G is in fact contractible, and in particular $\mathbb{F}_p$-acyclic.

We also recall one place where $\mathbb{F}_p$-acyclicity of K/G played a more "active" role than just as an indexing set: namely in Lemma 7.3.17, where this acyclicity is required for sharpness of the normalizer decomposition for K.

Results on the structure of K/G. As the above discussion indicates, there are important reasons for trying to show that the topological structure of the orbit complex K/G is essentially trivial—e.g. that K/G is at least $\mathbb{F}_p$-acyclic.

Contractibility for K/G. Indeed many of the results in the literature aim at establishing *contractibility* for K/G.

For example in the case of $\mathcal{S}_p(G)$, we had observed at Lemma 3.2.8 that contractibility of the orbit *poset* $\mathcal{S}_p(G)/G$ is immediate from Sylow's Theorem; but the problem is more subtle for the orbit CW-complex $K(\mathcal{S}_p(G))/G$: where Webb's conjecture of contractibility was eventually established by Symonds in Theorem 3.2.10.

Many important G-complexes K in group theory are known to admit flag-transitive action by G; and this in particular establishes contractibility of K/G. In the case of the building K for G of Lie type, we mentioned the corresponding standard result of Tits at Remark 2.1.4.2. For sporadic geometries K, this is typically established on an individual basis; for example, many such proofs are given in [**BS08**].

Structural results for the case of $L(G)/G$. There is also some literature[1] on the orbit poset and complex for the case of K given by the full subgroup lattice $L(G)$.

First we mention some work just on the reduced Euler characteristic: Hawkes, Isaacs, and Özaydin in [**HIÖ89**, Sec 7] consider the orbit *poset* $L(G)/G$ for solvable G; and they establish the relation:

$$|G'|\,\tilde{\chi}(\,\Lambda(G)/G\,) = \tilde{\chi}(\Lambda(G)),$$

where G' is the commutator subgroup of G. This holds also for the orbit complex—and the authors ask whether this might extend to general G.

But this coincidence is perhaps just a consequence of the fact that the orbit poset and complex are homotopy equivalent for solvable G. Indeed Thévenaz later established[2] the relationship in general for the orbit complex—while for the orbit poset, other authors had found nonsolvable counterexamples using sporadic Mathieu groups.

The deeper structure of $L(G)/G$, including the equivalence of orbit posets and complexes mentioned above, is considered by Welker in [**Wel95b**].

11.2. Orbit-poset formulations of Alperin's Conjecture

Alperin's Conjecture 11.2.2 below is a major open problem in p-modular representation theory. We briefly examine several equivalent versions of the Conjecture, which involve orbit posets such as $\mathcal{B}_p(G)/G$. (So far at least, it seems that these orbit structures appear only as indexing sets.)

---

[1] I thank Jacques Thévenaz for comments in this area.

[2] I believe that this result has remained unpublished.

The original form of the Conjecture. In the spring of 1985, Jon Alperin announced a conjecture on the number of irreducible p-modular representations, for any finite group G. It is expressed using the following language:

NOTATION 11.2.1. We write $l_p(G)$ for the number of non-isomorphic irreducible p-modular representations of G. (We work over a splitting field, e.g. an algebraically closed field of characteristic p.)

We write $z_p(G)$ for the number of such irreducibles which are also *projective*. (Recall Definition 6.1.3, and blocks of defect 0 in Remark 6.2.3.) ♢

The conjecture asserts that the "global" parameter $l_p(G)$ should be computable via summing the values of z_p on appropriate p-local subgroups $N_G(P)$; indeed the sum is indexed essentially by the *members* (as opposed to the poset structure) of the orbit poset $\mathcal{B}_p(G)/G$. (However, notice that the right-hand side also includes a global term for $P = 1$—which can be nonzero.)

CONJECTURE 11.2.2 (Alperin's Conjecture [**Alp87**]).

$$l_p(G) \quad \textit{should be given by} \qquad \sum_{P \in [1 \cup \mathcal{S}_p(G)]/G} z_p\big(N_G(P)/P\big).$$

Equivalently the sum on the right can be taken over $[1 \cup \mathcal{B}_p(G)]/G$.

To deduce the form over $\mathcal{B}_p(G)$: If $N_G(P)/P$ has a nontrivial normal p-subgroup, that subgroup lies in the kernel of any irreducible, which in turn cannot be projective (since that requires free action by a Sylow p-subgroup). That is, z_p takes the value 0, unless P lies in $\mathcal{B}_p(G)$.

As a quick summary, we note that Alperin himself established the conjecture for certain classes of groups; and subsequently other authors have provided considerable further positive evidence.

(Option B:) We quickly mention the case of G of Lie type in characteristic p: Here Cabanes in [**Cab88**] verified the Conjecture, using the well-known natural-characteristic representation theory for such G; the calculation reduces inductively to parabolic subgroups—which provide the normalizers of $P \in \mathcal{B}_p(G)$, as we had noted in Remark 4.4.9. Here is a small example:

EXAMPLE 11.2.3 (Alperin's Conjecture for $SL_3(2)$). Take $p = 2$ and $G = L_3(2)$. Thus G is of Lie type, with Lie rank 2.

For general G of Lie type, defined over $\mathbb{F}_q$ and with rank r, the highest-weight theory (recall Remark 10.2.5) shows that $l_p(G) = q^r$; so here $l_2(G) = 2^2 = 4$.

On the other hand, each $N_G(P)/P$ is itself of Lie type—possibly extended by diagonal automorphisms; and the only irreducibles which are projective are in fact just diagonal-conjugates of the Steinberg module St. Since $q = 2$ here, there are no nontrivial diagonal outer automorphisms, so the value of z_p is 1; hence the sum on the right reduces in this case to just the *size* of the index set $[1 \cup \mathcal{B}_p(G)]/G$. This is just number of *types* of parabolics—in general 2^r, giving $2^2 = 4$ in the present case with types $\emptyset, P, L, PL$. This of course agrees with our value of $l_2(G) = 4$ from the previous paragraph. ♢

EXERCISE 11.2.4. Similarly analyze the symplectic group $G = Sp_4(2)$. ♢

Other orbit-structure variants of the Conjecture. Already by the Paris meeting in June of 1985, Geoff Robinson had started to explore some equivalent forms of the Conjecture—exploiting combinatorial techniques related to orbit posets and complexes. This led to a considerable further literature.

An approach via normal chains. We now sketch some directions developed in an influential joint work [**KR89**] of Robinson with Reinhard Knörr.

The authors initially observed a similarity of the formula in Alperin's Conjecture 11.2.2 with the combinatorial situation for Möbius inversion (cf. [**Sta86**, Sec 3.7]): In the combinatorial language, the left-side function l_p is expressed via the partial-summand function z_p on the right side.[3] So Möbius inversion suggests an alternate form of the Conjecture, which would conversely express $z_p(G)$ in terms of l_p, namely as:

$$(11.2.5) \qquad \sum_{P \in [1 \cup \mathcal{S}_p(G)]/G} \mu(1,P)\, l_p\big(N_G(P)\big),$$

where μ is the usual Möbius function on $L(G)$ (e.g. (5.3.16)). Recall we saw at (5.3.17) that μ can be interpreted as a suitable Euler characteristic; this Euler characteristic viewpoint is even more suggestive, when two additional developments are introduced:

First, Knörr and Robinson replace $\mathcal{S}_p(G)$ with the equivalent poset (indeed complex) $\mathcal{N}_p(G)$ of "normal chains" of p-subgroups, which we had indicated in Remark 4.6.2 (and where we now allow chains to start with 1). Then a term on the right side above corresponds to a chain c, and the role of $N_G(P)$ is played by the stabilizer G_c; in particular the term $\mu(1,P)$ is just $(-1)^{\dim c}$—so that the conjectural formula for z_p in (11.2.5) becomes an ordinary alternating sum over the orbit complex $\mathcal{N}_p(G)/G$.

Second, Knörr and Robinson consider a "blockwise" version of the Conjecture: where characters are counted separately for each block B (recall our indication of blocks after Example 6.1.1) of the group algebra; so the term on the right uses $l_p(B_c)$—where B determines (via the standard correspondence of Brauer) a suitable block B_c of G_c. And then the conjectural formula for $z_p(B)$ suggested by (11.2.5) takes the form:

$$(11.2.6) \qquad \sum_{c \in \mathcal{N}_p(G)/G} (-1)^{\dim c}\, l_p\big(B_c\big).$$

At this point, we recall from Remark 6.2.3 that a projective irreducible arises as the unique modular irreducible in a block of defect 0. And correspondingly, for B of defect 0, Lemma 3.7 of [**KR89**] shows that the expression in (11.2.6) does indeed take the value 1, as desired for $z_p(B)$.

This observation changes the focus to blocks B of *positive* defect: That is, Alperin's Conjecture—as re-stated in the equivalent form in [**KR89**, Thm 3.8]—takes the form:

(11.2.7) For B with defect group $D > 1$, the expression (11.2.6) should vanish.

We mention that the equivalence proof requires "for *all* finite G" on both sides.

[3] Or more precisely, by functions $\bar{z}_p$ defined on quotient groups $N_G(P)/P$. That is, it fails to correspond to actual Möbius inversion, in that there is no poset common to all terms on the right side; in particular, the orbit poset for $\mathcal{S}_p\big(N_G(P)/P\big)$ need not be a natural subposet of $\mathcal{S}_p(G)/G$.

The reader may have noticed that the expression in (11.2.6) resembles an Euler characteristic—indeed reduced, in view of the term for the trivial chain 1 in $\mathcal{N}_p(G)$. We now sketch the development of a corresponding (augmented) chain complex in [**KR89**]:

Knörr and Robinson in fact define a local system $\mathcal{F}$ on $\mathcal{N}_p(G)$, whose term $\mathcal{F}_c$ at c is B_c—which has the additional structure (beyond just that of a G_c-module) of being a sum of block subalgebras of the group algebra kG_c.

Of course the system $\mathcal{F}$ as usual defines an augmented chain complex $\tilde{C}_*(\mathcal{F})$, with the term B for the trivial chain $c = 1$ in dimension -1. This is not yet the complex for (11.2.6)—but it is interesting to pursue it for just a moment longer: At [**KR89**, p 51], the authors define $\tilde{L}(B)$ as the (reduced) Lefschetz module of $\tilde{C}_*(\mathcal{F})$, and call it the *Lefschetz conjugation module*; in their Theorem 4.2, they show that $\tilde{L}(B)$ is a virtual projective module (in the representation ring for G). This result is an analogue (for the non-constant system $\mathcal{F}$) of Webb's Projectivity Theorem 6.3.4—where Webb's result in effect uses the constant system on a complex K like $\mathcal{N}_p(G)$.

However to return to our original purpose: It is more relevant to instead consider (in the spirit of Bredon cohomology and G-equivariant cohomology) the homology of the G-fixed chain subcomplex $\tilde{C}_*(\mathcal{F})^G$. Indeed the results [**KR89**, 4.3, 4.5] show that the reduced Euler characteristic corresponding to this fixed subcomplex is in fact the expression in (11.2.6).

But this intriguing approach has not yet led to a proof of Alperin's Conjecture.

An approach via G-equivariant K-theory. The above notion, namely of expressions which formally resemble Euler characteristics of (unknown) coefficient systems, plays a role in other formulas appearing in [**KR89**]; and also in work of Thévenaz [**Thé92a**], where they are called *locally determined functions*. One feature of these expressions is that they take the same value, when computed over the orbit complex of any of the members L of the standard homotopy type in Remark 4.6.1—see for example [**KR89**, 4.3, 4.5] and [**Thé92a**, 2.4].

A further paper of Thévenaz [**Thé93**] observes that, if we replace the function $l_p(G)$ by the function $k(G)$ giving the number of *complex* irreducible characters of G, then the appropriate Euler-characteristic-like combination actually computes the "equivariant Euler characteristic" $\chi_G(L)$ of the underlying complex L; see [**Thé93**, 2.1]. This usage of $\chi_G(L)$ is different from that of Brown, which we had indicated in Section 5.6; instead it is connected with the G-equivariant K-theory of L, in the sense of Atiyah-Segal—see [**Thé93**, p 186]. In this context, Thévenaz gives an equivalent version of Alperin's Conjecture: namely that $\chi_G(\mathcal{S}_p(G))$ should be $k(G) - z_p(G)$.

Thévenaz also mentions as [**Thé93**, Cor 2.2] another way of computing $\chi_G(L)$, summing over orbit posets for fixed-point sets, of the form $L^g/C_G(g)$, as g ranges over the conjugacy classes of G. This result has been used in several ways: Thévenaz provides at [**Thé92a**, Cor 4.4] a proof of the vanishing result $\tilde{\chi}(K(\mathcal{S}_p(G))/G) = 0$, originally established by Webb in [**Web87a**, Prop. 8.2]. Also Lehrer and Thévenaz in [**LT92**] compute with the orbit complexes in the corresponding formula, to give a new proof of the Alperin Conjecture in the special case of groups G of Lie type in characteristic p (originally established by Cabanes).

We end the section by mentioning the variant approach to Alperin's Conjecture given by Dade—whose work, it is hoped, will reduce the problem to verification essentially for simple groups (and hence invoking the CFSG for that list). This in turn has motivated a program of verification of the Conjecture for simple groups—which in particular entails the *determination* of $\mathcal{B}_p(G)$, for each simple group G. For the latter, see for example various papers of An-O'Brien starting with [**AO98**], and the summary for sporadic groups in Yoshiara [**Yos05b**][4] and [**Yos05a**].

[4]A minor correction for J_2 appears in [**Yos06**].

Bibliography

[AA93] Herbert Abels and Peter Abramenko, *On the homotopy type of subcomplexes of Tits buildings*, Adv. Math. **101** (1993), no. 1, 78–86. MR1239453 (95a:55013)

[AB08] Peter Abramenko and Kenneth S. Brown, *Buildings*, Graduate Texts in Mathematics, vol. 248, Springer, New York, 2008, Theory and applications. MR2439729

[AG94] J. L. Alperin and George Glauberman, *Coverings and coverings*, J. Austral. Math. Soc. Ser. A **57** (1994), no. 1, 125–128. MR1279291 (95d:20044)

[AK90] Michael Aschbacher and Peter B. Kleidman, *On a conjecture of Quillen and a lemma of Robinson*, Arch. Math. (Basel) **55** (1990), no. 3, 209–217. MR1075043 (91j:20040)

[Ale37] P. S. Aleksandrov, *Diskrete Räume*, Mat. Sbornik (N.S.) **2** (1937), 501–518.

[Alp67] J. L. Alperin, *Sylow intersections and fusion*, J. Algebra **6** (1967), 222–241. MR0215913 (35 #6748)

[Alp87] ———, *Weights for finite groups*, The Arcata Conference on Representations of Finite Groups (Arcata, Calif., 1986), Proc. Sympos. Pure Math., vol. 47, Amer. Math. Soc., Providence, RI, 1987, pp. 369–379. MR933373 (89h:20015)

[Alp89] J. Alperin, *Group representations and complexes*, (manuscript lecture notes), 1989, (Mathematics 407–408, U. Chicago).

[Alp90] J. L. Alperin, *A Lie approach to finite groups*, Groups—Canberra 1989, Lecture Notes in Math., vol. 1456, Springer, Berlin, 1990, pp. 1–9. MR1092218 (92a:20011)

[ALSS11] Michael Aschbacher, Richard Lyons, Stephen D. Smith, and Ronald Solomon, *The classification of finite simple groups: Groups of characteristic* 2 *type*, Surveys and Monographs of the A. M. S., vol. 172, Amer. Math. Soc., Providence, RI, 2011.

[Alv79] Dean Alvis, *The duality operation in the character ring of a finite Chevalley group*, Bull. Amer. Math. Soc. (N.S.) **1** (1979), no. 6, 907–911. MR546315 (81e:20012)

[AM95a] Alejandro Adem and R. James Milgram, *The cohomology of the Mathieu group* M_{22}, Topology **34** (1995), no. 2, 389–410. MR1318884 (96c:20099)

[AM95b] ———, *The mod* 2 *cohomology rings of rank* 3 *simple groups are Cohen-Macaulay*, Prospects in topology (Princeton, NJ, 1994), Ann. of Math. Stud., vol. 138, Princeton Univ. Press, Princeton, NJ, 1995, pp. 3–12. MR1368650 (96m:20084)

[AM97] ———, *The cohomology of the McLaughlin group and some associated groups*, Math. Z. **224** (1997), no. 4, 495–517. MR1452047 (98e:20057)

[AM04] ———, *Cohomology of finite groups*, second ed., Grundlehren der Mathematischen Wissenschaften [Fundamental Principles of Mathematical Sciences], vol. 309, Springer-Verlag, Berlin, 2004. MR2035696 (2004k:20109)

[AMM91] Alejandro Adem, John Maginnis, and R. James Milgram, *The geometry and cohomology of the Mathieu group* M_{12}, J. Algebra **139** (1991), no. 1, 90–133. MR1106342 (92g:20083)

[AO98] Jianbei An and E. A. O'Brien, *A local strategy to decide the Alperin and Dade conjectures*, J. Algebra **206** (1998), no. 1, 183–207. MR1637276 (99d:20015)

[AS83] Michael Aschbacher and Stephen D. Smith, *Tits geometries over* GF(2) *defined by groups over* GF(3), Comm. Algebra **11** (1983), no. 15, 1675–1684. MR702413 (84g:51009)

[AS92a] M. Aschbacher and Y. Segev, *Locally connected simplicial maps*, Israel J. Math. **77** (1992), no. 3, 285–303. MR1194797 (94k:57036)

[AS92b] Michael Aschbacher and Yoav Segev, *Extending morphisms of groups and graphs*, Ann. of Math. (2) **135** (1992), no. 2, 297–323. MR1154595 (93d:20034)

[AS93] Michael Aschbacher and Stephen D. Smith, *On Quillen's conjecture for the p-groups complex*, Ann. of Math. (2) **137** (1993), no. 3, 473–529. MR1217346 (94g:20073)

[AS04a] ———, *The classification of quasithin groups. I*, Mathematical Surveys and Monographs, vol. 111, American Mathematical Society, Providence, RI, 2004, Structure of strongly quasithin K-groups. MR2097623

[AS04b] ———, *The classification of quasithin groups. II*, Mathematical Surveys and Monographs, vol. 112, American Mathematical Society, Providence, RI, 2004, Main theorems: the classification of simple QTKE-groups. MR2097624

[AS09] Michael Aschbacher and John Shareshian, *Restrictions on the structure of subgroup lattices of finite alternating and symmetric groups*, J. Algebra **322** (2009), no. 7, 2449–2463. MR2553689 (2010h:20048)

[Asc86] Michael Aschbacher, *Overgroups of Sylow subgroups in sporadic groups*, Mem. Amer. Math. Soc. **60** (1986), no. 343, iv+235. MR831891 (87e:20037)

[Asc93] ———, *Simple connectivity of p-group complexes*, Israel J. Math. **82** (1993), no. 1-3, 1–43. MR1239044 (94j:20012)

[Asc94] ———, *Sporadic groups*, Cambridge Tracts in Mathematics, vol. 104, Cambridge University Press, Cambridge, 1994. MR1269103 (96e:20020)

[Asc00] M. Aschbacher, *Finite group theory*, second ed., Cambridge Studies in Advanced Mathematics, vol. 10, Cambridge University Press, Cambridge, 2000, 274 pages. MR1777008 (2001c:20001)

[Ass91] Amir H. Assadi, *Permutation complexes and modular representation theory*, J. Algebra **144** (1991), no. 2, 467–495. MR1140617 (92h:20080)

[Bar95] Matthew Kirby Bardoe, *On the universal embedding of the near-hexagon for* $\mathrm{U}_4(3)$, Geom. Dedicata **56** (1995), no. 1, 7–17. MR1338314 (96f:51016)

[Bar96a] ———, *The universal embedding for the involution geometry of the Suzuki sporadic simple group*, J. Algebra **186** (1996), no. 2, 447–460. MR1423271 (98e:51016)

[Bar96b] ———, *The universal embedding for the $U_4(3)$ involution geometry*, J. Algebra **186** (1996), no. 2, 368–383. MR1423267 (98e:51015)

[Bar96c] Laurence Barker, *Möbius inversion and the Lefschetz invariants of some p-subgroup complexes*, Comm. Algebra **24** (1996), no. 8, 2755–2769. MR1393282 (97d:20007)

[Bar99] Matthew Kirby Bardoe, *The universal embedding for the involution geometry of* Co_1, J. Algebra **217** (1999), no. 2, 555–572. MR1700514 (2000h:51021)

[BC87] David J. Benson and Jon F. Carlson, *Diagrammatic methods for modular representations and cohomology*, Comm. Algebra **15** (1987), no. 1-2, 53–121. MR876974 (87m:20032)

[Ben84] David Benson, *Modular representation theory: new trends and methods*, Lecture Notes in Mathematics, vol. 1081, Springer-Verlag, Berlin, 1984. MR765858 (86g:20013)

[Ben91] D. J. Benson, *Representations and cohomology. II*, Cambridge Studies in Advanced Mathematics, vol. 31, Cambridge University Press, Cambridge, 1991, Cohomology of groups and modules. MR1156302 (93g:20099)

[Ben94] Dave Benson, *Conway's group* Co_3 *and the Dickson invariants*, Manuscripta Math. **85** (1994), no. 2, 177–193. MR1302871 (95h:55018)

[Ben98] D. J. Benson, *Representations and cohomology. I*, second ed., Cambridge Studies in Advanced Mathematics, vol. 30, Cambridge University Press, Cambridge, 1998, Basic representation theory of finite groups and associative algebras. MR1644252 (99f:20001a)

[BGS82] A. Björner, A. M. Garsia, and R. P. Stanley, *An introduction to Cohen-Macaulay partially ordered sets*, Ordered sets (Banff, Alta., 1981), NATO Adv. Study Inst. Ser. C: Math. Phys. Sci., vol. 83, Reidel, Dordrecht, 1982, pp. 583–615. MR661307 (83i:06001)

[Bir40] Garrett Birkhoff, *Lattice Theory*, American Mathematical Society, New York, 1940. MR0001959 (1,325f)

[Bjö80] Anders Björner, *Shellable and Cohen-Macaulay partially ordered sets*, Trans. Amer. Math. Soc. **260** (1980), no. 1, 159–183. MR570784 (81i:06001)

[Bjö84] ———, *Some combinatorial and algebraic properties of Coxeter complexes and Tits buildings*, Adv. in Math. **52** (1984), no. 3, 173–212. MR744856 (85m:52003)

[Bjö95] A. Björner, *Topological methods*, Handbook of combinatorics, Vol. 1, 2, Elsevier, Amsterdam, 1995, pp. 1819–1872. MR1373690 (96m:52012)

[BL97] Robert Baddeley and Andrea Lucchini, *On representing finite lattices as intervals in subgroup lattices of finite groups*, J. Algebra **196** (1997), no. 1, 1–100. MR1474164 (98j:20022)

[BLO03] Carles Broto, Ran Levi, and Bob Oliver, *The homotopy theory of fusion systems*, J. Amer. Math. Soc. **16** (2003), no. 4, 779–856 (electronic). MR1992826 (2004k:55016)

[Bor09] David Bornand, *Counterexamples to a fiber theorem*, J. Algebra **322** (2009), no. 7, 2428–2432. MR2553687 (2010i:20019)

[Bou] S. Bouc, *Complexes de p-permutations*, (preprint, ENS Paris, 1989).

[Bou84a] Serge Bouc, *Homologie de certains ensembles ordonnés*, C. R. Acad. Sci. Paris Sér. I Math. **299** (1984), no. 2, 49–52. MR756517 (85k:20150)

[Bou84b] ______, *Modules de Möbius*, C. R. Acad. Sci. Paris Sér. I Math. **299** (1984), no. 1, 9–12. MR756298 (85g:20064)

[Bou91] S. Bouc, *Projecteurs dans l'anneau de Burnside, projecteurs dans l'anneau de Green, et modules de Steinberg généralisés*, J. Algebra **139** (1991), no. 2, 395–445. MR1113783 (92k:20016)

[Bou92] S. Bouc, *Homologie de certains ensembles de 2-sous-groups des groups symmetriques*, J. Algebra **150** (1992), 158–186.

[Bre67] Glen E. Bredon, *Equivariant cohomology theories*, Lecture Notes in Mathematics, No. 34, Springer-Verlag, Berlin, 1967. MR0214062 (35 #4914)

[Bro74] Kenneth S. Brown, *Euler characteristics of discrete groups and G-spaces*, Invent. Math. **27** (1974), 229–264. MR0385007 (52 #5877)

[Bro75] ______, *Euler characteristics of groups: the p-fractional part*, Invent. Math. **29** (1975), no. 1, 1–5. MR0385008 (52 #5878)

[Bro94] ______, *Cohomology of groups*, Graduate Texts in Mathematics, vol. 87, Springer-Verlag, New York, 1994, Corrected reprint of the 1982 original. MR1324339 (96a:20072)

[Bro98] ______, *Buildings*, Springer Monographs in Mathematics, Springer-Verlag, New York, 1998, Reprint of the 1989 original. MR1644630 (99d:20042)

[BS08] David J. Benson and Stephen D. Smith, *Classifying spaces of sporadic groups*, Mathematical Surveys and Monographs, vol. 147, American Mathematical Society, Providence, RI, 2008.

[BT88] Kenneth S. Brown and Jacques Thévenaz, *A generalization of Sylow's third theorem*, J. Algebra **115** (1988), no. 2, 414–430. MR943266 (89e:20048)

[BT08] S. Bouc and Jacques Thévenaz, *The poset of elementary abelian subgroups of rank at least* 2, Guido's book of conjectures, Enseignement Mathématique, vol. 54, University of Geneva, 2008, pp. 41–45.

[Bue79] Francis Buekenhout, *Diagrams for geometries and groups*, J. Combin. Theory Ser. A **27** (1979), no. 2, 121–151. MR542524 (83f:51003)

[Bux99] Kai-Uwe Bux, *Orbit spaces of subgroup complexes, Morse theory, and a new proof of a conjecture of Webb*, Proceedings of the 1999 Topology and Dynamics Conference (Salt Lake City, UT), vol. 24, 1999, pp. 39–51. MR1802676 (2002d:20031)

[BW83a] Anders Björner and Michelle Wachs, *On lexicographically shellable posets*, Trans. Amer. Math. Soc. **277** (1983), no. 1, 323–341. MR690055 (84f:06004)

[BW83b] Anders Björner and James W. Walker, *A homotopy complementation formula for partially ordered sets*, European J. Combin. **4** (1983), no. 1, 11–19. MR694463 (84f:06003)

[BW95] D. J. Benson and C. W. Wilkerson, *Finite simple groups and Dickson invariants*, Homotopy theory and its applications (Cocoyoc, 1993), Contemp. Math., vol. 188, Amer. Math. Soc., Providence, RI, 1995, pp. 39–50. MR1349127 (96d:55009)

[BW96] Anders Björner and Michelle L. Wachs, *Shellable nonpure complexes and posets. I*, Trans. Amer. Math. Soc. **348** (1996), no. 4, 1299–1327. MR1333388 (96i:06008)

[BW97] ______, *Shellable nonpure complexes and posets. II*, Trans. Amer. Math. Soc. **349** (1997), no. 10, 3945–3975. MR1401765 (98b:06008)

[Cab88] Marc Cabanes, *Brauer morphism between modular Hecke algebras*, J. Algebra **115** (1988), no. 1, 1–31. MR937599 (89e:20019)

[Car89] Roger W. Carter, *Simple groups of Lie type*, Wiley Classics Library, John Wiley & Sons Inc., New York, 1989, Reprint of the 1972 original, A Wiley-Interscience Publication. MR1013112 (90g:20001)

[CCN+85] J. H. Conway, R. T. Curtis, S. P. Norton, R. A. Parker, and R. A. Wilson, *Atlas of finite groups*, Oxford University Press, Eynsham, 1985, Maximal subgroups and ordinary characters for simple groups, With computational assistance from J. G. Thackray. MR827219 (88g:20025)

[CL81] C. W. Curtis and G. I. Lehrer, *Homology representations of finite groups of Lie type*, Papers in algebra, analysis and statistics (Hobart, 1981), Contemp. Math., vol. 9, Amer. Math. Soc., Providence, R.I., 1981, pp. 1–28. MR655971 (83g:20042)

[Coh73] Marshall M. Cohen, *A course in simple-homotopy theory*, Springer-Verlag, New York, 1973, Graduate Texts in Mathematics, Vol. 10. MR0362320 (50 #14762)

[Coh94] Steve M. Cohen, *A geometric presentation of the* λ_2*-modules of* $C_n(q)$ *and* $D_n(q)$, J. Algebra **170** (1994), no. 2, 470–486. MR1302851 (96a:20019)

[Con68] S. B. Conlon, *Decompositions induced from the Burnside algebra*, J. Algebra **10** (1968), 102–122. MR0237664 (38 #5945)

[Con71] J. H. Conway, *Three lectures on exceptional groups*, Finite simple groups (Proc. Instructional Conf., Oxford, 1969), Academic Press, London, 1971, pp. 215–247. MR0338152 (49 #2918)

[CR90] Charles W. Curtis and Irving Reiner, *Methods of representation theory. Vol. I*, Wiley Classics Library, John Wiley & Sons Inc., New York, 1990, With applications to finite groups and orders, Reprint of the 1981 original, A Wiley-Interscience Publication. MR1038525 (90k:20001)

[CR06] ———, *Representation theory of finite groups and associative algebras*, AMS Chelsea Publishing, Providence, RI, 2006, Reprint of the 1962 original. MR2215618 (2006m:16001)

[CS90] Steve M. Cohen and Stephen D. Smith, *A geometric presentation of the 26-dimensional module for* $F_4(q)$, J. Algebra **132** (1990), no. 2, 474–487. MR1061492 (91h:20072)

[CST89] Peter J. Cameron, Ron Solomon, and Alexandre Turull, *Chains of subgroups in symmetric groups*, J. Algebra **127** (1989), no. 2, 340–352. MR1028457 (91a:20008)

[Cur70] C. W. Curtis, *Modular representations of finite groups with split* (B, N)*-pairs*, Seminar on Algebraic Groups and Related Finite Groups (The Institute for Advanced Study, Princeton, N.J., 1968/69), Springer, Berlin, 1970, pp. 57–95. MR0262383 (41 #6991)

[Cur71] ———, *Chevalley groups and related topics*, Finite simple groups (Proc. Instructional Conf., Oxford, 1969), Academic Press, London, 1971, pp. 135–189. MR0340442 (49 #5197)

[Das95] Kaustuv Mukul Das, *Simple connectivity of the Quillen complex of* $\mathrm{GL}_n(q)$, J. Algebra **178** (1995), no. 1, 239–263. MR1358264 (97b:20025)

[Das98] ———, *Some results about the Quillen complex of* $\mathrm{Sp}_{2n}(q)$, J. Algebra **209** (1998), no. 2, 427–445. MR1659907 (99i:20066)

[Das00] ———, *The Quillen complex of groups of symplectic type: the characteristic* 2 *case*, J. Algebra **223** (2000), no. 2, 556–561. MR1735162 (2000j:20038)

[DF99] D. Dummit and R. Foote, *Abstract algebra*, Prentice-Hall, Upper Saddle River NJ, 1999.

[DGM09] Alice Devillers, Ralf Gramlich, and Bernhard Mühlherr, *The sphericity of the complex of non-degenerate subspaces*, J. Lond. Math. Soc. (2) **79** (2009), no. 3, 684–700. MR2506693 (2010c:51018)

[DK78] Gopal Danaraj and Victor Klee, *A representation of* 2*-dimensional pseudomanifolds and its use in the design of a linear-time shelling algorithm*, Ann. Discrete Math. **2** (1978), 53–63, Algorithmic aspects of combinatorics (Conf., Vancouver Island, B.C., 1976). MR500688 (80b:57013b)

[DK01] James F. Davis and Paul Kirk, *Lecture notes in algebraic topology*, Graduate Studies in Mathematics, vol. 35, American Mathematical Society, Providence, RI, 2001. MR1841974 (2002f:55001)

[DL76] P. Deligne and G. Lusztig, *Representations of reductive groups over finite fields*, Ann. of Math. (2) **103** (1976), no. 1, 103–161. MR0393266 (52 #14076)

[Dre69] Andreas Dress, *A characterisation of solvable groups*, Math. Z. **110** (1969), 213–217. MR0248239 (40 #1491)

[DS87] Andreas W. M. Dress and Rudolf Scharlau, *Gated sets in metric spaces*, Aequationes Math. **34** (1987), no. 1, 112–120. MR915878 (89c:54057)

[Dwy97] W. G. Dwyer, *Homology decompositions for classifying spaces of finite groups*, Topology **36** (1997), no. 4, 783–804. MR1432421 (97m:55016)

[Dwy98] ———, *Sharp homology decompositions for classifying spaces of finite groups*, Group representations: cohomology, group actions and topology (Seattle, WA, 1996), Proc. Sympos. Pure Math., vol. 63, Amer. Math. Soc., Providence, RI, 1998, pp. 197–220. MR1603159 (99b:55033)

[Dwy01] W. G. Dwyer, *Classifying spaces and homology decompositions*, Homotopy theoretic methods in group cohomology (W. G. Dwyer and H.-W. Henn, eds.), Advanced Courses in Mathematics—CRM Barcelona, Birkhäuser, 2001, pp. 3–53.

[Fei82] Walter Feit, *The representation theory of finite groups*, North-Holland Mathematical Library, vol. 25, North-Holland Publishing Co., Amsterdam, 1982. MR661045 (83g:20001)

[Fei93] ———, *Extending Steinberg characters*, Linear algebraic groups and their representations (Los Angeles, CA, 1992), Contemp. Math., vol. 153, Amer. Math. Soc., Providence, RI, 1993, pp. 1–9. MR1247494 (94k:20012)

[FH64] W. Feit and G. Higman, *The nonexistence of certain generalized polygons*, J. of Algebra **1** (1964), 114–131.

[Fis93] Tony J. Fisher, *Weight operators and group geometries*, J. Algebra **158** (1993), no. 1, 61–108. MR1223668 (94h:20038)

[Fol66] J. Folkman, *The homology groups of a lattice*, J. Math. Mech. **15** (1966), 631–636.

[Fro03] G. Frobenius, *Über einen fundamentalsatz der gruppentheorie*, Berliner Sitzungsberichte (1903), 987–991.

[FS73] Paul Fong and Gary M. Seitz, *Groups with a (B, N)-pair of rank 2. I, II*, Invent. Math. **21** (1973), 1–57; ibid. 24 (1974), 191–239. MR0354858 (50 #7335)

[FS92] Daniel E. Frohardt and Stephen D. Smith, *Universal embeddings for the ${}^3D_4(2)$ hexagon and J_2 near-octagon*, European J. Combin. **13** (1992), no. 6, 455–472. MR1193554 (94a:51004)

[GAP00] GAP, *(The GAP group), GAP—Groups, Algorithms, and programming*, http://www-gap.idcs.st-and.ac.uk/~gap, 2000.

[Gar79] Peter Freedman Garst, *Cohen-Macaulay complexes and group actions*, ProQuest LLC, Ann Arbor, MI, 1979, Thesis (Ph.D.)–The University of Wisconsin - Madison. MR2628756

[Gar80] A. M. Garsia, *Combinatorial methods in the theory of Cohen-Macaulay rings*, Adv. in Math. **38** (1980), 229–266.

[GL83] Daniel Gorenstein and Richard Lyons, *The local structure of finite groups of characteristic 2 type*, Mem. Amer. Math. Soc. **42** (1983), no. 276, vii+731. MR690900 (84g:20025)

[GLS94] Daniel Gorenstein, Richard Lyons, and Ronald Solomon, *The classification of the finite simple groups*, Mathematical Surveys and Monographs, vol. 40, American Mathematical Society, Providence, RI, 1994. MR1303592 (95m:20014)

[GLS96] ———, *The classification of the finite simple groups. Number 2. Part I. Chapter G*, Mathematical Surveys and Monographs, vol. 40, American Mathematical Society, Providence, RI, 1996, General group theory. MR1358135 (96h:20032)

[GLS98] ———, *The classification of the finite simple groups. Number 3. Part I. Chapter A*, Mathematical Surveys and Monographs, vol. 40, American Mathematical Society, Providence, RI, 1998, Almost simple K-groups. MR1490581 (98j:20011)

[Glu81] David Gluck, *Idempotent formula for the Burnside algebra with applications to the p-subgroup simplicial complex*, Illinois J. Math. **25** (1981), no. 1, 63–67. MR602896 (82c:20005)

[Gol70] David M. Goldschmidt, *A conjugation family for finite groups*, J. Algebra **16** (1970), 138–142. MR0260869 (41 #5489)

[Gor80] Daniel Gorenstein, *Finite groups*, second ed., Chelsea Publishing Co., New York, 1980. MR569209 (81b:20002)

[Gra04] R. Gramlich, *Phan theory*, Habilitationsschrift, TU Darmstadt, 2004.

[Gri09] R. Philip Grizzard, *On nonprojective block components of Lefschetz characters for sporadic geometries*, Comm. Algebra **37** (2009), no. 12, 4489–4502. MR2588862 (2011a:20018)

[Gro02] Jesper Grodal, *Higher limits via subgroup complexes*, Ann. of Math. (2) **155** (2002), no. 2, 405–457. MR1906592 (2003g:55025)

[GS06] Jesper Grodal and Stephen D. Smith, *Propagating sharp group homology decompositions*, Adv. Math. **200** (2006), no. 2, 525–538. MR2200855 (2007e:20110)

[Hal36] P. Hall, *The Eulerian functions of a group*, Oxford Quarterly J. Math. **7** (1936), 134–151.

[Hal59] Marshall Hall, Jr., *The theory of groups*, The Macmillan Co., New York, N.Y., 1959. MR0103215 (21 #1996)

[Hat02] Allen Hatcher, *Algebraic topology*, Cambridge University Press, Cambridge, 2002. MR1867354 (2002k:55001)

[Hen97] Hans-Werner Henn, *Centralizers of elementary abelian p-subgroups, the Borel construction of the singular locus and applications to the cohomology of discrete groups*, Topology **36** (1997), no. 1, 271–286. MR1410475 (98g:57060)

[Her75] I. N. Herstein, *Topics in algebra*, second ed., Xerox College Publishing, Lexington, Mass., 1975. MR0356988 (50 #9456)

[HI88] Trevor Hawkes and I. M. Isaacs, *On the poset of p-subgroups of a p-solvable group*, J. London Math. Soc. (2) **38** (1988), no. 1, 77–86. MR949083 (89g:20083)

[HIÖ89] T. Hawkes, I. M. Isaacs, and M. Özaydin, *On the Möbius function of a finite group*, Rocky Mountain J. Math. **19** (1989), no. 4, 1003–1034. MR1039540 (90k:20046)

[Hum72] J. E. Humphreys, *Introduction to lie algebras and representation theory*, Graduate Texts in Mathematics, vol. 9, Springer-Verlag, Berlin-New York, 1972.

[Hun80] Thomas W. Hungerford, *Algebra*, Graduate Texts in Mathematics, vol. 73, Springer-Verlag, New York, 1980, Reprint of the 1974 original. MR600654 (82a:00006)

[Hup67] B. Huppert, *Endliche Gruppen. I*, Die Grundlehren der Mathematischen Wissenschaften, Band 134, Springer-Verlag, Berlin, 1967. MR0224703 (37 #302)

[HW60] P. J. Hilton and S. Wylie, *Homology theory: An introduction to algebraic topology*, Cambridge University Press, New York, 1960. MR0115161 (22 #5963)

[IS94a] A. A. Ivanov and S. V. Shpectorov, *The flag-transitive tilde and Petersen-type geometries are all known*, Bull. Amer. Math. Soc. (N.S.) **31** (1994), no. 2, 173–184. MR1256977 (95a:51021)

[IS94b] ———, *Natural representations of the P-geometries of* Co_2*-type*, J. Algebra **164** (1994), no. 3, 718–749. MR1272112 (95h:20020)

[Isa06] I. Martin Isaacs, *Character theory of finite groups*, AMS Chelsea Publishing, Providence, RI, 2006, Corrected reprint of the 1976 original [Academic Press, New York; MR0460423]. MR2270898

[Iva91] Alexander A. Ivanov, *Geometric presentations of groups with an application to the Monster*, Proceedings of the International Congress of Mathematicians, Vol. I, II (Kyoto, 1990) (Tokyo), Math. Soc. Japan, 1991, pp. 1443–1453. MR1159328 (93d:20036)

[Iva99] A. A. Ivanov, *Geometry of sporadic groups. I*, Encyclopedia of Mathematics and its Applications, vol. 76, Cambridge University Press, Cambridge, 1999, Petersen and tilde geometries.

[IY91] Nobuo Iiyori and Hiroyoshi Yamaki, *On a conjecture of Frobenius*, Bull. Amer. Math. Soc. (N.S.) **25** (1991), no. 2, 413–416. MR1102753 (92e:20014)

[Jac80] N. Jacobson, *Basic algebra I,II*, Freeman, San Francisco, 1980.

[JLPW95] C. Jansen, K. Lux, R. Parker, and R. Wilson, *An Atlas of Brauer characters*, London Math. Soc. Monographs (new series), vol. 11, Oxford U. Press, Oxford, 1995, 327 pages.

[JM89] Stefan Jackowski and James E. McClure, *Homotopy approximations for classifying spaces of compact Lie groups*, Algebraic topology (Arcata, CA, 1986), Lecture Notes in Math., vol. 1370, Springer, Berlin, 1989, pp. 221–234. MR1000379 (90m:55012)

[JM92] Stefan Jackowski and James McClure, *Homotopy decomposition of classifying spaces via elementary abelian subgroups*, Topology **31** (1992), no. 1, 113–132. MR1153240 (92k:55026)

[JMO92] Stefan Jackowski, James McClure, and Bob Oliver, *Homotopy classification of self-maps of BG via G-actions. I*, Ann. of Math. (2) **135** (1992), no. 1, 183–226. MR1147962 (93e:55019a)

[Kan81] William M. Kantor, *Some geometries that are almost buildings*, European J. Combin. **2** (1981), no. 3, 239–247. MR633119 (83a:51020)

[Kan85] ———, *Some exceptional 2-adic buildings*, J. Algebra **92** (1985), no. 1, 208–223. MR7724804(86j:51007)

[Kan86] ———, *Generalized polygons, SCABs and GABs*, Buildings and the geometry of diagrams (Como, 1984), Lecture Notes in Math., vol. 1181, Springer, Berlin, 1986, pp. 79–158. MR843390 (87k:51014)

[KLT87] W. M. Kantor, R. A. Liebler, and J. Tits, *On discrete chamber-transitive automorphism groups of affine buildings*, Bull. Amer. Math. Soc. (N.S.) **16** (1987), no. 1, 129–133. MR866031 (87m:20124)

[KMW84] Peter Köhler, Thomas Meixner, and Michael Wester, *The affine building of type $\tilde{A}_2$ over a local field of characteristic two*, Arch. Math. (Basel) **42** (1984), no. 5, 400–407. MR756691 (85j:20043)

[KMW90] W. M. Kantor, T. Meixner, and M. Wester, *Two exceptional 3-adic affine buildings*, Geom. Dedicata **33** (1990), no. 1, 1–11. MR1042619 (91d:20050)

[KÖzaydin] S. Kutin and M. Özaydin, *On the shellability of the poset of p-subgroups of a finite group*, U. Okla. preprint 1995.

[KR89] Reinhard Knörr and Geoffrey R. Robinson, *Some remarks on a conjecture of Alperin*, J. London Math. Soc. (2) **39** (1989), no. 1, 48–60. MR989918 (90k:20020)

[Kso04] Rached Ksontini, *The fundamental group of the Quillen complex of the symmetric group*, J. Algebra **282** (2004), no. 1, 33–57. MR2095571 (2005f:20039)

[KT84] Charles Kratzer and Jacques Thévenaz, *Fonction de Möbius d'un groupe fini et anneau de Burnside*, Comment. Math. Helv. **59** (1984), no. 3, 425–438. MR761806 (86k:20011)

[KT85] ———, *Type d'homotopie des treillis et treillis des sous-groupes d'un groupe fini*, Comment. Math. Helv. **60** (1985), no. 1, 85–106. MR787663 (87b:06017)

[Lak72] Harry Lakser, *The homology of a lattice*, Discrete Math. **1** (1971/1972), no. 2, 187–192. MR0288755 (44 #5951)

[Lan65] S. Lang, *Algebra*, Addison-Wesley, Reading MA, 1965.

[Li01] Paul Li, *On the universal embedding of the $\mathrm{Sp}_{2n}(2)$ dual polar space*, J. Combin. Theory Ser. A **94** (2001), no. 1, 100–117. MR1816249 (2002a:51018)

[Lib07] Assaf Libman, *Minami-Webb splittings and some exotic p-local finite groups*, Math. Z. **255** (2007), no. 3, 515–548. MR2270287 (2007k:55025)

[Lin09] Markus Linckelmann, *The orbit space of a fusion system is contractible*, Proc. Lond. Math. Soc. (3) **98** (2009), no. 1, 191–216. MR2472165 (2009i:55025)

[LP89] Guglielmo Lunardon and Antonio Pasini, *A result on C_3-geometries*, European J. Combin. **10** (1989), no. 3, 265–271. MR1029173 (90m:51002)

[LR93] G. I. Lehrer and L. J. Rylands, *The split building of a reductive group*, Math. Ann. **296** (1993), no. 4, 607–624. MR1233486 (94g:20064)

[LT92] Gustav Isaac Lehrer and Jacques Thévenaz, *Sur la conjecture d'Alperin pour les groupes réductifs finis*, C. R. Acad. Sci. Paris Sér. I Math. **315** (1992), no. 13, 1347–1351. MR1199000 (93k:20067)

[Luc95] Maria Silvia Lucido, *On the partially ordered set of nilpotent subgroups of a finite group*, Comm. Algebra **23** (1995), no. 5, 1825–1836. MR1323703 (96h:20092)

[Luc03] ———, *On the poset of non-trivial proper subgroups of a finite group*, J. Algebra Appl. **2** (2003), no. 2, 165–168. MR1980405 (2004d:20021)

[Lus74] George Lusztig, *The discrete series of GL_n over a finite field*, Princeton University Press, Princeton, N.J., 1974, Annals of Mathematics Studies, No. 81. MR0382419 (52 #3303)

[Lus75] G. Lusztig, *On the discrete series representations of the classical groups over finite fields*, Proceedings of the International Congress of Mathematicians (Vancouver, B. C., 1974),Vol. 1, Canad. Math. Congress, Montreal, Que., 1975, pp. 465–470. MR0419628 (54 #7646)

[Mac63] George W. Mackey, *The mathematical foundations of quantum mechanics: A lecture-note volume*, W., A. Benjamin, Inc., New York-Amsterdam, 1963. MR0155567 (27 #5501)

[Mag95] J. Maginnis, *Local control of group cohomology*, J. Pure and Appl. Algebra **103** (1995), 81–90.

[Mat94] Andrew Mathas, *A q-analogue of the Coxeter complex*, J. Algebra **164** (1994), no. 3, 831–848. MR1272115 (95c:20055)

[Mat09] Francesco Matucci, *Finite solvable groups whose Quillen complex is Cohen-Macaulay*, J. Algebra **322** (2009), no. 4, 969–982. MR2537666 (2010h:20036)

[Mil56] J. Milnor, *Construction of universal bundles, II*, Ann. Math. **63** (1956), 430–436.

[MO08] John Maginnis and Silvia Onofrei, *New collections of p-subgroups and homology decompositions for classifying spaces of finite groups*, Comm. Algebra **36** (2008), no. 7, 2466–2480. MR2422498 (2009d:20040)

[MO09] ———, *On fixed point sets and Lefschetz modules for sporadic simple groups*, J. Pure Appl. Algebra **213** (2009), no. 6, 901–912. MR2498784 (2010b:20023)

[MO10] ———, *On fixed point sets of distinguished collections for groups of parabolic characteristic*, J. Combin. Theory Ser. A **117** (2010), no. 7, 872–883. MR2652100

[MS95] Bernhard Mühlherr and Peter Schmid, *The extended Steinberg character*, J. Algebra **175** (1995), no. 1, 373–384. MR1338984 (96h:20013)

[MT95] R. James Milgram and Michishige Tezuka, *The geometry and cohomology of M_{12}. II*, Bol. Soc. Mat. Mexicana (3) **1** (1995), no. 2, 91–108. MR1387653 (97f:20064)

[Mun84] James R. Munkres, *Elements of algebraic topology*, Addison-Wesley Publishing Company, Menlo Park, CA, 1984. MR755006 (85m:55001)

[Neu84] A. Neumaier, *Some sporadic geometries related to* PG(3, 2), Arch. Math. (Basel) **42** (1984), no. 1, 89–96. MR751477 (86a:51028)

[Oli75] Robert Oliver, *Fixed-point sets of group actions on finite acyclic complexes*, Comment. Math. Helv. **50** (1975), 155–177. MR0375361 (51 #11556)

[Oli76] ———, *A proof of the Conner conjecture*, Ann. of Math. (2) **103** (1976), no. 3, 637–644. MR0415650 (54 #3730)

[OS09] Fumihito Oda and Masato Sawabe, *A collection of subgroups for the generalized Burnside ring*, Adv. Math. **222** (2009), no. 1, 307–317. MR2531376 (2010k:19001)

[Pah93] Herbert Pahlings, *On the Möbius function of a finite group*, Arch. Math. (Basel) **60** (1993), no. 1, 7–14. MR1193087 (94c:20039)

[Pah95] H. Pahlings, *Character polynomials and the Möbius function*, Arch. Math. (Basel) **65** (1995), no. 2, 111–118. MR1338241 (96g:20029)

[Pul] J. Pulkus, *Das poset der p-untergruppen einer endlichen gruppe*, (Diplomarbeit, U. Erlangen, 1994).

[PW00] Jürgen Pulkus and Volkmar Welker, *On the homotopy type of the p-subgroup complex for finite solvable groups*, J. Austral. Math. Soc. Ser. A **69** (2000), no. 2, 212–228. MR1775179 (2001i:20046)

[PY01] Jonathan Pakianathan and Ergün Yalçın, *On commuting and noncommuting complexes*, J. Algebra **236** (2001), no. 1, 396–418. MR1808361 (2002c:20036)

[Qui71] Daniel Quillen, *The spectrum of an equivariant cohomology ring. I, II*, Ann. of Math. (2) **94** (1971), 549–572; ibid. (2) 94 (1971), 573–602. MR0298694 (45 #7743)

[Qui78] ———, *Homotopy properties of the poset of nontrivial p-subgroups of a group*, Adv. in Math. **28** (1978), no. 2, 101–128. MR493916 (80k:20049)

[Ram05] Daniel A. Ramras, *Connectivity of the coset poset and the subgroup poset of a group*, J. Group Theory **8** (2005), no. 6, 719–746. MR2179666 (2006g:20039)

[Rob88] Geoffrey R. Robinson, *Some remarks on permutation modules*, J. Algebra **118** (1988), no. 1, 46–62. MR961325 (89k:20018)

[Rob89] ———, *On projective summands of induced modules*, J. Algebra **122** (1989), no. 1, 106–111. MR994938 (90c:20011)

[Ron81] Mark A. Ronan, *Coverings of certain finite geometries*, Finite geometries and designs (Proc. Conf., Chelwood Gate, 1980), London Math. Soc. Lecture Note Ser., vol. 49, Cambridge Univ. Press, Cambridge, 1981, pp. 316–331. MR627510 (83b:51015)

[Ron84] M. A. Ronan, *Triangle geometries*, J. Combin. Theory Ser. A **37** (1984), no. 3, 294–319. MR769219 (86f:51008)

[Ron87] ———, *Embeddings and hyperplanes of discrete geometries*, European J. Combin. **8** (1987), no. 2, 179–185. MR896131 (88h:51009)

[Ron89a] ———, *Duality for presheaves on chamber systems, and a related chain complex*, J. Algebra **121** (1989), no. 2, 263–274. MR992762 (90c:17014)

[Ron89b] Mark Ronan, *Lectures on buildings*, Perspectives in Mathematics, vol. 7, Academic Press Inc., Boston, MA, 1989. MR1005533 (90j:20001)

[Rot64] Gian-Carlo Rota, *On the foundations of combinatorial theory. I. Theory of Möbius functions*, Z. Wahrscheinlichkeitstheorie und Verw. Gebiete **2** (1964), 340–368 (1964). MR0174487 (30 #4688)

[Rot88] Joseph J. Rotman, *An introduction to algebraic topology*, Graduate Texts in Mathematics, vol. 119, Springer-Verlag, New York, 1988. MR957919 (90e:55001)

[RS80] M. A. Ronan and S. D. Smith, *2-local geometries for some sporadic groups*, The Santa Cruz Conference on Finite Groups (Univ. California, Santa Cruz, Calif., 1979), Proc. Sympos. Pure Math., vol. 37, Amer. Math. Soc., Providence, R.I., 1980, pp. 283–289. MR604595 (82c:20033)

[RS84] M. A. Ronan and G. Stroth, *Minimal parabolic geometries for the sporadic groups*, European J. Combin. **5** (1984), no. 1, 59–91.

[RS85] Mark A. Ronan and Stephen D. Smith, *Sheaves on buildings and modular representations of Chevalley groups*, J. Algebra **96** (1985), no. 2, 319–346. MR810532 (87h:20087)

[RS86] ———, *Universal presheaves on group geometries, and modular representations*, J. Algebra **102** (1986), no. 1, 135–154. MR853235 (87m:20022)

[RS89] ———, *Computation of 2-modular sheaves and representations for $L_4(2), A_7, 3S_6$, and M_{24}*, Comm. Algebra **17** (1989), no. 5, 1199–1237. MR993399 (90c:20020)

[RSY90] A. J. E. Ryba, Stephen D. Smith, and Satoshi Yoshiara, *Some projective modules determined by sporadic geometries*, J. Algebra **129** (1990), no. 2, 279–311. MR1040940 (91a:20019)

[RT87] M. A. Ronan and J. Tits, *Building buildings*, Math. Ann. **278** (1987), no. 1-4, 291–306. MR909229 (89e:51005)

[Saw03] Masato Sawabe, *On a p-local geometry associated with a complex of non-trivial p-subgroups*, Bull. London Math. Soc. **35** (2003), no. 2, 196–202. MR1952395 (2003j:20044)

[Saw05] ———, *On the reduced Lefschetz module and the centric p-radical subgroups*, Tokyo J. Math. **28** (2005), no. 1, 79–90. MR2149625

[Sch92] Peter Schmid, *Extending the Steinberg representation*, J. Algebra **150** (1992), no. 1, 254–256. MR1174899 (93k:20016)

[Seg94] Yoav Segev, *Simply connected coset complexes for rank 1 groups of Lie type*, Math. Z. **217** (1994), no. 2, 199–214. MR1296394 (95j:20012)

[Sha97] John Shareshian, *On the Möbius number of the subgroup lattice of the symmetric group*, J. Combin. Theory Ser. A **78** (1997), no. 2, 236–267. MR1445416 (98c:20039)

[Sha01] ———, *On the shellability of the order complex of the subgroup lattice of a finite group*, Trans. Amer. Math. Soc. **353** (2001), no. 7, 2689–2703. MR1828468 (2002k:06006)

[Sha03] ———, *Topology of order complexes of intervals in subgroup lattices*, J. Algebra **268** (2003), no. 2, 677–686. MR2009327 (2004h:20032)

[Sha04] ———, *Hypergraph matching complexes and Quillen complexes of symmetric groups*, J. Combin. Theory Ser. A **106** (2004), no. 2, 299–314. MR2059000 (2005e:20029)

[Smi82] Stephen D. Smith, *Irreducible modules and parabolic subgroups*, J. Algebra **75** (1982), no. 1, 286–289. MR650422 (83g:20043)

[Smi85] ———, *Sheaf homology and complete reducibility*, J. Algebra **95** (1985), no. 1, 72–80. MR797655 (87e:20077)

[Smi88a] ———, *Construction of group representations from geometries*, Computers in algebra (Chicago, IL, 1985), Lecture Notes in Pure and Appl. Math., vol. 111, Dekker, New York, 1988, pp. 143–149. MR1060765 (91h:20019)

[Smi88b] ———, *Geometric methods in group theory*, Algebraic, extremal and metric combinatorics, 1986 (Montreal, PQ, 1986), London Math. Soc. Lecture Note Ser., vol. 131, Cambridge Univ. Press, Cambridge, 1988, pp. 219–237. MR1052669 (91k:20044)

[Smi90] ———, *On decomposition of modular representations from Cohen-Macaulay geometries*, J. Algebra **131** (1990), no. 2, 598–625. MR1058568 (91h:20025)

[Smi94a] ———, *The dual-parapolar space for the Monster embeds in the Griess module* Mod 2, European J. Combin. **15** (1994), no. 1, 57–65, Algebraic combinatorics (Vladimir, 1991). MR1257675 (95d:51013)

[Smi94b] ———, *Universality of the 24-dimensional embedding of the* ·1 *2-local geometry*, Comm. Algebra **22** (1994), no. 13, 5159–5166. MR1291025 (95e:20028)

[Sol67] Louis Solomon, *The Burnside algebra of a finite group*, J. Combinatorial Theory **2** (1967), 603–615. MR0214679 (35 #5528)

[Sol69] ———, *The Steinberg character of a finite group with BN-pair*, Theory of Finite Groups (Symposium, Harvard Univ., Cambridge, Mass., 1968), Benjamin, New York, 1969, pp. 213–221. MR0246951 (40 #220)

[Spa81] Edwin H. Spanier, *Algebraic topology*, Springer-Verlag, New York, 1981, Corrected reprint. MR666554 (83i:55001)

[SS86] Yoav Segev and Stephen D. Smith, *Apartments and the Cayley-algebra module for* $G_2(k)$, Comm. Algebra **14** (1986), no. 3, 493–506. MR823350 (87h:20076)

[Sta82] Richard P. Stanley, *Some aspects of groups acting on finite posets*, J. Combin. Theory Ser. A **32** (1982), no. 2, 132–161. MR654618 (83d:06002)

[Sta86] ———, *Enumerative combinatorics. Vol. I*, The Wadsworth & Brooks/Cole Mathematics Series, Wadsworth & Brooks/Cole Advanced Books & Software, Monterey, CA, 1986, With a foreword by Gian-Carlo Rota. MR847717 (87j:05003)

[SU96] Stephen D. Smith and Kristin L. Umland, *Stability of cohomology via double-coset products and suborbit diagrams*, J. Algebra **182** (1996), no. 3, 627–652. MR1398114 (98a:20057)

[Sur85] David B. Surowski, *A character theoretic proof of the homological Sylow theorem*, Comm. Algebra **13** (1985), no. 4, 987–989. MR776871 (86f:20061)

[Suz56] Michio Suzuki, *Structure of a group and the structure of its lattice of subgroups*, Ergebnisse der Mathematik und ihrer Grenzgebiete, Neue Folge, Heft 10, Springer-Verlag, Berlin, 1956. MR0083487 (18,715e)

[SV89] Stephen D. Smith and Helmut Völklein, *A geometric presentation for the adjoint module of* $\mathrm{SL}_3(k)$, J. Algebra **127** (1989), no. 1, 127–138. MR1029407 (91e:20033a)

[SW94] Yoav Segev and Peter Webb, *Extensions of G-posets and Quillen's complex*, J. Austral. Math. Soc. Ser. A **57** (1994), no. 1, 60–75. MR1279286 (95d:20042)

[SW09] John Shareshian and Michelle L. Wachs, *Top homology of hypergraph matching complexes, p-cycle complexes and Quillen complexes of symmetric groups*, J. Algebra **322** (2009), no. 7, 2253–2271. MR2553198 (2010j:20081)

[Swe09] Daniel E. Swenson, *The Steinberg complex of an arbitrary finite group in arbitrary positive characteristic*, ProQuest LLC, Ann Arbor, MI, 2009, Thesis (Ph.D.)–University of Minnesota. MR2713697

[SY95] Stephen D. Smith and Satoshi Yoshiara, *Group geometries and error-correcting codes*, J. Algebra **175** (1995), no. 1, 199–211. MR1338975 (97c:94016)

[SY97] ———, *Some homotopy equivalences for sporadic geometries*, J. Algebra **192** (1997), no. 1, 326–379. MR1449965 (98e:20024)

[Sym98] Peter Symonds, *The orbit space of the p-subgroup complex is contractible*, Comment. Math. Helv. **73** (1998), no. 3, 400–405. MR1633359 (99f:20033)

[Sym05] ———, *The Bredon cohomology of subgroup complexes*, J. Pure Appl. Algebra **199** (2005), no. 1-3, 261–298. MR2134305 (2006e:20093)

[Sym08] ———, *A relative version of Webb's theorem on the equivariant chain complex of a subgroup complex*, J. Pure Appl. Algebra **212** (2008), no. 8, 1984–1986. MR2414698 (2009d:20125)

[tD79] Tammo tom Dieck, *Transformation groups and representation theory*, Lecture Notes in Mathematics, vol. 766, Springer, Berlin, 1979. MR551743 (82c:57025)

[Thé85] Jacques Thévenaz, *The top homology of the lattice of subgroups of a soluble group*, Discrete Math. **55** (1985), no. 3, 291–303. MR802667 (86j:20045)

[Thé86] ———, *Idempotents de l'anneau de Burnside et caractéristique d'Euler*, Séminaire sur les groupes finis, Tome III, Publ. Math. Univ. Paris VII, vol. 25, Univ. Paris VII, Paris, 1986, pp. iii, 207–217. MR907481 (88i:20040)

[Thé87] ———, *Permutation representations arising from simplicial complexes*, J. Combin. Theory Ser. A **46** (1987), no. 1, 121–155. MR899904 (88f:20025)

[Thé92a] ———, *Locally determined functions and Alperin's conjecture*, J. London Math. Soc. (2) **45** (1992), no. 3, 446–468. MR1180255 (94f:20025)

[Thé92b] ———, *On a conjecture of Webb*, Arch. Math. (Basel) **58** (1992), no. 2, 105–109. MR1143156 (93d:20095)

[Thé93] ———, *Equivariant K-theory and Alperin's conjecture*, J. Pure Appl. Algebra **85** (1993), no. 2, 185–202. MR1207508 (94c:20022)

[Tho67] John G. Thompson, *Defect groups are Sylow intersections*, Math. Z. **100** (1967), 146. MR0213432 (35 #4296)

[Tho68] ———, *Nonsolvable finite groups all of whose local subgroups are solvable*, Bull. Amer. Math. Soc. **74** (1968), 383–437. MR0230809 (37 #6367)

[Tim83] Franz G. Timmesfeld, *Tits geometries and parabolic systems in finitely generated groups. I*, Math. Z. **184** (1983), no. 3, 377–396. MR716284 (85g:20050a)

[Tim01] Franz Georg Timmesfeld, *Abstract root subgroups and simple groups of Lie type*, Monographs in Mathematics, vol. 95, Birkhäuser Verlag, Basel, 2001. MR1852057 (2002f:20070)

[Tit74] Jacques Tits, *Buildings of spherical type and finite BN-pairs*, Springer-Verlag, Berlin, 1974, Lecture Notes in Mathematics, Vol. 386. MR0470099 (57 #9866)

[Tit81] J. Tits, *A local approach to buildings*, The geometric vein, Springer, New York, 1981, pp. 519–547. MR661801 (83k:51014)

[Tit86] Jacques Tits, *Immeubles de type affine*, Buildings and the geometry of diagrams (Como, 1984), Lecture Notes in Math., vol. 1181, Springer, Berlin, 1986, pp. 159–190. MR843391 (87h:20077)

[Tit92] ———, *Twin buildings and groups of Kac-Moody type*, Groups, combinatorics & geometry (Durham, 1990), London Math. Soc. Lecture Note Ser., vol. 165, Cambridge Univ. Press, Cambridge, 1992, pp. 249–286. MR1200265 (94d:20030)

[TW91] J. Thévenaz and P. J. Webb, *Homotopy equivalence of posets with a group action*, J. Combin. Theory Ser. A **56** (1991), no. 2, 173–181. MR1092846 (92k:20049)

[TW02] Jacques Tits and Richard M. Weiss, *Moufang polygons*, Springer Monographs in Mathematics, Springer-Verlag, Berlin, 2002. MR1938841 (2003m:51008)

[TY96] M. Tezuka and N. Yagita, *On odd prime components of cohomologies of sporadic simple groups and the rings of universal stable elements*, J. Algebra **183** (1996), no. 2, 483–513. MR1399037 (97e:20073)

[Vog82] K. Vogtmann, *A Stiefel complex for the orthogonal group of a field*, Comment. Helv. Math. **57** (1982), 11–21.

[Völ89a] Helmut Völklein, *The 1-cohomology of the adjoint module of a Chevalley group*, Forum Math. **1** (1989), no. 1, 1–13. MR978973 (90b:20038)

[Völ89b] ———, *On the geometry of the adjoint representation of a Chevalley group*, J. Algebra **127** (1989), no. 1, 139–154. MR1029408 (91e:20033b)

[Wal81a] J. W. Walker, *Topology and combinatorics of ordered sets*, Ph.D. thesis, M. I. T., 1981.

[Wal81b] James W. Walker, *Homotopy type and Euler characteristic of partially ordered sets*, European J. Combin. **2** (1981), no. 4, 373–384. MR638413 (83g:06002)

[Web87a] P. J. Webb, *A local method in group cohomology*, Comment. Math. Helv. **62** (1987), no. 1, 135–167. MR882969 (88h:20065)

[Web87b] ———, *Subgroup complexes*, The Arcata Conference on Representations of Finite Groups (Arcata, Calif., 1986), Proc. Sympos. Pure Math., vol. 47, Amer. Math. Soc., Providence, RI, 1987, pp. 349–365. MR933372 (89f:20057)

[Web91] ———, *A split exact sequence of Mackey functors*, Comment. Math. Helv. **66** (1991), no. 1, 34–69. MR1090164 (92c:20095)

[Web00] Peter Webb, *A guide to Mackey functors*, Handbook of algebra, Vol. 2, North-Holland, Amsterdam, 2000, pp. 805–836. MR1759612 (2001f:20031)

[Wei03] Richard M. Weiss, *The structure of spherical buildings*, Princeton University Press, Princeton, NJ, 2003. MR2034361 (2005b:51027)

[Wel92] Volkmar Welker, *The poset of conjugacy classes of subgroups in a finite solvable group*, J. Algebra **148** (1992), no. 1, 203–218. MR1161572 (93e:20033)

[Wel94] ———, *Combinatorial properties of intervals in finite solvable groups*, Rocky Mountain J. Math. **24** (1994), no. 4, 1567–1590. MR1322248 (96b:20024)

[Wel95a] ———, *Direct sum decompositions of matroids and exponential structures*, J. Combin. Theory Ser. B **63** (1995), no. 2, 222–244. MR1320168 (96e:05042)

[Wel95b] ———, *Equivariant homotopy of posets and some applications to subgroup lattices*, J. Combin. Theory Ser. A **69** (1995), no. 1, 61–86. MR1309152 (95j:05172)

[Wil09] Robert A. Wilson, *The finite simple groups*, Graduate Texts in Mathematics, vol. 251, Springer-Verlag London Ltd., London, 2009. MR2562037 (2011e:20018)

[Woo08] Russ Woodroofe, *An EL-labeling of the subgroup lattice*, Proc. Amer. Math. Soc. **136** (2008), no. 11, 3795–3801. MR2425717 (2009f:06010)

[WW93] Michael Weidner and Volkmar Welker, *The combinatorics and the homology of the poset of subgroups of p-power index*, J. Pure Appl. Algebra **90** (1993), no. 3, 253–274. MR1255714 (94i:05112)

[WW97] ———, *On the poset of conjugacy classes of subgroups of π-power index*, Comm. Algebra **25** (1997), no. 5, 1347–1364. MR1444007 (98f:06003)

[Yos83] Tomoyuki Yoshida, *Idempotents of Burnside rings and Dress induction theorem*, J. Algebra **80** (1983), no. 1, 90–105. MR690705 (85d:20004)

[Yos89] Satoshi Yoshiara, *Some geometries for J_3 and O'N*, European J. Combin. **10** (1989), no. 5, 499–506. MR1014558 (90j:20031)

[Yos90] S. Yoshiara, *Embeddings of geometries and related codes*, (proceedings of 1990 conference on algebraic combinatorics), Suuriken-Kokyuuroku, vol. 768, Kyoto Institute of Mathematical Sciences and Analysis, 1990.

[Yos05a] Satoshi Yoshiara, *Odd radical subgroups of some sporadic simple groups*, J. Algebra **291** (2005), no. 1, 90–107. MR2158512 (2006d:20029)

[Yos05b] ———, *Radical 2-subgroups of the Monster and the Baby Monster*, J. Algebra **287** (2005), no. 1, 123–139. MR2134261

[Yos06] ———, *Corrigendum to: "Radical 2-subgroups of the Monster and the Baby Monster" [J. Algebra 287 (2005), no. 1, 123–139*, J. Algebra **303** (2006), no. 1, 447. MR2253672 (2007c:20037)

[Yuz87] Sergey Yuzvinsky, *Cohen-Macaulay rings of sections*, Adv. in Math. **63** (1987), no. 2, 172–195. MR872352 (88d:13039)

Index

Page locations for definitions, as well as for references which are particularly fundamental, are indicated in **boldface**.